Atom movements

Diffusion and mass transport in solids

Jean PHILIBERT

Professor of Materials Science
Université de Paris-Sud

Translated from the French by

Steven J. Rothman

Metallurgist, Argonne National Laboratory

PREFACE by

David Lazarus

University of Illinois

les éditions de physique

Avenue du Hoggar,
Zone Industrielle de Courtaboeuf,
B.P. 112,
F-91944 Les Ulis Cedex A, France

"To explain that which is visible but complicated
by that which is invisible but simple..."

Jean Perrin,
in preface to *Les Atomes* (1912)

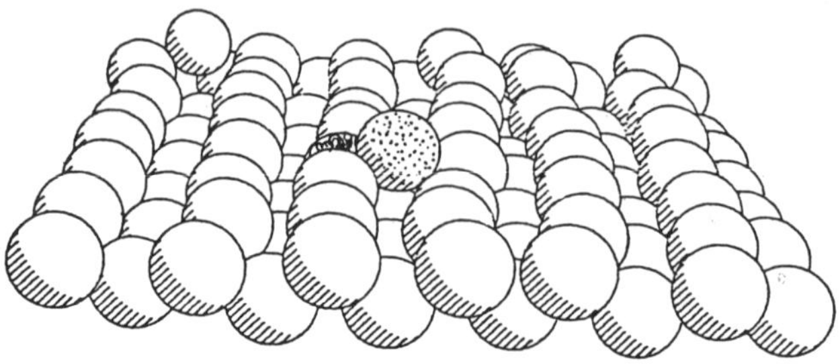

a

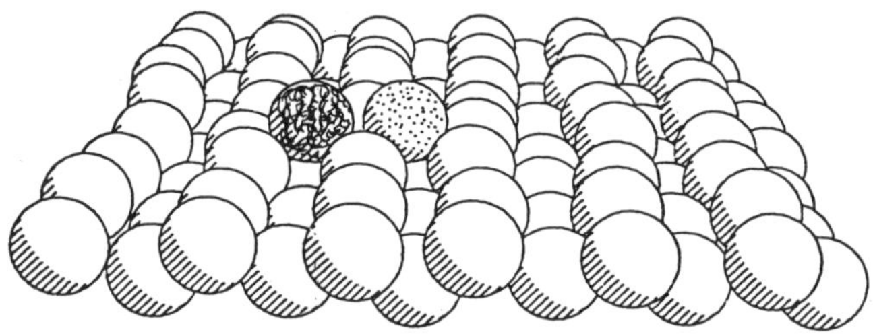

b

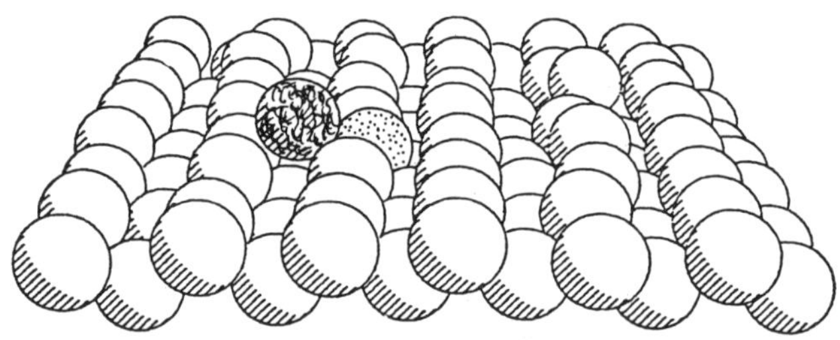

c

Diffusion of an adatom on a (110) surface of a fcc crystal by the exchange mechanism:

a) Adatom in initial position $t = t_0$

b) Saddle point position $t = t_0 + 6 \times 10^{-12}$ s

c) Final position $t = t_0 + 10 \times 10^{-12}$ s

The figures show an "instantaneous" view of two atomic layers, viewed along a direction that makes an angle of 20° with the (110) plane ; each plane contain six <110> strings of eight atoms.

The atom coordinates were calculated by a molecular dynamics simulation using a Lennard-Jones potential with parameters corresponding to solid argon at 0.4 T_m. (see G. de Lorenzi *et al.*, reference at end of Ch. VI.)

The author thanks Drs. Madeleine Meyer and Vassili Pontikis for preparing the figure.

Preface

As I write this preface, in January 1989, it is hard for me to believe that a full 23 years have passed since the publication of "LA DIFFUSION DANS LES SOLIDES" (Presses Universitaires de France, Paris, 1966). This glorious two-volume work by Yves Adda and Jean Philibert was, until very recently, the basic "bible" for all serious scientists working in the field of diffusion in solids. In 1985 Professor Philibert published a condensed, updated version, suitable as a textbook for advanced students of materials science or solid-state physics : "DIFFUSION ET TRANSPORT DE MATIERE DANS LES SOLIDES" (Monographies de Physique, les Editions de Physique, Paris, 1985).

Unfortunately, the world includes fewer francophones than persons who wish to, or should, enter into the serious study of the field of solid-state diffusion- an area which is absolutely fundamental to understanding a virtual cornucopia of important phenomena in materials science: nucleation, crystal growth, sintering, hardening, alloying, phase transformations, oxidation, plastic flow, fracture, photography...... the list is almost endless. Thus, many not raised with a sufficient knowledge of French, (including most of my own graduate students over two decades) have either had to learn enough French to wade slowly and painfully through the Adda-Philibert "bible," or, far worse, had no access at all to this most important reference.

Finally, a miracle has occured : Dr. S. J. Rothman of Argonne National Laboratory, not only a fluent francophone but also a scientist who himself has made enormous contributions to the field of solid-state diffusion, has made an English-language translation of Professor Philibert's 1985 text, now entitled "ATOM MOVEMENTS". Moreover, the new edition has been updated in important ways and includes an extensive set of extremely practical homework exercises to help the serious reader master the field in a professional manner. This, if I may steal a line from Shakespeare, is "...a consummation devoutly to be wished."

The most wonderful aspects of the original Adda-Philibert "bible" are faithfully preserved in Professor Philibert's French-language 1985 book and again in this English-language edition. This is a work of love by a scientist who understands the field thoroughly and deeply, from its fundamental atomistic aspects to the most practical of its "real-world" applications. The selection of topics is superb, and the treatment of each subject is thorough and complete, appropriate in level for advanced undergraduate or graduate students, as well as active research workers, who demand a thorough grounding in this vital area.

Thus, through the joint efforts of Jean Philibert and Steve Rothman, we finally have available "ATOM MOVEMENTS", a superb basic text in English, which should be "required reading" for serious students of diffusion throughout the world. My one sadness is that it comes too late for my own graduate students (I have now retired from active research), but then, I can always console myself with the thought that by forcing them to learn enough French to read the "bible," I also made it possible for them to enjoy much more fruitful visits to France themselves in their post-student lives!

As a final and personal note, I want to express my own sincere thanks to my old and dear friends and colleagues, Jean Philibert, who wrote the new book, and Yves Adda, who joined with Jean in writing the original "bible," for all that they have done for the field of solid-state diffusion, in general, and for me and my own research programs over the past decades. Their books, as well as their own vital and basic scientific work in this field, will endure for generations. I am delighted that their work, through this English-language edition, will now be more widely available.

David Lazarus
Loomis Laboratory of Physics
The University of Illinois
Urbana, Illinois, USA

Foreword

This book was written to remedy a deficiency: at this time, an elementary text on diffusion in solids does not exist either in French or in English. On the other hand, literature for specialists at an advanced level is abundant ; during the last fifteen years, a number of colloquia and workshops have resulted in publications, many of which resemble review articles. Still, there is no first book that would prepare a graduate student or beginning researcher to use these review articles or the original literature fruitfully.

The present book is the result of diverse courses on diffusion. It is intended to give as complete an overview as possible of diffusion in solid media, while relating the processes of diffusion to both their physical bases and their applications. In this spirit, certain fundamental aspects of these processes, such as the calculation of correlation factors or the theory of the atomic jump, which require long mathematical derivations, have been considered only on an elementary level, with the important results given without proof. However, when a simple approach was possible, the important relations have been derived, but concentrating more on the physics than on the mathematical formalism.

A series of a real situations is covered in this account, from self-diffusion of radiotracers to the more complex cases of mass flow under chemical or thermal gradients or under electric fields, or diffusion in structures of lower dimensionality (surfaces and interfaces). In all these analyses, no category of materials was favored ; metals, ionic crystals, oxides, and semiconductors all had their turn. Only polymers were not specifically touched. One chapter is specifically devoted to techniques for studying diffusion, including methods of numerical simulation, and a last and long chapter gives a number of metallurgical phenomena in which diffusion plays a fundamental role.

In the spirit of the book, neither a review of experimental results nor an exhaustive bibliography has been given. Only a few typical results, with their references, are given to illustrate important points. The rest of the bibliography lists references to books and review articles which allow the reader to penetrate the subject more deeply before going to the original literature.

This work is addressed first of all to graduate students, but may serve a larger audience in allowing researchers to refresh their memories on some points of diffusion. They will grasp that the points of view, the approaches, of this apparently classical subject have recently experienced a significant evolution, as shown in the series of colloquia held over the last fifteen years and cited in the general bibliography. The background is classical ; the new perspectives open with new materials.

May this small book inspire the reader to futher research and renewal in a field in which several laboratories in our country have long been active.

* *
*

The author thanks all those who, by reading a part of the manuscript and by discussion have helped him to clarify a number of points. His thanks go equally to the secretaries who had to face a difficult stenographic task, and especially to Mrs. Marie-Claire Dolou, who took care of a large part of this reproduction, as well as to the publisher, Editions de Physique, who have lavished much care on the production of this volume.

J. Philibert

Foreword to the English Edition

The good reception given this work in the scientific community and the urging of several colleagues have encouraged the author to prepare an English edition.

The title chosen for this edition, evoking that of the ASM seminar published in 1952, is intended to indicate the aim of this book: to understand the processes encountered in Materials Science which are governed by the movement of atoms. As for the text itself, it has been revised, expanded, and corrected, and, last but not least, a set of exercices of various levels of difficulty has been added.

The author wishes to thank all those who have made suggestions about the book, and especially the translator ; his many suggestions have considerably improved the original text, so that it may be of even better service to its readers.

J. Philibert, October 1988

Translator's Acknowledgments

Dr. Charles Wiley and Prof. Jean Philibert read the translation manuscript ; I thank them for their many excellent suggestions, which helped greatly to improve the clarity of the translation. I especially thank Prof. Philibert for his constant friendly encouragement. I am grateful to Dr. David Price for reading and correcting the parts on neutron diffraction, and to Drs. Alex McKale and Nestor Zalucec for assistance with word processing. I thank my wife, Ms. Barbara Rothman, for her frequent suggestions of the right word or the correct grammar, and for her patience and support during the course of this work.

S. J. Rothman, October 1988

TABLE OF CONTENTS

CHAPTER VI: DIFFUSION AND DRIFT IN ALLOYS AND COMPOUNDS 203

CHAPTER VII: DIFFUSION IN MEDIA OF LOWER DIMENSIONALITY 249

Part 1. — Internal short-circuits (dislocations, interfaces)

General Bibliography

I. — BOOKS

Atom Movements, J. H. Hollomon ed., ASM, Cleveland (1951).

SHEWMON P. G., *Diffusion in Solids* (McGraw-Hill, New York) 1963.

Diffusion in BCC Metals, ASM, Cleveland (1965).

ADDA Y. and PHILIBERT J., *La Diffusion dans les Solides*, 2 vols. (P.U.F., Paris) 1966.

QUÉRÉ Y., *Défauts Ponctuels dans les Métaux* (Masson et Cie, Paris) 1967.

MANNING J. R., *Diffusion Kinetics for Atoms in Crystals* (Van Nostrand, Princeton) 1968.

Atomic Transport in Solids and Liquids, A. Lodding and T. Lagerwall eds., *Zeits. Naturforsch.*, Tübingen (1970).

Diffusion Processes, J. N. Sherwood, A. V. Chadwick, W. M. Muir and F. L. Switon eds., 2 vols (Gordon and Breach, London) 1971.

Diffusion ASM Seminar, ASM, Cleveland (1972).

FLYNN C. P., *Point Defects and Diffusion* (Clarendon Press, Oxford) 1972.

Atomic Diffusion in Semiconductors, D. Shaw ed. (Plenum, New York) 1973.

TUCK B., *Introduction to Diffusion in Semiconductors* (IEE Monograph Series 16, Inst. Electr. Eng.) 1974.

Diffusion in Solids, Recent Developments, A. S. Nowick and J. J. Burton eds., (Academic Press, New York) 1975.

Point Defects in Solids, J. H. Crawford and L. M. Slifkin eds. (Plenum Press, New York)
Vol. 1, *General and Ionic Crystals* (1972)

Vol. 2, *Semiconductors and Molecular Crystals* (1975)
Vol. 3, *Defects in Metals*

Low Temperature Diffusion and Application to Thin Films, A. Gangulee, P. S. Ho, and K. -N. Tu eds., *Thin Solid Films* **25** #1–2 (1975) 1 vol. (Elsevier Sequoia, Lausanne)

La Diffusion dans les Milieux Condensés, Théorie et Applications 19e Coll. Métallurgie INSTN Saclay (1976).

STARK J. P., *Solid State Diffusion* (Wiley, New York) 1976.

Eléments de Métallurgie Physique, Y. Adda, J. M. Dupouy, J. Philibert and Y. Quéré eds. (La Documentation Française, Paris) 1973–1983.

Défauts Ponctuels dans les Solides, Ecole de Confolant (1977) (Les Editions de Physique, Les Ulis) 1978.

DIMETA–82 *Diffusion in Metals and Alloys*, F. J. Kedves and D. L. Beke eds., Diff. and Defects Monograph Series #7, Trans. Tech. Pub. Aedermannsdorf, Switzerland (1983).

Mass Transport in Solids, F. Bénière and C. R. A. Catlow eds., NATO series, Series B, vol. 97 (Plenum Press, London, New York) 1983.

Physical Metallurgy, R. W. Cahn and P. Haasen eds., 3rd ed., 2 vols (North Holland) 1984.

Diffusion in Crystalline Solids, G. E. Murch and A. S. Nowick eds. (Academic Press, New York) 1984.

Nontraditional Methods in Diffusion, G. E. Murch, H. K. Birnbaum and J. Cost eds. (The Metallurgical Society of AIME, Warrendale, PA) 1984.

Diffusion in solids, Recent Developments, M. A. Dayananda and G.E. Murch eds. (The Metallurgical Society of AIME, Warrendale,PA) 1985.

Interstitials and Vacancies in Metals and Alloys, C. Abromeit and H. Wollenberger, eds. *Materials Sci. Forum* **15–18** (Trans. Tech. Pub., Aedermannsdorf, Switzerland) 1987.

GHEZ R., *A Primer of Diffusion Problems* (Wiley, New York) 1988.

Diffusion Processes in High-Technology Materials, D. Gupta, A. D. Romig, jr., and M. A. Dayananda eds., *Def. and Diff. Forum* **A59** (Sci-Tech Publishers,

Vaduz, Lichtenstein) 1988.

KIRKALDY J. S. and YOUNG D. J., *Diffusion in the Condensed State* (Institute of Metals, London) 1988.

BORG R. J. and DIENES G. J., *An Introduction to Solid State Diffusion* (Academic Press, Boston) 1988.

TUCK B., *Atomic Diffusion in III-V Semiconductors* (A. Hilger, London) 1988.

SHEWMON P. G., Diffusion in Solids, 2nd ed. (TMS, Warrendale, PA) 1989.

Diffusion in Materials, A. L. Laskar, G. Brebec, J.-L. Bocquet and C. Monty eds., Nato Series B (Kluwer Academic, Dordrecht, Netherlands) 1990.

DIMETA-88, Diffusion in Metals and Alloys, F. J. Kedves and D. L. Beke eds., Material Science Forum, **66-69**, 1990.

II. — BIBLIOGRAPHIES

DIFFUSION DATA, 1967–1973, vol. 1–7.
Diffusion Information Center, (Cleveland).

DIFFUSION AND DEFECT DATA (DDD) 1974–, vol.8–, F. H. Wohlbier and D. J. Fisher eds. (Trans. Tech. Pub., Aedermannsdorf, Switzerland)

BUTRYMOWICZ D. B., *Diffusion Rate Data and Mass Transport Phenomena in Copper Systems* vol.VIII, INCRA Series, New York (1983).

DIFFUSION IN SOLID METALS AND ALLOYS, H. Mehrer ed. Landolt-Börnstein, New Series, Group III vol. 26, Springer Verlag, Berlin (1990)

Notation

a	lattice parameter ($2a$ for ionic crystals)
c	concentration (generally mol. cm^{-3})
D	diffusion coefficient
D^*	diffusion coefficient of a tracer
D_A^B	solute diffusion coefficient of solute B in solvent A
D_A	intrinsic diffusion coefficient
$\tilde{D}$	interdiffusion coefficient or chemical diffusion coefficient (N. B. for diffusion in a couple, say, metal A/ metal B, we use the term "interdiffusion" rather than "chemical diffusion".)
D'	grain boundary diffusion coefficient
D_s	surface diffusion coefficient
F	Helmholtz free energy
f	correlation factor
G	Gibbs free energy
H	enthalpy
ΔH^f	enthalpy of formation
ΔH^m	enthalpy of motion
J	flux (particles cm^{-2} s^{-1})
k	Boltzmann's constant
K	equilibrium constant
L	lattice (subscript)
L_{ij}	phenomenological coefficient (constant of proportionality between flux and force)
Log	logarithm to base e (natural logarithm)
m	mass
n_i	concentration (number of particles of species i per unit volume)
n	total number of particles per unit volume
N_i	mole fraction of species i
$\bar{N}_v$	mole fraction of vacancies at equilibrium
Q	activation energy
Q^*, Q^{**}	heat of transport
q	electric charge
S	entropy
T	absolute temperature
t	time
u	mobility

v	velocity ; as subscript, vacancy
V	volume
V_m	molar volume
V_i	partial molar volume of species i
$[V]$	mole fraction of vacancies $\equiv N_v$
w	jump frequency
X	thermodynamic "force"
Z	valence
Z^*	effective valence
β	geometric factor in diffusion coefficient (value depends on the lattice)
Γ	total number of jumps per unit time
Γ_s	number of jumps toward a given site per unit time
γ_i	thermodynamic activity coefficient of species i
γ_b	surface tension of grain boundary
γ_s	surface tension
δ	grain-boundary width
ζ	coordination number
κ	segregation coefficient
λ	jump distance ; Boltzmann transformation variable $\equiv x/\sqrt{t}$
μ_i	chemical potential of species i
$\tilde{\mu}_i$	electrochemical potential
μ	shear modulus
ν	vibration frequency
ν_0	attempt frequency
ρ	density (g cm^{-3})
ρ_d	density of dislocations
σ	electrical conductivity, stress, or rate of entropy production
τ	mean time of stay ($\tau = \Gamma^{-1}$)
τ_R	relaxation time
Φ	electrical potential, or thermodynamic factor ($1 + \partial \mathrm{Log}\, \gamma_i / \partial \mathrm{Log}\, N_i$)
$\Phi(r)$	interaction potential between two particles
Ω	atomic volume

CHAPTER I

Diffusion and drift

From a phenomenological point of view, the processes of diffusion and, more generally, of atomic migration, can be approached in two ways: 1) in a somewhat formal manner, e.g., some of the transport laws in physics, or 2) in a physically much better grounded manner, based on the thermodynamics of irreversible processes. The latter justifies, within an approximation to be given below, the expressions that follow naturally from the first type of derivation. This is the framework of the present chapter; derivations from the thermodynamics of irreversible processes are the subject of chapter VIII.

I. Flux of particles. Fick's equation.

Let us consider the flux of particles of a certain species in a one dimensional system. The particles can be molecules, atoms, ions, point defects, free electrons, electron holes, etc... Let $c(x, t)$ be their concentration, expressed in number of particles or moles per unit volume.

One assumes that in the presence of a concentration gradient $\partial c/\partial x$, a flux of particles is established, in direction *down* the concentration gradient, and that the flux is proportional to the concentration gradient:

$$J = - D \frac{\partial c}{\partial x}, \tag{I.1}$$

where D is called the diffusion coefficient or diffusivity.

The flux is expressed in number of particles (or moles) traversing unit area per unit time (1). Thus D has dimensions of $\mathrm{L}^2\ \mathrm{T}^{-1}$, generally $\mathrm{cm}^2\ \mathrm{s}^{-1}$ or, in SI units, $\mathrm{m}^2\ \mathrm{s}^{-1}$.

This law, called Fick's law, is formally identical to Fourier's law of heat flow:

$$J_q = - k \frac{\mathrm{d}T}{\mathrm{d}x},$$

where k is the thermal conductivity and J_q is the flux of heat. It is also identical

(1) Rigorously, J is a flux density, but is traditionally, and here, called a flux.

to Ohm's law for the flow of electric charge:

$$i = \sigma E = -\sigma \frac{\mathrm{d}V}{\mathrm{d}x},$$

where i is the current density, and σ is the electric conductivity.

Fick's law is easily generalized to three dimensions:

$$\mathbf{J} = -\mathbf{D}\,\nabla \mathrm{c}. \tag{I.1a}$$

$\mathbf{D}$ is thus a second-rank tensor.

The flow of particles in a given direction can, in fact, be due to two causes: one is the above concentration gradient, and the other is the action of an *external force* or *driving force*. Under the influence of this force, the particles move with a certain average velocity $\langle v \rangle$, which gives rise to a flux $\langle v \rangle c$. Thus the general expression for the flux is:

$$\boxed{J = -D\frac{\partial c}{\partial x} + \langle v \rangle c.} \tag{I.2}$$

The first term on the right hand side is called the *Fickian or diffusional flux*, and the second term the *drift* (2). This distinction is more practical than profound, as can be shown by the thermodynamics of irreversible processes (cf. Chap. VIII), but is of interest because it corresponds to the experimental conditions. We shall describe the "external forces" more precisely below; here we only remark that the second term in (I.2) is formally analogous to the first, in that it gives a flux proportional to a gradient, because the force can be seen as a gradient (the electric field in Ohm's law, for example, is the gradient of the electric potential).

II. Time-dependent case.

Equations (I.1) and (I.2), or their three-dimensional equivalents, can only be used in a *steady-state* regime, that is, when the flux is independent of time. For example, in the case of the diffusion of a gas through a foil, a distribution $c(x)$ independent of t is established, and $J = -D(\mathrm{d}c/\mathrm{d}x) = \text{const}$.

In the *time-dependent* case, i.e., when the flux at every point varies with time, one must combine Fick's law with an equation of material balance. For species which obey a conservation law, this is the *equation of continuity*:

$$\frac{\partial J}{\partial x} = -\frac{\partial c}{\partial t}, \tag{I.3}$$

(2) The terminology varies; "drift" is also called "mass flow".

which is easily derived. Consider a cylinder parallel to the diffusion direction (Fig. I.1), of unit cross sectional area, with a section bounded by the planes P and P', having coordinates x, $x + \mathrm{d}x$, respectively. Let the flux of particles moving from left to right be $J(x)$ and $J(x + \mathrm{d}x)$ at x and $x + \mathrm{d}x$, respectively.

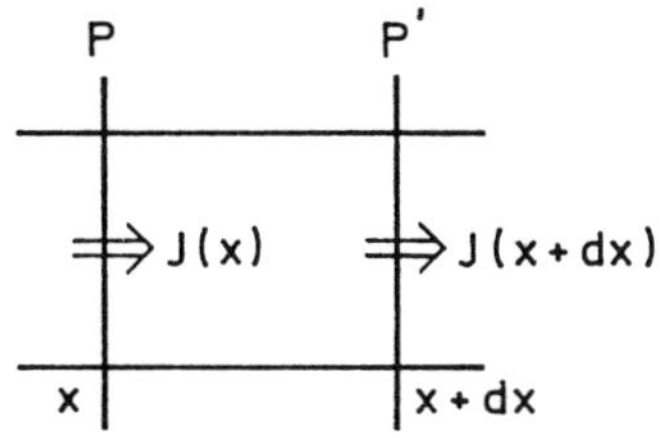

Fig. I.1. — One-dimensional time-dependent diffusion.

The quantity of material which accumulates in the cylinder between P and P' in the time $\mathrm{d}t$ is then obtained by applying the formula for a differential. This quantity must be equal to the change of the number of particles in the volume $1.\mathrm{d}x$, which is $[c(t + \mathrm{d}t) - c(t)]\mathrm{d}x = \partial c/\partial t \; \mathrm{d}t \; \mathrm{d}x$. Setting the two expressions equal:

$$\frac{\partial c}{\partial t} = -\frac{\partial J}{\partial x}. \tag{I.3}$$

Combining (I.2) and (I.3):

$$\frac{\partial c}{\partial t} = \frac{\partial}{\partial x}\left(D \frac{\partial c}{\partial x}\right) - \frac{\partial}{\partial x}\left(\langle v \rangle c\right). \tag{I.4}$$

Thus the general diffusion equation is a second-order partial differential equation. It cannot be solved analytically if D and $\langle v \rangle$ depend on concentration, and thus on x and t. The solution becomes simpler if D and $\langle v \rangle$ are independent of concentration, which is the case for systems that remain chemically homogeneous (e.g., self-diffusion). Then equation (I.4) can be written:

$$\frac{\partial c}{\partial t} = D \frac{\partial^2 c}{\partial x^2} - \langle v \rangle \frac{\partial c}{\partial x}. \tag{I.5}$$

If there is no drift term, equation (I.5) reduces to:

$$\boxed{\frac{\partial c}{\partial t} = D \frac{\partial^2 c}{\partial x^2}} \tag{I.6}$$

an equation often, and improperly, called Fick's second law. There exist different analytical solutions for equations (I.5) and (I.6) corresponding to different initial and boundary conditions. These solutions express the form of the *concentration profile* or *depth profile* $c(x, t)$ and allow the calculation of the diffusion coefficient D from the experimentally obtained concentration profile.

The evolution of a concentration profile $c(x, t)$ with time can be predicted qualitatively from the form of equation (I.6): the concentration increases as a function of time in regions where the profile is concave upward (Fig. I.2a), and decreases with increasing time where the profile is convex (Fig. I.2b).

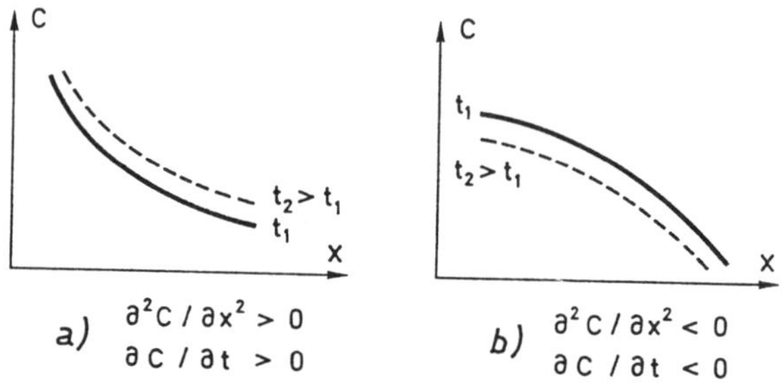

Fig. I.2. — Evolution of a concentration profile in non-steady state; the profile tends to smooth out.

Note that this analysis holds only for $D > 0$. See chapter IV, § II.1.

This derivation is identical to the theory of the conduction of heat. One begins with Fourier's equation, and for the time-dependent case, writes an equation for the conservation of heat energy. In the differential volume considered in figure I.1, the change in temperature is written $(\partial T/\partial t)\mathrm{d}t$. The heat capacity of the slice is $1.\mathrm{d}x.\rho C$, where ρ is the mass density and C the specific heat. The heat balance due to the flux entering and leaving the volume is, as above, $-\ (\partial J/\partial x)\ \mathrm{d}x\ \mathrm{d}t$, whence:

$$-\ \frac{\partial J}{\partial x}\ \mathrm{d}x\ \mathrm{d}t\ =\ \rho C\ \mathrm{d}x\ \frac{\partial T}{\partial t}\ \mathrm{d}t.$$

Combining this with Fourier's equation and assuming k is constant gives:

$$\frac{\partial T}{\partial t} = \frac{k}{\rho C}\,\frac{\partial^2 T}{\partial x^2},$$

an equation formally analogous to Fick's second equation with thermal diffusivity $k/\rho C$. Many practical cases of heat flow have been studied and solutions obtained; these can be directly used in diffusion problems by simply replacing T with c. These solutions are collected in the classic work of Carslaw and Jaeger (1959); a similar but not exactly overlapping book on diffusion was written later by J. Crank (1956, 2nd ed. 1975).

III. Solutions of the diffusion equation (or Fick's second law).

We limit ourselves here to a few simple cases, in order to illustrate the kinetics of diffusion. For the mathematical derivations, the reader is referred to the Appendices and the works cited as references. The experimental diffusion couples that correspond to these solutions are discussed in chapter IX.

III.1 THIN LAYER OR INSTANTANEOUS SOURCE. — The diffusing species is deposited as a "thin" layer on the surface of the sample or in a "sandwich" of two identical samples (Fig. I.3). The initial condition is:

$$t = 0, \quad c(x,0) = Q\delta(x),$$

where δ indicates the Dirac delta function, defined by:

$$\delta(x) = 0, \; x \neq 0,$$

and

$$\forall t, \; \int_0^{+\infty} c \, \mathrm{d}x = Q \quad \text{or} \quad \int_{-\infty}^{+\infty} c \, \mathrm{d}x = Q,$$

where Q is the quantity of atoms deposited per unit area. The limits of integration are 0, $+\infty$ for the surface layer (semi-infinite sample) and $-\infty$, $+\infty$ for the sandwich (infinite sample). The analytical solution for the sandwich geometry is:

$$c(x,t) = \frac{Q}{2\sqrt{\pi D t}} \exp\left(-\frac{x^2}{4\,Dt}\right). \tag{I.7}$$

Fig. I.3. — Thin-layer geometry (surface deposit or "sandwich").

For the surface deposit, the factor of 1/2 in the pre-exponential is removed. This is the equation of a *Gaussian* (bell curve), which decays as the square root of time and broadens as a function of $x/2\sqrt{Dt}$ (Fig. I.4). The Gaussian can also be written as $\exp(-\,x^2/2w^2)$; its "width" is thus $w = \sqrt{2Dt}$. The quantity $\sqrt{2Dt}$ is a measure of the penetration depth. The Gaussian solution is valid only if the thickness of the deposited layer is small with respect to $\sqrt{2Dt}$.

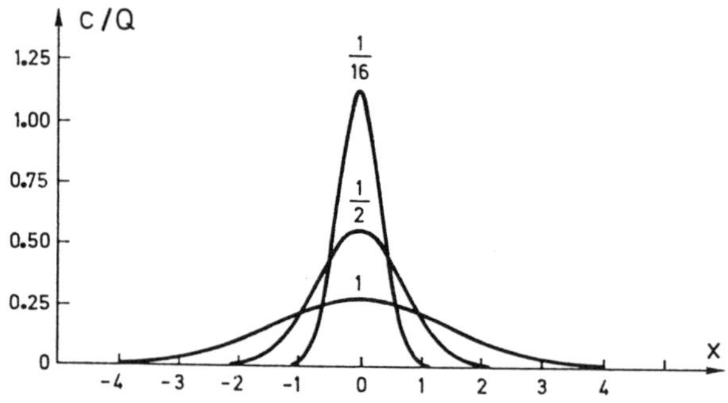

Fig. I.4. — Evolution of concentration profiles from a thin layer as a function of time. The numbers on the curves indicate values of Dt.

In practice, one plots the logarithm of the concentration against x^2 and calculates the value of D from the slope of the straight line. The validity of the experiment is indicated by how well the experimental points follow the analytical solution; a quantitative measure of the goodness of fit is the correlation coefficient (see the section on least-squares analysis, Ch. IX, part 1, § IV).

It is interesting to see how the concentration profile is modified by a driving force. The solution of the corresponding equation (Eq. (I.5)) is:

$$c = \frac{Q}{2\sqrt{\pi Dt}} \exp\left[-\,\frac{(x - \langle v\rangle\, t)^2}{4\,Dt}\right] . \qquad \text{(I.8)}$$

One obtains the same Gaussian, the center of which translates with velocity $\langle v\rangle$. This result justifies the distinction we have drawn between the two terms in Fick's equation; the first or Fickian term leads to a broadening of the distribution of the diffusing atoms, while the drift term leads to a translation of the entire distribution.

The fact that diffusion leads to a Gaussian distribution of the diffusing atoms reveals the fundamentally random movement of the diffusing atoms, as opposed to a directed migration under the effect of a force. But both processes involve the elementary jumps of the particles, so that $\langle v\rangle$ and D must be related to each other (see the Einstein relation, below).

III.2 CONSTANT SURFACE CONCENTRATION (DIFFUSION IN A SEMI-INFINITE SOLID).

Initial conditions: $t = 0, \; x > 0, \; c(x,0) = c_0$.
Boundary conditions: all $t > 0, \; x = 0, \; c(0,t) = c_s$.

The solution is:

$$\frac{c - c_s}{c_0 - c_s} = \operatorname{erf}\left(\frac{x}{2\sqrt{Dt}}\right), \qquad \text{(I.9)}$$

where "erf" means the error function or error integral of Gauss:

$$\operatorname{erf} z = \frac{2}{\sqrt{\pi}} \int_0^z e^{-u^2} \, du.$$

Values of this function are tabulated in many places (see Table p. 566-567). Two cases of equation (I.9) are of particular interest:

1) $$c_s = 0$$

$$c/c_0 = \operatorname{erf}(x/2\sqrt{Dt}). \qquad \text{(I.9a)}$$

From this, the outward flux at the surface is $D\, c_0/\sqrt{\pi D t}$ and the total quantity of substance that escapes by diffusion across a unit area of surface in time t is:

$$M(t) = \int_0^t \frac{Dc_0}{\sqrt{\pi D t}} \, dt = 2c_0 \sqrt{\frac{Dt}{\pi}}. \qquad \text{(I.10)}$$

Examples of this case are the decarburization of a steel in an oxidizing atmosphere or the dezincification of brass under vacuum (Fig. I.5a).

2) $$c_0 = 0$$

$$\frac{c}{c_s} = \operatorname{erfc}(x/2\sqrt{Dt}), \qquad \text{(I.9b)}$$

where "erfc" is the complementary error function, = 1-erf. An example of this case is diffusion from an atmosphere which, at equilibrium, keeps the surface concentration constant (Fig. I.5b).

The quantity of material which diffuses into the solid per unit area is:

$$M(t) = 2\, c_s \sqrt{Dt/\pi}. \qquad \text{(I.10a)}$$

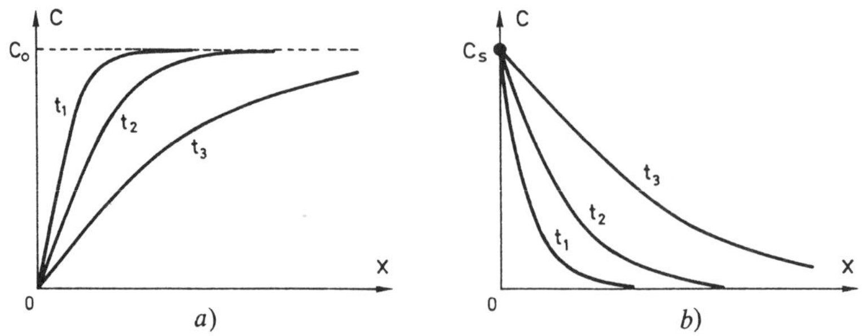

Fig. I.5. — Evolution of concentration profiles with time of diffusion: a) out-diffusion, b) in-diffusion. $t_1 < t_2 < t_3$.

In practice, instead of c one plots $\text{erf}^{-1}\,(c/c_0)$ or $\text{erfc}^{-1}\,(c/c_s)$ *vs.* x or *vs.* $x/\sqrt{t}$. One hopes to obtain a straight line, in which case D can be calculated from the slope ([3]). There is a practical difficulty in that c_s is not always known with the necessary precision, and one must vary both D and c_s. The boundary condition c_s = constant cannot always be proven since the introduction of the diffusing species may take place by means of chemical reactions or the process of absorption. These processes may be rate determining, in which case the flux into the solid is given by:

$$J(0,t) = K\left[c_s^0 - c_s(t)\right],$$

where K is the reaction rate constant (the reaction is assumed to be first order), and $c_s(t)$ and c_s^0 are the real and equilibrium surface concentrations. The solution of Fick's equation then becomes:

$$\frac{c - c_0}{c_s^0 - c_0} = \text{erfc}\left(\frac{x}{2\sqrt{Dt}}\right) - \exp(hx + h^2Dt)\,\text{erfc}\left(\frac{x + 2hDt}{2\sqrt{Dt}}\right), \quad (\text{I.11})$$

with $h = K/D$. The reader can verify that for the limiting case $h \longrightarrow 0$, equation (I.11) predicts no diffusion ($c - c_0 = 0$), whereas for $h \longrightarrow \infty$, the erfc solution is obtained, as the second term of (I.11) goes to zero ([4]).

([3]) A straight line is simply obtained by plotting c/c_0 *vs.* x on "probability paper" (see Exercice 1).

([4]) Using the approximation

$$\sqrt{\pi}z\,\exp\left(z^2\right)\,\text{erfc}\,(z) = 1 + \sum_{\text{m}=1}^{\infty}(-1)^{\text{m}}\,\frac{1\cdot 3\ldots(2\text{m} - 1)}{\left(2z^2\right)^{\text{m}}}.$$

A solution is also available for the case when the surface is displaced parallel to itself with velocity v due to evaporation or condensation of the base material:

$$c(x,t) = \frac{1}{2} c_s \left[\operatorname{erfc} \frac{x + vt}{2\sqrt{Dt}} + \exp\left(-\frac{vx}{D}\right) \operatorname{erfc} \frac{x - vt}{2\sqrt{Dt}} \right], \qquad (I.12)$$

where $v > 0$ for evaporation and $v < 0$ for condensation. Thus in these last two cases, the solution depends on two kinetic parameters, D and K or v, in addition to the surface concentration, and this must be taken into account in fitting the data to determine D. We emphasize that these solutions are valid only for D independent of concentration.

In the solution for constant surface concentration (Eq. (I.9)), the x coordinate of a plane parallel to the surface and at which the concentration remains constant with time is given by $x/2\sqrt{Dt}$ = Constant, i.e., this plane moves as the square root of time. This result is characteristic of diffusion: *the penetration depth varies with the square root of the time.* The same is true for the amount of diffusant introduced or lost (Eqs. (I.10) and (I.10a)).

In the case of a sheet of thickness l comparable to $\sqrt{Dt}$, there is reflection of the flux at the back surface, and the solution (I.9) is no longer valid. Equation (I.6) is solved by the method of separation of variables (Appendix I). Equation (A.4) gives for the case $c_s = 0$:

$$\frac{c}{c_0} = \frac{4}{\pi} \sum_{n=0}^{\infty} \frac{1}{2n + 1} \sin\left[\frac{(2n + 1)\pi x}{l}\right] \exp\left[-\frac{(2n + 1)^2 \pi^2}{l^2} Dt\right].$$

III.3 Infinite initial distribution (Fig. I.6). — The initial condition is a step function at $x = 0$:

$$t = 0 \begin{cases} x < 0 & c = c_2 \\ x > 0 & c = c_1 \end{cases}.$$

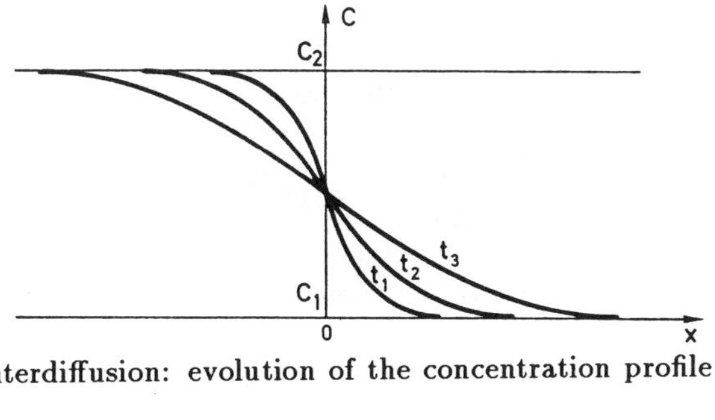

Fig. I.6. — Interdiffusion: evolution of the concentration profile with time. $t_1 < t_2 < t_3$.

For reasons of symmetry, the concentration at the initial interface ($x = 0$) remains constant at the value of $\frac{1}{2}(c_1 + c_2)$. Substituting this value for c_s and c_1 for c_0 in equation (I.9), we obtain:

$$\frac{c - c_1}{c_2 - c_1} = \frac{1}{2} \operatorname{erfc}\left[\frac{x}{2\sqrt{Dt}}\right]. \tag{I.13}$$

The two branches of the solution on either side of the original interface can be brought into coincidence by a 180° rotation around the point at $x = 0$.

III.4 The Boltzmann transformation. — The fact that many of the previous solutions contain the combination $x/2\sqrt{Dt}$ suggests the choice of $\lambda = x/\sqrt{t}$ as a useful variable. This change of variable, known as the Boltzmann transformation, can be applied only when the initial and boundary conditions are themselves functions of only the variable λ. Since

$$\frac{\partial}{\partial x} = \frac{\mathrm{d}}{\mathrm{d}\lambda}\frac{\partial \lambda}{\partial x} = \frac{1}{\sqrt{t}}\frac{\mathrm{d}}{\mathrm{d}\lambda},$$

and

$$\frac{\partial}{\partial t} = \frac{\mathrm{d}}{\mathrm{d}\lambda}\frac{\partial \lambda}{\partial t} = -\frac{\lambda}{2t}\frac{\mathrm{d}}{\mathrm{d}\lambda},$$

the Boltzmann transformation converts the partial differential diffusion equation

$$\frac{\partial c}{\partial t} = \frac{\partial}{\partial x}\left(D\frac{\partial c}{\partial x}\right) \tag{I.14}$$

into a second order ordinary differential equation:

$$-\frac{\lambda}{2}\frac{\mathrm{d}c}{\mathrm{d}\lambda} = \frac{\mathrm{d}}{\mathrm{d}\lambda}\left(D\frac{\mathrm{d}c}{\mathrm{d}\lambda}\right), \tag{I.14a}$$

the solution of which is some function $c(\lambda)$. When D is a constant, i.e., independent of λ, equation (I.14a) simplifies to:

$$\lambda\frac{\mathrm{d}c}{\mathrm{d}\lambda} = -2D\frac{\mathrm{d}^2c}{\mathrm{d}\lambda^2}. \tag{I.14b}$$

The integration of (I.14b) is straightforward:

$$c(\lambda) = A + B \operatorname{erf}\left(\lambda/2\sqrt{D}\right),$$

or

$$c(x,t) = A + B \operatorname{erf}\left(x/2\sqrt{Dt}\right),$$

where A and B are constants to be determined from the initial and boundary conditions. This solution applies e.g., to the case of the constant surface concentration (§ III.2) and the infinite distribution (§ III.3), but not the instantaneous source (§ III.1), since here the initial condition is given in terms of x, and not λ.

III.5 CONCENTRATION-DEPENDENT DIFFUSION COEFFICIENT. — In many practical cases, where the composition varies over a large range of concentrations, or in semiconductors, where the diffusing species changes the Fermi level of the base material, the diffusion coefficient is not constant, and a graphical or numerical integration of the diffusion equation is required. The Boltzmann transformation is very useful in this case also. To calculate D for a given value of λ, equation (I.14a) is integrated from one end of the diffusion couple, i.e., from $\lambda = \infty$, to the value of λ of interest, or between the corresponding values of the concentration, c_1 and c:

$$-\frac{1}{2}\int_{c_1}^{c} \lambda \, \mathrm{d}c = D \left.\frac{\mathrm{d}c}{\mathrm{d}\lambda}\right|_c - D \left.\frac{\mathrm{d}c}{\mathrm{d}\lambda}\right|_{c_1} = D \left.\frac{\mathrm{d}c}{\mathrm{d}\lambda}\right|_c, \qquad \text{(I.15)}$$

since for c_1 the concentration gradient is zero. Thus

$$D = -\frac{1}{2}\frac{\int_{c_1}^{c} \lambda \, \mathrm{d}c}{(\mathrm{d}c/\mathrm{d}\lambda)_c},$$

or transforming back to the original variables:

$$\boxed{D(c) = -\frac{1}{2t}\frac{\int_{c_1}^{c} x \, \mathrm{d}c}{(\mathrm{d}c/\mathrm{d}x)_c}}. \qquad \text{(I.16)}$$

This is the well-known *Boltzmann-Matano equation*, which allows the calculation of D for the entire range of concentration provided that the initial and boundary conditions are functions of λ (see above). Equation (I.16) can be used for a semi-infinite medium with a constant surface concentration, the origin on the x axis being taken at the surface. For an interdiffusion problem (infinite medium), one has to choose the origin on the x axis carefully; if the integration is carried out from c_1 to c_2, the right hand side of equation (I.15) is zero, whence:

$$\int_{c_1}^{c_2} \lambda \, \mathrm{dc} = 0,$$

or

$$\int_{c_1}^{c_2} x \, \mathrm{dc} = 0. \tag{I.17}$$

This equation determines the origin on the x or λ axis; this origin is traditionally called the *Matano interface.* The Matano interface fixes two equal areas on the $c(x)$ profile (Fig. I.7a); it is the plane through which equal amounts of material have moved in the positive and negative directions. To calculate D, one first determines the location of the Matano interface by successive approximations, and then determines the area corresponding to the integral (shown hatched in Fig. I.7b) and the slope of the tangent to the $c(x)$ curve at the point in question. It makes no difference at which end of the diffusion couple the integration is started, because the two differently hatched areas in figure I.7b are equal by equation (I.17). It is useful in practice to verify that the concentration profiles from diffusion anneals carried out at the same temperature but for varying times superpose when plotted as c vs. λ. Equation (I.16) is easily programmed even on a small computer. As a first step, one fits the data, by the method of least squares, to a polynomial of sufficiently high degree to represent the function $c(x)$ or better by spline interpolation. See chapter IX, part IV. See also exercice 17 to calculate $\tilde{D}$ without determining the Matano interface. The case of variable molar volume will be discussed in chapter V, appendix I.

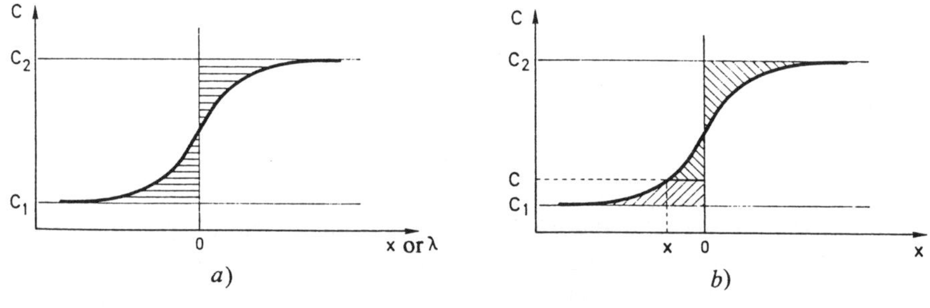

Fig. I.7. — Interdiffusion: a) determination of the Matano interface; b) calculation of D.

IV. Relation between drift and diffusion. The Nernst-Einstein equation.

The combined effect of a concentration gradient and a driving force can lead to a steady state if the corresponding fluxes are equal and of opposite sign (Fig. I.8):

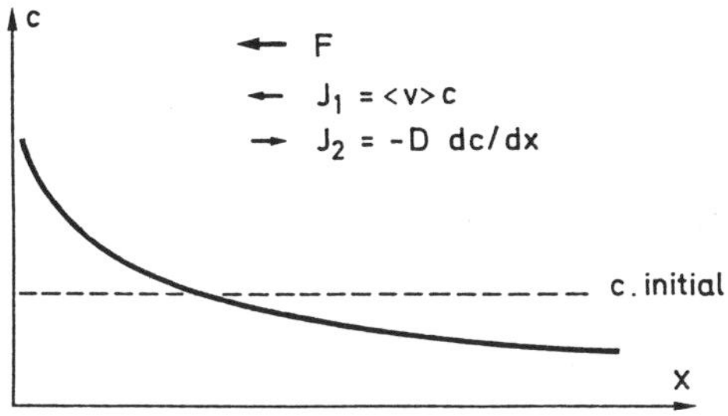

Fig. I.8. — Concentration profile under a driving force.

$$J = 0 \longrightarrow D\,\frac{\mathrm{d}c}{\mathrm{d}x} = \langle v\rangle\, c. \qquad \text{(I.18)}$$

Assume that the applied force is the derivative of a potential, i.e., $F = -\mathrm{d}\Phi/\mathrm{d}x$. At thermodynamic equilibrium, the distribution of the species under consideration should follow the Boltzmann distribution:

$$c(x) \;=\; \text{const.}\,\exp\left[-\,\Phi(x)/kT\right]. \qquad \text{(I.19)}$$

This expression should be the solution of equation (I.18) at steady state. Differentiating with respect to x yields:

$$\frac{\mathrm{d}c}{\mathrm{d}x} \;=\; -\,\frac{c}{kT}\,\frac{\mathrm{d}\Phi}{\mathrm{d}x} \;=\; \frac{cF}{kT}.$$

Substituting in (I.18), we obtain:

$$\boxed{\frac{\langle v\rangle}{D} \;=\; \frac{F}{kT}}\,. \qquad \text{(I.20)}$$

This is the *Nernst-Einstein equation.* If the driving force is due to an *electric field*, then for particles with charge q,

$$F \;=\; qE.$$

One introduces the mobility u, defined by:

$$\langle v \rangle = uE.$$

Equation (I.20) then becomes:

$$\boxed{\frac{u}{D} = \frac{q}{kT}}. \qquad \text{(I.21)}$$

Fick's equation can then be written in general form as:

$$J = -D\frac{\partial c}{\partial x} + \frac{D}{kT}Fc. \qquad \text{(I.22)}$$

Under the effect of the electric field, the particles of charge q migrate with velocity $\langle v \rangle$, from which the drift flux is:

$$J = \langle v \rangle c = \frac{FDc}{kT} = \frac{qcD}{kT}E,$$

and the current density

$$i = qJ = \frac{cq^2D}{kT}E.$$

But this is just *Ohm's law*, with the electrical conductivity given by:

$$\sigma = \frac{cq^2D}{kT}. \qquad \text{(I.23)}$$

V. The nature of the driving force.

The nature of, and the analytical expression for, the driving force can only be determined from the thermodynamics of irreversible processes. We have so far considered explicitly only the case of the electric field, a case that appears quite clear at first glance. The table below compares five different kinds of driving force, with the corresponding analytical expressions for the driving force, for introduction into equation (I.20) or (I.22):

The first case concerns charged particles; when these are atoms or ions, their motion under an electric field is called *electromigration* or *electrotransport*. The charge q is indeed that of the ion or of the point defect in ionic crystals; however, in the case of metals, this charge can be very different from the charge on the ion: $q^* \neq q$ (see Ch. VIII, § III.4). The same equations apply to electrons and holes in semiconductors and insulators; the electric field can be that of a *pn* junction or a Schottky barrier. The terms for diffusion and mass flow or drift are additive.

Table I. — *Driving Forces*

Nature of Force	Expression	Remarks
Gradient of electrical Potential, $E = -\mathrm{d}\Phi/\mathrm{d}x$	$q^* E$	q^* is the effective charge
Temperature Gradient $\mathrm{d}T/\mathrm{d}x$	$-(Q^*/T)\,\mathrm{d}T/\mathrm{d}x$	Q^* is the heat of transport
Gradient of Chemical Potential (non – ideal part only)	$-kT\,\partial \operatorname{Log}\gamma/\partial x$	γ is the thermo–dynamic activity coefficient
Stress Gradient	$-\mathrm{d}U/\mathrm{d}x$	U is the elastic interaction energy in the stress field $\sigma(x)$
Centrifugal force	$m^* \omega^2 r$	m^* is the effective molecular mass, ω is the angular velocity

Thermotransport, ***thermomigration***, or ***thermal diffusion*** is an important phenomenon in certain practical cases where large temperature gradients are encountered, such as nuclear fuel elements. It is a result of processes physically so complex that the significance of the heat of transport is not always obvious (see Ch.VIII, § III.5).

In the case of a *non-ideal solution*, the gradient of the activity coefficient gives rise to a driving force. It is easily shown that the sum of this term and the term due to the concentration gradient expresses the total effect of the gradient of thermodynamic activity or chemical potential – even though in this case the precise distinction between the two terms is at first glance not as simple as it is for the other forces ([5]).

A uniform stress cannot generate a flux (Curie-Neumann principle); thus the driving force must be the *stress gradient* and its effect must be considered

([5]) In fact, the flux is in all cases proportional to the gradient of the generalized chemical potential (electrochemical, mechano-chemical...) chapter VIII.

whenever the elastic interaction energy of the particle with the stress field is large enough. Typical cases are 1) *the Gorski effect*: migration of atoms in an elastically deformed sample (this effect is particularly noticeable for atoms with a strong size effect and high diffusivity such as interstitials, especially hydrogen in metals), and 2) formation of *Cottrell atmospheres* around dislocations, where the stress field varies as the reciprocal of the distance from the dislocation line (Ch. X, § IV.5 and Exercise 71).

The effect of a *centrifugal force* requires extremely rapid rotation to be observed. Only a very small number of systems have been studied in order to demonstrate the interstitial nature of anomalously fast diffusers (equilibrium sedimentation of Au in Na, K, and Pb, Barr and Le Claire (1969)).

VI. A variety of diffusion processes and generalization of Fick's law.

At the beginning of this chapter, the flux of the diffusing species was separated into two parts: a Fickian diffusion flux, and drift or mass flow under the action of a driving force (see Tab. I). The equation expressing this, equation (I.2), was subsequently combined with the equation of continuity to yield the general diffusion equation (I.4). However, the equation of continuity in the form (I.3) is valid only for species or particles that obey conservation laws. For species that do not obey the conservation laws, terms must be added to equation (I.3) for the number of particles annihilated or created per unit volume per unit time (*source* or *sink* terms), or for particles that can be *trapped* or *de-trapped.*

The creation or annihilation of particles can be due to various processes. A distinction is drawn between terms due to an external process, e.g., the rate of generation $G(x, y, z, t)$ of Frenkel defects in crystals subjected to irradiation, or of electron-hole pairs in a semiconductor irradiated by light or by an electron beam, and terms resulting from the migration of the particles themselves: annihilation of point defects at surfaces, grain boundaries, or dislocations, recombination of interstitials and vacancies, or defect reactions that lead to the formation of complex defects (divacancies...) etc.

Consider the migration of point defects to the sinks where they annihilate. Fick's equation, with the appropriate boundary conditions (especially on the sinks), allows the concentration of the defects $c(x, y, z, t)$ to be calculated. Consider further a simple geometry, e.g., spherical geometry, a uniform concentration c_0 at $t = 0$, with the surface $r = R_0$ a perfect sink. The solution of Fick's equation is (see Ch. X, § IV.4):

$$\frac{c}{c_0} = \frac{2 R_0}{\pi r} \sum_n \frac{(-1)^n}{n} \sin\left(\frac{n \pi r}{R_0}\right) \exp\left(- \frac{n^2 \pi^2 D t}{R_0^2}\right).$$

The exponential terms decay very rapidly for $n \geq 2$. The transient regime is described principally by the term $n = 1$; the number of defects in the sphere thus follows an exponential time dependence:

$$n(t) = n_0 \exp(-Kt) \tag{I.24}$$

with

$$\begin{aligned} K &= \alpha D, \text{ and} \\ \alpha &= \pi^2/R_0^2 \text{ for a sphere,} \\ &= (2.405/R_0)^2 \text{ for a cylinder, and} \\ &= \pi^2/4\,R_0^2 \text{ for a foil of thickness } 2\,R_0. \end{aligned}$$

Other geometries lead to an identical time dependence, but a different value of α, depending on whether the sinks are on the surface or internal. (One considers a sphere or cylinder around each point or line sink, respectively, and solves the diffusion equation in that region.) The expression $n(t)$ can be considered as the solution of a chemical rate equation for first-order kinetics:

$$\mathrm{d}n/\mathrm{d}t = -Kn = -n/\tau. \tag{I.25}$$

Describing the "reaction" of defect annihilation at unsaturated sinks:

$$\text{Point defect} + \text{Sink} \rightleftharpoons \text{Sink}.$$

The equation of continuity is thus replaced by a *balance equation* for the number of the particles of the species under consideration. The time derivative in equation (I.3) is replaced by the one from equation (I.25), and substitution of equation (I.1) leads to a generalized Fick's equation. Following the above remark, one can often replace diffusional terms by kinetic terms. A remarkable example of this is the dissociative diffusion mechanism (the diffusing species equilibrates between substitutional and interstitial sites and diffuses by two coupled processes, Ch. V, § V). Further examples are given here to illustrate the method.

Example 1: creation of point defects by irradiation. In the simple model, the concentration of vacancies obeys the balance equation:

$$\partial c_v/\partial t = G - K_1\, c_v\, c_i - K_2\, c_v, \tag{I.26}$$

where G is the rate of production of Frenkel pairs per unit volume, the term $K_1\, c_v\, c_i$ describes annihilation with interstitials (mutual recombination) and the term $K_2\, c_v$ describes annihilation at sinks. In order to write equation (I.26) without introducing spatial derivatives, the sinks are modeled by a continuous distribution (6) and the annihilation reactions are assumed to follow first order kinetics. The formation of divacancies, on the other hand, leads to second order kinetics: $- K_3\, c_v^2$.

An analogous equation can be written for the self-interstitials:

$$\partial\, c_i/\partial t \;=\; G \;-\; K_1\, c_v\, c_i \;-\; K_2'\, c_i. \tag{I.27}$$

These equations have different solutions for different conditions. For steady state, $dc_v/dt \;=\; dc_i/dt \;=\; 0$, whence $K_2\, c_v \;=\; K_2'\, c_i$, and $D_v\, c_v \;=\; D_i\, c_i$, where D_v and D_i are the diffusion coefficients of vacancies and interstitials, respectively. In general, the simultaneous non-linear differential equations (I.26) and (I.27) cannot be solved analytically, and numerical solutions must be found for each case. Because $D_i \;>>\; D_v$ in most cases, stability problems are encountered in the solutions.

Example 2: The equation of continuity also does not apply to a system in which the diffusing species can change to another species by a ***chemical reaction*** or by ***trapping*** of the diffusing species, e.g., at some defect. Trapping differs from annihilation in that the diffusing particle is not annihilated, but the concentrations of trapped and untrapped particles follow the laws of chemical equilibrium. The kinetics of trapping and de-trapping, or of chemical reaction, must then be introduced into the mass balance equation. Trapping is especially common in the case of the diffusion of hydrogen in metals, for which non-Fickian behavior is often observed both at steady state or in the time-dependent regime (7) (cf. Exercise 31).

Assume that there exists only one type of trap, of which there are n_t per unit volume, each trap being occupied by 0 or 1 particle. Let Θ be the fraction of traps that is occupied. The number of particles per unit volume is then $n_L + \Theta\, n_t$, where n_L is the concentration of particles in solution in the lattice and N_L their mole fraction. The diffusion equation then becomes:

$$\partial n_L/\partial t \;+\; n_t\; \partial\Theta/\partial t \;=\; \frac{\partial}{\partial x}\; (D \partial n_L/\partial x)\,. \tag{I.28}$$

(6) To calculate K_2, the diffusion equation (I.5) or (I.6) must be solved under the boundary conditions appropriate to an isolated sink of the correct geometry (e.g., a cylinder for an isolated straight dislocation line, see Eq. (A.18)).

(7) Consider permeation through a sheet of thickness l, with concentrations c_0 and 0 at the entry and exit faces, respectively. According to the diffusion equation, the steady-state flux is Dc_0/l, and the average quantity of hydrogen in the medium is $Q_0 \;=\; c_0\; l/2$, and $Q_1 \;=\; c_0\; l$ is needed to saturate the sheet uniformly. The ratio Q_0/Q_1 is thus 0.5, but higher values (0.8) are found experimentally, suggesting the presence of traps.

If trapping and detrapping follow first order kinetics with rate constants k and k' respectively,

$$\partial\Theta/\partial t \;=\; kN_L\;(1\;-\;\Theta)\;-\;k'\Theta, \tag{I.28a}$$

where the first term on the RHS is the trapping rate, proportional to the mole fraction of free particles and the fraction of empty traps, and the second term is the detrapping rate. Equations (I.28 and I.29) (the equation of McNabb and Foster (1963)) cannot be easily solved analytically, although they have been solved for certain boundary conditions with the use of the Laplace transformation (cf. App. I, § 3), and numerical solutions have been given by Caskey and Pillinger (1975). However, if the trapped atoms are in local equilibrium with the atoms in solution in the lattice, solution of Fick's equation will give an apparent diffusion coefficient that is smaller than the true D. This is also discussed by Crank (Ch. 14).

In effect, for steady state $\partial\Theta/\partial t = 0$, i.e., following (I.28a),

$$K \;=\; k/k' \;=\; \Theta/[N_L\;(1-\Theta)], \tag{I.28b}$$

which is just an expression of the law of mass action. The total concentration is $n_L \;+\; n_t\Theta$. Equating the flux to the gradient of total concentration, an apparent diffusion coefficient is obtained such that:

$$\begin{aligned} J \;&=\; -\,D_L\;\partial n_L/\partial x \;=\; -D(\partial/\partial x)\;(n_L\;+\;n_t\,\Theta)\\ &=\; -D\;(\partial n_L/\partial x\;+\;n_t\;\partial\Theta/\partial x)\,. \end{aligned}$$

From equation (I.28b), we obtain:

$$\frac{\partial N_L}{\partial x} \;=\; \frac{1}{K}\,\frac{\partial\Theta}{\partial x}\;(1\;-\;\Theta)^{-2},$$

from which:

$$\boxed{D \;=\; \frac{D_L}{1\;+\;K\;N_t\;(1\;-\;\Theta)^2},} \tag{I.29}$$

or, with the trap concentration written in terms of the mole fraction N_t,

$$D \;=\; D_L\;N_L/\,[N_L\;+\;N_t\,\Theta\;(1\;-\;\Theta)]\,.$$

This equation can be simplified for the limiting conditions of small Θ (few traps occupied):

$$D \;\approx\; D_L/\,(1\;+\;K\;N_t)\,,$$

which is identical to the equation obtained by Oriani following a different approach (cf. Ch. III, § VII.2)). This apparent diffusivity does not depend on concentration so that the usual analytical solutions for gas permeation can be used, which would not be the case for saturated traps (i.e., $\Theta \lesssim 1$).

Example 3: Injection of minority carriers into a semiconductor by electron bombardment. This is the case of a semiconductor in a scanning electron microscope (Fig. I.9). Consider that the semiconductor is covered by a thin layer of metal, forming a Schottky barrier and collecting the minority carriers (electrons in a p-type material, holes in n-type). The electron beam produces $g(\mathbf{r})$ electron-hole pairs per unit volume at the point $\mathbf{r}$ in the "generation volume". Let n be the concentration of minority carriers and neglect the zone near the surface where the carriers migrate under the influence of the electric field (depth of depleted zone << depth of generation volume). The charge carriers migrate by diffusion, and can also be trapped and, while trapped, recombine with carriers of the opposite sign. The diffusion equation is then:

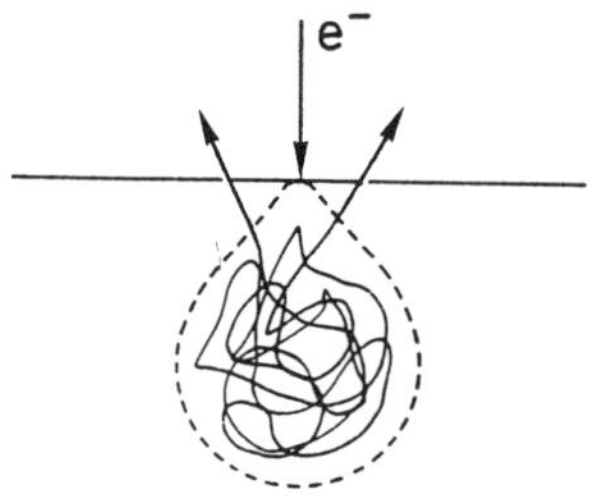

Fig. I.9 — Penetration and spreading of electrons in a target. The dotted line defines the generation volume.

$$\frac{\partial n}{\partial t} = g(\mathbf{r}) + D\nabla^2 n(r) - Kn(r). \tag{I.30}$$

The kinetic equation for the elimination of excess carriers with uniform concentration is:

$$\frac{\partial n}{\partial t} = -Kn(r),$$

from which:

$$\begin{aligned} n(t) &= n_0 \exp(-Kt) \\ &= n_0 \exp(-t/\tau). \end{aligned} \tag{I.31}$$

In general, the rate constant K is replaced in the balance equation by the *mean lifetime*, τ, of the carriers ($\tau(\mathbf{r})$ in the general case), with τ dependent on the trap density.

A simple solution exists for the case of a point source of constant strength on the surface: $g(\mathbf{r}) = g_0\,\delta(\mathbf{r})$. At steady state, equation (I.30) becomes:

$$D\,\nabla^2 n(\mathbf{r}) + g_0\,\delta(\mathbf{r}) - n(\mathbf{r})/\tau = 0,$$

or putting the Laplacian into spherical coordinates (cf. App. II),

$$D\,\frac{1}{r^2}\,\frac{\mathrm{d}}{\mathrm{d}r}\left(r^2\,\frac{\mathrm{d}n}{\mathrm{d}r}\right) - \frac{n}{\tau} = 0.$$

The source term is expressed as a boundary condition; if the diffusion flux across the surface of a sphere of radius r is g_0, this means that for a point source:

$$-\,4\,\pi r^2 \cdot D\,\frac{\mathrm{d}n}{\mathrm{d}r} = g_0, \quad r \rightarrow 0.$$

An easily verified solution of the diffusion equation is:

$$n_0(r) = \text{Const}\,\exp\,(-r/L)\,/Dr,$$

with $L = \sqrt{D\tau}$, the *diffusion length* of the minority carriers. The constant is determined from the boundary condition:

$$n_0(r) = \frac{g_0}{4\,\pi D}\,\frac{\exp\left(-\,r/L\right)}{r}. \tag{I.32}$$

This solution allows us to reach the case of more realistic boundary conditions (e.g., the generation volume is a sphere of radius R), and further to calculate the current collected by the junction, which is equal to the integral of the flux across its area S:

$$I = qD \int_s \nabla\mathbf{n}.\mathrm{d}\mathbf{S},$$

where q is the charge of the carriers. Consider the case of a lateral Schottky junction, with the electron beam bombarding the surface at a distance x_0 from the junction (Fig. I.10). The above expression for $n(r)$ then yields:

$$I = q g_0 \exp\left(-\,x_0/L\right). \tag{I.33}$$

The diffusion length can thus be determined experimentally from the variation of current collected with distance from the junction x_0.

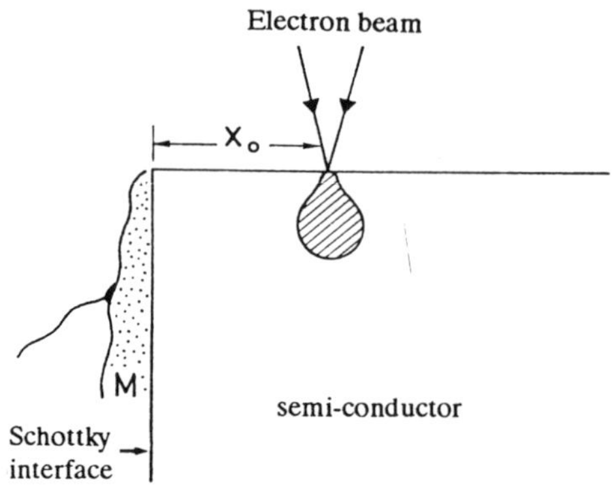

Fig. I.10. — Schematic of experiment to measure the diffusion length of charge carriers in a semiconductor by means of a scanning electron microscope.

VII. Diffusion with phase change. Multiphase diffusion.

In the case of interdiffusion, the concentration range of the couple often extends over two or more phases; two examples are given in figures I.11a and I.11b. The same is true for the growth of a new phase at the surface due to chemical reactions such as oxidation, sulfidation, etc. (Fig. I.11c), or to the evaporation of constituent B from a two-phase alloy with precipitates of a B-rich phase (Fig. I.11d).

The Boltzmann-Matano method is applicable in all these cases because the function $c(x)$ is differentiable and integrable at all points (at discontinuities, there exists a derivative on the right and one on the left) and because the phase boundaries are points of constant concentration. It is, however, useful to solve the equations assuming that the interdiffusion coefficient in each phase is constant, i.e., independent of composition. The problem is then attacked in the following manner:

1) In each phase, the concentration follows an error function:

$$c_i = A_i + B_i \operatorname{erf}\left(x/2\sqrt{D_i t}\right). \tag{I.34}$$

2) The interface, being a plane of constant concentration, moves as $\sqrt{t}$:

$$\xi = 2\,\gamma\sqrt{D_i t}, \tag{I.35}$$

where D_i is the diffusion coefficient in one of the phases, and γ is a constant.

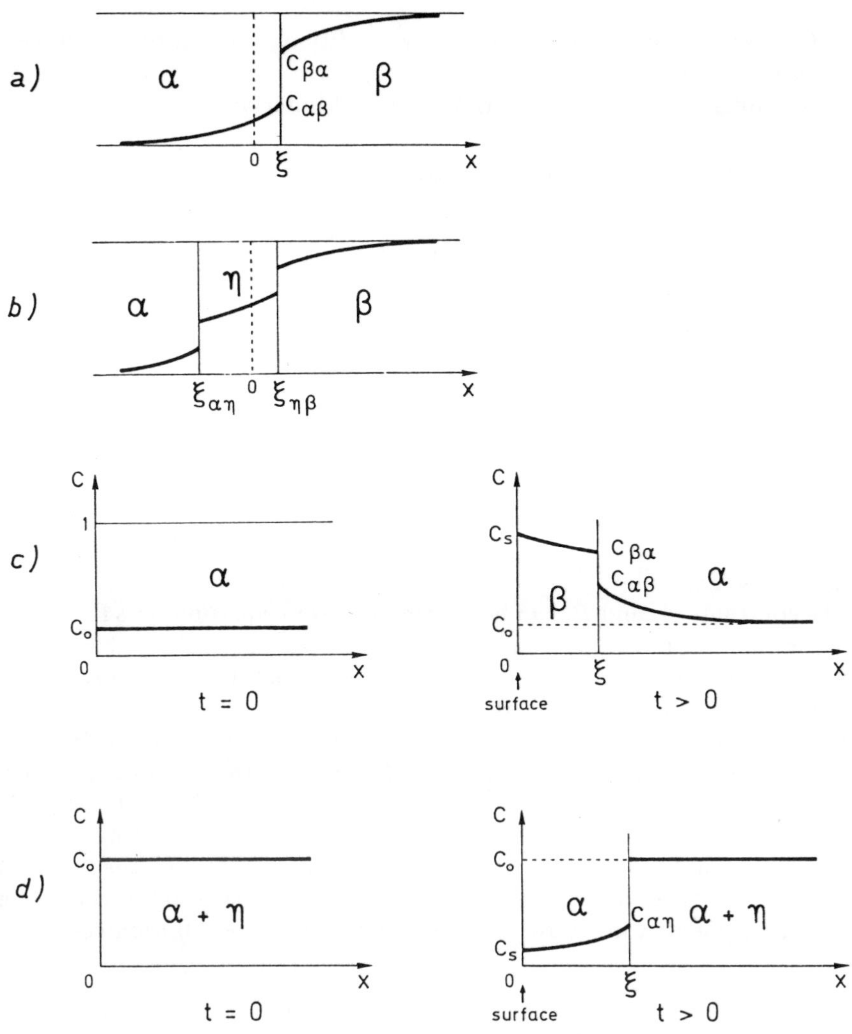

Fig. I.11. — Multiphase diffusion: a) two solid solutions; b) two terminal solid solutions with an intermediate phase; c) growth of a second phase from the surface; d) two-phase alloy, evaporation of one component. The concentration of component B is plotted along the ordinate.

3) A materials balance equation is written for the moving interface: the quantity of material that diffuses across the plane of the discontinuity during time dt is equal to the difference between the diffusion fluxes in the two phases at the interface (see App. III):

$$(c_{\alpha\beta} - c_{\beta\alpha}) \cdot \frac{\mathrm{d}\xi}{\mathrm{d}t} = \left[D_\beta \frac{\partial c_\beta}{\partial x}\right]_{\xi+} - \left[D_\alpha \frac{\partial c_\alpha}{\partial x}\right]_{\xi-} . \tag{I.36}$$

4) One writes the initial and boundary conditions relevant to the problem. In the case of several phase boundaries, an equation similar to equation (I.35) but with a different value of γ is written for each interface.

SOLUTION. — Differentiate equation (I.35):

$$d\xi/dt = \gamma (D_\alpha/t)^{1/2} . \tag{I.37}$$

Calculate the gradients at ξ^+ and ξ^-

$$\begin{aligned} \left(\frac{\partial c}{\partial x}\right)_{\xi^-} &= B_\alpha \frac{2}{\sqrt{\pi}} \frac{1}{2\sqrt{D_\alpha t}} \exp\left(-\frac{x^2}{4 D_\alpha t}\right) = B_\alpha \frac{\exp(-\gamma^2)}{(\pi D_\alpha t)^{1/2}}, \\ \left(\frac{\partial c}{\partial x}\right)_{\xi^+} &= B_\beta \frac{\exp(-\gamma^2/\varphi)}{(\pi D_\beta t)^{1/2}}, \end{aligned} \tag{I.38}$$

with $\varphi = D_\beta/D_\alpha$.

The constants A_i and B_i can be determined from equations (I.34 and I.38) and the initial and boundary conditions. The balance equation (1.36) is then used to determine the growth parameter γ which is obtained from an implicit equation of the form $f(\gamma, \varphi) = 0$.

Knowing the concentrations at the interface (e.g., from the phase diagram) and the diffusion coefficients in each phase (or, rather, their ratio φ), one can calculate γ. But in most cases, the measurement of $\xi(t)$ and knowledge of the solubility limits are not sufficient for the determination of the two diffusivities D_α and D_β because there are only two equations ((I.35) and the equation $f(\gamma, \varphi) = 0$) for three unknowns (D_α, D_β, γ). An exception to this conclusion is the case represented in figure I.11d, where only the one diffusion coefficient D_α enters.

Examples:

1) *Two terminal solid solutions* (Fig. I.11a).
The initial and boundary conditions are:

$$\begin{array}{lll} t = 0 & x < 0 & c_\alpha = 0 \\ & x > 0 & c_\beta = 1 \\ \\ t > 0 & x = \xi^- & c = c_{\alpha\beta} \\ & x = \xi^+ & c = c_{\beta\alpha} . \end{array}$$

One finds easily:

$$\begin{cases} c_\alpha = \dfrac{c_{\alpha\beta}}{1+\operatorname{erf}\gamma}\left(1+\operatorname{erf}\dfrac{x}{2\sqrt{D_\alpha t}}\right) \\ c_\beta = \dfrac{1}{\operatorname{erfc}\left(\gamma/\varphi^{1/2}\right)}\left[c_{\beta\alpha}-\operatorname{erf}\left(\gamma/\varphi^{1/2}\right)+(1-c_{\beta\alpha})\operatorname{erf}\dfrac{x}{2\sqrt{D_\beta t}}\right] \end{cases} \tag{I.39}$$

$$\frac{c_{\alpha\beta}}{c_{\beta\alpha}-c_{\alpha\beta}}\,\frac{\exp\left(-\gamma^2\right)}{\gamma\sqrt{\pi}\,(1+\operatorname{erf}\gamma)}-\frac{1-c_{\beta\alpha}}{c_{\beta\alpha}-c_{\alpha\beta}}\,\frac{\varphi^{1/2}\exp\left(-\gamma^2/\varphi\right)}{\gamma\sqrt{\pi}\operatorname{erfc}\left(\gamma/\varphi^{1/2}\right)}=1. \tag{I.40}$$

2) *Formation of a surface phase* β (Fig. I.11c).
Initial conditions:

$$t=0 \qquad x>0 \qquad c=c_0.$$

Boundary conditions:

$$\begin{array}{lll} t>0 & x=0 & c=c_s \\ & x=\xi^- & c=c_{\beta\alpha} \\ & x=\xi^+ & c=c_{\alpha\beta}. \end{array}$$

One obtains easily:

$$\begin{cases} \text{phase } \beta: \; c_\beta = c_s - \dfrac{c_s-c_{\beta\alpha}}{\operatorname{erf}\gamma}\operatorname{erf}\dfrac{x}{2\sqrt{D_\beta t}} \\ \text{phase } \alpha: \; c_\alpha = c_0 + \dfrac{c_{\alpha\beta}-c_0}{\operatorname{erfc}\left(\gamma\varphi^{1/2}\right)}\operatorname{erfc}\left(\dfrac{x}{2\sqrt{D_\alpha t}}\right). \end{cases} \tag{I.41}$$

Finally:

$$\frac{c_s-c_{\beta\alpha}}{\gamma\sqrt{\pi}\operatorname{erf}\gamma}\exp\left(-\gamma^2\right)-\frac{(c_{\alpha\beta}-c_0)\exp\left(-\gamma^2\varphi\right)}{\sqrt{\pi}\,\gamma\,\varphi^{1/2}\operatorname{erfc}\left(\gamma\,\varphi^{1/2}\right)}=c_{\beta\alpha}-c_{\alpha\beta}. \tag{I.42}$$

Recall that here $\xi = 2\gamma\sqrt{D_\beta t}$, $\varphi = D_\beta/D_\alpha$. These equations apply, for example, to oxidation (β is the oxide), with diffusion of oxygen in the metal substrate; in this case $c_0 \approx 0$. The unknowns are D_α, D_β, and γ; one can measure ξ, and $c_\beta(x)$ and $c_\alpha(x)$, if one can determine the depth profiles in both phases.

3) *Evaporation from an initially two-phase mixture* $\alpha+\eta$ (Fig. I.11d).

Initial conditions:

$$t = 0 \qquad c = c_0 .$$

Boundary conditions:

$$t > 0 \qquad \begin{array}{ll} x = 0 & c = c_s \\ x = \xi^- & c = c_{\alpha\eta} . \end{array}$$

This yields:

$$c_\alpha = c_s + \frac{c_0 - c_{\alpha\eta}}{\exp(-\gamma^2)} \sqrt{\pi}\gamma \operatorname{erf}\left(\frac{x}{2\sqrt{D_\alpha t}}\right) \tag{I.43}$$

and

$$\sqrt{\pi}\gamma \exp\left(\gamma^2\right) \cdot \operatorname{erf}(\gamma) = \frac{c_{\alpha\eta} - c_s}{c_0 - c_{\alpha\eta}} . \tag{I.44}$$

If c_s and $c_{\alpha\eta}$ are known, γ can be calculated, and from it, D_α via the movement of the interface $\xi = 2\gamma \sqrt{D_\alpha t}$.

APPENDIX I

Methods for solving the diffusion equation.

1. METHOD OF SUPERPOSITION AND REFLECTION. — The first of these leads to the solution for a thick or semi-infinite layer by addition (in fact, integration) of the solutions for a series of thin layers. The second yields solutions for a finite sample, under the assumption that the boundaries are impermeable, i.e., the diffusing material is "reflected" at the boundary, with no loss of the diffusing material by evaporation.

2. METHOD OF SEPARATION OF VARIABLES. — A solution of the form

$$c(x, t) = X(x) \cdot T(t) \tag{A.1}$$

is assumed. The diffusion equation then becomes:

$$\begin{aligned} \frac{1}{T}\frac{\mathrm{d}T}{\mathrm{d}t} &= -\lambda^2 D \\ \frac{1}{X}\frac{\mathrm{d}^2X}{\mathrm{d}x^2} &= -\lambda^2 . \end{aligned} \tag{A.2}$$

The solutions of these equations are:

$$\begin{aligned} T &= \exp\,(-\,\lambda^2\,Dt) \\ X &= A\sin\lambda x + B\cos\lambda x. \end{aligned} \tag{A.3}$$

The initial and boundary conditions are used to determine a series of values of λ_m, A_m, and B_m so that the general solution is:

$$c\,(x,\,t) = \sum_{m=0}^{\infty} (A_m \sin\lambda_m\, x + B_m \cos\lambda_m\, x)\,\exp\,(-\,\lambda_m^2\,Dt)\,. \tag{A.4}$$

This method is used for the problem of the plane sheet (§ III.2).

3. THE LAPLACE TRANSFORM. — By definition:

$$\bar{c}(s) = \int_0^\infty e^{-st}\,c(t)\,dt, \tag{A.5}$$

where s is large enough so the integral converges. Multiplying both sides of Fick's equation by $e^{-st}\,dt$ and integrating, we obtain:

$$D\int_0^\infty e^{-st}\,\frac{\partial^2 c}{\partial x^2}\,dt = \int_0^\infty e^{-st}\,\frac{\partial c}{\partial t}\,dt. \tag{A.6}$$

Reversing the order of differentiation and integration yields:

$$\int_0^\infty e^{-st}\,\frac{\partial^2 c}{\partial x^2}\,dt = \frac{\partial^2}{\partial x^2}\int_0^\infty e^{-st}\,c\,dt = \frac{\partial^2 \bar{c}}{\partial x^2}.$$

Integration by parts of the second term of equation (A.6) gives:

$$\int_0^\infty e^{-st}\,\frac{\partial c}{\partial t}\,dt = \left[c\,e^{-st}\right]_0^\infty + s\,\bar{c}.$$

Let the initial and boundary conditions correspond to the semi-infinite solid with constant surface concentration:

$$\begin{cases} t = 0 & x > 0 \quad c = 0 \\ t > 0 & x = 0 \quad c = c_0. \end{cases}$$

then the term in brackets is zero, and the diffusion equation becomes:

$$D\,\frac{\partial^2 \bar{c}}{\partial x^2} = s\bar{c}, \tag{A.7}$$

the solution of which has the form:

$$\bar{c} = A \exp\left(\sqrt{\frac{s}{D}}\, x\right) + B \exp\left(-\sqrt{\frac{s}{D}}\, x\right). \tag{A.8}$$

Inserting the boundary conditions yields:

$$\bar{c} = \frac{c_0}{s} \exp\left(-\sqrt{\frac{s}{D}}\, x\right). \tag{A.9}$$

From a table of Laplace transforms:

$$c = c_0 \operatorname{erfc}\left(x/2\sqrt{Dt}\right). \tag{A.10}$$

4. Fourier transform. — By definition of the Fourier transform:

$$\left.\begin{aligned} \bar{c}(X,\, t) &= \int_{-\infty}^{\infty} c(x,\, t)\, e^{-2\pi i x X}\, \mathrm{d}x \\ c(x,\, t) &= \int_{-\infty}^{\infty} \bar{c}(X,\, t)\, e^{2\pi i x X}\, \mathrm{d}X. \end{aligned}\right\} \tag{A.11}$$

Differentiating the second equation twice with respect to x:

$$\frac{\partial^2 c}{\partial x^2} = \int_{-\infty}^{\infty} (-\,4\,\pi^2\, X^2)\; \bar{c}\,(X,\, t)\; \mathrm{e}^{2\pi i x X}\, \mathrm{d}X,$$

which shows that $(-\,4\,\pi^2\, X^2)\,\bar{c}$ is the transform of $\partial^2 c/\partial x^2$. Evidently $\partial \bar{c}/\partial t$ is the transform of $\partial c/\partial t$. Fick's equation then becomes:

$$\frac{\partial \bar{c}(X,\, t)}{\partial t} = -\,4\,\pi^2\, X^2\, D\bar{c}(X,\, t)$$

or

$$\partial \log \bar{c}/\partial t = -\,4\,\pi^2\, X^2\, D. \tag{A.12}$$

The solution is a Gaussian. With the thin-layer initial condition:

$$\begin{aligned} c(x,\, 0) &= \delta(x), \\ \bar{c}(X,\, 0) &= 1. \end{aligned}$$

From which

$$\bar{c}(X,\, t) = \exp\left(-\,4\,\pi^2\, Dt\, X^2\right). \tag{A.13}$$

Since the transform of $\exp[-\pi\ (x/l)^2]$ is $(1/|l|)\ \exp[-\pi\ (lX)^2]$,

$$c(x,\ t) = \frac{1}{2\sqrt{\pi D t}} \exp\left(-\frac{x^2}{4\ Dt}\right). \tag{A.14}$$

Note that the Fourier integral exists provided that $\int_{-\infty}^{\infty} |c(x)|\ dx$ converges, which excludes certain functions from this treatment.

5. NUMERICAL SOLUTIONS. — With the development of advanced computers, there is less interest in analytical solutions for complicated cases, as numerical integration of the diffusion equation is often simpler, e.g., the *method of finite differences* (Crank, 1975, Ch. 8). Analytical and numerical methods appear as complementary tools for the solution of diffusion problems (Ghez, 1988).

APPENDIX II

Diffusion in three dimensions.

When diffusion in a medium is isotropic, the flux is:

$$\mathbf{J} = -D\ \nabla \mathbf{c} \tag{A.15}$$

and the diffusion equation is given by:

$$\frac{\partial c}{\partial t} = D\ \nabla^2 c. \tag{A.16}$$

This equation is, in cartesian coordinates:

$$\frac{\partial c}{\partial t} = D \left[\frac{\partial^2 c}{\partial x^2} + \frac{\partial^2 c}{\partial y^2} + \frac{\partial^2 c}{\partial z^2}\right], \tag{A.17}$$

in cylindrical coordinates:

$$\frac{\partial c}{\partial t} = \frac{D}{r}\frac{\partial}{\partial r}\left(r\ \frac{\partial c}{\partial r}\right), \tag{A.18}$$

in spherical coordinates:

$$\frac{\partial c}{\partial t} = \frac{D}{r^2}\frac{\partial}{\partial r}\left(r^2\ \frac{\partial c}{\partial r}\right). \tag{A.19}$$

Cylindrical coordinates. — Separation of variables leads to a series of terms of the type $J_0\ (\alpha_n\ r)\exp\left(-\ \alpha_n^2\ Dt\right)$, where J_0 is the Bessel function of zeroth order. For steady state, equation (A.18) becomes:

$$\frac{\mathrm{d}}{\mathrm{d}r}\left(r\ \frac{\mathrm{d}c}{\mathrm{d}r}\right) = 0 \tag{A.20}$$

the general solution of which has the form:

$$c = A + B \log r, \tag{A.21}$$

with the constants A and B determined from the boundary conditions.

Spherical coordinates. — Let $u = rc$. Equation (A.19) then becomes:

$$\frac{\partial u}{\partial t} = D\ \frac{\partial^2 u}{\partial r^2}. \tag{A.22}$$

Equation (A.22) is easily solved by separation of variables if u goes to zero as r becomes very large. For steady state:

$$\frac{\mathrm{d}}{\mathrm{d}r}\left(r^2\ \frac{\mathrm{d}c}{\mathrm{d}r}\right) = 0, \tag{A.23}$$

the general solution of which is:

$$c = (A/r) + B, \tag{A.24}$$

where the constants A and B are again determined from the boundary conditions.

APPENDIX III

Conservation at a moving boundary.

The total quantity of the diffusing species is given by:

$$M = \int_{-\infty}^{\infty} c(x,\ t)\mathrm{d}x = \int_{-\infty}^{\xi^-} c(x,\ t)\mathrm{d}x + \int_{\xi^+}^{+\infty} c(x,\ t)\mathrm{d}x$$

and must satisfy the condition $\mathrm{d}M/\mathrm{d}t = 0$. With $\xi = \xi(t)$, we obtain:

$$\mathrm{d}M/\mathrm{d}t = \int_{-\infty}^{\xi^-} \frac{\partial c}{\partial t}\,\mathrm{d}x + \frac{\mathrm{d}\xi}{\mathrm{d}t}\,c(\xi^-, t) + \int_{\xi^+}^{\infty} \frac{\partial c}{\partial t}\,\mathrm{d}x - \frac{\mathrm{d}\xi}{\mathrm{d}t}\,c(\xi^+, t)$$

$$= -\int_{-\infty}^{\xi^-} \frac{\partial J}{\partial x}\,\mathrm{d}x - \int_{\xi^+}^{\infty} \frac{\partial J}{\partial x}\,\mathrm{d}x + \frac{\mathrm{d}\xi}{\mathrm{d}t}\left[c(\xi^-, t) - c(\xi^+, t)\right].$$

For an infinite medium, $J(\pm\infty) = 0$. Then:

$$\mathrm{d}M/\mathrm{d}t = J(\xi^+) - J(\xi^-) + \frac{\mathrm{d}\xi}{\mathrm{d}t}\left[c(\xi^-, t) - c(\xi^+, t)\right] = 0,$$

or, with the notation of figure I.11a:

$$\frac{\mathrm{d}\xi}{\mathrm{d}t}\,(c_{\alpha\beta} - c_{\beta\alpha}) = \left(D_\beta \frac{\partial c_\beta}{\partial x}\right)_{\xi^+} - \left(D_\alpha \frac{\partial c_\alpha}{\partial x}\right)_{\xi^-}.$$

Bibliography

I. — General references

Adda Y. and Philibert J., *op. cit.* Ch. III.
Carslaw H. S. and Jaeger J. C., *Conduction of Heat in Solids* (Clarendon Press, Oxford) 1959.
Crank J., *The Mathematics of Diffusion* (Clarendon Press, Oxford) 1st ed. 1956, 2nd ed. 1975.
Ghez R., *A Primer of Diffusion Problems* (J. Wiley & Sons, New York) 1988.

II. — Specific references

Barr L. W. and LeClaire A. D., *Phil. Mag.* **20** (1969) 1289–1291.
Caskey G. R. and Pillinger W. L., *Metall. Trans.* **6A** (1975) 467–476.
McNabb A. and Foster P. K., *Trans. Metall. Soc. AIME* **227** (1963) 618–627.

CHAPTER II

Atomic theory of diffusion

Diffusion and mass flow or drift result from individual jumps of atoms and/or point defects in the solid. Thus the parameters in Fick's equations, the diffusivity, D, and the velocity, v, can reasonably be expressed in terms of the physical quantities that describe these elementary processes: the jump frequency and the jump distance. The simplest case is that of the crystalline lattice, since the jump distances and frequencies can take on only a small number of values.

I. A simplified model.

To show the relationship of diffusion to the underlying physical parameters, consider the very simple case of unidirectional diffusion, restricted to the flux of atoms between two neighboring planes, 1 and 2, in a simple cubic lattice with interplanar spacing λ (Fig. II.1). The net flux from left to right is:

$$\begin{aligned} J &= j_{12} - j_{21} \\ &= n_1\,\Gamma_{12} - n_2\,\Gamma_{21}, \end{aligned} \tag{II.1}$$

where Γ_{ij} is the jump frequency from plane i to plane j and n_i is the number of atoms of diffusing species per unit area on the plane i ($n_i \,/\, \lambda = c_i$, the concentration in atoms per unit volume). The quantity n_i can be expressed as a function of the average concentration, n, and its gradient:

$$\begin{aligned} n_1 &= n - \frac{1}{2}\,\lambda\,\frac{\mathrm{d}n}{\mathrm{d}x} \\ n_2 &= n + \frac{1}{2}\,\lambda\,\frac{\mathrm{d}n}{\mathrm{d}x}. \end{aligned} \tag{II.2}$$

The net flux is then:

$$J = -\frac{1}{2}\,\lambda\,(\Gamma_{12} + \Gamma_{21})\,\frac{\mathrm{d}n}{\mathrm{d}x} + n\,(\Gamma_{12} - \Gamma_{21}).$$

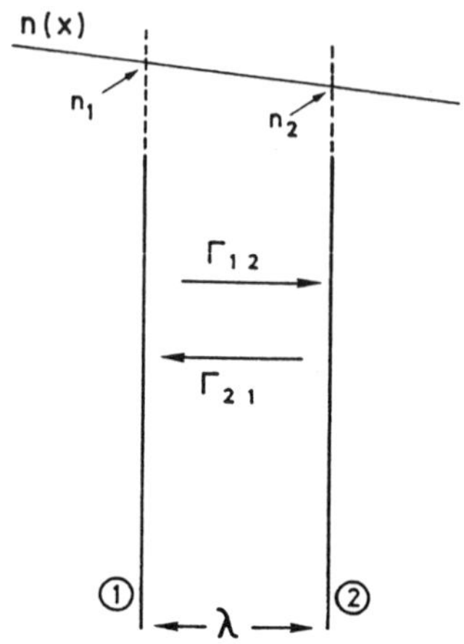

Fig. II.1. — Simplified model of diffusion in a shallow concentration gradient $n(x)$.

Changing to units of volume concentration:

$$J = -\lambda^2 \Gamma_s \frac{dc}{dx} + \lambda c (\Gamma_{12} - \Gamma_{21}), \tag{II.3}$$

with:

$$\Gamma_s = \frac{1}{2}(\Gamma_{12} + \Gamma_{21}).$$

But this is just Fick's first law, with:

$$D = \lambda^2 \Gamma_s \tag{II.4}$$

$$\langle v \rangle = \lambda (\Gamma_{12} - \Gamma_{21}). \tag{II.5}$$

In a simple cubic lattice, each atom can jump to one of six nearest neighbor sites; its total jump frequency is $\Gamma = 6\,\Gamma_s$, Γ_s being the frequency of a jump to any one of the six neighboring sites:

$$\boxed{D = \frac{1}{6} \Gamma \cdot \lambda^2}. \tag{II.6}$$

This expression is true for any cubic lattice as long as only *nearest neighbor jumps* are allowed. All the ζ jump frequencies (ζ is the coordination number) are equal, but the projection of each jump distance along the x axis is $\lambda \cos \gamma_i$ (λ is the nearest-neighbor distance and γ_i is the angle between the ith jump direction and the x axis). The above derivation is then repeated, but with the fluxes j_{12} and j_{21} being due to the different types of jumps; the jump frequencies

Γ_{12} and Γ_{21} are replaced by the ζ frequencies Γ_i and the corresponding jump distances projected on the x axis, $\lambda \cos \gamma_i$, are used:

$$J_x = -\frac{1}{2}\,\Gamma_s\,\lambda^2 \sum_{i=1}^{\zeta} \cos^2 \gamma_i\, \frac{\mathrm{d}c}{\mathrm{d}x}.$$

In the absence of a driving force and in an isotropic lattice, all the Γ_i are equal. Since $\sum \cos^2 \gamma_i = \zeta/3$,

$$D = \frac{1}{6}\,\Gamma\,\lambda^2,$$

with $\Gamma = \zeta\,\Gamma_s$. For all cubic lattices, this expression can be written as:

$$D = \Gamma_s\, a^2\ ;$$

here a is the lattice parameter.

In the presence of a driving force:

$$\Gamma_{12} \neq \Gamma_{21}$$

and expanding to first order:

$$\left.\begin{aligned}\Gamma_{12} &= \Gamma_s\,(1 + \varepsilon)\\ \Gamma_{21} &= \Gamma_s\,(1 - \varepsilon)\,,\end{aligned}\right\} \qquad \text{(II.7)}$$

whence:

$$\langle v\rangle = 2\,\varepsilon\lambda\Gamma_s. \qquad \text{(II.8)}$$

The thermal energy kT per net jump in the direction of the applied force equals the work done by this force over the distance λ:

$$\begin{aligned}kT\cdot(\Gamma_{12} - \Gamma_{21})/\Gamma_s &= F\cdot\lambda\\ \varepsilon &= \frac{1}{2}\,F\cdot\lambda/kT\end{aligned} \qquad \text{(II.9)}$$

and

$$\langle v\rangle = \frac{F\lambda^2\,\Gamma_s}{kT} = \frac{FD}{kT}. \qquad \text{(II.10)}$$

But this is just the Nernst-Einstein equation. It is valid only if $F\,\lambda \ll kT$. (For the case of F large, see Ch. III, § VII.2).

II. General theory of random walk.

Random walk means that each jump of a particle is independent of all its preceding jumps, both in length and in direction. In a crystalline medium, where the jump distance is the same for all jumps, the migration of a particle can be described by two numbers drawn at random for each jump: the first decides the direction of the jump, the second decides if the particle jumps or not. Only one number is needed in a cubic lattice if only nearest-neighbor jumps are allowed because all the jump directions are equally probable.

A good example of random movement is the *Brownian motion* of small particles in suspension in a liquid (Brown, 1828). Figure II.2, borrowed from Jean Perrin (Les Atomes, 1936), shows the paths of gamboge particles in suspension in water. The particles were observed in a microscope and projected with a camera lucida at equal time intervals. Figure II.2 shows that the concept of a mean velocity of agitation, which one had tried to define for a long time, clearly had no significance since each segment in figure II.2 is the resultant of a very large number of very small jumps forming a tangled path. In 1906, Einstein and Smoluchowski proposed independently that the course of the motion be characterized by the vector $\mathbf{R}(t)$ which joins the initial location of the particle to its final one. For motion that is truly random, $\langle \mathbf{R}(t) \rangle = 0$ is expected; the angular brackets here designate the average over a very large number of particles. This is illustrated in figure II.3, where the ends of the vectors $\mathbf{R}$, measured for the same time of observation, are plotted referred to the same origin. The vectors $\mathbf{R}$ are distributed isotropically, with no favored direction: $\langle \mathbf{R} \rangle = 0$.

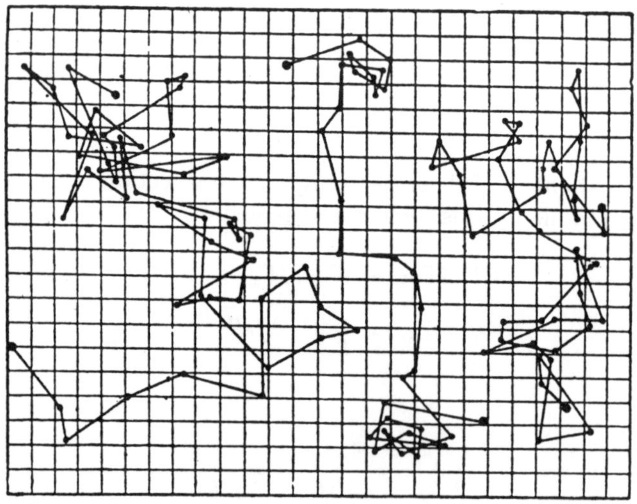

Fig. II.2. — Brownian motion of gamboge particles in an aqueous emulsion, projected at equal time intervals in a camera lucida. One division represents approximately 3 μm (Jean Perrin, Les Atomes, 1936).

What then is the magnitude characteristic of a random walk at the end of the time t? This will be shown to be the mean-square displacement $\langle R^2 \rangle$. To keep the derivation general, the medium is taken as not necessarily isotropic, and the component of the flux along the x axis, as well as the component along the x axis, $X(t)$, of the displacement $\mathbf{R}(t)$, will be considered.

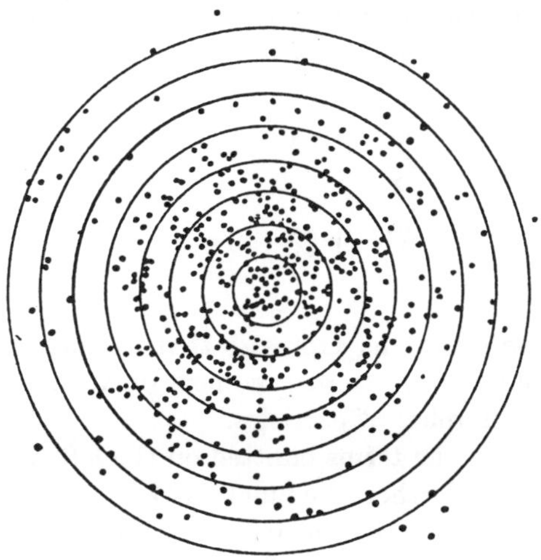

Fig. II.3. — Perrin's verification of the laws of Brownian motion. Successive displacements **R** of 0.37 μm diameter particles during a 30 second period are plotted from a common origin. The radii of the circles are equal to 1/4, 1/2, 3/4, 1,... times the root-mean-square displacement $\langle R^2 \rangle^{1/2} = 7.84$ μm.

We introduce a distribution function $W(X, \tau)$ which gives the probability that at time τ the particle will have travelled a path with projection X. It is assumed that W depends neither on x nor on τ. (This is a restrictive hypothesis that does not include, for example, the case of interdiffusion ([1])). The precise analytical form of W is not needed.

Consider next a balance for the number of particles of the diffusing species located in the plane x at the time $t + \tau$. At time t, these particles were located in the planes $x - X$. Thus:

$$c(x,\ t + \tau) = \sum_X c(x - X,\ t) \cdot W(X,\ \tau),$$

([1]) The hypothesis requires that D, $\langle v \rangle$, and F be independent of x and t.

the summation being carried out over all values of X. Expanding c to second order in x and first order in t, and dropping higher order terms:

$$c(x,\ t)+\tau\frac{\partial c}{\partial t}+\ldots=\sum_X\left[c(x,\ t)-X\frac{\partial c}{\partial x}+\frac{X^2}{2}\frac{\partial^2 c}{\partial x^2}+\ldots\right]W(X,\ \tau),\quad \text{(II.11)}$$

where the derivatives of c are defined for the plane x at time t.

By definition of the moments of X:

$$\sum_X W(X,\ \tau)=1$$
$$\sum_X X^m\, W(X,\ \tau)=\langle X^m\rangle, \quad \text{(II.12)}$$

where the first of these equations expresses that the probabilities W are normalized, and the second defines the mth moment of X: the average value of X^m taken over a very large number of particles.

For small values of τ, the terms omitted on the left hand side of equation (II.11) become negligible. Further, the function $W\ (X,\ \tau\ \rightarrow\ 0)$ becomes more and more narrow around $x\ =\ 0$, and terms higher than second order on the right hand side can also be neglected. Thus:

$$\frac{\partial c}{\partial t}=-\frac{\langle X\rangle}{\tau}\frac{\partial c}{\partial x}+\frac{\langle X^2\rangle}{2\tau}\frac{\partial^2 c}{\partial x^2}. \quad \text{(II.13)}$$

For a random walk in the absence of a driving force $\langle X\rangle\ =\ 0$:

$$\frac{\partial c}{\partial t}=\frac{\langle X^2\rangle}{2\,\tau}\frac{\partial^2 c}{\partial x^2}, \quad \text{(II.13a)}$$

and one obtains the diffusion equation with:

$$\boxed{D_x=\frac{\langle X^2\rangle}{2\,\tau}}. \quad \text{(II.14)}$$

The expression (II.14) is due to Einstein (1904).

Note that:

1) The limit $\tau\ \rightarrow\ 0$ has been omitted from the Einstein equation because $\langle X^2\rangle/\tau$ is independent of τ for a random walk.

In the presence of a driving force, $\langle X\rangle\ \neq\ 0$, and $\langle X^2\rangle/\tau$ is not independent of τ. Since no restriction in this regard was included in the preceding derivation (cf. the definition of W, which does not assume a random walk, so that the above derivation is quite general), the Einstein equation remains valid on the condition that ([2]):

$$D_x = \lim_{\tau \to 0} \frac{\langle X^2 \rangle}{2\,\tau}.$$

2) For an isotropic medium, $\langle X^2 \rangle = \langle Y^2 \rangle = \langle Z^2 \rangle$ and $\langle R^2 \rangle = \langle X^2 \rangle + \langle Y^2 \rangle + \langle Z^2 \rangle$, whence:

$$D = \frac{\langle R^2 \rangle}{6\,\tau} \quad (D_x = D_y = D_z = D). \tag{II.14a}$$

3) In the presence of a driving force, $\langle X \rangle \neq 0$. The transport term is then given by equation (II.13), with:

$$\langle v_x \rangle = \lim_{\tau \to 0} \frac{\langle X \rangle}{\tau}. \tag{II.15}$$

III. Expressions for the mean-square displacement $\langle X^2 \rangle$ and for the diffusion coefficient.

Let a given particle carry out n jumps during the time τ. The projection of its path on the x axis is:

$$X = x_1 + x_2 + \cdots + x_n,$$

where x_i is the projection of the ith jump, and

$$X^2 = \sum_{i=1}^{n} x_i^2 + 2 \sum_{i=1}^{n-1} \sum_{j=i+1}^{n} x_i\, x_j.$$

([2]) The condition $\tau \to 0$ looks very restrictive. Actually it is related to the present derivation. If the materials balance were derived on the basis of the number of particles, an equation:

$$\partial c / \partial t = \left[\langle X^2 \rangle / 2t \right] \partial^2 c / \partial x^2$$

would have been obtained without the above condition on time. A very rigorous derivation (Le Claire and Rabinovitch, 1989) does not introduce the condition $\tau \to 0$.

Taking the average over a large number of particles, and using the fact that the average of a sum of terms equals the sum of the averages of the terms:

$$\langle X^2 \rangle = \sum_{i=1}^{n} \langle x^2 \rangle + 2 \sum_{i=1}^{n-1} \sum_{j=i+1}^{n} \langle x_i \, x_j \rangle . \qquad \text{(II.16)}$$

The double sum contains $n(n-1)$ terms; it equals $n(n-1)$ times the average value of the products $x_i \, x_j$. But for a random walk, this must be zero, as for every pair $x_i \, x_j$ one can find for another particle a pair $x_i \, x_j$ equal and opposite in sign. Thus:

$$\langle X^2 \rangle = \sum_{i=1}^{n} \langle x_i^2 \rangle . \qquad \text{(II.17)}$$

In a crystalline medium, the x_i can take on only a few definite values, corresponding to the ζ possible jumps (ζ is the nearest neighbor coordination number, if only nearest neighbor jumps are allowed). Among the n jumps, there are n_s jumps of the type s with $1 \le s \le \zeta$, with projected length on the x axis x_s

$$\begin{aligned} \langle X^2 \rangle &= \langle n_1 \, x_1^2 \rangle + \cdots + \langle n_\zeta \, n_\zeta^2 \rangle \\ &= \langle n_1 \rangle \, x_1^2 + \cdots + \langle n_\zeta \rangle \, x_\zeta^2 . \end{aligned}$$

Introduction of the corresponding average jump frequencies defined as:

$$\langle n_s \rangle = \Gamma_s \, \tau$$

transforms the above equation into:

$$\langle x^2 \rangle = \tau \sum_{s=1}^{\zeta} \Gamma_s \, x_s^2 . \qquad \text{(II.18)}$$

Application of the Einstein equation (Eq. (II.14)) yields:

$$\boxed{D_x = \frac{1}{2} \sum_{s=1}^{\zeta} \Gamma_s \, x_s^2} . \qquad \text{(II.19)}$$

APPLICATIONS.

Crystals with cubic structure. — Γ_s is a constant, independent of s, if only nearest-neighbor jumps are allowed. $D_x = D_y = D_z = D$.

$$D = \frac{1}{2} \Gamma_s \sum_{s=1}^{\zeta} x_s^2, \text{ where}$$

$$D = \frac{1}{6} \Gamma_s \sum \lambda^2 = \frac{1}{6} \zeta \, \Gamma_s \lambda^2 \quad (\lambda \text{ is the jump distance}). \tag{II.19a}$$

N. B. :
Do not confuse the jump frequency towards a given site, Γ_s, with the total jump frequency $\Gamma = \zeta \Gamma_s$, which expresses the average number of jumps, (toward any nearest-neighbor site) of the particle per unit time: $D = (1/6) \, \Gamma \, \lambda^2 = (1/6)\lambda^2/\tau$, where $\tau = \Gamma^{-1}$ is the mean time of residence of a particle on a site. (τ has a different meaning here than in § II of this chapter.)

For a *BCC structure, with lattice parameter a*, there are 8 jump directions, each with a projected length of $\pm a/2$ (Fig. II.4):

$$D = \frac{1}{2} \Gamma_s \, 8 \, \frac{a^2}{4} = \Gamma_s \, a^2. \tag{II.20}$$

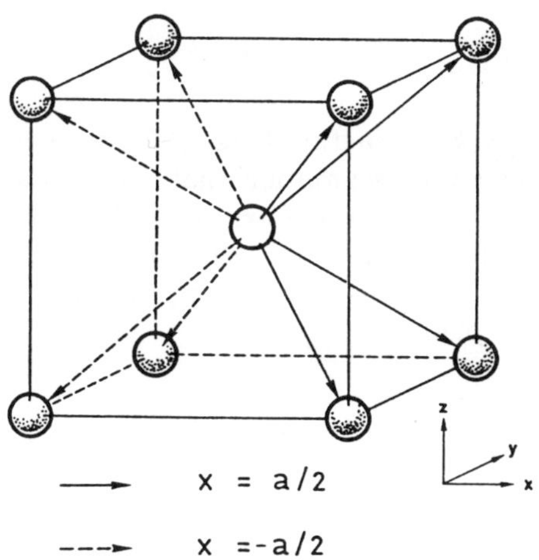

Fig. II.4. — BCC structure; jumps to the 8 nearest neighbors.

For a *FCC structure* there are 4 jumps with projected length $-a/2$, 4 with zero projected length, and 4 with projected length $+a/2$ (Fig. II.5):

$$D = \frac{1}{2} \Gamma_s \left(4 \, \frac{a^2}{4} + 4 \cdot 0 + 4 \, \frac{a^2}{4}\right) = \Gamma_s \, a^2. \tag{II.20a}$$

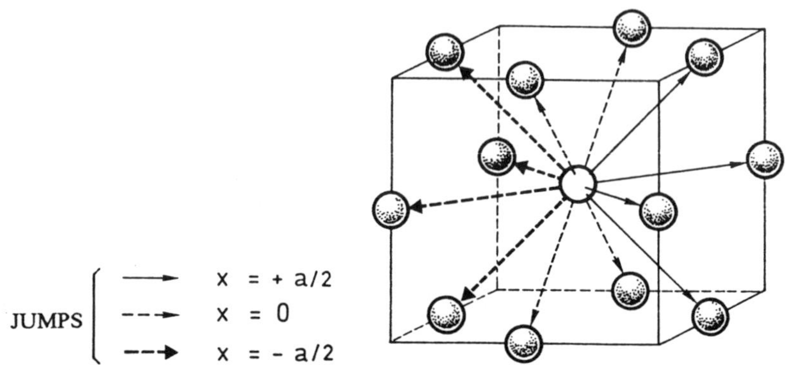

Fig. II.5. — FCC structure; jumps toward the 12 nearest neighbors.

For a *diamond cubic structure*, (Fig. II.6) the coordination number is 4, and there are thus 4 jumps of length λ such that $\lambda^2 = 3\,(a/4)^2$. Hence:

$$D = \frac{1}{6}\, 4\Gamma_s\, 3\,(a/4)^2 = \frac{1}{8}\, \Gamma_s\, a^2. \tag{II.21}$$

In the *hexagonal close-packed* structure (Fig. II.7), the symmetry is such that $D_x = D_y \neq D_z$, under the assumption that the z axis is parallel to the c direction of the lattice. Further, there are two values of the jump frequency Γ_s, one for jumps in the basal plane (Γ_1), the other for jumps out of the basal plane (Γ_2).For the x or y directions, considering 6 jumps in the basal plane and 6 jumps out of the basal plane:

$$\begin{aligned} D_x &= \frac{1}{2}\left[2\,\Gamma_1\, a^2 + 4\,\Gamma_1\, \frac{a^2}{4} + 4\,\Gamma_2\, \frac{a^2}{4} + 2\,\Gamma_2 \cdot 0\right] \\ &= \frac{a^2}{2}\,(3\,\Gamma_1 + \Gamma_2)\,. \end{aligned} \tag{II.22}$$

For the z axis, only 6 jumps with projected length $\pm\, c/2$ contribute to the difffusion:

$$D_z = \frac{1}{2}\, 6\,\Gamma_2\, \frac{c^2}{4} = \frac{3}{4}\, \Gamma_2\, c^2. \tag{II.22a}$$

For an ideal HCP structure, the axial ratio $c/a = \sqrt{8/3}$, $\Gamma_1 = \Gamma_2$, and

$$D_x = D_z\,.$$

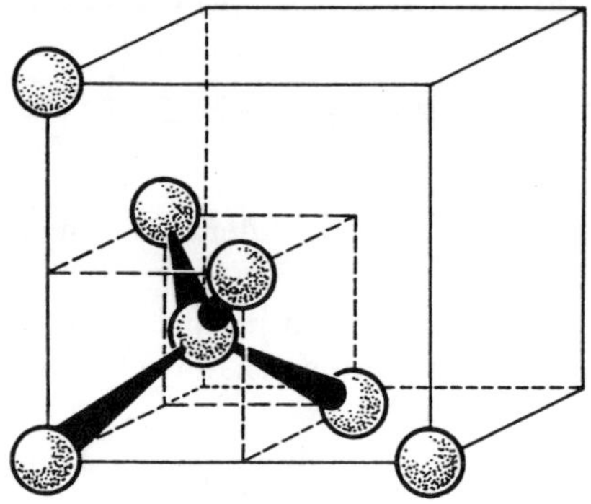

Fig. II.6. — Diamond cubic structure; jumps toward the 4 nearest neighbors.

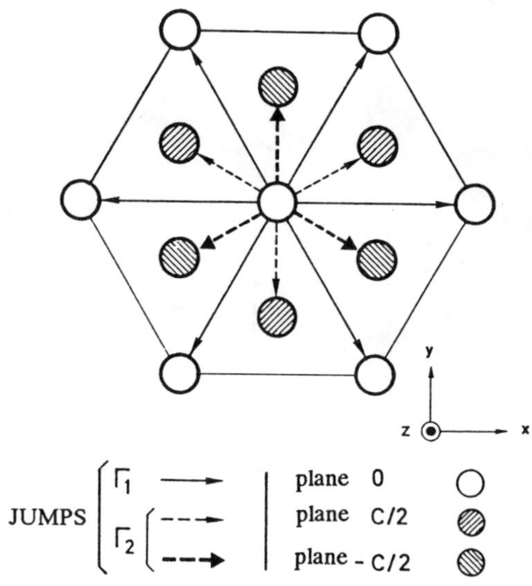

Fig. II.7. — Hexagonal close-packed cell, projected on the basal plane; jumps toward the 12 nearest neighbors, of which 6 are in the plane of the figure, 3 in the plane at $+c/2$ and 3 in the plane at $-c/2$.

Interstitial diffusion. — The above cases concerned particles on normal or substitutional sites in the lattice. The general formula also encompasses particles moving over interstitial sites, and is applied below to particles confined to *octahedral sites*:

FCC structure: Each interstitial can jump toward 12 neighboring interstitial sites (Fig. II.8a). Then $\Gamma = 12\,\Gamma_s$, $\lambda = a/\sqrt{2}$, and from equation (II.19a):

$$D = \frac{1}{12}\,\Gamma\,a^2 = \Gamma_s a^2 = \frac{1}{12}\,\frac{a^2}{\tau_m}, \qquad \text{(II.23)}$$

where $\tau_m = \Gamma^{-1}$ is the mean time of residence on one site.

BCC structure: Each interstitial can jump toward 4 neighboring sites at a distance $\lambda = a/2$. Then from equation (II.19a) (Fig. II.8b):

$$D = \frac{1}{24}\,\Gamma\,a^2 = \frac{1}{6}\,\Gamma_s\,a^2 = \frac{1}{24}\,\frac{a^2}{\tau_m}. \qquad \text{(II.24)}$$

The case of diffusion over octahedral and tetrahedral sites is considered in exercice # 37.

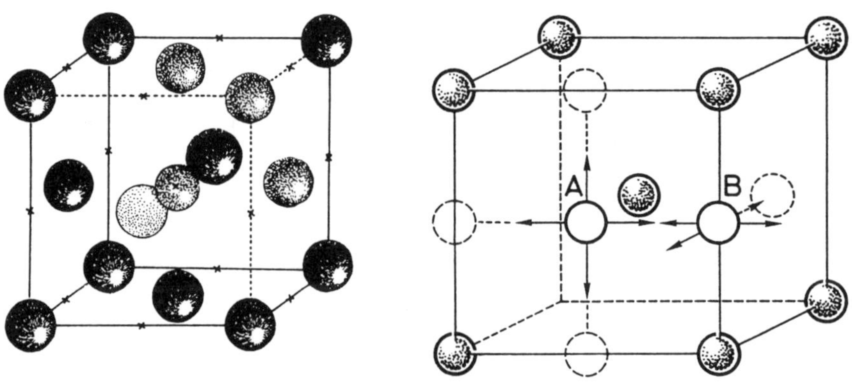

Fig. II.8. — a) FCC structure: the interstitial atom at the center of the cube can jump to octahedral sites (X). b) BCC structure: jumps on octahedral interstitial sites.

Note that the calculation of the mean-square displacement, $\langle R^2 \rangle$, or its x component, $\langle X^2 \rangle$, has been carried out under the assumption of a *purely random walk* of the diffusing particles. This presupposes two conditions:

1). For the calculation of the averages indicated by $\langle\ \rangle$, the movement of each particle must be independent of the motion of all the other particles. This is not always the case; for example, diffusion on a lattice of interstitial sites, with interaction between diffusing atoms (or ions) such that the sites neighboring one atom are forbidden to the others (cf. Ch. III, § VI).

2). The successive jumps of a particle must be independent of each other. If this condition is not met, the jumps are said to be correlated. In this case, the averages of the products in equation (II.16) are not zero. Calculation of the effect of correlation between jumps is given in chapter III.

IV. Diffusion in the presence of a driving force.

In the above notation:

$$X = x_1 + x_2 + \cdots + x_n .$$

$$\langle X \rangle = \sum_{s=1}^{\zeta} \langle n_s \rangle \, x_s$$

$$\frac{\langle X \rangle}{\tau} = \sum_{s=1}^{\zeta} \Gamma_s \, x_s ,$$

with

$$\Gamma_s = \Gamma_s^0 \, (1 + \varepsilon_s) ,$$

and following equation (II.9):

$$\varepsilon_s = \frac{1}{2} F_x \, x_s / kT,$$

where F_x is the component of the driving force along the x axis. By symmetry:

$$\sum \Gamma_s^0 \, x_s = 0,$$

whence

$$\frac{\langle X \rangle}{\tau} = \frac{1}{2} \frac{F_x}{kT} \sum \Gamma_s^0 \, x_s^2 . \qquad \text{(II.25)}$$

Calculating $\langle X^2 \rangle$:

$$\langle X^2 \rangle = \sum_{i=1}^{n} \langle x_i^2 \rangle + 2 \sum_{i \neq j} \langle x_i \, x_j \rangle .$$

Consider a simple cubic lattice. The jumps of length λ take place to the left with a probability $p \leftarrow$ and to the right with probability $p \rightarrow$. Then for n jumps:

$$\langle X \rangle = n\lambda \; (p \rightarrow - p \leftarrow) .$$

In the expression for $\langle X^2 \rangle$, arrange the terms $x_i \, x_j$ in a table with the subscript i denoting the row and the subscript j denoting the column. There are $n\,(n-1)$ products; they can be positive with probability $(p \leftarrow)^2$ or $(p \rightarrow)^2$ (two jumps with the same sign), or negative with probability $(p \leftarrow)(p \rightarrow)$ or $(p \rightarrow)(p \leftarrow)$.

It then follows:

$$\langle X^2 \rangle = n\lambda^2 + n(n-1)\ \lambda^2\ [p \rightarrow^2 + p \leftarrow^2 - 2p \rightarrow\ p \leftarrow]$$
$$\langle X^2 \rangle = n\lambda^2 + \frac{n-1}{n}\ \langle X \rangle^2 \approx n\lambda^2 + \langle X \rangle^2 .$$

In the limit $\tau \rightarrow 0$, $\langle X \rangle$ becomes negligible and $\langle X^2 \rangle \rightarrow n\lambda^2$, independent of the driving force. The value of the diffusion coefficient D is therefore not affected by the driving force. Thus:

$$\langle X^2 \rangle = \langle X^2 \rangle_{F=0} + \langle X \rangle^2$$
$$D = \frac{\langle X^2 \rangle_{F=0}}{2\,\tau} = \frac{1}{2\,\tau}\ \left[\langle X^2 \rangle - \langle X \rangle^2\right]. \tag{II.26}$$

This result is exactly the translation of the Gaussian under an applied electric field that was noted in the preceding chapter (Eq. (I.8)). The diffusion coefficient is obtained from the width of the Gaussian (the same as in the absence of F, given by $\langle X^2 \rangle_{F=0}$). It is clear, however, that $\langle X^2 \rangle$ is modified due to the translation $\langle X \rangle$ of the entire Gaussian.

Equation (II.25) above, when combined with equation (II.18) shows that:

$$\langle X \rangle = \frac{1}{2}\ \frac{F_x}{kT}\ \langle X^2 \rangle_{F=0}, \tag{II.27}$$

or once again

$$\langle v_x \rangle = \frac{F_x D}{kT}, \tag{II.27a}$$

the Nernst-Einstein equation.

V. Explicit form of the function $W(X,\ \tau)$.

For one dimensional diffusion, the n jumps taken by a particle can be separated into the $n \rightarrow$ jumps to the right and the $n \leftarrow$ jumps to the left. Then:

$$n \rightarrow = \frac{1}{2}\ (n+m)$$
$$n \leftarrow = \frac{1}{2}\ (n-m),$$

where the net displacement is $m\lambda = (n \rightarrow - n \leftarrow)\,\lambda$.

Let $W(m,\ n)$ be defined as the probability that in a sequence of n jumps, the particle undergoes a net displacement $m\lambda$. If there is no driving force, the probabilities of jumps to the left and to the right are equal:

$$p \rightarrow = p \leftarrow = \frac{1}{2}.$$

The probability of a sequence of $n \rightarrow +n \leftarrow$ jumps is then:

$$\left(\frac{1}{2}\right)^{n\rightarrow + n\leftarrow} = \left(\frac{1}{2}\right)^{n}.$$

But the net displacement $m\lambda$ can be accomplished by various distinguishable permutations of $n \rightarrow$ and $n \leftarrow$ jumps:

$$\frac{n!}{\left(\frac{n+m}{2}\right)!\left(\frac{n-m}{2}\right)!},$$

whence:

$$W(m,\ n) = \left(\frac{1}{2}\right)^{n} \frac{n!}{\left(\frac{n+m}{2}\right)!\left(\frac{n-m}{2}\right)!}. \tag{II.28}$$

This expression can be simplified by using Stirling's approximation of $n!$ for large n:

$$n! \approx \left(\frac{n}{e}\right)^{n} \sqrt{2\ \pi n}\left[1 + \frac{1}{12n} + 0\left(n^{-2}\right)\right].$$

If terms of order n^{-2} and higher are neglected:

$$W(m,\ n) = \sqrt{\frac{2}{\pi n}} \exp\left(-\frac{m^2}{2\ n}\right). \tag{II.29}$$

This function is discontinuous; nevertheless, the points are situated on a Gaussian. It is possible to pass to a continuous variable, X, with $X = m\lambda$. Care must be taken that the function W is normalized:

$$W(X,\ n)\ \mathrm{d}X = W(m,\ n)\ \frac{\mathrm{d}X}{2\lambda},$$

the probability that X is between $m\lambda$ and $(m+2)\lambda$, since m must be even. Then:

$$W(X,\ n) = \frac{1}{\sqrt{2\ \pi n \lambda^2}} \exp\left(-\frac{X^2}{2\ n\lambda^2}\right),$$

or, since for one-dimensional diffusion $D = \Gamma_s \lambda^2 = \frac{1}{2}\,\Gamma\lambda^2$, we obtain:

$$W(X,\ \tau) = \frac{1}{2\sqrt{\pi D\tau}}\ \exp\left(-\ \frac{X^2}{4\ D\tau}\right). \tag{II.30}$$

VI. Variable jump distance.

Consider a sequence of n_1 jumps of projected length λ_1 and n_2 jumps of projected length λ_2. One can arbitrarily choose first all the jumps of type 1, followed by all the jumps of type 2, to calculate the displacement $X = m_1\lambda_1 + m_2\lambda_2$:

$$W(m_1\ \lambda_1 + m_2\ \lambda_2,\ n) = W(m_1,\ n_1)\cdot W(m_2,\ n_2),$$

where $n = n_1 + n_2$. Changing to continuous variables:

$$X = X_1 + X_2 = m_1\ \lambda_1 + m_2\ \lambda_2,$$

$$W(X,\ n_1,\ n_2)\,\mathrm{d}X = \int_{-\infty}^{+\infty} \left[W(X_1,\ n_1)\cdot W(X - X_1,\ n_2)\,\mathrm{d}X_1\right]$$

$$= \frac{1}{\sqrt{2\pi\,\langle X^2\rangle}}\ \exp\left(-\frac{X^2}{2\,\langle X^2\rangle}\right),$$

with:

$$\langle X^2\rangle = n_1\ \lambda_1^2 + n_2\ \lambda_2^2,$$

that is:

$$W(X,\ \tau) = \frac{1}{2\sqrt{\pi D\tau}}\ \exp\left(-\ \frac{X^2}{4\ D\tau}\right),$$

with:

$$D = \frac{\langle X^2\rangle}{2\tau} = \frac{1}{2}\left[\frac{n_1}{\tau}\ \lambda_1^2 + \frac{n_2}{\tau}\ \lambda_2^2\right].$$

This result can be generalized to any number of types of jump:

$$2D = (\Gamma_1\ \lambda_1^2 + \Gamma_2\ \lambda_2^2 + \cdots) \tag{II.31}$$

an expression which is often written

$$D = \sum_i D_i, \tag{II.31a}$$

using the *partial diffusion coefficients* $D_i = (1/2)\,\Gamma_i\,\lambda_i^2$ ([3])

This expression will be found later in the case of the "effective" diffusion in a crystal containing dislocations, with the index 1 refering to the "good" crystal and the index 2 to the "bad" crystal or the dislocation cores; this is Hart's model for dislocation enhancement of diffusion (see Ch. VII, § I.2). The expression (II.31a) is encountered again in the case where several diffusion mechanisms are operating due to the presence of several types of point defects (vacancies and divacancies, for example); in this case, one simply adds the diffusion coefficients calculated for each mechanism. Equation (II.31) also applies to the case considered above of jumps of the same length but making different angles with the x axis:

$$\langle X^2 \rangle = \sum_{j=1}^{\varsigma} N_j\,\lambda^2\,\cos^2\psi_j,$$

where ψ_j is the angle of the vector of jump j with the x axis; equation (II.20) and the following equations are easily found from this.

For a non-cubic structure possessing at least one axis of three fold or higher symmetry (e.g. the hexagonal system), the diffusion coefficient along a direction making an angle α with that axis is given by:

$$D_\alpha = D_{\parallel}\,\cos^2\alpha + D_{\perp}\,\sin^2\alpha.$$

For a continuous distribution of the lengths of the jump distances, if we denote the jump frequency for jumps of length λ to $\lambda + \mathrm{d}\lambda$ by $\Gamma(\lambda)\mathrm{d}\lambda$, the expression for D becomes:

$$D = \frac{1}{2}\int_0^{\infty} \Gamma(\lambda)\,\lambda^2\,\mathrm{d}\lambda, \tag{II.32}$$

which is applicable to amorphous structures.

VII. Correlation functions.

To describe diffusion in liquids or gases, a formalism different from the above is used because of the variable structure of the medium. This formalism

([3]) Note that in this paragraph, the subscripted λ s are projected lengths.

is also used for solids in the interpretation of certain experiments (neutron scattering, NMR) and in simulation by molecular dynamics. For that reason, some elements of it are given here.

VII.1. CHARACTERIZATION OF THE STRUCTURE IN A NON-CRYSTALLINE MEDIUM. — The structure is characterized by a *radial distribution*: let an atom be at point j (Fig. II.9); the volume element Δr, located at a distance r, contains n_{ij} atoms at the instant of time t. The average of the n_{ij} is calculated for the N atoms j contained in the volume I. At equilibrium this average is independent of time, and the *radial distribution function* is given by:

$$\begin{aligned} \langle n_{r,\ r+\Delta r} \rangle &= \frac{1}{N} \sum_{j=1}^{N} n_{ij} \\ &= 4\ \pi r^2\ \Delta r \rho(r), \end{aligned} \tag{II.33}$$

where $\rho(r)$ is called the *radial density*. $\rho(r)$ is zero for $r <$ the atomic diameter, goes through a maximum at $r \approx r_0$, the first nearest-neighbor distance, and tends to $\rho_0 = N/V$, the atomic density, for $r \to \infty$ (Fig. II.10). More often, a dimensionless variable is used:

$$g(r) = \rho(r)/\rho_0 \qquad g(r) \to 1 \qquad r \to \infty$$

which is called the *pair distribution function*, or $G(r) = g(r) - 1$, called the *pair correlation function*.

If the particles are mobile, these functions must be functions of time, t, as well as of the space coordinates, $\mathbf{r}$.

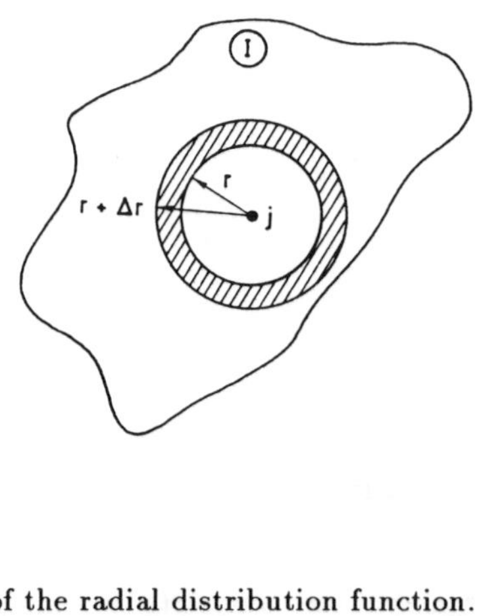

Fig. II.9. — Definition of the radial distribution function.

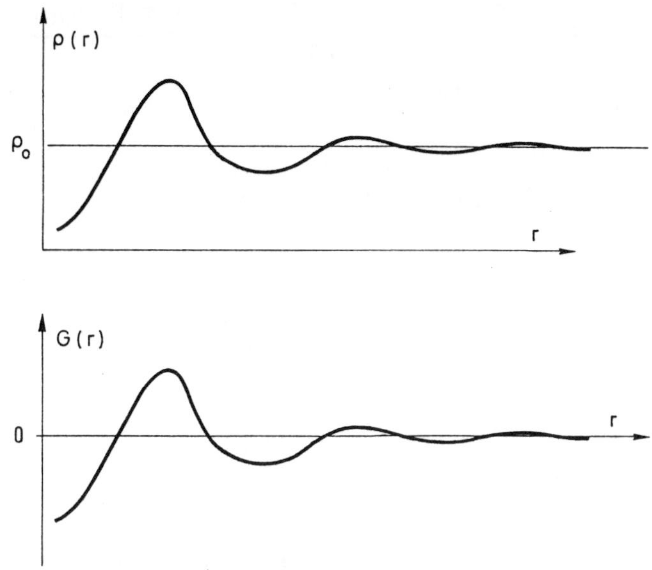

Fig. II.10. — Radial density and the pair correlation function.

The Van Hove self-correlation function $G_s(\mathbf{r},\ t)$. — This function gives the probability that a particle originally at position $\mathbf{r} = 0$ at time $t = 0$ is located at position $\mathbf{r}$ at time t; it expresses the *correlation* between the positions of the same particle at times τ and $\tau + t$. It resembles the function $W(X,\ t)$ introduced previously. It is often written in the form of an average:

$$G_s(\mathbf{r},\ t) = \langle \delta\left\{\mathbf{r} - (\mathbf{r}_i(t) - \mathbf{r}_i(0))\right\}\rangle\,, \tag{II.34}$$

where the Dirac δ function is zero except for $\mathbf{r} = \mathbf{r}_i\ (t) - \mathbf{r}_i\ (0)$.

One can also define the function $G_d(\mathbf{r},\ t)$, where the subscript d signifies "distinct", as opposed to the subscript s (self):

$$G_d(\mathbf{r},\ t) = \frac{n(\mathbf{r},\ t)}{4\ \pi r^2\ \Delta r}. \tag{II.35}$$

Here $n(\mathbf{r},\ t)$ is the number of particles situated at time t at a distance between $\mathbf{r}$ and $\mathbf{r} + \Delta\mathbf{r}$ from the position of a certain atom at $t = 0$. $G_d(\mathbf{r},\ t)$ thus represents the evolution with time of the pair distribution function $g(\mathbf{r})$:

$$G_d(\mathbf{r},\ 0) \equiv \rho_0 g(\mathbf{r}) \quad \text{and} \quad G_s(\mathbf{r},\ 0) = \delta(\mathbf{r}).$$

The functions G_s and G_d are shown schematically in figure II.11 for a crystalline solid and for a liquid. Their definitions and principal properties are given in the appendix.

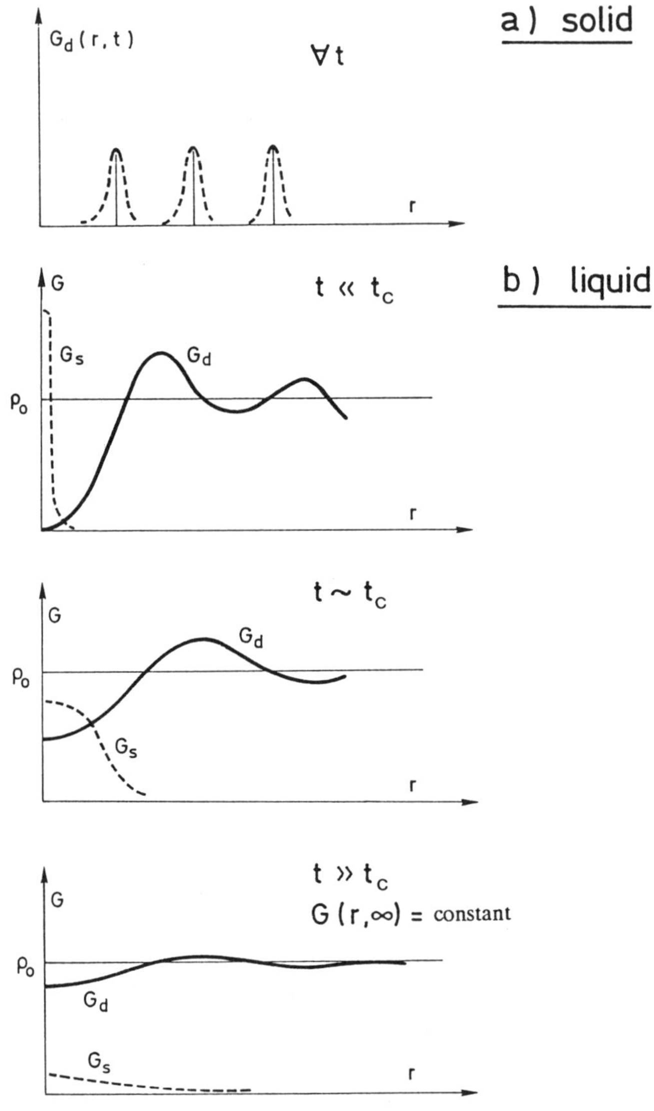

Fig. II.11. — Examples of pair correlation functions: a) for a solid, b) for a liquid, for different times.

VII.2 Diffusion. — The self-correlation function $G_s(\mathbf{r}, t)$ must obey Fick's equation for times t much longer than the time between collisions, t_c. In effect, the relative concentration, c/n_0, of the diffusing species can be interpreted as the probability density that a marked particle, situated at $\mathbf{r} = 0$ at $t = 0$, is located at point $\mathbf{r}$ at time t.

$$\frac{\partial G_s}{\partial t} = D\, \nabla^2 G_s(\mathbf{r},\, t) \qquad t >> t_c, \tag{II.36}$$

with the initial condition $G_s(\mathbf{r},\, 0) = \delta(\mathbf{r})$.

This equation is solved by use of the Fourier transform. The transform of G_s with respect to $\mathbf{r}$, called the intermediate scattering function, is given by:

$$F_s(\mathbf{k},\, t) = \int e^{-i\mathbf{k}.\mathbf{r}}\, G_s(\mathbf{r},\, t)\, d\mathbf{r}, \tag{II.37}$$

or, with the definition of G_s given in equation (II.34):

$$F_s(\mathbf{k},\, t) = \langle \exp i\mathbf{k} \cdot [\mathbf{r}(t+\tau) - \mathbf{r}(\tau)] \rangle\, . \tag{II.38}$$

Recall that the transform of $\delta\,(\mathbf{r} - \mathbf{r}_0)$ is just $\exp\,(-i\mathbf{k}{\cdot}\mathbf{r}_0)$.

Note that $F_s(k,\, 0) = 1$. The solution of the transformed Fick's equation is (cf. Ch. I, Appendix 1):

$$F_s\,(\mathbf{k},\, t) = \exp\left(-k^2\, Dt\right) \tag{II.39}$$

and performing the inverse transformation:

$$G_s(\mathbf{r},\, t) = (4\,\pi Dt)^{-3/2}\, \exp\left(-\frac{r^2}{4\, Dt}\right). \tag{II.40}$$

This Gaussian can be characterized by its width, that is, by the second moment of the function G_s:

$$\begin{gathered} \left\langle |\mathbf{r}(t) - \mathbf{r}(0)|^2 \right\rangle = \langle r^2 \rangle = \int_0^t r^2\, G_s(r,\, t)\, dt \\ \langle \mathbf{r}^2 \rangle = 6\, Dt \qquad t >> \tau_c. \end{gathered} \tag{II.41}$$

The diffusion coefficient D is then a measure of the time dependence of the mean-square displacement.

Dynamic structure factor. — The double Fourier transform of $G(\mathbf{r},\, t)$ is the dynamic structure factor $S(\mathbf{k},\, \omega)$. A part, S_s, of this factor corresponds to G_s:

$$S_s(\mathbf{k},\, \omega) = \frac{1}{2\,\pi} \int_{-\infty}^{\infty} e^{i\omega t}\, F_s(\mathbf{k},\, t)\, dt. \tag{II.42}$$

If F_s is exponential in time (cf. Eq. (II.39)), its transform is a Lorentzian:

$$S_s(\mathbf{k},\, \omega) = \frac{1}{\pi} \frac{Dk^2}{(Dk^2)^2 + \omega^2}. \tag{II.43}$$

Also, in the limit for long wavelengths:

$$\lim_{k\to 0} (1/k^2)\ S_s(\mathbf{k},\ \omega) = \frac{D}{\pi\omega^2}$$

$$\boxed{D = \pi \lim_{\omega\to 0} \lim_{k\to 0} \frac{\omega^2}{k^2} S_s(\mathbf{k},\ \omega)}\ . \tag{II.44}$$

This relation is little used, except for quasi-elastic neutron scattering (Ch. IX, pt. 2, § **II**.1). A more common starting point is the velocity autocorrelation function:

$$\langle \mathbf{v}(0)\cdot\mathbf{v}(t)\rangle = \frac{1}{N}\left\langle \sum_{i=1}^{N} \mathbf{v}_i(\tau)\cdot\mathbf{v}_i(\tau+t)\right\rangle_\tau\ . \tag{II.45}$$

The average $< - >$ is calculated over various values of τ. In particular, for the components along an axis:

$$Z(t) = \langle v_z(t)\cdot v_z(0)\rangle\ .$$

Rewriting equation (II.41) in the form:

$$D = \lim_{t\to\infty} \frac{1}{2\,t}\ \langle z^2(t)\rangle\ , \tag{II.46}$$

with

$$\langle z(t)\rangle = \int_0^t v_z(t)\ \mathrm{d}t$$

$$\langle z^2(t)\rangle = \int_0^t \mathrm{d}t' \int_0^t \mathrm{d}t''\ \langle v_z(t')\cdot v_z(t'')\rangle\ ,$$

it is easy to show that:

$$\boxed{D = \int_0^\infty \langle v_z(t)\ v_z(0)\rangle\ \mathrm{d}t}\ . \tag{II.47}$$

As the diffusion coefficient is the integral of the velocity autocorrelation function, it contains less information than the latter. The velocity autocorrelation function is also closely related to the dynamic structure factor $S_s(\mathbf{k},\ \omega)$ and its Fourier transform $F_s(\mathbf{k},\ t)$. If the vector $\mathbf{k}$ is parallel to the z axis,

$$Z(t) = -\lim_{k\to 0} \frac{1}{k^2}\frac{\mathrm{d}^2}{\mathrm{d}t^2} F_s(\mathbf{k},\ t) \tag{II.48}$$

and

$$Z(\omega) = \omega^2 \lim_{k \to 0} \frac{S_s(\mathbf{k},\ \omega)}{k^2}.$$

Information on the velocity autocorrelation function can be obtained from measurements of incoherent inelastic neutron scattering near $k \to 0$ (cf. Ch. IX, pt. 2, § II.1).

Note that:

1) If there is no diffusion, then (cf. Appendix, Eq. (A.17)):

$$G_d = \frac{1}{\rho_0} \langle \rho(\mathbf{r}) \cdot \rho(\mathbf{r}') \rangle \text{ and } G_s = \delta(\mathbf{r}). \qquad \text{(II.49)}$$

This is the Patterson function of x-ray crystallography. In a crystal, $G_d(\mathbf{r})$ is zero except near certain values of $\mathbf{r}$ (Fig. II.11a, note the broadening due to thermal agitation). In an amorphous material, G_d approaches the density ρ_0 for large distances (Fig. II.11b).

2) The intensity of the scattered radiation is proportional to $S(\mathbf{k},\ \omega)$, the double Fourier transform of $G(\mathbf{r},\ t)$.

VIII. Limitations of Fick's law.

Fick's law was written for a continuous medium, neglecting the discontinuous structure of real materials. That is why the solutions $c(x)$ yield non-zero concentrations at any distance, no matter how large, for any time $t > 0$, which has no physical meaning. On the other hand, for short times, or in the presence of large gradients, a linear equation seems a gross oversimplification.

In the simple model of § I, the balance equation for plane n can be written, assuming a constant jump frequency:

$$\frac{\mathrm{d}c_n}{\mathrm{d}t} = \Gamma_s\,(c_{n+1} + c_{n-1} - 2c_n) \qquad \text{(II.50)}$$

(the factor 2 multiplying c_n takes account of the fact that jumps from plane n occur to both right and left). We compare a solution of this equation (Martin and Benoist 1977) with the solution predicted for Fickian behavior, for the same initial and boundary conditions. For a sinusoidal initial condition with wavelength Λ:

$$c(x,\ 0) = c_1 + (c_2 - c_1)\sin(2\ \pi x/\Lambda),$$

the solution has the form:

$$c(x,\ t) = c_1 + (c_2 - c_1)\sin(2\ \pi x/\Lambda)\exp(-\alpha t). \qquad \text{(II.51)}$$

The concentration is uniform only for infinite time. Substituting equation (II.51) into the classic diffusion equation, one obtains:

$$\alpha_F = (2\ \pi/\Lambda)^2 D,$$

but the same substitution in equation (II.50) leads to:

$$\alpha_B = 2\ \Gamma_s\,[1 - \cos(2\ \pi\lambda/\Lambda)]\,,$$

with λ the interplanar spacing. Then expanding cos $2\ \pi\lambda/\Lambda$, with $D = \Gamma_s\lambda^2$, we obtain:

$$\alpha_B = \left(\frac{2\pi}{\Lambda}\right)^2 D\left[1 - \frac{2}{4\,!}\left(\frac{2\ \pi\lambda}{\Lambda}\right)^2 + \frac{2}{6\,!}\left(\frac{2\ \pi\lambda}{\Lambda}\right)^4 + \cdots\right].$$

The solutions for the Fickian regime and the balance equation are identical only if $2\ \pi\lambda/\Lambda << 1$, i.e., for small concentration gradients ($<< (c_2 - c_1)/2\pi\lambda$); α_B and α_F differ by 3% for $\Lambda = 10\lambda$. Since all concentration profiles can be expanded in Fourier series, it follows that the short-wavelength components will not behave in a Fickian manner.

The case of a concentration gradient so steep that the free energy is a function of the concentration gradient, a condition applicable to synthetic multilayer structures (cf. ch. VIII, § V and ch. IX, part 1, § III.5) or to spinodal decomposition, originally treated by Cahn and Hilliard (1958), has been reviewed by Greer and Spaepen (1985).

In all cases where very short distances come into play, even if only in one direction, an approach similar to equation (II.50) is *a priori* recommended. Examples are given in chapter VII: grain boundary diffusion (pt. 1, § III) and surface diffusion (pt. 2, § III). Other examples that should be mentioned are the kinetics of ordering or clustering in solids, and all the relaxation and resonance phenomena directly sensitive to atomic jumps (Ch. IX, pt. 2).

APPENDIX

Some definitions.

1) The time correlation function of two dynamic variables $A(t)$ and $B(t)$, which are functions of the variables of position $\mathbf{r}_n$ and momentum $\mathbf{p}_n$, is written as:

$$C_{BA}(t) \equiv \langle B\,(t+s)\ \ A\,(s)\rangle\,. \qquad \text{(A.1)}$$

If $B = A$, $C_{AA}(t)$ is the *self-correlation function* of the variable A.

The averages can be defined either over the ensembles:

$$C_{BA}(t) = \int B\,(t+s)\ f_0^N\,(s)\,A\,(s)\ \mathrm{d}\Gamma_N\,, \tag{A.2}$$

where $\mathrm{d}\Gamma_N$ is the element of volume in phase space, and f_0 the distribution function of microscopic states at equilibrium, or over time:

$$C_{BA}(t) = \lim_{T\to\infty} \int_0^T B(t+s+t')\ A(s+t')\ \mathrm{d}t'. \tag{A.3}$$

In a system at *equilibrium*, the correlation function is *stationary*, i.e., independent of the origin of time:

$$C_{BA}(t) = \langle B(t+s)\ A(s)\rangle = \langle B(t)\ A\rangle\,. \tag{A.4}$$

2) The density of particles at the point $\mathbf{r}$:

$$\rho(\mathbf{r},\ t) = \sum_{i=1}^{N} \delta\,(\mathbf{r} - \mathbf{r}_i(t))\,. \tag{A.5}$$

The *time-dependent density autocorrelation function* is defined as:

$$G(\mathbf{r},\ \mathbf{r}',\ t) \ = \frac{1}{N}\ \langle \rho(\mathbf{r}'+\mathbf{r},\ t)\cdot\rho(\mathbf{r}',\ 0)\rangle \tag{A.6}$$

$$= \frac{1}{N}\left\langle \sum_i \sum_j {}^{\prime}\delta\,(\mathbf{r}+\mathbf{r}'-\mathbf{r}_j(t))\cdot\delta\,(\mathbf{r}_i(0)-\mathbf{r}')\right\rangle. \tag{A.7}$$

The dependence on the location of the origin is eliminated by integrating over $\mathbf{r}'$

$$G(\mathbf{r},\ t) = \frac{1}{N}\left\langle \sum_i \sum_j \int \delta\,[\mathbf{r}'+\mathbf{r}-\mathbf{r}_j(t)]\cdot\delta\,[\mathbf{r}_i(0)-\mathbf{r}']\ \mathrm{d}r'\right\rangle, \tag{A.8}$$

which is the correlation function in the form introduced by Van Hove (1954). If the operators $\mathbf{r}_i(0)$ and $\mathbf{r}_j(t)$ commute:

$$G(\mathbf{r},\ t) = \frac{1}{N}\left\langle \sum_{i=1}^{N} \sum_{j=1}^{N} \delta\,[\mathbf{r}+\mathbf{r}_i(0)-\mathbf{r}_j(t)]\right\rangle, \tag{A.9}$$

which can be separated into two parts:

$$G_s(\mathbf{r},\ t) = \frac{1}{N}\left\langle \sum_{i=1}^{N} \delta\,[\mathbf{r}+\mathbf{r}_i(0)-\mathbf{r}_i(t)]\right\rangle \tag{A.10}$$

and

$$G_{\mathrm{d}}(\mathbf{r},\ t) = \frac{1}{N}\left\langle \sum\sum_{i \neq j} \delta\left[\mathbf{r} + \mathbf{r}_i(0) - \mathbf{r}_j(t)\right] \right\rangle. \tag{A.11}$$

Clearly:

$$G_{\mathrm{s}}(\mathbf{r},\ 0) = \delta(\mathbf{r}) \tag{A.12}$$

$$G_{\mathrm{d}}(\mathbf{r},\ 0) = \rho_0\, g(\mathbf{r}) \quad \text{with} \quad \rho_0 = N/V. \tag{A.13}$$

The normalization is given by:

$$\left.\begin{aligned} \int G_{\mathrm{s}}(\mathbf{r},\ t)\ \mathrm{d}\mathbf{r} &= 1 \\ \int G(\mathbf{r},\ t)\ \mathrm{d}\mathbf{r} &= N \end{aligned}\right\} \tag{A.14}$$

for all t. The limits are (Fig. II.11):

$$\lim_{\mathbf{r}\to\infty} G_{\mathrm{s}}(\mathbf{r},\ t) = \lim_{t\to\infty} G_{\mathrm{s}}(\mathbf{r},\ t) = \frac{1}{V} \sim 0 \tag{A.15}$$

$$\lim_{\mathbf{r}\to\infty} G(\mathbf{r},\ t) = \lim_{t\to\infty} G(\mathbf{r},\ t) = \rho_0 = \frac{N}{V}. \tag{A.16}$$

We note again:

$$\begin{aligned} G(\mathbf{r},\ t) &= \frac{\langle \rho(\mathbf{r},\ t)\cdot\rho(0,\ 0)\rangle}{\rho_0} \\ G_{\mathrm{s}}(\mathbf{r},\ t) &= V\,\langle \rho_i(\mathbf{r},\ t)\cdot\rho_i(0,\ 0)\rangle \end{aligned} \tag{A.17}$$

where ρ_i means the number density of i particles, $\rho_i = V^{-1}$.

Bibliography

BROWN R., *Philos. Mag.* **4** (1828) 161.

CAHN J. W. and HILLIARD J. E., *J. Chem. Phys.* **28** (1958) 258; *Acta Metall.*, **19** (1971) 151–161.

EINSTEIN A., *Investigations on the Theory of Brownian Motion*, ed. R. Furth, (Dover) 1956.

GREER A. L. and SPAEPEN F., *Synthetic Modulated Structures*, eds. L. L. Chang and B. C. Giessen, (Academic Press, 1985) pp. 419–486.

HANSEN J. P. and MCDONALD I. R., *Theory of Simple Liquids* (Academic

Press) 1976.
LE CLAIRE A. D. and RABINOVITCH A., *DIMETA-88, op. cit.*, 3-10.
MANNING J., *Diffusion Kinetics of Atoms in Crystals* (Van Nostrand) 1968.
MARTIN G. and BENOIST P., *Scr. Metall.* **11** (1977) 503–508.
PERRIN J., *Les Atomes* (F. Alcan, Paris) 1936.
VAN HOVE L., *Phys. Rev.* **95** (1954) 249–262.
WEISS G. H., *Am. Sci.* **71** (1983) 65–71.

CHAPTER III

Diffusion mechanisms and correlation effects

The hypothesis of random walk (independence of successive jumps) is not always appropriate; the successive jumps of a particle can be independent or not, depending on the nature of the particle and the jump mechanism. When they are not independent, the jumps are said to be *correlated*; the jump frequencies of the particle in the different jump directions deviate from the probabilities calculated on the basis of a purely random walk, and the deviation depends on the nature of the preceding jump.

I. Mechanisms of diffusion.

How does an atom migrate in a lattice? Figure III.1 recalls the most simple mechanisms by which this migration can take place.

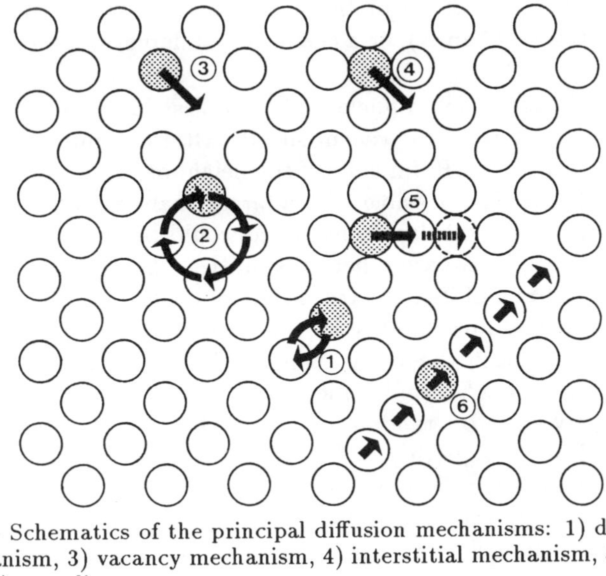

Fig. III.1. — Schematics of the principal diffusion mechanisms: 1) direct exchange, 2) ring mechanism, 3) vacancy mechanism, 4) interstitial mechanism, 5) interstitialcy mechanism, 6) crowdion.

I.1 DIRECT INTERCHANGE. — For an atom on a lattice site, it is easy to conceive a mechanism of atomic motion by direct interchange with a neighboring atom, or by the rotation of a ring of three or more atoms. The first of these is, however, very improbable, at least in close-packed structures, in view of the strong short-range repulsion of the atoms, which prohibits the occupation of the intermediate position where the two atoms would be at mid-path. In the ring mechanism, on the other hand, the repulsive forces play a positive role, each atom "pushing" its neighbor in the course of a sort of circular permutation. However, this mechanism requires the coordination of several atomic jumps, which makes it somewhat improbable. Even if a tracer could migrate by such a mechanism, it could not be responsible for net mass transport in a given direction, such as that observed in chemical diffusion, creep, sintering, etc. Ring mechanisms are invoked for tracer diffusion in ordered alloys, in which certain ring cycles preserve the order, contrary to mechanisms that depend on the motion of point defects (see Ch. V, § VII).

I.2. MECHANISMS INVOLVING POINT DEFECTS. — Most mechanisms of diffusion require the presence of point defects:

1) *Vacancy Mechanism.* — If a lattice site is not occupied, a nearest neighbor atom can jump onto this site, and the vacancy will appear on the site that the atom has just vacated. The vacancies are conserved, and the process can be viewed as the migration of either the vacancy or the atom, but the behavior of the two is not identical. In a crystal of a pure element, the vacancy can jump toward any of its nearest neighbors at any time; its successive jumps are indeed independent of each other. This is, however, not true of a tagged atom, be it a different isotope of the atoms that make up the crystal (self-diffusion of the tracer A^* in A), or a chemically different atom (solute diffusion of B in A).

Consider the case of self-diffusion. Let the first jump of the tagged atom, located on site A, be toward the vacant site B. After this jump (Fig. III.2), the atom, now situated on site B, finds itself the neighbor of the vacancy on site A. A second jump BA is then possible with a larger-than-random probability, and in the calculation of the displacement, there would be two jumps with no net displacement. Clearly there are other possibilities: the vacancy located on site A can jump toward any of its ζ nearest neighbors with the same probability; it thus jumps into B with a probability $1/\zeta$. The probability that the second jump of the tagged atom cancels the first is then $1/\zeta$. In n jumps, there will be $n(1/\zeta)$ such pairs of jumps, that is, $2n(1/\zeta)$ lost jumps, or $n(1-2/\zeta)$ effective jumps, i.e., jumps that give rise to a random walk. This leads to the idea of an *effective jump frequency* $\Gamma_{\text{eff}} = f\Gamma$, with $f = 1-2/\zeta$, i.e., the mean square displacement and the diffusion coefficient are reduced by this factor with respect to the value calculated for a purely random walk. The quantity f is called the *correlation factor* to denote that it is due to the non-independence (correlation) between successive jumps. Its value is always between zero and unity. Note that this type of correlation is connected with the fact that the tracer atom is tagged (Bardeen and Herring, 1952).

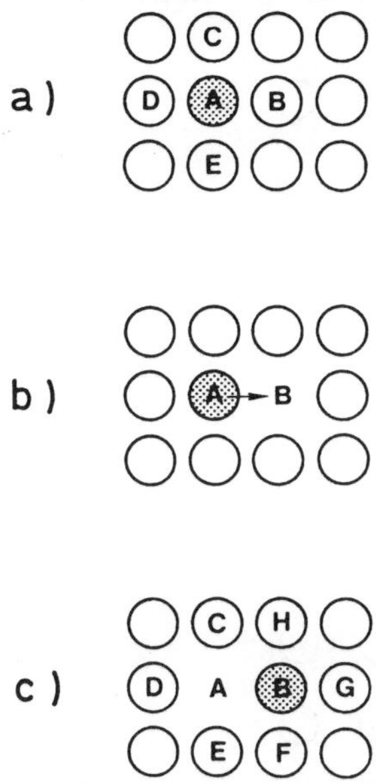

Fig. III.2. — Correlation between successive jumps for the vacancy mechanism: a) The tracer atom located on site A can jump only if the vacancy arrives on a neighboring site B, C, D, or E. b) A vacancy has arrived on site B; the tracer can now carry out the jump AB. c) After the AB jump of the tracer, the vacancy is still a nearest neighbor of the tracer, with equal probability of jumping toward B, C, D, or E. But the tracer can only jump to A, at least until a vacancy arrives at F, G, or H; the tracer thus has a higher-than-random probability of making the jump BA after first making the jump AB.

The preceding description is considerably simplified since the vacancy on site A can migrate over a succession of diverse sites and find itself again, after n jumps, on one of the nearest-neighbor sites to the tagged atom, and exchange with it again; the second jump of the atom is then not independent of the first, even though it is not necessarily its inverse. The value of $1 - 2/\zeta$ is an overestimate in the argument above because the reverse jump of the atom does not necessarily cancel the first one 100% of the time.

If the tagged atom is chemically different (solute diffusion of B in A at infinite dilution), the situation is still more complicated, since the probabilities of the exchange of the vacancy with atoms A or B, respectively, are not equal. However, because the B atoms are present at infinite dilution, the total vacancy movement is not affected. The situation is different in a concentrated alloy, where the jumps of the vacancy are also correlated. Say that the vacancy exchanges more frequently with an A atom than with a B atom; then after a first exchange with an A atom, the inverse exchange will be favored, and thus the successive jumps of the vacancy will not be independent.

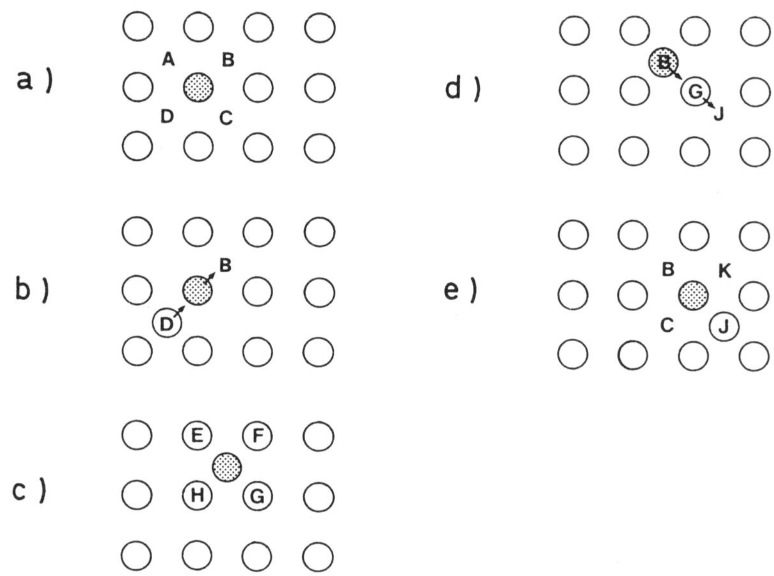

Fig. III.3. — Correlation of successive jumps for an interstitialcy mechanism: a) The 4 interstitial sites A, B, C, and D around the tagged atom are equivalent; the direction of the jump is thus completely random. b) A self-interstitial arrives at D; the tracer ends up, for example, at B if the jump is collinear. The *interstitial defect moves* from D to B. c) The tracer atom, which is now in the interstitial position, has equal probability of jumping to any of the 4 neighboring sites E, F, G, or H, whatever the direction of the previous jump. d) Let the interstitial defect make the jump BJ, with the tracer moving to G. e) Now the tracer again has the self-interstitial for a neighbor and therefore has a larger than random probability to jump back into B. The sequence of the two jumps BG, GB ($A_i^* \longrightarrow A_L^* \longrightarrow A_i^*$) is correlated. ($A_i^*$ and A_L^* are the interstitial and substitutional positions).

2) *Interstitial mechanism.* — An interstitial atom jumping from interstitial site to interstitial site is a point defect, the successive jumps of which are not correlated. This is the *interstitial* (or *direct interstitial*) mechanism (Fig. III.1, # 4); it is typical for atoms that are normally found in interstitial solid solution, such as carbon in iron.

A variant of this mechanism is possible if the tagged atom can occur on either the substitutional or the interstitial position; this is the *interstitialcy* or *indirect interstitial* mechanism (Fig. III.1, # 5).

Consider the tagged atom first located on a substitutional site (Fig. III.3a). The tagged atom can jump only if an interstitial atom arrives at a neighboring position and "pushes" it from its site, placing the tracer in the interstitial position. The tagged atom can then jump into any of the neighboring substitutional sites, creating by this act an interstitial defect. At this point, the configuration is similar to that in the case of the vacancy mechanism, since a point defect (an interstitial) is next to the tagged atom, and the inverse jump of the interstitial takes the configuration back to its preceding state: tagged atom in the interstitial position. Thus these two jumps are correlated. But the effect of correlation stops here, for once the tagged atom is in the interstitial position, no memory of the previous jump is left in the system, and all the jumps are equally probable. The correlation effect is limited to pairs of jumps "interstitial → substitutional → interstitial", while in the vacancy mechanism all the succeding jumps are correlated.

The real situation is somewhat more complex than the drawing of figure III.3 suggests because the jumps of the two atoms involved in the elementary jump process are not necessarily collinear. Several types have to be considered depending upon the arrangement of the interstitial sites around the substitutional site.

The configuration of the interstitial defect can also be a source of complication if it appears as a *dumbbell interstitial*, which is just two atoms centered at one site (Fig. III.4). As this figure shows, three "jumps" are necessary to go from one site to the neighboring one: the first moves the atom into the dumbbell, the second correspond to the jump of the dumbbell, and on the third the tagged atom returns to a substitutional site. If one of the atoms in the dumbbell is an impurity, favored energetically to be in an interstitial site (e.g., because of its small size), the dumbbell may become trapped, and able to execute only so-called "caging" and "looping" motions, which lead to no long-range displacement (Johnson and Lam, 1976). This can be an important phenomenon under irradiation because the interstitial, which is normally the faster moving defect, becomes the slower moving one.

3) *Divacancy mechanism.* — The atom exchanges with one of the vacancies in the vacancy pair. In the FCC structure, this can be done without dissociating the divacancy, but in the BCC structure, such a jump necessarily involves a modification of the distance between the two vacancies because in this structure there is no group of three atoms which are all nearest neighbors of one another.

4) *Other mechanisms.* — More complex mechanisms are possible. The point defect can be less localized than in the classic description. One such concept is the *extended interstitial*, which, for a string of atoms, is just the "crowdion". Another is the *delocalized vacancy*, which in the limit is just a small disordered or even non-crystalline zone, known for a long time as a "relaxion". The latter defect has been described as a zone of local melting; the atoms can then migrate by a "shuffling" motion similar to that of the atoms in a liquid. Complexes of point defects are also invoked for particular cases, e.g., the interstitial-vacancy pair, in which the interstitial can also be

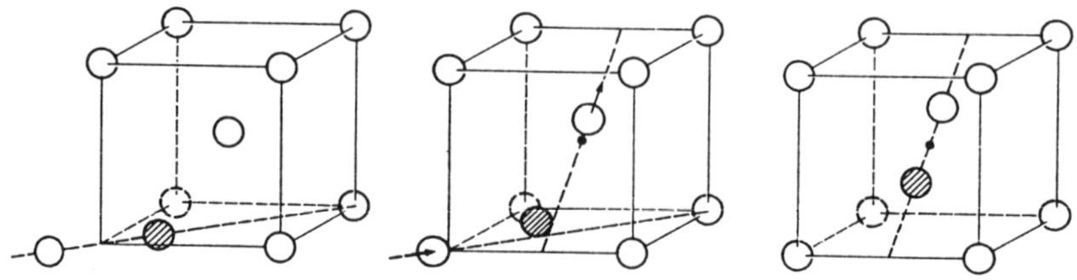

(a) (b) (c)

BCC lattice, <110> dumbbell

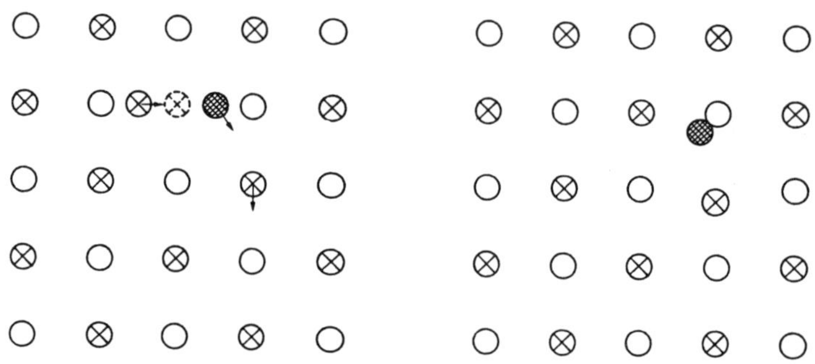

(d) (e)

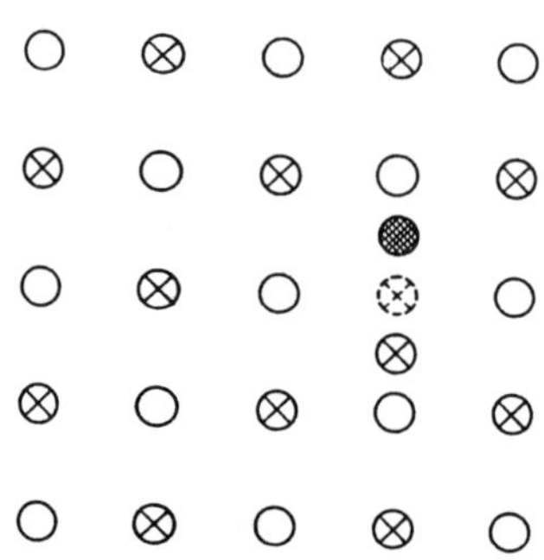

(f)

FCC lattice, <100> dumbbell

o atoms at level 0 and a

⊗ atoms at level a/2 , • tracer atom

Fig. III.4. — Migration of a tracer atom by the dumbbell interstitial: -a, b, c, BCC structure, <110> dumbbell (b is the saddle point). -d, e, f, FCC structure, <100> dumbbell (e is the saddle point). o atoms at levels 0 and a, ⊗ atoms at level $a/2$, • tracer atom.

in the dissociated state. More complex mechanisms have been proposed for high temperatures, such as the transient Frenkel defects proposed by Doan and Adda (1987),which just "open the lock" for a ring exchange to operate with the help of the transient vacancy. The correlation between jumps has to be studied for each case individually.

N. B.:

1) Several elementary mechanisms can be operative simultaneously; the tagged atom can jump via one or several of the mechanisms described above.

2) The effect of correlation between two successive jumps appears as soon as 3 different types of particles participate in the diffusion process, for example: host atoms, tracer atom, and point defect. Successive jumps of an atom are correlated if and only if ***at least 3 types of particles*** are involved in the process. In the application of this rule, lattice atoms, tagged atoms, vacancies, interstitials (including self-interstitials) are considered as different species.

II. Definition of the correlation factor.

The theory of random walk leads to the reference value of the mean square displacement $\langle X^2 \rangle_\mathrm{r}$:

$$\langle X^2 \rangle_\mathrm{r} = \sum \langle x_i^2 \rangle .$$

If the jumps are correlated, the averages of the products $\langle x_i \, x_j \rangle$ are no longer zero. One writes as a definition:

$$\langle X^2 \rangle = f \, \langle X^2 \rangle_\mathrm{r}$$

with:

$$\boxed{f = \frac{\langle X^2 \rangle}{\langle X^2 \rangle_\mathrm{r}} = 1 + \frac{2 \sum_{i=1}^{n-1} \sum_{j=i+1}^{n} \langle x_i \, x_j \rangle}{\sum_{i=1}^{n} \langle x_i^2 \rangle}}, \qquad \text{(III.1)}$$

where the general expression for $\langle X^2 \rangle$ calculated in the preceding chapter (Eq. (II.16)) has been used. For a purely random walk, $\langle x_i \, x_j \rangle = 0$ and $f = 1$. In fact, f is defined in the *limit as* $n \to \infty$, in order to take account of all the effects of correlation. If all the successive jumps are correlated (vacancy mechanism), it follows that the jth jump is correlated to the ith jump across the entire chain of the intervening $j - i$ jumps. All the correlations must be taken into account in the products $\langle x_i \, x_j \rangle$, hence the necessity for passing to the limit $n \to \infty$. In fact, a detailed examination of the sequence of jumps leads to a slightly different statement of this assertion (see § III "the encounter model"). It should be noted here that if the successive jumps of a tagged atom are caused by *different* defects, they are *not correlated.* The terms $\langle x_i \; x_j \rangle$ are zero if the jumps i and j are due to two different defects.

APPLICATION TO CUBIC LATTICES. — In the *interstitialcy mechanism*, only two immediately successive jumps can be correlated; $\langle x_i \; x_j \rangle = 0$ if $j > i+1$, and further, correlation is found only in the sequence $A_i^* \longrightarrow A_L^* \longrightarrow A_i^*$ and not in the sequence $A_L^* \longrightarrow A_i^* \longrightarrow A_L^*$, where A_i^* and A_L^* are the interstitial and substitutional (lattice) positions of the tracer, respectively. The double sum reduces to $(1/2)\; 2 \sum \langle x_i \; x_{i+1} \rangle$.

In a cubic lattice, all the jumps have the same length (if only one type of interstitial site comes into play); the averages of the products can be written $\left\langle x_i^2 \cos\theta \right\rangle$, where θ is the angle between the jumps i and $i+1$. This product is independent of i:

$$\sum \langle x_i^2 \rangle = n \langle x_i^2 \rangle \quad \text{and} \quad \sum \langle x_i \; x_j \rangle = n \langle \cos\theta \rangle \langle x_i^2 \rangle$$

from which:

$$\boxed{f = 1 + \langle \cos\theta \rangle} \,. \tag{III.2}$$

Consider next the *vacancy mechanism.* Since in a cubic structure $\langle x_i \; x_j \rangle = \left\langle x_i^2 \cos\theta_{ij} \right\rangle$, where θ_{ij} is the angle between the vectors of jumps i and j, and since $\sum \left\langle x_i^2 \right\rangle = n \left\langle x_i^2 \right\rangle$, equation (III.1) becomes:

$$f = 1 + \lim_{n\to\infty} \frac{2}{n} \sum_{i=1}^{n-1} \sum_{j=i+1}^{n} \langle \cos\theta_{ij} \rangle \,.$$

Following a calculation given in the appendix ([1]):

$$\langle \cos\theta_{i,\; i+m} \rangle = (\langle \cos\theta \rangle)^m \,.$$

If θ is the angle between two successive jumps i and $i+1$, then:

$$\begin{aligned} f &= 1 + \lim_{n\to\infty} \frac{2}{n} \sum_{i=1}^{n-1} \langle \cos\theta_{i,\; i+1} \rangle + \langle \cos\theta_{i,\; i+2} \rangle + \cdots \\ &= 1 + \lim_{n\to\infty} \frac{2}{n} (n-1) \left\{ \langle \cos\theta \rangle + \langle \cos\theta \rangle^2 + \cdots + \langle \cos\theta \rangle^{n-1} \right\} \\ &= 1 + 2 \langle \cos\theta \rangle / \left[1 - \langle \cos\theta \rangle\right] , \end{aligned}$$

([1]) This relation is valid only if the jump vector is a two-fold or higher axis of rotational symmetry, a condition realized for self-diffusion in a pure crystal. This is not the case for diffusion by divacancies, nor for non-cubic structures. It is valid for the diffusion of a solute in a cubic lattice, but not for a solvent atom next to the solute.

from which:

$$\boxed{f = \frac{1 + \langle \cos\theta \rangle}{1 - \langle \cos\theta \rangle}.} \qquad \text{(III.3)}$$

Note that $\langle \cos\theta \rangle$ is always negative for a vacancy mechanism so that f is < 1. Note also that in a non-cubic crystal, the correlation factor depends on the direction of diffusion in the crystal. In the same way that generally $D_x \neq D_y \neq D_z$, $f_x \neq f_y \neq f_z$ in the calculation of $\langle X^2 \rangle$, $\langle Y^2 \rangle$, or $\langle Z^2 \rangle$ with the use of formulae of the type (III.1).

III. The encounter model.

Computer simulation has shown that for self-diffusion by a vacancy mechanism, the situation is in some ways simpler than the reasoning and general formulae above would indicate.

Consider the configuration after the 1st tagged atom-vacancy exchange (Fig. III.5). Following this first jump, the vacancy will migrate in the lattice, eventually returning to a site nearest-neighbor to the tagged atom, and exchange with it. In reality, it is not probable that the vacancy will return a large number of times since another vacancy can intervene; its arrival destroys the chain of correlation.

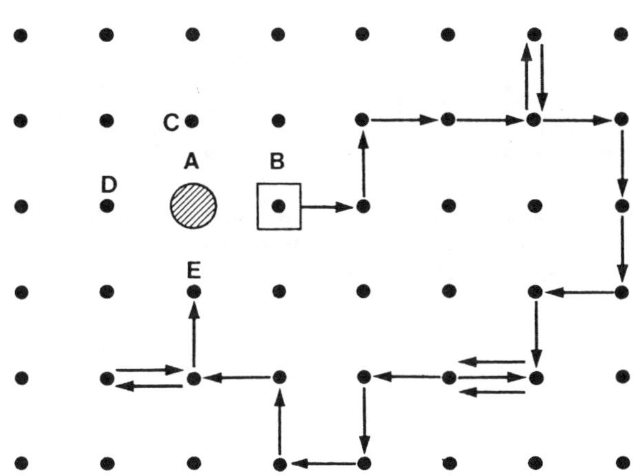

Fig. III.5. — Vacancy mechanism, correlation between jumps. (□) = position of the vacancy after the first exchange with the tracer atom (◍). At the end of a random walk consisting of $n_v - 1$ jumps, the vacancy finds itself on site E, next to the tracer; its n_v^{th} jump can be toward site A. The second jump of the tracer in this case is AE; it is not independent of the first jump BA because the same vacancy was involved in both jumps.

The sequence of exchanges of the tagged atom with the same vacancy is called the *encounter*. Let Z_e be the number of these exchanges which give rise to a path $\mathbf{R}_e$. The only jumps correlated with each other are those inside the sequence corresponding to a given encounter. The path can be described by the sum of the $\mathbf{R}_e$ vectors (Figs. III.6 and III.7). These jumps are uncorrelated if the interaction between vacancies is negligible. This is usually the case at thermal equilibrium, when the concentration of vacancies is small (less than 1 vacancy per 10,000 sites) ([2]), although divacancies can still form. In a cubic crystal,

$$D = (1/6)\left\langle R_e^2 \right\rangle / \tau_e, \tag{III.4}$$

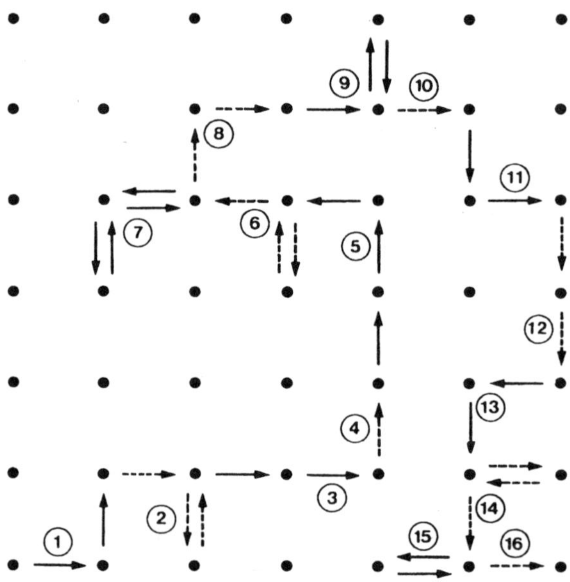

Fig. III.6. — The encounter model. Path of a tracer atom following encounters with 16 different vacancies. The jumps corresponding to each vacancy are indicated alternately by solid and dashed arrows. (After D. Wolf, 1983).

with $\tau_e = Z_e \tau$ and $\tau = \Gamma_0^{-1}$ the residence time of an atom on a site. But in terms of the jump length, λ,

$$D = \frac{1}{6} f \frac{\lambda^2}{\tau},$$

with the correlation factor f. From which:

$$f = (1/Z_e)\left\langle R_e^2 \right\rangle / \lambda^2. \tag{III.5}$$

([2]) This is not true for certain non-stoichiometric compounds in which the vacancy concentration can be very high.

The quantities $\langle R_e^2 \rangle$ and Z_e can be evaluated as a function of the number of jumps of the vacancy, ν_v, by computer simulation (see below). For $\nu_v \to \infty$ in a FCC lattice, $Z_e = 1.3446$ and $\langle R_e^2 \rangle = 1.0509\ \lambda^2$. Sequences of more than three exchanges of the tagged atom with the same vacancy (correlated jumps) are highly improbable. Unfortunately, the values of Z_e and $\langle R_e^2 \rangle$ approach their final values very slowly with increasing ν_v, and values of $\nu_v \geq 100$ must be taken. However, their ratio, which gives f, converges significantly more rapidly. Figure III.6 gives a picture of the diffusion of an atom: its path is made up of the vectors $\mathbf{R}_e$, which are not correlated, but each of these vectors is the sum of a small number of elementary jumps, which are correlated. The calculation of f is then simplified because it is enough to study the exchanges of the tagged atom with only one vacancy. The number of tagged atom-vacancy exchanges in the course of one encounter is given by:

$$Z_e = 1 + p_2 + p_3 + \cdots, \tag{III.6}$$

where p_2 is the probability of the first return of the vacancy (i.e., the second jump of the tagged atom), etc. Since $1 - p_2$ is the probability that after any number of returns, the vacancy never comes back, one can write on the average:

$$Z_e = \sum_{k=1}^{\infty} k p_k (1 - p_2), \tag{III.7}$$

since the probability of k exchanges ($Z_e = k$) equals $p_k (1 - p_2)$ (p_k returns times a non-return $(1 - p_2)$). From this:

$$Z_e = (1 - p_2)\left[1 + 2p_2 + 3p_3 + \cdots\right].$$

The recursion formula is:

$$p_k = p_2^{k-1}$$

whence:

$$\begin{aligned} Z_e &= (1 - p_2)\left(1 + 2p_2 + 3p_2^2 + \cdots + k p_2^{k-1} + \cdots\right) \\ &= \frac{1}{1 - p_2}. \end{aligned}$$

Consider an ultra-simplified example: $\nu_v = 1$ and the encounter stops on the second exchange:

$$Z_e = 1 + p_2.$$

The probability of the return of the vacancy, p_2, equals $1/\zeta$ (where ζ is the number of nearest neighbors):

$$R^2 = \frac{1}{\zeta}\left[(\zeta - 1)\lambda^2 + 1 \cdot 0\right] = \left(1 - \frac{1}{\zeta}\right)\lambda^2,$$

since there is 1 case in which the atom makes two jumps which cancel each other for $(\zeta - 1)$ cases where it makes only one lone jump. Then from equation (III.1):

$$f = \frac{1 - 1/\zeta}{1 + 1/\zeta}, \tag{III.8}$$

an expression which overestimates f.

Calculation of the probability p_2 leads back to the classic expression (III.3).

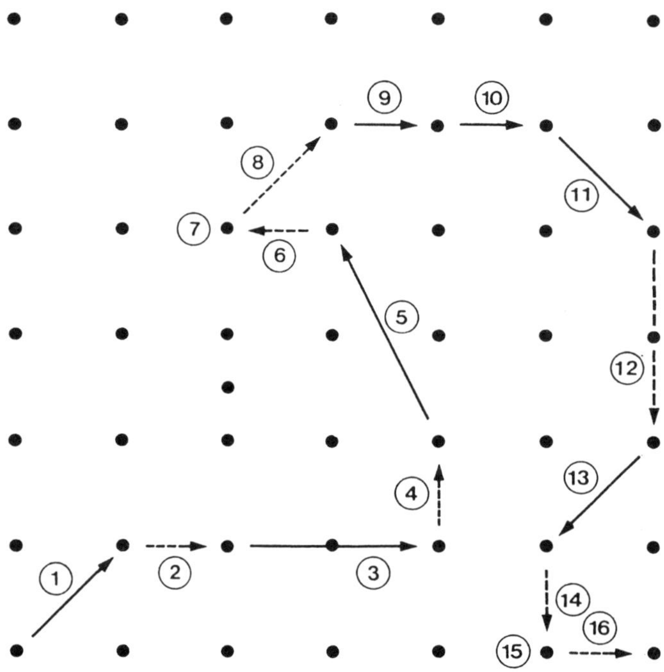

Fig. III.7. — The same path as in the preceding figure, showing the paths $\mathbf{R}_e$ resulting from each encounter. The vectors $\mathbf{R}_e$ are not correlated.

The correlation factor defined above is sometimes called the *spatial correlation* factor, for obvious reasons. Following the jumps over the course of time, it is seen that the second jump of the tagged atom most probably takes place after a few jumps of the vacancy. There is thus an effect of *bunching* in time of the jumps of the tagged atom, forming a packet of a small number of jumps near the beginning of the encounter, followed by a "dead time" awaiting another encounter (Fig. III.8). Certain experimental techniques (NMR, Mössbauer effect, incoherent quasi-elastic neutron scattering) which are called microscopic (they measure not the flux but jump frequencies or times between jumps) give access to the individual movements of the atoms or to the relative movement of a pair of atoms. In fact, because of the "bunching effect", these methods do not see the successive atomic jumps, but only a random succession of encounters, and measure the mean time τ_e between two "bunches". This is called *temporal correlation*, following Eisenstadt and Redfield (1963). It differs from spatial correlation in that $\tau/\tau_e = f\lambda^2/\langle R_e^2\rangle$; that is, $\tau/\tau_e < f$.

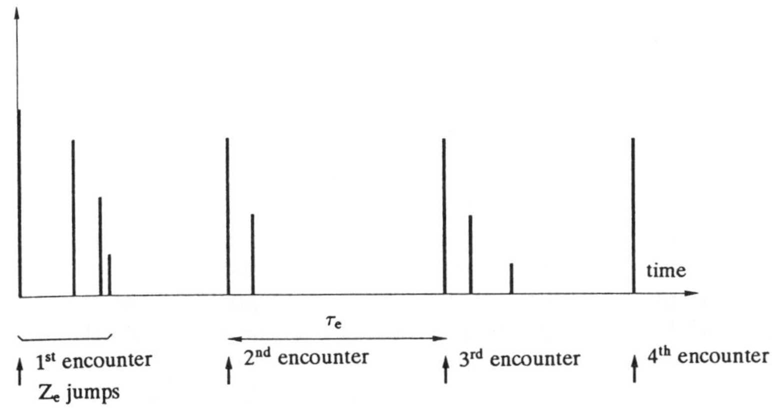

Fig. III.8. — Temporal correlation effect: "bunching" of tracer jumps due to several successive exchanges with the same vacancy. (After D. Wolf, 1983.)

OTHER MECHANISMS. — The encounter model is applicable just as well to other mechanisms, especially to the interstitialcy. It is not of great interest for the simple mechanism, in which the chain of correlation is destroyed as soon as the tracer moves into the interstitial position. On the other hand, it is very useful for dumbbell interstitials. The description of the process is quite complicated and beyond the scope of this book. The encounter model also can be applied to diffusion by several mechanisms operating simultaneously, each encounter taking place at random with one or another type of defect.

The above ideas cannot be applied to a *plane lattice*, where the probability of the return of the vacancy p_2 is unity, and thus the number of jumps a tagged atom makes with a vacancy is infinite, although a correlation factor can still be calculated for this case.

IV. A simple simulation of self-diffusion and electromigration.

As an example, we give an outline of a very simple Monte Carlo simulation, written for pedagogical purposes by G. E. Murch (1979).

The simulation is for a simple cubic crystal made up of $n \times n \times n$ atoms, into which a vacancy is introduced; in the simulation, the motion of the vacancy is followed. To keep the vacancy concentration constant, every time that the vacancy leaves the crystal, it is reintroduced on the opposite face. (These are the so-called "periodic boundary conditions", which reduce to considering a crystal of infinite dimensions, with a vacancy concentration equal to n^{-3}). All the atoms on the $n^3 - 1$ sites are numbered (labeled).

The random motion of the vacancy is followed by drawing a number for each jump; a subroutine is used that generates random numbers between 0 and 1. For the drawing, the segment [0, 1] is divided into 6 equal parts, corresponding to the 6 possible jump directions (x, $-x$, y, $-y$, z, $-z$). The number drawn fixes the jump direction. The new coordinates of the vacancy and the atom involved in the jump are recorded, and the procedure is repeated. At the end of the experiment, the vacancy has jumped N times. As each atom is labeled, the number of jumps m^+ and m^- it carried out in the x and $-x$ directions is known, and one can obtain the distribution function $W(m,\ N)$ by counting the number of atoms which have undergone a displacement $m\lambda = (m^+ - m^-)\lambda$. Even though there was no attempt to simulate the thin layer geometry directly, an equivalent result has been obtained because each labeled atom can arbitrarily be considered to have begun its random walk at the coordinate (0, 0, 0).

Figure III.9 gives the distribution function $W(N,\ m)$ for $N = 640,000$ jumps of the vacancy in a $20 \times 20 \times 20$ crystal. The Gaussian nature of the distribution can easily be checked by replotting in the coordinates $\log W$ *vs.* x^2 (with $x = m\lambda$), or simply *vs.* m^2. The correlation effect can also be demonstrated by a calculation of $\langle X^2 \rangle$, averaging the squares of the displacements of 7999 atoms. Each atom has carried out on the average $m = 1/3\ (640,000/7,999) = 26.67$ jumps in the direction x or $-x$. The calculation shows that $\langle X^2 \rangle \neq m\lambda^2$, and that $f = \langle X^2 \rangle / m\lambda^2 = 17.46/26.67 = 0.654$ (λ was taken as 1 in all the calculations). (The exact value of f is 0.65311...).

It is easy to introduce a driving force into such a program; for each jump of the vacancy, two numbers have to be drawn (random numbers on the segment [0, 1]). The first, as before, indicates the direction of the jump. The second indicates whether the jump really takes place. If the force is applied along the x axis, take $R = \exp(-W'/kT)$, with $R_x \neq R_{-x} \neq R_{\pm y} = R_{\pm z}$. If the number drawn is greater than R, the jump does not take place; if it is less than or equal to R, the jump takes place and the coordinates of the vacancy and the atom it exchanged with are updated. The displacement $\langle X \rangle$ of the distribution W under the action of an electric field (or any other constant force) is shown in figure III.10. The probability of a jump is proportional to $\exp(-U/kT)$ in the absence of a field; in the presence of a field, it is proportional to $\exp\left(-\dfrac{(U \pm \frac{1}{2}\lambda qE)}{kT}\right)$.

Taking for instance $\exp(-U/kT) = 0.9$ and $\exp(-\lambda qE/kT) = 0.9$ gives $R_x = 1$, $R_{-x} = 0.81$, $R_y = R_z = 0.9$. The Nernst-Einstein equation is also verified (see Eq. (II.27)), by comparing $\langle X \rangle$ with $\langle X^2 \rangle$, the latter evaluated in the absence

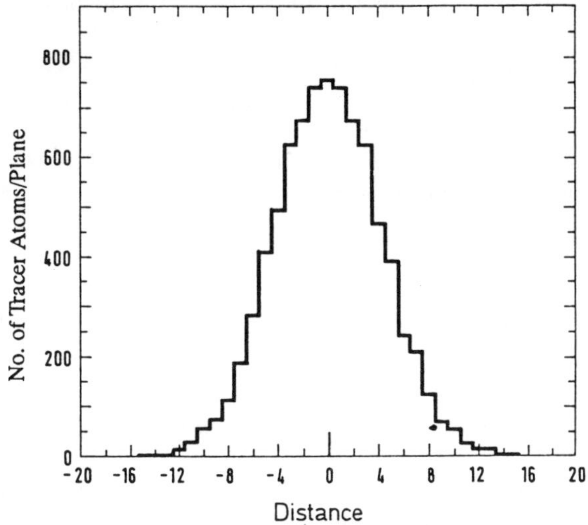

Fig. III.9. — Numerical simulation of a concentration profile for unidirectional diffusion from an instantaneous source $\delta(x)$ of tracer at time $t = 0$. The units of distance are lattice parameters. 640,000 jumps on a $20 \times 20 \times 20$ lattice containing one vacancy (After Murch, 1979).

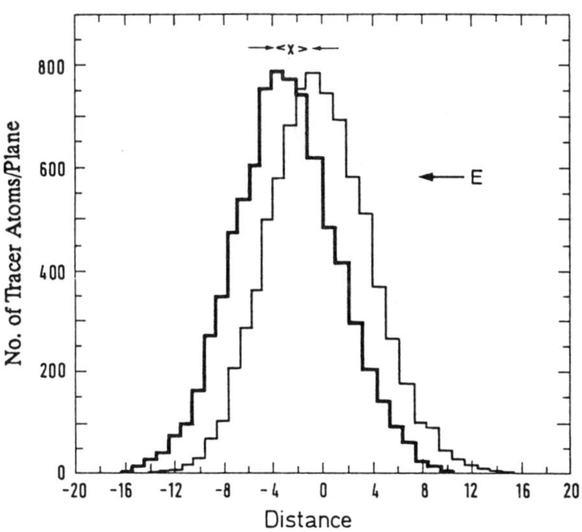

Fig. III.10. — Same as figure III.9, but with an electric field E giving rise to a displacement $\langle X \rangle$ from the right to the left (After Murch, 1979).

of a field. The value of $\langle X^2 \rangle$ in the presence of a field is given by equation (II.26) as:

$$\langle X^2 \rangle_{E=0} = \langle X^2 \rangle_{E\neq 0} - \langle X \rangle^2 .$$

With the values found above, $\langle X \rangle = \langle X^2 \rangle \, |\log 0.9| / f = 17.341 \; |\log 0.9| / 0.654$ $= 2.793$, while the simulation of figure III.10 gives 2.784.

V. Methods of calculating the correlation factor.

A detailed discussion of these methods is beyond the scope of this book. Some of the more important methods are sketched below.

1) *The random walk method.* — This is a direct calculation based on the general equation (III.1), from which it is easy to arrive at the relations (III.2) and (III.3) if certain symmetry conditions are fulfilled. The calculation of $\langle \cos\,\theta \rangle$, where θ is the angle between two successive jumps, is carried out based on the probabilities $p_N(s)$ of the return of the defect to each of the ζ sites nearest neighbor to the tagged atom, after all the possible sequences of N jumps of the vacancy. This is the series of Bardeen and Herring:

$$\langle \cos\,\theta \rangle = \psi_2 \sum_{s=1}^{\zeta} \sum_{N=0}^{\infty} p_N(s) \cos\,\theta_s , \qquad \text{(III.9)}$$

where s is the site to which the vacancy returns and ψ_2 is the probability that it exchanges with the tagged atom. The calculation is difficult because its accuracy increases only slowly with N. A matrix method due to Le Claire and Lidiard (1955) can be used to calculate $\langle \cos\,\theta \rangle$. Mullen (1961) and Howard (1966) developed a matrix method based on equation (III.1) for the more general case of anisotropic crystals and jump directions without any particular symmetry. All these methods tend to underestimate the correlation effect somewhat, because they are limited to a finite volume such that any vacancy which jumps outside it is considered lost. A more precise method is due to Benoist, Bocquet, and Lafore (1977); it consists of solving the transport equation for the vacancy by a double Laplace and Fourier transformation.

2) *The kinetic method,* so named by analogy to the kinetic theories of transport in liquids [Lidiard (1955), (1983); Franklin and Lidiard (1983)]. For the diffusion of an impurity, the flux of atoms in a stationary concentration gradient is calculated directly from the kinetic equations for different impurity-defect pairs; the flux is obtained as a function of the different jump frequencies and jump lengths, and the value of f by comparison with Fick's law. The method is often used for the analysis of relaxation by pair reorientation.

3) *Analog simulation.* — Compaan and Haven (1956, 1958) have devised an elegant method based on the similarity of Ohm's and Fick's laws: electric currents in an electrical network of equivalent resistances are used to simlulate the diffusional flux. In some cases it is possible to solve the equations and calculate f; in more complex cases, the equivalent network is built and the necessary electrical measurements are carried out.

4) *Numerical simulation.* — This method has developed greatly with the availability of powerful computers. Random numbers are drawn (Monte Carlo method) to simulate atomic paths and to calculate $\langle X^2 \rangle$ and then f by the equation $f = \langle X^2 \rangle / \langle X^2 \rangle_r$. A simple example of this has been given in § IV. This method is very powerful, but cannot yield analytical expressions, only numerical results.

N. B.: The correlation factor is a purely geometric numerical factor, uniquely determined by the structure and is therefore independent of the temperature only if the defect has *one unique jump frequency.* This condition is realized only for self-diffusion (where the atom-defect exchange frequencies are identical for lattice atom and tracer isotope, within the accuracy of the isotope effect, cf. § III of the following chapter) and in an isotropic lattice. This condition is not realized in the HCP structure (see Eqs. (II.22) and (II.22a)). In non-cubic structures, the character of f has been considered either as a tensor (Manning, 1968) or a second rank matrix (Flynn, 1972).

VI. Types of correlation factors.

The effects described until now are based, following Bardeen and Herring (1952), on the local configuration around a tagged atom or ion migrating via point defects. When the mean square displacement is of interest, this leads to the introduction of a *spatial* correlation factor. If, however, effects connected to the jump frequency are measured, a *time-dependent correlation* has to be introduced, to connect the jump frequency with the effective frequency. Other effects which have different origins can give rise to deviations from the relations which give $\langle X^2 \rangle$ or the jump frequencies.

VI.1 Dynamic correlations. — In a gas, Bardeen-Herring type effects clearly do not exist. Nevertheless, there exists a correlation between successive "jumps" of a molecule because of the memory of its motion before the collision that produced the new "jump". In a crystalline solid, the atomic jump results from a superposition of phonons. Once the jump is effected, the energy of the jumping atom is dissipated into the lattice. The atom is "thermalized" in its new site, and its nearest neighbors are also thermalized (residence time

>>> period of atomic vibration). Successive jumps of the particle are thus *dynamically independent.*

But dynamic effects are possible if the motion of the atom after its first jump is not sufficiently damped; it can then cross a second energy barrier, even a third..., or make a reverse jump. Such effects were discovered for crystaline solids in the course of a molecular dynamics simulation. Recall that in this method, the energies are not calculated for 0 K, but the thermal energy is partitioned according to a Maxwellian distribution over all the degrees of freedom of the atoms, and the equations of motion are solved with these initial conditions on the velocities and the appropriate interatomic pair potentials (see Ch. IX, part 3, § II).

For Al and Na, Bennett (1976) showed that for very high jump frequencies ($\approx 10^{12}$ s^{-1}), the jumps of the vacancy are no longer independent events. When the mean residence time Γ^{-1} between two jumps is the same order of magnitude as the time it takes the atom to travel the interatomic distance λ at thermal velocity (3), (i.e., $\Gamma^{-1} \approx (m\lambda^2/kT)^{1/2} \approx 10^{-12}$ s), the distribution of "forward" and "reverse" jumps after a first jump has taken place no longer corresponds to a random distribution. The forward jumps are far more frequent, as if the vacancy carried out ***double jumps*** at one time. Figure III.11 pertains to sodium at 298 K; among the 8 possible jumps, two in the forward direction ($\theta = 0$ or 72°) and two in the reverse direction ($\theta = 109$ or 180°) are singled out. The 240 jumps of the vacancy are divided into four groups according to the mean time between two jumps. For the shortest times ($<$ 0.52 ps), the distribution is very different from random and strongly favors the forward jumps. Such double jumps of the vacancy decrease the correlation between tracer jumps and increase the correlation factor of an isotope because they increase the separation of the vacancy from the tracer.

On certain close-packed surfaces, adatoms show an even more spectacular effect (see Ch. VII, part 2). Other examples are the motion of interstitials in structures with channels (see, e.g., the anisotropy of the diffusion of Co in TiO_2, where there are channels parallel to the *c*-axis (Sasaki, Peterson, and Hoshino, (1985)), or preferential diffusion on a sublattice such as is found in superionic conductors (Ch. V, § VIII), or, another example, diffusion along grain boundaries (Ch. VII, part 1, § IV).

Such dynamic effects occur in the course of ***coupled jumps,*** i.e., when the "simultaneous" motion of several atoms is brought into play, such as the migration of the dumbbell interstitial. "Simultaneous" here means that the time between jumps is much shorter than the mean residence time and comparable to the time of thermalization after a jump.

---

(3) The thermal velocity is defined by $(1/2)mv^2 = (3/2)kT$

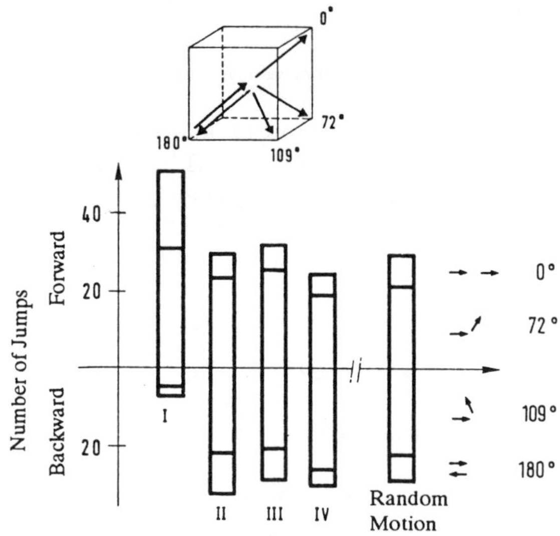

Fig.III.11. — Dynamic correlations in a BCC lattice. Calculated distribution of "forward" jumps (the angle between two successive jumps is 0° or 72°) and "reverse" jumps (109 or 180°), compared to a distribution given by purely random motion (rectangle farthest right). Average times between two successive jumps in picoseconds: Group I $t < 0.52$; Group II $0.52 < t < 1.25$; Group III $1.25 < t < 2.50$; Group IV $t > 2.5$. Calculation for sodium by C. H. Bennett (1976).

VI.2 PHYSICAL CORRELATION (SATO AND KIKUCHI 1971, MURCH 1980). — Consider the conductivity σ of an ionic crystal, which is due to the migration of one type of ion of charge q (i.e., the corresponding point defect). Then from the Nernst-Einstein equation (I.23):

$$D_\sigma = \left(kT/nq^2\right)\sigma \tag{III.10}$$

(n ions of charge q per unit volume). In effect, *this equation defines a quantity D_σ which has the dimensions of a diffusion coefficient.* It is known that $D_\sigma \neq D^*$, where D^* is measured by a tracer of the same ion. This difference originates not only in the correlation between the jumps of the tracer, but also in other effects such as the different jump lengths of the tracer atoms and the point defects, even if only one mechanism of diffusion is operative. The *Haven ratio* H is defined as the ratio:

$$H = D^*/D_\sigma = \left(nq^2/kT\right)\left(D^*/\sigma\right). \tag{III.11}$$

The conductivity is easily calculated as a function of the jump frequencies and jump lengths of the point defects. If there is only one kind of defect, the current density i is given by:

$$i = \sigma E = n_d q \langle v \rangle ,$$

where n_d is the number per unit volume of point defects of charge q. Following the simple calculation of chapter II (§ 1), with force $F = qE$, equation (II.10) becomes, for the defects d with jump frequency w_d:

$$<v> = w_d \; Eq\lambda_d^2/kT,$$

whence

$$\sigma = (1/kT) \; n_d \; q^2 \; w_d \; \lambda_d^2, \tag{III.12}$$

and according to the definition (III.10):

$$D_\sigma = N_d \; w_d \; \lambda_d^2.$$

This clearly corresponds to a diffusion coefficient of a defect with uncorrelated jumps. But in fact, theoretical calculations predict a conductivity which may be smaller than the above. For this reason, Sato and Kikuchi have introduced a factor $f_{\mathrm{I}} \leq 1$ into the conductivity; keeping the definition (III.10), this factor appears in D_σ:

$$D_\sigma = f_{\mathrm{I}} \; N_d \; w_d \; \lambda_d^2. \tag{III.13}$$

The form of this expression explains the origin of the term *physical correlation factor* used to designate f_{I}. The modifier "physical" indicates that the origin of this factor is different from the "geometric" correlation factor introduced earlier in this chapter.

The self-diffusion coefficient for a tracer of these ions via the defects d is given by:

$$D^* = f_d \; \Gamma \; \lambda^2,$$

with the corresponding geometric correlation factor f_d. Then the Haven ratio is given by:

$$H = \frac{D^*}{D_\sigma} = \frac{f_d}{f_{\mathrm{I}}} \; \frac{\Gamma}{N_d \; w_d} \; \frac{\lambda^2}{\lambda_d^2}. \tag{III.14}$$

In a simple case, such as the vacancy mechanism, $\Gamma = N_d \; w_d$ and $\lambda = \lambda_d$, so that the Haven ratio is equal to the geometric correlation factor only if $f_{\mathrm{I}} = 1$. The interstitialcy mechanism presents other complications because $\lambda \neq \lambda_d$ (see Ch. IV, § X).

Since the Haven ratio is not always equal to unity, it follows that the

Nernst-Einstein equation is not confirmed, because it corresponds, in effect, to equation (III.11) with $H = 1$.

VI.3 MEANING OF THE PHYSICAL CORRELATION FACTOR. — The expression "physical correlation factor" is somewhat ambigous. This factor was introduced in the expression for the conductivity calculated for a model of non-ideal diffusion (interaction among the diffusing particles, i.e., the paths of the diffusing particles are not independent, independence being a necessary condition for random motion). It follows that the same type of effect can enter into the calculation of the self-diffusion coefficient for the same models and can modify f, which then includes both geometric and physical correlations. The quantities f and H are very sensitive to these interactions.

The physical correlation factor expresses, in reality, a departure from the laws of random motion. This appears as differences in the jump frequencies in different directions, for reasons connected with the local structure, so that the migration of a defect, vacancy or interstitial, is no longer a random process in the strictest sense. Some possible causes of these deviations are briefly analyzed below:

– *Alloys.* — In an alloy, the jump frequency of a vacancy depends on the nature of the atom with which it exchanges, as well as on the configuration of the neighboring atoms. In a very simple model of a binary alloy, one considers only two jump frequencies for the vacancy, w_A and w_B, depending on whether the exchange is with an A or a B atom. Thus the local configuration is neglected (Manning's model of a random alloy, Ch. V, § VII, and "mean field" models).

– *The vacancy wind.* — If an exterior force (a gradient of chemical potential, an electric field, a temperature gradient) gives rise to a flux of vacancies in one direction, the availability of vacancies for a given tracer atom is biased. The equilibrium configuration (probability of the presence of a vacancy on a given site) is locally distorted by the applied field of force. For example, consider an impurity having an exchange frequency with vacancies less than that of the solvent atoms (Fig. III.12). The availability of vacancies near the solute atom is *ipso facto* higher on the side from which the vacancy flux comes, or the "windward" side, in contrast to the "leeward" side. The solute atom will thus be able to jump more frequently toward the windward side, that is, against the vacancy wind. Figure III.12 shows the distortion of the vacancy flux around the atom of slowly diffusing solute. This effect was first considered by Manning in his analysis of interdiffusion (see Ch. VI, § II).

– *Interaction among point defects.* — Any attraction or repulsion between the diffusing particles, vacancies or interstitials, can cause a deviation from random behavior. Such effects are expected in the case of a high concentration of point defects: compounds that deviate strongly from stoichiometry (certain oxides, for example), or structures in which a one, two, or three dimensional sublattice offers many more positions for diffusing particles than there are particles ("superionic" conductors (See Ch. V, § VIII), and, in general, diffusion of interstitial impurities). In the last two cases, a factor f_I appears in the expression for the conductivity; f_I can vary with the composition.

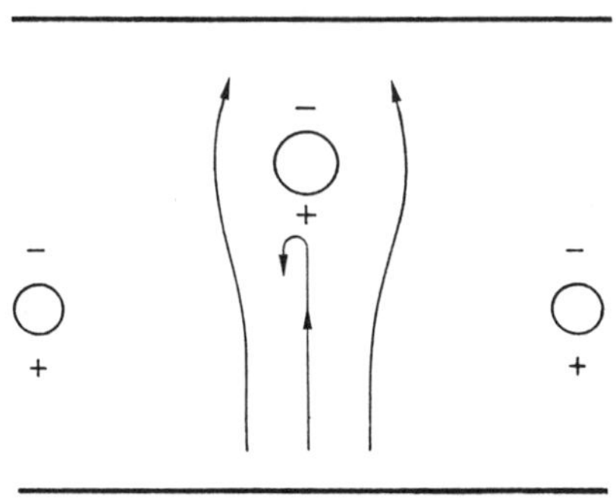

Fig. III.12. — Effect of the vacancy wind: the availability of vacancies is higher on the (+) side than the (−) side of the atoms of slowly diffusing solute (denoted by o).

VI.4 COMPOUNDS WITH A HIGH CONCENTRATION OF DEFECTS. — In compounds with a strong departure from stoichiometry or in the superionic conductors, the correlation factor for self-diffusion can be modified for different reasons:

- *Decorrelation.* — When many vacancies are present, the sequence of exchanges of one tracer atom with the same vacancy is often broken up by an exchange with another vacancy; the number of jumps in an encounter is the more reduced, the higher the concentration of vacancies. Thus the correlation factor increases, and tends to unity in the limit of very few particles with a majority of vacancies. For very high concentrations of defects, a description in terms of *percolation effects* can be interesting (see App. II).

- *Interactions among point defects.* — or among diffusing particles. Locally, the jump frequencies can be increased according to the environment of the defect. A simple case is that of the neighboring sites being excluded due to repulsion between defects. In other words, in a potential-barrier model, the energy minima are not always on the same level, so a supplementary term is introduced into the jump frequency. This term is given by:

$$w = \tilde{\nu} \exp(-\Delta H/kT) \exp(\zeta\Phi/kT),$$

where ζ is the number of sites that are nearest neighbors to the site to which the particle can jump, and are occupied by identical particles with which the particle has an interaction Φ; for simplicity, this interaction is limited to nearest neighbors. These effects manifest themselves by a change in the correlation factor f which becomes concentration dependent, and by the appearence of a factor f_I in the expression for the conductivity.

– *Distortion of the local equilibrium of defects* of the same type as those that give rise to the *vacancy wind.* This is the case for the electrotransport of impurity ions.

Numerical calculation of these effects is difficult, and analytical methods are no easier. Most of the calculations are carried out either by a method of statistical thermodynamics, the "Path Probability Method" of Sato and Kikuchi (1971) which is an expansion of the Cluster Variation Method (CVM) to time dependent processes, or by a Monte Carlo simulation, or by molecular dynamics with the self-correlation function (see Ch. IX, 3rd part). The last are based on a "lattice gas" model in which the accessible sites are on a two- or three-dimensional lattice. In the molecular dynamics formalism, a maximum of one particle can occupy each site, the sites may have different energy levels, and interactions among defects can be introduced.

Interpretation of the Haven ratio is simple only in the absence of interactions among the diffusing particles.

VII. Migration of point defects. Effect of temperature.

To talk about the migration of point defects is equivalent to talking about the migration of atoms, because the former process, the jumping of a vacancy or interstitial, necessarily implies the latter. We consider then the jump of a material particle from one site to a nearest-neighbor site, and attempt to evaluate the probability of the jump. It is assumed in this model that it is possible to treat separately the movements of atoms around their sites and the atomic jump from one site toward the neighboring site. The condition for this is that the average residence time of an atom on a site is much longer than the time of flight between two sites. This case, as well as intermediate cases, can be studied by numerical simulations (see § VII.4).

VII.1 THE POTENTIAL-BARRIER MODEL. — Consider two equivalent nearest neighbor sites M and M′, at a distance λ from each other (Fig. III.13); in each of these sites the particle is in a "well" of potential energy. In effect, due to thermal agitation, the particle at M carries out small vibrational motions at various frequencies (a single ν_E in the Einstein model of a harmonic solid) which are functions of the mass of the particles and of the interatomic force constants. The amplitude of the vibrations depends on the thermal energy of the particle. When the particle moves from the average position, its potential energy $\Phi(\mathbf{r})$ increases. Let us follow the particle in a thought experiment, evaluating $\Phi(\mathbf{r})$ for each position in the course of the jump from M to M′. $\Phi(\mathbf{r})$ must pass through a maximum somewhere between M and M′, but this maximum is the least high for a certain path which defines the *saddle point* of the function $\Phi(\mathbf{r})$ and the height of the *potential barrier* $\Delta\Phi^{\mathrm{m}}$.

For values of $r << \lambda$, $\Phi(\mathbf{r})$ is a quadratic function of the distance from the average position $\Delta\mathbf{r}$. Near the equilibrium position in the parabolic potential well, $\Phi = 1/2\ K\ (\Delta r)^2$; since the restoring force is proportional to the distance $\Delta\mathbf{r}$, or equals $K\Delta\mathbf{r}$, the oscillations are harmonic with frequency $\nu = (1/2\ \pi)\sqrt{K/m}$. However, this form of $\Phi(\mathbf{r})$ is not valid all the way to the saddle point, as this would lead to too high a barrier (Fig. III.13).

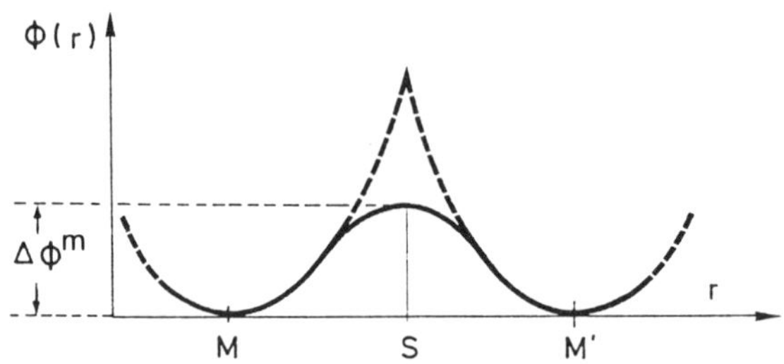

Fig. III.13. — Thermal activation: the potential-barrier model. $MM' = \lambda$.

The position of the saddle point is suggested by the geometry of the lattice. In a hard-sphere model, the saddle point is determined by the least narrow gate defined by the atoms that are nearest neighbors of both sites M and M'. For example, for an $a/2 < 110 >$ jump in the FCC structure, the gate is the four atoms at the halfway point (Fig. III.14). For a $a/2 < 111 >$ jump in a BCC structure, there is a double saddle point since the six common neighbors are not coplanar.

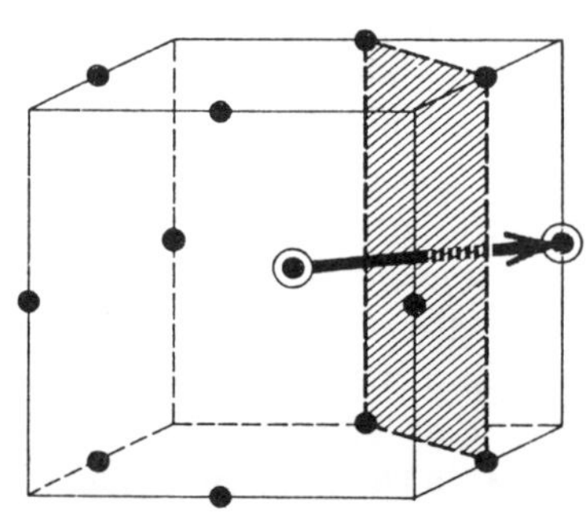

Fig. III.14. — Position of the saddle point between nearest-neighbor sites in the FCC structure.

The classic calculation of the jump frequency follows the same line as the calculation of a rate in chemical kinetics, the "reaction" being written as $A_M \longrightarrow A_S$, where A_S corresponds to the "activated state" of Eyring's theory of absolute reaction rates. One can, in principle, calculate the partition functions Z_M and Z_S of the ground state and activated state, respectively, and the partition function $Z=Z_M+Z_S$ of the system of the two states M and S. For a system at constant volume, the probability of each state is equal to Z_M/Z and Z_S/Z, hence the ratio of the probabilities is:

$$\frac{P(\mathrm{S})}{P(\mathrm{M})} = \frac{Z_S}{Z_M} = \exp\left[-\left(F_S - F_M\right)/kT\right],$$

since in statistical thermodynamics, the Helmholtz free energy is given by $F = -kT \log Z$. The quantity $\Delta F^m = F_S - F_M$ is called the Helmholtz free energy of activation. For a system at constant pressure instead of constant volume, the Gibbs free energy of activation is used:

$$\frac{P(\mathrm{S})}{P(\mathrm{M})} = \frac{\exp-\left(pV_S + F_S\right)/kT}{\exp-\left(pV_M + F_M\right)/kT} = \exp\left(-\Delta G^m/kT\right), \tag{III.15}$$

with a *volume of activation* $\Delta V = V_S - V_M$.

The jump frequency is then written:

$$\boxed{w = \tilde{\nu}\exp\left(-\Delta G^m/kT\right),} \tag{III.16}$$

where $\tilde{\nu}$ is the effective frequency (or attempt frequency) of the particle, introduced for consistency in the dimensions.

This approach suffers from several limitations:

– 1) the particle is considered to move reversibly from M to S, all of its states being defined by equilibrium thermodynamic quantities. In the framework of a pair-interaction model, ΔF^m can be considered as the isothermal, elastic work done against the interatomic forces. Or, the movements of the electrons and nuclei (or, better, the ions) being completely decoupled, the energy of the electronic system can be calculated for any coordinate $\mathbf{r}$ of the moving particle;

– 2) in fact, the parameters $\tilde{\nu}$ and ΔF^m cannot be calculated theoretically in a rigorous way.

VII.2 More refined models. — In reality, the jumping particle cannot be isolated from the system of the surrounding atoms. The problem is thus a many-body problem and very difficult to solve. Two approaches have been proposed:

– 1) Vineyard (1957) has worked out a many-body calculation based on equilibrium statistical mechanics. The top of the potential barrier defines a certain state of the system, with a given configuration and normal modes different from those of the ground state.

– 2) Rice (1958) and Manley (1960), in a "dynamic" approach based on theorems of random variables, calculated the probability of a superposition of phonons such that the migrating particle and the neighboring atoms simultaneously undergo displacements, the amplitudes and directions of which are such that the jump is permitted.

It can be shown that these two approaches are, in fact, equivalent. The reader is referred to Flynn's book (1972) or to a review article by Glyde (1967) for the details. The latter author's result is:

$$w = \frac{|A|_{3n}^{1/2}}{|A^+|_{3n-1}^{1/2}} \frac{1}{2\pi m^{1/2}} \exp\left[-\Delta\Phi^{\mathrm{m}}/kT\right], \qquad \text{(III.17)}$$

where m is the mass of the particle, $|A|_{3n}$ and $|A^+|_{3n-1}$ are the determinants of the force constant matrix for the ground state and the saddle point (S), respectively. The matrix elements are simply the second derivatives of the potential energy of the crystal in each of these states. If $\mathbf{\Phi}$ $(\mathbf{r}_1, \mathbf{r}_2, \ldots, \mathbf{r}_N)$ is the potential written as a function of the coordinates of N atoms:

$$A_{nn'} \equiv \left(\partial^2\Phi/\partial\mathbf{r}_n\, \partial\mathbf{r}_{n'}\right).$$

The potential $\mathbf{\Phi}$ can be expanded to second order:

$$\Phi = \Phi_0 + \frac{1}{2}\sum_{nn'} A_{nn'}\mathbf{u}_n \cdot \mathbf{u}_{n'}$$

as a function of the displacements $\mathbf{u}_n$ and $\mathbf{u}_{n'}$ of the atoms n, n', with respect to their zero of position in the ground state or at the saddle point. Note that A^+ corresponds to an unstable state with $3n - 1$ degrees of freedom, whereas the stable state A has $3n$ degrees of freedom.

The theory thus predicts that the *jump frequency is inversely proportional to the square root of the mass of the particle.* The activation energy is calculable theoretically: it is the difference between the potential energy of the crystal with an atom at the saddle point and the crystal with that atom at M and a vacancy at M′.

The entropy term does not appear explicitly in the preexponential factor, but the entropy of vibration is implicity there. Introducing $\nu_0 = \frac{1}{2\pi}\sqrt{\frac{A_0}{m}}$, a vibration frequency of the ground state (the so-called attempt frequency), one obtains:

$$\boxed{w = \nu_0 \exp\frac{\Delta S^{\mathrm{m}}}{k} \exp\left(-\frac{\Delta\Phi^{\mathrm{m}}}{kT}\right)} \qquad \text{(III.18)}$$

with

$$\frac{\Delta S^{\mathrm{m}}}{k} = \frac{1}{2}\left[\log|A|_{3n} - \log A_0 - \log\left|A^+\right|_{3n-1}\right],$$

an expression analogous to one for a difference in entropies of vibration. (Recall that the frequencies are proportional to the square roots of the force constants.) This term can be calculated quite simply with a pair-interaction model, restricting the calculation to nearest neighbors, 12 in the ground state, 4 at the saddle point for a FCC structure. However, the sign of ΔS^{m} is not guaranteed!

In the framework of the dynamic theory, after some simplifying hypotheses (the Debye model), Flynn arrived at the following expression:

$$w = \sqrt{\frac{3}{5}}\nu_{\mathrm{D}} \exp\left(-\frac{C\Omega\delta^2}{kT}\right), \tag{III.19}$$

with ν_{D} the Debye frequency, Ω the atomic volume, and C a function of the elastic constants:

$$\frac{15}{2C} = \frac{3}{C_{11}} + \frac{2}{C_{11} - C_{12}} + \frac{1}{C_{44}},$$

that is, an elastic modulus which averages over the longitudinal and transverse branches of the phonon spectrum in the ratio 3/1/1 for the moduli C_{11}, $(C_{11} - C_{12})/2$, and C_{44}. δ is the fraction of the displacement toward the saddle point necessary for the jump to be completed ($\delta = 0.32$ in the FCC structure).

N. B.:

1) All the theories cited above are for temperatures high enough so that quantum effects can be neglected. In fact, according to quantum mechanics, even in the absence of assistance from thermal activation, a particle has non-zero probability of crossing a potential barrier by the ***tunnel effect.*** Such an effect is possible at very low temperatures for very light particles (cf. Ch. VII, diffusion of hydrogen and of muons, and Exercice 36).

2) For particles only slightly coupled to the lattice (see the diffusion of light interstitials), the jumps can be described by the theory of Brownian motion, especially the ***Langevin equation:***

$$m\left(\mathrm{d}^2y/\mathrm{d}t^2\right) + \beta\left(\mathrm{d}y/\mathrm{d}t\right) = F(t),$$

which describes the motion of a damped oscillator (the damping term β is inversely proportional to D, $D = kT/\beta$) and a fluctuating force $F(t)$ such that $\langle F(t)\rangle = 0$.

VII.3 THE ISOTOPE EFFECT. — Let m_1 be the mass of the jumping particle and m_0 the mass of the host atoms in the crystal (isotopic substitution). It is considered that this substitution does not affect the potential energy and its effect is limited to the entropy term, i.e., only the preexponential terms have to be compared. For example in the framework of the dynamic theory, one considers the n atoms constituting the gate and one can show that (see, e.g., Flynn 1972, § 7.4):

$$w \propto \frac{1}{\mu^{1/2}} = \left(\frac{1}{m_1} + \frac{1}{nm_0}\right)^{1/2}. \tag{III.20}$$

For light particles, such as the case of the diffusion of interstitials in the transition metals, $\frac{1}{m_1} >> \frac{1}{nm_0}$, the probability of the jump is closely proportional to $m^{-1/2}$, a simple isotope effect, verified by experiment. On the other hand, if $m_1 \approx m_0$, the diffusing atom appears to have a mass μ different from m_1 and given by equation (III.20). Let

$$m_1 = m_0 + \delta m.$$

The change of the jump frequency of the atom with mass m_1 due to the participation in the jump of the n atoms with mass m_0 is given by:

$$\frac{\Delta w}{w} = \frac{w(m_0 + \delta m) - w(m_0)}{w(m_0)} = -\frac{n}{n+1}\frac{\delta m}{2m_0},$$

or

$$\boxed{\frac{\Delta w}{w} = -\Delta K \frac{\delta m}{2m_0}}. \tag{III.21}$$

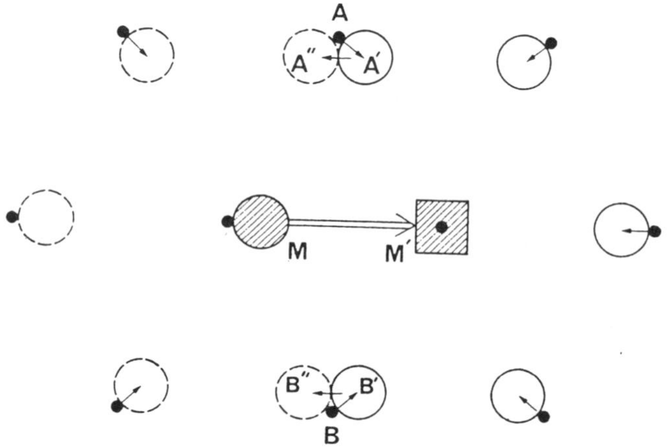

Fig. III.15. — Relaxation effects. The atoms A and B which form the gate relax to A′ and B′ due to the presence of the vacancy, and to A″ and B″ after the jump; hence the components of their movements A′A″ and B′B″ are antiparallel to the jump of the atom MM′.

The simple isotope effect is reduced by a mass-effect factor $\Delta K = n/(n+1)$ so that for a FCC structure, $\Delta K \approx 4/5 = 0.80$.

In Vineyard's theory, which is based on absolute reaction rates, ΔK is viewed as the fraction of the kinetic energy of the entire jump process that is carried by the jumping atom. In effect, as shown in figure III.15, in a vacancy mechanism, the atoms nearest neighbor to the two sites and the saddle point have components of displacement and velocity anti-parallel to the jump.

VII.4 NUMERICAL SIMULATION. — The most probable path of the atom, the position of the saddle point, the energy of migration, and the relaxation of the neighboring atoms can all be calculated by numerical simulation. Molecular-dynamics simulations can follow the trajectories of a large number of atoms. This technique has revealed the possibility of ***aborted jumps***, i.e., jumps in which the atom returns immediately from the saddle point (20% of the jumps at a temperature near T_M in the crystal model studied by Bennett (1976)). These aborted jumps result from anharmonic interferences between normal modes, which manifest themselves as an "obstruction" of the saddle point by a neighboring atom. This results in a value of ΔK sensibly lower than that predicted from a harmonic model. Molecular dynamics should give important insights into processes taking place at temperatures near the melting point, where the theory of thermal activation may no longer be applicable (see Ch. IX, part 3, § II.3, and the reference to Marchese *et al.* in Ch. IV).

VII.5 SOME SIMPLE APPLICATIONS OF THE POTENTIAL-BARRIER MODEL. — This simple model is often used to compare jump frequencies. Figure III.16 shows the effects of a driving force and of a frequency that depends on position.

In the first case, the barrier $\Delta\Phi$ is lowered by $\varepsilon/2$ for jumps from left to right, and raised by $\varepsilon/2$ for jumps from right to left. ε equals the work done by the constant driving force F over the distance λ: $\varepsilon = F\lambda$.

One can write:

$$w \rightarrow = \nu \exp\left(-\frac{\Delta\Phi - \varepsilon/2}{kT}\right) = w_0 \exp\left(\frac{\varepsilon/2}{kT}\right),$$

$$w \leftarrow = \nu \exp\left(-\frac{\Delta\Phi + \varepsilon/2}{kT}\right) = w_0 \exp\left(-\frac{\varepsilon/2}{kT}\right).$$

The difference of the jump frequencies is then:

$$\begin{aligned} w \rightarrow - w \leftarrow &= w_0 \left[\exp\left(F\lambda/2kT\right) - \exp\left(-F\lambda/2kT\right)\right] \\ &= 2w_0 \sinh\left(F\lambda/2kT\right). \end{aligned}$$

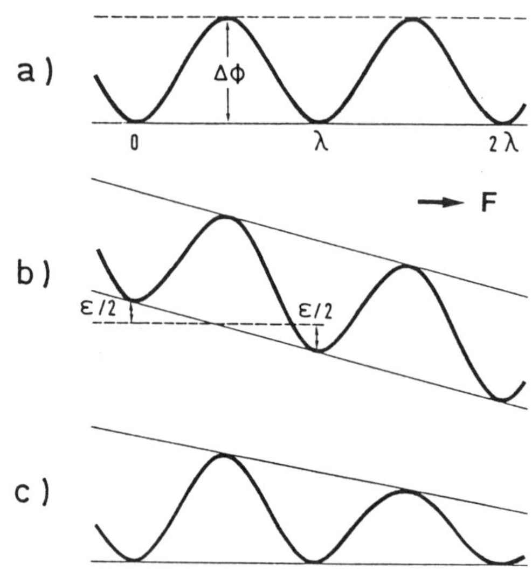

Fig.III.16. — The potential barrier (a) is modified by the presence of an applied force F (b). In (c), the barrier height varies with position (example: variation of jump frequency with chemical composition).

From this, the velocity of mass flow is:

$$\langle v \rangle = (w \rightarrow - w \leftarrow)\, \lambda = 2 w_0 \lambda \, \sinh\left(F\lambda/2kT\right),$$

or

$$\langle v \rangle = (2D/\lambda) \, \sinh\left(F\lambda/2kT\right). \qquad \text{(III.22)}$$

Expanding the hyperbolic sine to third order:

$$\langle v \rangle = DF/kT \left[1 + 1/24\,(F\lambda/kT)^2 + ...\right],$$

which reduces to the Nernst-Einstein equation only for $F\lambda << kT$ (see also Exercice 34).

The potential-barrier model is equally useful for describing the effect of *traps*. Figure III.17 shows a simple trap, allowing the calculation of the trapping frequency $\nu_1 \exp(-\Delta\Phi_1/kT)$ and the detrapping frequency $\nu_2 \exp(-\Delta\Phi_2/kT)$ with $\Delta\Phi_2 > \Delta\Phi_1$. We can then compare these with the jump frequency for normal migration, $\nu \exp(-\Delta\Phi^{\mathrm{m}}/kT)$.

The model has also been applied to interstitials, with numerous refinements, especially the introduction of non-uniform distributions of traps with varying depths, following Oriani's original model (1970). These turn out to be a particular case of McNabb's and Foster's general equations (see Ch. I, § VI, example 3; see also the kinetics of detrapping, Ch. IX, Appendix I).

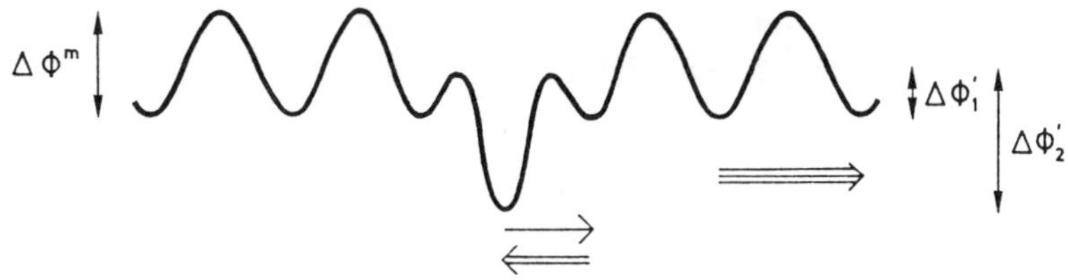

Fig. III.17. — Migration, trapping, and detrapping. Normal jump: ≡>, trapping: ⟹, detrapping: ⟶.

The effect of trapping is to reduce the diffusion coefficient at low temperatures as compared to values extrapolated from higher temperatures (on an Arrhenius plot, the curvature is convex upwards).

APPENDIX I

Calculation of $\langle \cos\theta \rangle$.

Consider a simple cubic lattice with interatomic spacing λ. The values of $\cos\theta_{ij}$ are 0, ± 1. The calculation of its average value is here limited to the jumps either parallel or antiparallel to the first jump: $\cos\theta_{ij} = \pm 1$. Let p_1 be the probability that the $(i+1)^{\text{st}}$ jump is parallel to the previous jump, p_2 that it is antiparallel:

$$\langle \cos\theta_{i,\ i+1} \rangle = (+1)p_1 + (-1)p_2 = p_1 - p_2.$$

For the jump $j = i + 2$, 4 cases are possible:

jump i	$i+1$	$i+2$	$\cos\theta_{ij}$	
$\longrightarrow$	$\longrightarrow$	$\longrightarrow$	$+1$	p_1^2
		$\longleftarrow$	-1	$p_1\ p_2$
	$\longleftarrow$	$\longrightarrow$	$+1$	p_2^2
		$\longleftarrow$	-1	$p_2\ p_1$

Whence:

$$\begin{aligned}\langle \cos\theta_{i,\ i+2} \rangle &= p_1^2 - p_1\ p_2 + p_2^2 - p_2\ p_1 = (p_1 - p_2)^2 \\ &= \langle \cos\theta_{i,\ i+1} \rangle^2 .\end{aligned}$$

As this is true for all i, we have the recurrence relation:

$$\langle \cos \theta_{i,\ i+k} \rangle = \langle \cos \theta_{i,\ i+1} \rangle^k .$$

This relation is applicable to all structures where the jump vector is an axis of at least two-fold symmetry (see e.g. Adda and Philibert, *op. cit.*, Ch. VII, Appendix I; Manning, *op. cit.*, § 3.4).

APPENDIX II

Percolation.

Consider a plane network, every element of which is an electrical resistor. Remove, at random, one, two, ... n elements of this network, continuously measuring the total electrical resistance of the network in the meantime with a multimeter. After a certain fraction of the resistors has been removed, the network no longer conducts current. This fraction is well defined for a network with an infinite number of resistors. Let us define the ***percolation threshold*** ([4]) by the fraction P_c of the resistors which are left in the network such that just above the percolation threshold, there exists a completely connected cluster of resistors that traverses through and through the network, but below the threshold, only finite clusters exist. Statistical analysis can be used to describe how the conductance of such a network varies in the neighborhood of the percolation threshold, i.e., the insulator-conductor transition (Fig. III. 18 for a ball model):

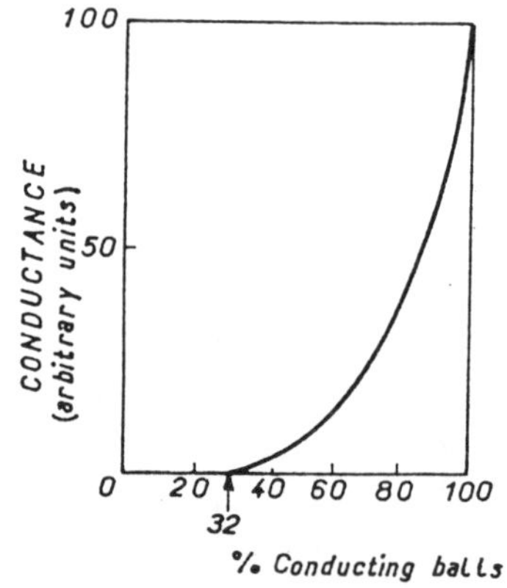

Fig. III.18. — Mixture of conducting and insulating units.

([4]) The name "percolation" is used because this phenomenon is analogous to the problem of a filter with some channels connected and others closed or obstructed.

The percolation transition is characterized by three quantities:

– the percolation threshold P_c,

– the correlation length, which is finite below the threshold and diverges at $P = P_c$ for an infinite lattice,

– the critical exponent in the power law $|P - P_c|^{\nu}$ which is followed by certain properties near the threshold. It is remarkable that this exponent has a "universal" value, i.e., for a given property, e.g., the correlation length, the conductance, ..., it depends only on the dimensionality of the system, but not on the details of this structure.

In diffusion, the idea of percolation clearly applies to all systems such that the sites on which an atomic or defect species can migrate are, in fact, not all accessible, some fraction of these sites being forbidden for various reasons. There exists then a *percolation threshold* which is the minimum concentration of forbidden sites – i.e., the maximum concentration of accessible sites – for which diffusion is not possible. Above this threshold, the flux is proportional to a fraction of the atomic jumps, determined by the *percolation efficiency* which increases in the same measure as the concentration of the forbidden sites decreases (Fig. III.19). The value of the percolation threshold depends only on the crystal structure.

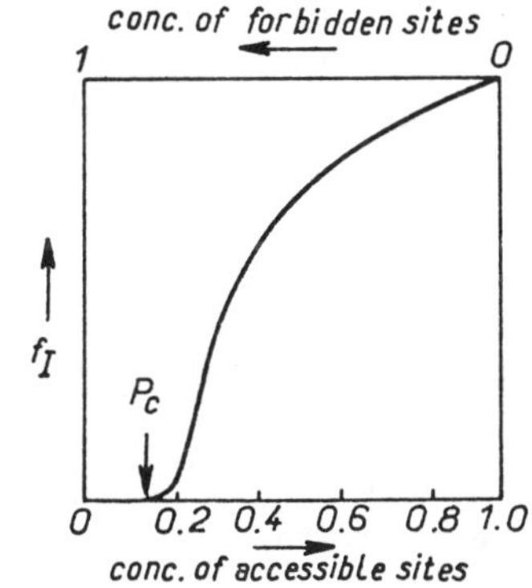

Fig. III.19. — Variation of the percolation efficiency $f_{\rm I}$ as a function of forbidden (accessible) sites for an FCC lattice. The percolation threshold is indicated by P_c (After Murch and Rothman (1981)).

The above is a very simple case. However, the idea of percolation extends to systems in which the forbidden sites are not distributed at random, or to the case in which there are no forbidden sites, but several types of sites with different energies, whether the differences are imposed by the crystal structure (*à priori* inequivalence of sites) or the differences are due to the interactions among the diffusing species.

All these effects cause departures from purely random motion, such that the flux is reduced. Various authors, depending on their point of view, have called the coefficient describing this reduction the *physical correlation factor* or the *percolation efficiency*. For the diffusion of a defect (or for true self-diffusion), this factor must thus be introduced, but for the diffusion of a tagged atom, the *geometric or tracer correlation factor* must also be taken into account (see, e.g., H. Sato, S. A. Akbar, and G. E. Murch (1981)).

A simple case of forbidden sites on a lattice is that of "blocking", e.g., immobile impurities on interstitial sites which close off these sites from the migration of other interstitials. A detailed theory of random walk on a lattice with blocked sites has been developped by Tahir-Kheli (1985).

Bibliography

I. — GENERAL REFERENCES

FLYNN C. P., *Point Defects and Diffusion*, (1972) op. cit, chap. 7.

HAUS J. W. and KEHR K. W., *Physics Reports* **150** (1987) 263–406.

LE CLAIRE A. D., *in Physical Chemistry, an Advanced Treatise*, eds. W. Jost, H. Eyring, O. Henderson (Academic Press) 1970, vol. **X**, p 261 ff.

MANNING J. R., *Diffusion Kinetics for Atoms in Crystals*, (1968) *op. cit.*

MURCH G. E., *Atomic Diffusion Theory in Highly Defective Solids*, Diff. and Defects Monograph Series, #6, Trans. Tech. Publ. Aedermannsdorf (1980).

II. SPECIFIC REFERENCES

BARDEEN J. and HERRING C., *Imperfections in Nearly Perfect Crystals*, ed. W. Shockley (Wiley, New York) 1952, p 261–288.

BENOIST P., BOCQUET J-L., and LAFORE P., *Acta Met.* **25** (1977) 265–275.

BENNETT C. H., Colloque Saclay, *op. cit.*, Vol II (1976) p 655–684.

COMPAAN K. and HAVEN Y., *Trans. Faraday Soc.* **52** (1956) 786–801, **54** (1958) 1498–1508.

DOAN N. V. and ADDA Y., *Philos. Mag.* **A56** (1987) 269–283.

EISENSTADT M. and REDFIELD A. G., *Phys. Rev.* **132** (1963) 635–643.

FRANKLIN A. D. and LIDIARD A. B., *Proc. Roy. Soc.* London **A389** (1983) 405–431.

GLYDE H. R., *Rev. Mod. Phys.* **39** (1967) 373–382.

HOWARD R. E., *Phys. Rev.* **144** (1966) 650–661.

JOHNSON R. A. and LAM N. Q., *Phys. Rev.* **B13** (1976) 4364–4375.

LE CLAIRE A. D. and LIDIARD A. B., *Philos. Mag.* **1** (8) (1955) 518–527.

LIDIARD A. B., *Philos. Mag.* **46** (1955) 1218; erratum HOWARD R. E. and LIDIARD A. B., *J. Phys. Soc. Japan* **18**, Supp. II (1963) 197–201.

LIDIARD A. B., *Mass Transport in Solids, op. cit.*, chap. 3 (1983) p 43–79.

MANLEY O. P., *J. Phys. Chem. Solids* **13** (1960) 244–250.
MULLEN J. G., *Phys. Rev.* **124** (1961) 1723–1730.
MURCH G. E., *Am. J. Phys.* **47** (1979) 78–80 and 958–960.
MURCH, G. E., ROTHMAN S. J., *Philos. Mag.* **A43** (1981) 229-238.
ORIANI R. A., *Acta Metall.*, **18** (1970) 147–157.
RICE S. A., *Phys. Rev.* **112** (1958) 804-811.
SASAKI J., PETERSON N. L., and HOSHINO K., *J. Phys. Chem. Solids* **46** (1985) 1267–1283.
SATO H. and KIKUCHI R., *J. Chem Phys.* **55** (1971) 677–702, and 702–715.
VINEYARD G. H., *J. Phys. Chem. Solids* **3** (1957) 121–127.
WOLF D., *Mass Transport in Solids, op. cit.*, chap 7 (1983) p 149–168.

CHAPTER IV

Self-diffusion

The preceding theories are now applied to the simplest case, a chemically homogenous system, either a pure element, or a binary (or ternary...) compound insofar as the two (resp. 3,...) sublattices can be considered as independent.

I. The self-diffusion coefficient.

Consider a body of pure A in which the diffusion of A atoms is studied: this is *self-diffusion*. In practice, tagged atoms are used as a tracer, i.e., an isotope A*, usually, but not necessarily, radioactive. The diffusion coefficients corresponding to the A atoms and A* tracers differ due to the correlation effect:

$$D^{\mathrm{A}}_{\mathrm{A}^*} = f D^{\mathrm{A}}_{\mathrm{A}}, \qquad \text{(IV.1)}$$

or more generally:

$$\boxed{D^* = f D_{\mathrm{random}}}. \qquad \text{(IV.2)}$$

By incorrect custom, $D^{\mathrm{A}}_{\mathrm{A}^*}$ is called the self-diffusion coefficient. In the case of self-diffusion, f_0 is a numerical factor independent of temperature, but only for cubic structures, in which, because of isotropy, the jumps toward all the nearest neighbors have the same frequency (Tab. I).

These definitions and values of f apply equally to compounds $\mathrm{A}_\alpha\mathrm{B}_\beta$, where the species A and B form two independent sublattices. Since the migration of the A* atoms is confined to their own sublattice, this case is the same as that of the elemental material.

II. Variation of the diffusion coefficient with temperature.

II.1 Vacancy mechanism. — Assume that diffusion takes place via the vacancy mechanism. It is clear that in any given time, the total number of vacancy jumps is equal to the total number of atomic jumps, or, per unit volume, $n_v \zeta w_0 = n_{\mathrm{A}} \zeta \Gamma_{\mathrm{A}}$.

Table I. — *Correlation factors for self-diffusion*

Mechanism	Crystal structure	f_0
Vacancy	Honeycomb Square Triangular Simple cubic BCC FCC Diamond cubic Ideal HCP	1/3 0.467 0.56 0.653 0.727 0.7815 1/2 $f_x = f_y = 0.7812$ $f_z = 0.7815$
Divacancy	FCC	0.458*
Interstitial	All structures (high dilution)	1
Interstitialcy	AgCl (diffusion of Ag) CaF_2 (diffusion of Ca)	2/3 (collinear jumps, $\Theta = 0°$) 0.9697 (non-collinear, $\Theta = 70.53°$) 0.9643 (non-collinear, $\Theta = 109.47°$) 4/5 (collinear jumps, $\Theta = 0°$) 0.9855 (non-collinear, $\Theta = 70.53°$) 1 (non-collinear, $\Theta = 90°$)
Dumbbell interstitial, <100>	FCC	0.88

* Or 0.468, depending on the method of calculation (see Murch, 1984).

Since $n_v << n_A$, n_v/n_A is practically equal to the mole fraction of vacancies, $\bar{N}_v$, at thermodynamic equilibrium,

$$\Gamma_A = \bar{N}_v w_0. \tag{IV.3}$$

The atomic jump frequency denoted by Γ_A is the same as the one denoted by Γ_s in previous chapters. The jump frequency of the vacancy in a nearest-neighbor site is denoted by w_0 for reasons which will appear later. The total jump frequency equals the coordination number ζ times the jump frequency toward a given site, Γ_s. At equilibrium, the atomic fraction of vacancies is given by:

$$\bar{N}_v = \exp\left(-\Delta G_v^f/kT\right), \tag{IV.4}$$

where ΔG_v^f is the Gibbs free energy of vacancy formation. The jump frequency

w_0 can be expressed in the form (see Ch. III, § VIII):

$$w_0 = \tilde{\nu} \exp\left(-\Delta G_v^{\mathrm{m}}/kT\right), \tag{IV.5}$$

where $\tilde{\nu}$ is a vibration frequency ($\approx$ the Debye frequency) and ΔG_v^{m} is called the Gibbs free energy of migration.

Since $D^* = \beta f_0 \Gamma_s\, a^2$ in a cubic structure, if the entropies and enthalpies of formation migration are substituted, one obtains:

$$D^* = \tilde{\nu}\beta a^2\, f \exp\left(\frac{\Delta S_v^{\mathrm{f}} + \Delta S_v^{\mathrm{m}}}{k}\right) \exp\left(-\frac{\Delta H_v^{\mathrm{f}} + \Delta H_v^{\mathrm{m}}}{kT}\right), \tag{IV.6}$$

with $\beta = 1$ for FCC and BCC and 1/8 for diamond cubic.

This expression has the form of the *Arrhenius equation*:

$$\boxed{D = D_0 \exp\left(-\Delta H/kT\right)}, \tag{IV.7}$$

or $D = D_0 \exp\left(-Q/RT\right)$; R is the gas constant and

$$\boxed{D_0 = \beta a^2\, \tilde{\nu} f \exp\left(\frac{\Delta S_v^{\mathrm{f}} + \Delta S_v^{\mathrm{m}}}{k}\right)} \tag{IV.8}$$

$$\boxed{\Delta H = -k\frac{\mathrm{d}\ \mathrm{Log}\ D}{\mathrm{d}\,(1/T)} = \Delta H_v^{\mathrm{f}} + \Delta H_v^{\mathrm{m}}}, \tag{IV.9}$$

even if ΔH and ΔS are temperature dependent(1). Note that the entropy terms appear in D_0.

The relation (IV.6) above can also be written:

$$D = f_0 N_v D_v, \tag{IV.10}$$

with

$$D_v = \beta a^2\, \tilde{\nu} \exp\left(-\Delta G_v^{\mathrm{m}}/kT\right), \tag{IV.11}$$

where D_v is the diffusion coefficient or diffusivity of the vacancies.

(1) *Note on units:* Boltzmann's constant: $k = 1.38 \times 10^{-16}$ erg K^{-1}, 1.38×10^{-23} J K^{-1}, or 0.8617×10^{-4} eV K^{-1}. ΔH is given in eV, with 1 eV $= 1.602 \times 10^{-12}$ erg or 1.602×10^{-19} J. Q can also be given in cal. mole^{-1} with $R = 1.9872$ cal. mole^{-1} K^{-1} or 8.314 J mole^{-1} K^{-1}. Recall 1 eV = 23061 cal. mole^{-1} = 96485 J mole^{-1}.

II.2 Divacancy mechanism. — The mole fraction of divacancies is given by:

$$\bar{N}_{2v} = \alpha \exp\left(-\Delta G^{\rm f}_{2v}/kT\right), \qquad \text{(IV.12)}$$

with $\alpha = 6$ for the FCC lattice, where there are 6 possible orientations for a divacancy (i.e., 2 vacancies on nearest-neighbor sites). The total number of atom jumps per second in the FCC structure is 12 $n_{\rm A}\Gamma_s$ because 12 Γ_s is the total jump frequency of one atom. The total number of vacancy jumps is 2 n_{2v} (number of vacancies) multiplied by 4 w_{2v} because each vacancy can jump to only 4 nearest neighbor sites without dissociating the divacancy. Thus:

$$\begin{aligned} n_{\rm A}\ 12\ \Gamma_s &= 2 n_{2v} \cdot 4\ w_{2v} \\ \Gamma_s &= \frac{2}{3}\ \bar{N}_{2v} w_{2v}. \end{aligned} \qquad \text{(IV.13)}$$

Hence the seif-diffusion coefficient of a tracer by the divacancy mechanism is:

$$D^*_{2v} = f^0_{2v}\ \Gamma_s\ a^2 = \frac{2}{3}\ f^0_{2v}\ \bar{N}_{2v}\ w_{2v}\ a^2, \qquad \text{(IV.14)}$$

with the proper correlation factor for this mechanism, f^0_{2v}=0.458, and $\alpha = 6$, or:

$$D^*_{2v} = 4\ a^2\ f^0_{2v}\ \tilde{\nu}_{2v} \exp\left[-\frac{\Delta G^{\rm f}_{2v} + \Delta G^{\rm m}_{2v}}{kT}\right]. \qquad \text{(IV.15)}$$

Although D^*_{2v} can be expressed as a function of the diffusivity of divacancies, D_{2v}, care must be taken as the length of the divacancy jump is half that of the tracer if the center of the divacancy is considered (Fig. IV.1).

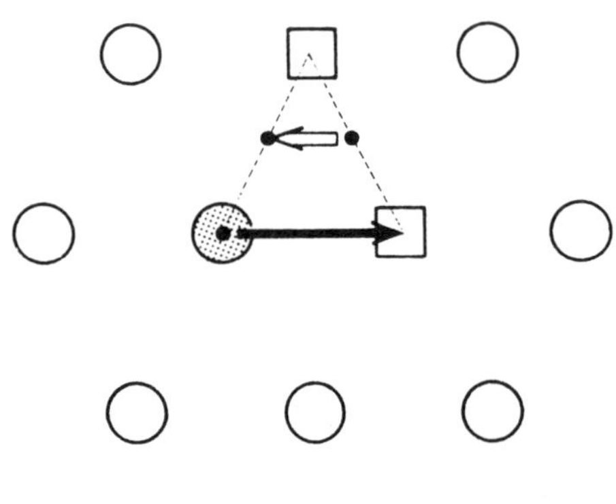

Fig. IV.1. — Divacancy mechanism: jump of the atom (→); jump of the "center" of the divacancy (⇐).

From this,

$$\frac{D^*_{2v}}{D_{2v}} = 2\, f^0_{2v}\ \bar{N}_{2v} = H_{2v}\ \bar{N}_{2v}\,, \tag{IV.16}$$

where H_{2v} is the *Haven ratio*, which differs from the correlation factor by a numerical factor.

II.3 Interstitial mechanism. — This mechanism is rather unlikely for self-diffusion in metals because of the high energy of formation of the self-interstitial in a close-packed structure, although it has been suggested for sodium, and it applies for silicon because of its rather open structure. Self-interstitials can be created by irradiation. They are found at equilibrium in silicon and certain compounds, e.g. Ag in AgCl. The equations established for the vacancy mechanism can be transformed to those for the interstitial mechanism directly. However, for the interstitialcy mechanism, the jump distance of the defect is not equal to that of the tracer. Thus:

$$D^*_i = H_i\ \bar{N}_i D_i\,, \tag{IV.17}$$

with the Haven ratio H_i. For a collinear jump, $H_i = f^0/2$ (see § X.2).

II.4 Several mechanisms operating simultaneously. — If several independent mechanisms are operative, then according to equation (II.31) the diffusion

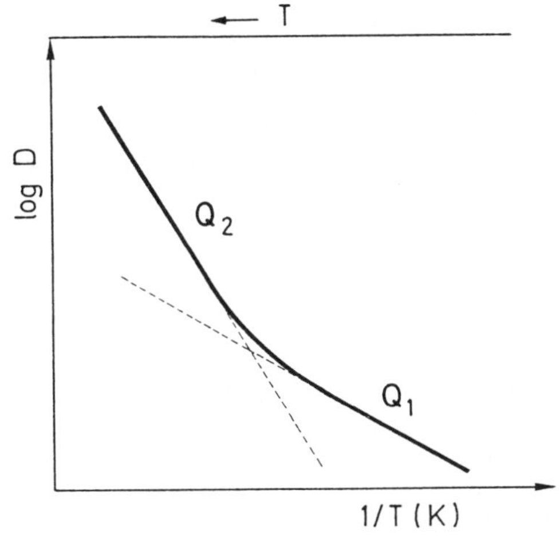

Fig. IV.2. — Arrhenius plot for the case of two diffusion mechanisms, with $Q_1 < Q_2$.

coefficients calculated for each type of defect are added:

$$D^* = \sum_d D_d^* = \sum_d H_d N_d D_d, \qquad \text{(IV.18)}$$

where D_d is the diffusion coefficient of the defect d, and H_d is the corresponding Haven ratio which contains the correlation factor. The sum can include vacancies, divacancies, self-interstitials, etc. In this case, the self-diffusion coefficient does not follow an Arrhenius dependence. The Arrhenius plot is curved, possibly with straight-line portions where one of the mechanisms is dominant (Fig. IV.2).

III. Anisotropy of diffusion.

It was shown in chapter II that a different diffusion coefficient has to be considered for each coordinate axis. In the HCP structure, $D_x = D_y \neq D_z$. There are two jump frequencies, w_a for jumps in the basal plane, and w_b for jumps out of the basal plane. Considering correlation effects, equations (II.22) become, with $\Gamma_i = \bar{N}_v w_i$ (i = a or b):

$$\begin{aligned} D_x &= \frac{a^2}{2}\bar{N}_v\ (3\ f_{\mathrm{a}x}\ w_\mathrm{a} + f_{\mathrm{b}x}\ w_\mathrm{b}) \\ D_z &= \frac{3}{4}\bar{N}_v f_{\mathrm{b}z}\ w_\mathrm{b}\ c^2. \end{aligned} \qquad \text{(IV.19)}$$

The anisotropy of diffusion is usually measured as $A = (c/a)^2\ (D_x/D_z)$(²):

$$A = \frac{2}{3}\,\frac{3\ f_{\mathrm{a}x}\ w_\mathrm{a} + f_{\mathrm{b}x}\ w_\mathrm{b}}{f_{\mathrm{b}z}\ w_\mathrm{b}}. \qquad \text{(IV.20)}$$

The partial correlation factors vary with $w_\mathrm{a}/w_\mathrm{b}$, as does the anisotropy. Figure IV.3 shows this variation compared to the variation expected when correlation is neglected:

$$A_\mathrm{random} = \left[2\frac{w_\mathrm{a}}{w_\mathrm{b}} + \frac{2}{3}\right].$$

Experimentally, D_x/D_z is found to be < 1 or > 1 as the c/a ratio is $>$ or $< \sqrt{8/3}$. The value of the ratio $w_\mathrm{a}/w_\mathrm{b}$ is obtained from the experimental value of the anisotropy. A more general discussion of the symmetry of the **D** tensor is given in chapter VIII, Appendix II.

(²) D_x/D_z or $D_\perp/D_\parallel$, as you like it. The values of the correlation factors given in Table I are defined as follows: $f_x\ (= f_y)\ = (3f_{\mathrm{a}x}w_\mathrm{a} + f_{\mathrm{b}x}w_\mathrm{b})\,/\,(3w_\mathrm{a} + w_\mathrm{b})$ and $f_z = f_{\mathrm{b}z}$. The numerical values correpond to the particular case $w_\mathrm{a} = w_\mathrm{b}$, so that the f's are temperature independent.

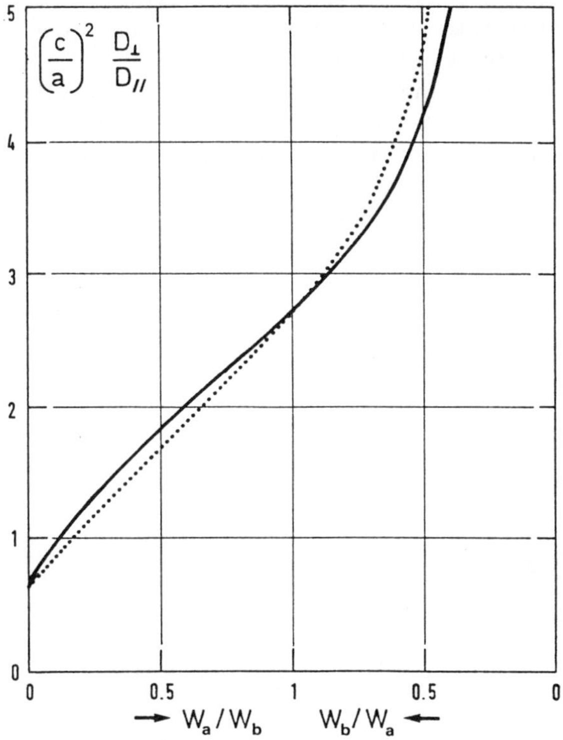

Fig. IV.3. — Anisotropy in the HCP structure. Variation of the anisotropy $(c/a)^2$ (D_x/D_z) of the self-diffusion coefficients perpendicular to and parallel to the c axis, as a function of the ratio of the jump frequency in the basal plane, w_a, to that out of the basal plane, w_b. Dotted line: same, without correlation. After Mullen (1961).

IV. Deviations from the Arrhenius law.

A non-linear Arrhenius plot can be interpreted in several ways:

1) *A single mechanism of diffusion,* but with ΔH and ΔS dependent on temperature. It can be shown in a general way that this is indeed the case. Zener (1951) had suggested that the Gibbs free energy of migration for interstitials (in this case, ΔG reduces to ΔG^{m}) represents the work to deform the lattice elastically during the course of an atomic jump. The variation of ΔG^{m} with T would then be the same as that of the shear modulus:

$$\frac{\partial\,(\Delta G^{\mathrm{m}}/\Delta G_0)}{\partial T} = \frac{\partial\,(\mu/\mu_0)}{\partial T}, \qquad \text{(IV.21)}$$

$$\mu = \mu_0 + T\,(\mathrm{d}\mu/\mathrm{d}T),$$

where the subscript 0 refers to a value at 0 K. ΔH^{m} and ΔS^{m} are obtained from the classical thermodynamic relations; these quantities are independent of T only if μ varies linearly with T.

In Flynn's model of the atomic jump (Ch. III, § VII), the Gibbs free energy of migration contains an average elastic modulus. If this modulus does not vary linearly with T, the entropy and enthalpy of migration are functions of the temperature.

Varotsos *et al.* (1978) considered that the energy for the formation and migration of a defect are drawn from the heat reservoir made up of the lattice vibration modes; they assumed that 1) only the dilatational modes were important, and 2) that these could be related to the static bulk modulus. On this basis, they obtained for the Gibbs free energy of isothermal dilatation per unit volume:

$$G = B\varepsilon\Omega, \tag{IV.22}$$

where B is the isothermal bulk modulus, ε a constant that depends on the type of crystal structure but is independent of temperature, and Ω the atomic volume. Then by differentiation:

$$\Delta S\,(T) = -\frac{H_0}{B_0}\left[B\alpha + \frac{\partial B}{\partial T}\right] \exp \int_0^T \alpha \,\mathrm{d}T + \text{vibration terms independent of T;} \tag{IV.23a}$$

$$\Delta H\,(T) = \frac{H_0}{B_0}\left[B - TB\alpha - T\frac{\partial B}{\partial T}\right] \exp \int_0^T \alpha \,\mathrm{d}T. \tag{IV.23b}$$

The subscript 0 indicates a value at 0 K; α is here the coefficient of volume thermal expansion. Clearly, ΔS and ΔH vary with temperature unless $\alpha = 0$ and $\partial B/\partial T$ is constant, and they can vary very strongly at high temperatures due to the anharmonicity (α varies with T). The ratio $\Delta S/\Delta H$ is determined by the anharmonic properties of the crystal; it should be the same for the formation and migration of all defects. The model predicts a linear temperature dependence of log D on $B\Omega/kT$, so that only the variations of B or Ω with temperature are responsible for the form of the temperature dependence of D.

Even though $(\Delta H^{\mathrm{m}}+\Delta H^{\mathrm{f}})$ should vary with temperature, it is still the enthalpy of activation. Let, for example:

$$\Delta H = \Delta H_0\,(1 - \alpha T)\,.$$

It is *incorrect* to call ΔH_0 the enthalpy of activation, and to ascribe the effect of temperature to an entropy term exp $(\Delta H_0\alpha/k)$. For if ΔH varies with temperature, so does ΔS, according to the thermodynamic relation:

$$T\left(\frac{\partial\,\Delta S}{\partial T}\right)=\frac{\partial\,\Delta H}{\partial T},$$

from which:

$$\begin{aligned} Q=-R\,\frac{\partial\,\mathrm{Log}\,D}{\partial\,(1/T)} &= -\frac{\partial\,\Delta S}{\partial\,(1/T)}+\Delta H_0 \\ &= -\Delta H_0\,\alpha T+\Delta H_0=\Delta H. \end{aligned}$$

1 a) *A single mechanism of diffusion, with dynamic correlation effects,* such as the vacancy double jumps described in chapter III, § VI, also exhibits a deviation from Arrhenius behavior because of the decorrelation of the tracer jumps at high temperature.

2) *Two (or more) mechanisms of diffusion,* the one with the higher enthalpy of activation being dominant at high temperature. We distinguish three cases:

a) several types of point defects are contributing to the transport. This is the reason for the gentle curvature of the Arrhenius plot noted near the melting point for several FCC metals (Cu, Ag, Fig. IV.7). The effect is ascribed to the presence of divacancies (concentration $\approx 10^{-6}$ near T_{m}), the single vacancy mechanism being determined by the measurements of D down to quite low temperatures ($\approx 0.4\ T_{\mathrm{m}}$). The same arguments apply to the isotope effect and to the volume of activation (see below). In other cases, the contribution of self-interstitials at high temperature should perhaps be considered;

b) a single type of point defect, but with an "intrinsic" region at high temperature, as opposed to an "extrinsic" region at low temperature where the concentration of defects no longer varies with the temperature, being fixed by some other factor (thus $\Delta H = \Delta H^{\mathrm{m}}$). This is the case of ionic crystals, in which the concentration of defects at low temperatures is controlled by the concentration of heterovalent impurities;

c) a single type of point defect, but with jumps to both *nn* (nearest neighbor) and *nnn* (next-nearest neighbor) sites.

3) *Structural heterogeneities*, notably grain boundaries, subgrain boundaries, and dislocations (Frank networks), which constitute "short circuiting" paths for diffusion, i.e., regions where the diffusion coefficient is higher and the activation energy for diffusion is lower than in the crystal. The effect of boundaries can be avoided experimentally by using single crystals free of substructure. The effect of isolated dislocations on self-diffusion is generally small; it can be noticeable in the case of impurities that segregate to the dislocation cores (and, *a fortiori*, to the boundaries). See chapter VII.

To decide which of these hypotheses is the correct one, the following criteria are used:

- the values of D_0 and Q,
- The value of the isotope effect $E = f\ \Delta K$ (§ V),
- The value of the activation volume (§ VI).

The criteria are based on the values of these parameters and on their variation with temperature. For example, the high-temperature contribution of divacancies manifests itself by a decrease of the correlation factor and hence of the isotope effect, and by an increase of the activation volume with increasing temperature. But it is still very difficult to decide between hypotheses 1 and 2.

V. The isotope effect.

Are the self-diffusion coefficients of two isotopes of the same element equal ? They can be expected to differ by a little, no matter what the mechanism of diffusion, owing to their difference in atomic mass. For a mechanism involving a single type of point defect and two isotopes α and β:

$$\frac{D_\alpha}{D_\beta} = \frac{f_\alpha}{f_\beta}\frac{w_\alpha}{w_\beta}, \tag{IV.24}$$

where $w_{\alpha,\beta}$ designates the exchange frequency of the isotopes α and β with the point defect. The idea that $D_\alpha/D_\beta = w_\alpha/w_\beta$ when $w_\alpha \neq w_\beta$ is correct only if $f = 1$. Consider the situation where the vacancy has one atom of tracer α as a neighbor. Among its ζ neighbors there are then $\zeta - 1$ atoms of mass m_0 and one atom of mass m_α; the jumps of the vacancy then cannot be described by one frequency w_0 alone. The tracer has to be considered in this context as a different atom and the expression for the correlation factor of an impurity (solute diffusion at infinite dilution) has to be used.

The isotope effect can then be expected to be very sensitive to the mechanism of the atomic jumps; this is the reason why the isotope effects is of such great interest. We next calculate the relative difference of the diffusion coefficients of two isotopes:

$$\frac{\Delta D}{D} = \frac{D_\alpha}{D_\beta} - 1.$$

1st *case: a single jump mechanism.*

$$\frac{\Delta D}{D} = \frac{w_\alpha f_\alpha - w_\beta f_\beta}{w_\beta f_\beta}. \tag{IV.25}$$

$f_{\alpha,\beta}$ is a function of $w_{\alpha,\beta}$ (defect-tracer exchange frequency) and the other exchange frequencies w_i (defect-host atom exchange frequencies); the latter are assumed not changed because the tracer is present. Then to first order:

$$f_\alpha = f_\beta + \frac{\partial f}{\partial w}\,(w_\alpha - w_\beta)\,, \tag{IV.26}$$

whence

$$\begin{aligned} \frac{\Delta D}{D} &= \frac{w_\alpha\,(f_\alpha/f_\beta) - w_\beta}{w_\beta} \\ \frac{\Delta D}{D} &= \left[\frac{w_\alpha - w_\beta}{w_\beta}\right]\left[1 + \frac{\partial \log f}{\partial \log w}\right], \end{aligned} \tag{IV.27}$$

since $f_\alpha \approx f_\beta$(³) and $w_\alpha \approx w_\beta$.

Examples :

1) FCC lattice, vacancy mechanism. The correlation factor is written (see Ch. V, Eq. (V.3)):

$$f_{\alpha,\beta} = \frac{\psi\,(w_i)}{\psi\,(w_i) + w_{\alpha,\beta}}, \tag{IV.28}$$

where $w_{\alpha,\beta}$ designate the frequency w_2 for the isotopes α and β, and the w_i are the other frequencies, assumed to be unchanged. After some arithmetic:

$$\boxed{\frac{\Delta D}{D} = f_\alpha \cdot \left[\frac{w_\alpha}{w_\beta} - 1\right]}\,. \tag{IV.29}$$

This well-known formula is valid *only* if the correlation factor can be expressed by equation (IV.28). If not, the proper expression for $f\,(w)$ and equation (IV.27) must be used.

2) *Several concurrent mechanisms.* — Let D_1, $D_2, \ldots$ be the diffusion coefficients calculated for each mechanism. From equation (II.31a) above:

$$D = D_1 + D_2 + \cdots$$

Applying this formula to each tracer, and introducing the corresponding isotope

(³) f_α and f_β are very close to the value of f_0 given in table I for the ideal situation in which the jump frequencies of the tracer and the host atoms are equal. In fact the numerical difference among f_α, f_β, and f_0 is generally not significant, but for this calculation their explicit variation with the jump frequency must be taken into account.

effects, one obtains:

$$\frac{\Delta D}{D} = \frac{D_1}{D_1 + D_2 + \cdots} \cdot \frac{\Delta D_1}{D_1} + \frac{D_2}{D_1 + D_2 + \cdots} \cdot \frac{\Delta D_2}{D_2} + \cdots \tag{IV.30}$$

This expression is often used with D_1 for diffusion by single vacancies ($D_1 = f_v N_v D_v$) and D_2 for diffusion by divacancies.
More complex expressions must be calculated if the tracer can make several types of jumps: e.g., vacancy in an anisotropic crystal, or an interstitialcy (collinear or non-collinear).

RELATION TO THE ATOMIC MASS OF THE TRACERS. — Assume that the enthalpy of migration is independent of the atomic mass, a reasonable hypothesis in the absence of quantum effects. Then:

$$\frac{w_\alpha}{w_\beta} = \frac{\tilde{\nu}_\alpha}{\tilde{\nu}_\beta}.$$

If the atom is likened to a small harmonic oscillator:

$$\nu = (1/2\ \pi)\ \sqrt{K/m},$$

whence:

$$w_\alpha / w_\beta = (m_\beta / m_\alpha)^{1/2}. \tag{IV.31}$$

But in fact, the jump of the vacancy does not involve only one atom. Because of the relaxation around the vacancy, a collective process must be considered; during the course of the jump, certain neighbors undergo a small displacement in the opposite direction (see Fig. III.15).

These many-body effects are difficult to analyze; Fig. III.15 is a very simplistic scheme. Theoretically one is led to write (Ch. III. § VII, Eq. (III.21)):

$$(w_\alpha / w_\beta) - 1 = \Delta K \left[(m_\beta / m_\alpha)^{1/2} - 1 \right] \tag{IV.32}$$

where ΔK expresses in a gross way the fraction of the kinetic energy of the configuration localized on the tracer (4). It is anharmonic effects that are

(4) Le Claire has proposed a simple approximate expression for ΔK, with a simple physical meaning for the case of the vacancy mechanism:

$$\Delta K \approx \left[1 + (n/3) \left(1 - \Delta V_v^{\mathrm{f}} \right) \right]^{-1},$$

where n is the number of nearest neighbor atoms of the initial and final sites of the jump ($n = 4$ for FCC, 6 for a BCC lattice), and ΔV_v^{f} is the volume of formation of a vacancy as a fraction of the atomic volume. Marchese *et al.* (1986, 1988) have carried out a first-principles calculation of ΔK for FCC and BCC metals that closely reproduces the experimental values.

responsible for the values of ΔK significantly different from unity ($\approx 0.8 - 0.9$ for vacancy diffusion in FCC structures).

For mechanisms involving the simultaneous jumps of several atoms, such as interstitialcy or ring mechanisms, the ratio of the average masses of the moving atoms must be considered:

$$\frac{w_\alpha}{w_\beta} - 1 = \Delta K \left\{ \left[\frac{(n-1)\ m_0 + m_\beta}{(n-1)\ m_0 + m_\alpha} \right]^{1/2} - 1 \right\} \tag{IV.33}$$

for n atoms (1 tracer of mass $m_{\alpha,\beta}$ and $n-1$ atoms of mass m_0). ***The isotope effect*** usually refers to the quantity $E\,(n=1)$, or more simply, E defined as:

$$E \equiv \frac{D_\alpha / D_\beta - 1}{(m_\beta / m_\alpha)^{1/2} - 1}, \tag{IV.34}$$

or an equivalent expression $E\,(n)$ derived from equation (IV.33) for the case of a mechanism involving n atoms:

$$E_n \equiv \frac{D_\alpha / D_\beta - 1}{\left(\dfrac{m_\beta + (n-1)\ m_0}{m_\alpha + (n-1)\ m_0} \right)^{1/2} - 1}. \tag{IV.34a}$$

The ideal value of the isotope effect, $E = 1$, is made smaller by two effects:

– the diffusion of the tracers is decreased by correlation,

– the variation of the mass affects the kinetic energy of atoms other than the jumping one (ΔK effect).

Keeping in mind equations (IV.29) and (IV.32), equation (IV.34) can be written:

$$\boxed{E = f \cdot \Delta K} \tag{IV.35}$$

which is valid for the same conditions as equations (IV.28) and (IV.29). Nevertheless, this equation has been used even in cases where its validity remains to be shown, a point that would merit further study (Ishioka *et al.*, 1987). ***The isotope effect measures the correlation factor within a factor of*** ΔK.

For two simultaneous mechanisms, e.g., vacancies and divacancies:

$$E = \frac{D_1}{D_1 + D_2} f_1 \, \Delta K_1 + \frac{D_2}{D_1 + D_2} f_2 \, \Delta K_2 . \tag{IV.36}$$

The effect to be measured is very small:

$$\frac{\Delta D}{D} \sim \frac{1}{2} f \frac{\Delta m}{m} < \frac{1}{2} \frac{\Delta m}{m} .$$

For ^{22}Na and ^{24}Na, $\Delta D/D \approx 4\%$; for ^{55}Fe and ^{59}Fe, $\Delta D/D \approx 4\%$ and $\Delta f/f < 1\%$.

Note that E *can depend on temperature only if two or more mechanisms are operating simultaneously*; this property is often used to verify the uniqueness of the diffusion mechanism. When vacancies and divacancies, with their respective values of the correlation factor, are operating simultaneously, E decreases with increasing temperature (Fig. IV.8). Because $E < 1$, the mass dependence of D is weaker than the inverse square root relation (see Exercise 42).

VI. Effect of pressure.

The effect of hydrostatic pressure p on diffusion can be deduced from the general expression for D:

$$D = \text{const } a^2 \, \tilde{\nu} f \exp(-\Delta G/kT) . \tag{IV.37}$$

From the Gibbs free energy of activation, a corresponding volume can be calculated from the thermodynamic relation:

$$\Delta V = (\partial \, \Delta G/\partial p)_T ,$$

whence

$$\left(\frac{\partial \log D}{\partial p}\right)_T = -\frac{\Delta V}{kT} + \left[\frac{\partial \log a^2}{\partial p}\right]_T + \left[\frac{\partial \log \tilde{\nu}}{\partial \, p}\right]_T , \tag{IV.38}$$

since for self-diffusion in an isotropic medium, f is a pure number. The term $(\partial \log a^2/\partial p)_T$ is zero, since the distance measurements from which D is obtained are made at ambient pressure. The term $(\partial \log \tilde{\nu}/\partial p)_T$ can be calculated with the help of Grüneisen's relation:

$$\frac{\partial \log \tilde{\nu}}{\partial p} = \frac{\partial \log \theta_{\mathrm{D}}}{\partial p} = \frac{\partial \log \theta_{\mathrm{D}}}{\partial \log V} \cdot \frac{\partial \log V}{\partial p} = \gamma \chi ,$$

where θ_{D} is the Debye temperature, γ the Grüneisen constant, and χ the isothermal compressibility. This term is always small. In practice, log D varies linearly (decreases, in most cases) with the pressure. Hence the equation for the measurement of the activation volume for self-diffusion is:

$$\boxed{\Delta V \simeq -kT\,(\partial \log D/\partial p)_T}\,.$$

The volume can be decomposed into a formation and a migration term for defects in the same way that the Gibbs free energy was:

$$\Delta V = \Delta V^{\mathrm{f}} + \Delta V^{\mathrm{m}}. \tag{IV.39}$$

The meaning of the first term, ΔV^{f}, is clear. To form a vacancy means to remove an atom from the crystal and place it on a surface, with a change of volume of $+\Omega$. In the same way, to form an interstitial means to remove an atom from the surface and place it in an interstitial position, with a change of volume of $-\Omega$. The change in volume due to the relaxation around the defect must be added in, corrected for the image forces due to the presence of the external surface (the Eshelby term):

$$\Delta V^{\mathrm{f}} = \pm\Omega + f_{\mathrm{Esh}}\Delta V_{\mathrm{rel}}.$$

Estimates of ΔV^{f} are given in table II, in units of the atomic volume Ω.

The meaning of ΔV^{m} is much less clear (cf. the theory of the atomic jump, Ch. III. § VII.1); ΔV^{m} is estimated to be positive and $\leq 0.2\Omega$.

It follows that the volume of activation is always positive or null, except perhaps in the case of diffusion via self-interstitials in the BCC structure. For diffusion by single vacancies, $\Delta V < \Omega$, a prediction that turns out to be correct for the noble metals. On the other hand, for aluminum $\Delta V \approx 1.3\Omega$ (outward relaxation around the vacancy in this metal ?). Negative volumes of activation have been observed in ε-Pu and in certain BCC rare earths, a result to be compared with the fact that the melting temperature of these metals decreases with increasing pressure.

If several mechanisms are operating simultaneously, the measurement gives a ΔV_{eff} which is a linear combination of the ΔV's corresponding to each type of defect. For example, for diffusion by single and di-vacancies:

$$D^* = D^*_{1v} + D^*_{2v},$$

and it is easily shown that:

$$\Delta V_{\mathrm{eff}} = (D^*_{1v}/D^*)\ \Delta V_{1v} + (D^*_{2v}/D^*)\,\Delta V_{2v}. \tag{IV.40}$$

The volume of activation is then expected to approach ΔV_{1v} at a sufficiently low temperature and to increase with the temperature (role of ΔV_{2v}). Figure IV.9 gives an example of this behavior for silver (cf. the Arrhenius plot (Fig. IV.7) and the variation of the isotope effect (Fig. IV.8) for the same metal).

The measurement of the activation volume is thus an additional method for identifying the diffusion mechanism.

Table II. — *Formation Volume of Defects (After Lazarus, 1983)*

Structure	Defect	V^f	$\Delta V_{rel.}$	f_{Esh}	ΔV^f
FCC	Vacancy	+ 1	− 1/5	1.5	0.7
	Divacancy	+ 2	− 2/3		~ 1
	Interstitial	− 1	+ 2/3		~ 0
BCC	Vacancy	+ 1	− 1/3	1.6	0.4
	Divacancy	+ 2	− 3/4		0.8
	Interstitial	− 1	+ 1/2		− 0.2 (?)
NaCl	Schottky	+ 1	+ 1/3	1.8	1.6
	Frenkel (cations)	0	+ 1/3		0.6
AgCl	Schottky	+ 1	+ 1/4	1.3	1.3
	Frenkel (cations)	0	0		0

VII. Empirical correlations.

The correlations between the self-diffusion parameters and the melting temperature are quite general:

1) $D\,(T_m)$ (extrapolation) is a constant for a given crystal structure and type of bonding, equally good for metals, diverse ceramics, semiconductors, and ionic crystals (Fig. IV.4).

2) The ratio Q/T_m is constant for a given crystal structure and type of bonding (Fig IV.5, Van Liempt's rule). The same is true for the ratio Q/L_m and Q/L_s, where L_m is the heat of fusion and L_s the heat of sublimation = the cohesive energy.

3) The volume of activation follows the relation:

$$\Delta V = (Q/T_m)\,(\mathrm{d}T_m/\mathrm{d}p)\,,$$

which is applicable even for the case $\Delta V < 0$ (ε-Pu...).

The "constants" defined in this manner cover a very large range of numerical values; Q/T_m varies over a factor of 3.5, $D\,(T_m)$ over a factor of 10^6.

The $Q - T_m$ correlation justifies the use of the reduced temperature T/T_m for comparison of thermally activated processes. Such correlations can also be useful when experimental data are not available, especially for compounds and alloys. For the latter, the solidus temperature is taken as T_m. For binary

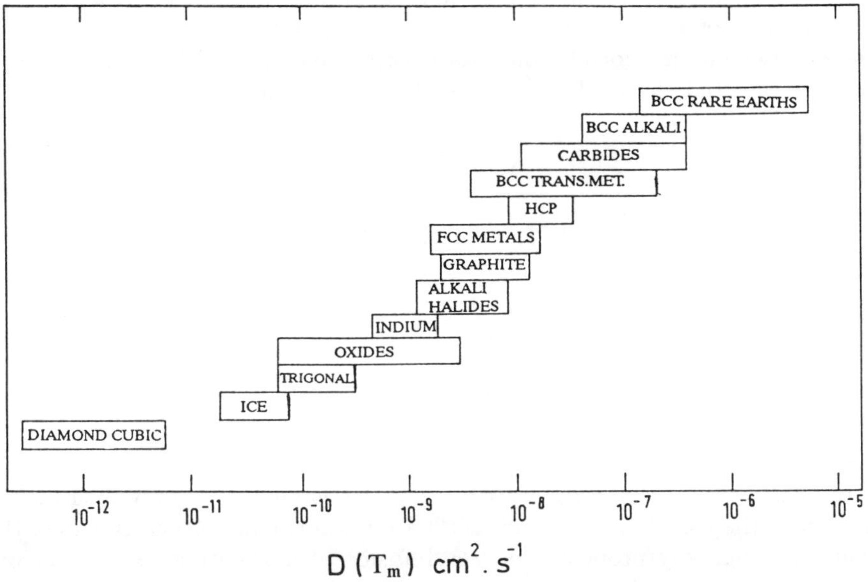

Fig. IV.4. — Comparison of self-diffusion coefficients extrapolated to the melting point for different classes of crystalline solids. After Brown and Ashby (1980).

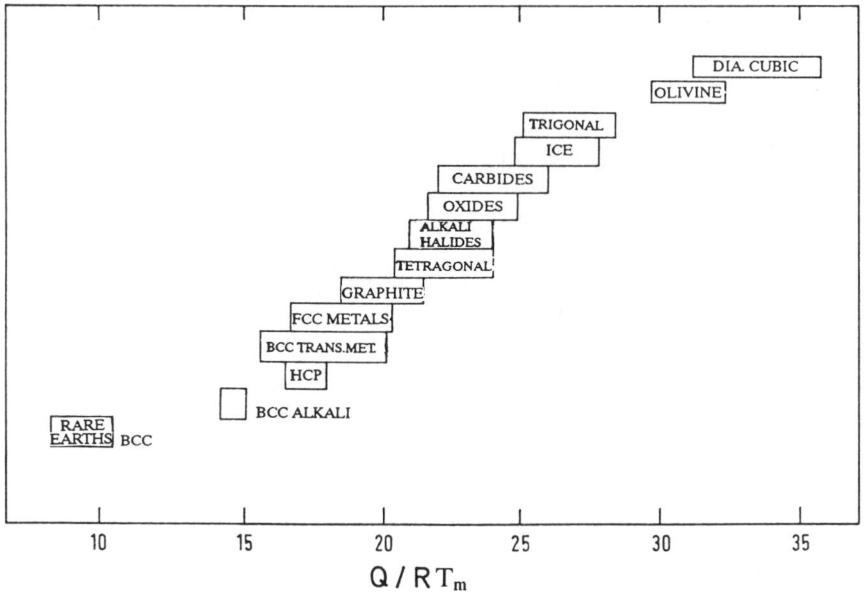

Fig. IV.5. — Comparison of reduced activation energies (Q/T_m) for self-diffusion in diverse classes of crystalline solids. After Brown and Ashby (1980).

systems, the values of D and Q for the more slowly diffusing species are used.

Another correlation often invoked in the literature connects D_0 and Q. The idea behind this correlation goes back to Zener's model for the diffusion of interstitials (c.f. § IV). If the thermodynamic equation

$$\Delta S = -\frac{\partial(\Delta G)}{\partial T}$$

is substituted in equation (IV.21), and with $\Delta H^{\mathrm{m}} \approx \Delta G_0$ (if ΔH^{m} is independent of T):

$$\begin{aligned}\Delta S^{\mathrm{m}} &= \Delta H^{\mathrm{m}} \frac{\mathrm{d}(\mu/\mu_0)}{\mathrm{d}T} \\ &= -\frac{\Delta H^{\mathrm{m}}}{T_{\mathrm{m}}} \frac{\partial(\mu/\mu_0)}{\partial(T/T_{\mathrm{m}})} = \theta\, \Delta H^{\mathrm{m}}/T_{\mathrm{m}}. \qquad \text{(IV.41)}\end{aligned}$$

As the derivative $\theta = -\partial(\mu/\mu_0)/\partial(T/T_{\mathrm{m}})$ varies but little (only 0.25 to 0.45 for most of the metals), ΔS^{m} and ΔH^{m} are proportional. Zener extended this argument, originally proposed for the diffusion of interstitials, to the entropy of activation for the diffusion of atoms on substitutional sites:

$$\Delta S = \lambda \frac{\Delta H}{T_{\mathrm{m}}}\,\theta,$$

whence

$$D_0 = \text{const}\; a^2\, \nu\, \exp\left(\frac{\lambda\theta Q}{RT_{\mathrm{m}}}\right), \qquad \text{(IV.42)}$$

with λ a universal (?) constant. The constant in D_0 is a well-known geometric factor.

This relation is often well fulfilled. One generally plots log $(D_0/a^2\nu)$ vs. Q/RT_{m}. Nevertheless, it can lead to mistakes, given that the errors in determining Q and D_0 from the experimental $D(T)$ curve are not independent (cf. Ch. IX, part 1, § IV.2).

Recall the relation proposed by Varotsos *et al.* (IV.22), which explains equally well the effects of temperature and pressure on D (since both B and Ω depend on T and p), especially the effects due to anharmonicity of the vibrational modes, and which also predicts a relation between ΔS and ΔH.

VIII. Self-diffusion in metals.

For most metals:

$$5 \times 10^{-2} < D_0 < 10\ \mathrm{cm^2 s^{-1}}$$
$$Q/R \approx 17\ T_{\mathrm{m}}\,(\mathrm{K}).$$

Extrapolating to the melting point:

$$D(T_m) \approx 10^{-8}\ \mathrm{cm^2s^{-1}}\ \text{for FCC, HCP, and BCC transition metals,}$$
$$\approx 10^{-7}\ \mathrm{cm^2s^{-1}}\ \text{for the alkali metals.}$$

At $T = T_m/2$, D is approximately 10^{-16} to $10^{-15}\ \mathrm{cm^2s^{-1}}$. The values of D_0 correspond to an entropy of activation $\Delta S \approx k$.

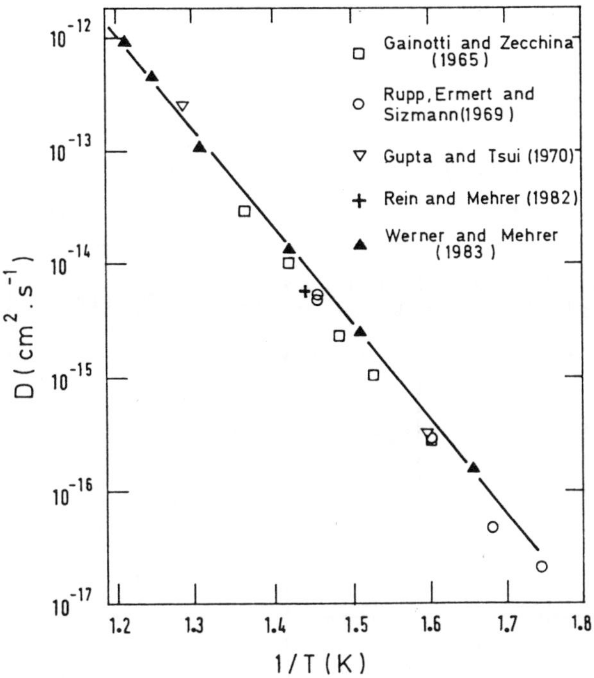

Fig. IV.6. — Self-diffusion in gold. Compilation of experimental results by Werner and Mehrer (1983). Measurements at very low temperatures carried out by electrochemical or sputter sectioning.

1) FCC *metals.* — Equation (IV.9) relating the activation enthalpy for diffusion to the enthalpies of formation and motion of vacancies holds well for FCC metals. This result is remarkable because it concerns the comparison of completely different experiments: the rapid quenching and the annealing of small specimens in one case, the migration of radiotracers in the other. In fact, it is necessary to couple the two types of measurements in order to interpret the results, especially those due to the presence of divacancies.

Some recently developed techniques (see Ch. IX) allow the measurement of very small diffusion coefficients and the extension of the Arrhenius plot of D^* *vs.* $1/T$ over a very wide range of temperature, down to $T/T_m = 0.35$.

A perfectly linear Arrhenius plot over a fairly broad temperature range is found only for some FCC metals. For the noble metals, a slight curvature in the Arrhenius plot at high temperature may exist for gold (Fig. IV.6) and clearly exists for copper and silver (Fig. IV.7). This deviation from linearity is usually ascribed to a divacancy contribution, an interpretation that is bolstered by the decrease of the isotope effect $E = f\ \Delta K$ [5] (Fig. IV.8) and by the increase of the volume of activation (Fig. IV.9) with increasing temperature.

This type of interpretation is also applied to other FCC metals (Al, Ni, γ-Fe). The preponderant role of single vacancies is well established on the basis of equation (IV.9); see table III.

Table III. — *Self-diffusion in some* FCC *and* HCP *metals*
(After G. Brébec (1983) and N. L. Peterson (1978))

	D_0 ($cm^2 . s^{-1}$)	Q (kcal/mole)	0,034 T_m (kcal/mole)	$\Delta H^f + \Delta H^m$	ΔV (atomic volume)	ΔK (f = 0,78)
Al (*)	0.035	28.75	31.7	31.8	1.3-0.71	
Ag (*)	0.44	44.3	42	44.4	0.7-0.9	0.92 Low T
Cu (*)	0.62	49.6	46.1	49.2	0.9	0.87
Ni (*)	1.9	68	58.7	66.8		
Au	0.091	41.7	45.4	40.7-42.1	0.72	
Pb	1.37	26	20.4	22.8	0.73	
Pd	0.21	63.6	62			1
γ Fe	0.49	67.9				0.68
Cd //	0.12	18.7	20.2	18	0.53	
Cd ⊥	0.18	19.6			0.585	
Zn //	0.13	21.9	23.5	22.6	0.40	0.95
Zn ⊥	0.18	23			0.47	
Mg //	1	32.2	31.4	29.2		
Mg ⊥	1.5	32.5				

* Indicates a slightly curved Arrhenius plot.

2) In metals with the *HCP structure,* diffusion is anisotropic, and the ratio of the diffusion coefficients correlates well with the c/a ratio of the lattice parameters: $D_\perp / D_\parallel > 1\ (< 1)$ if $c/a < \sqrt{8/3}\ \left(> \sqrt{8/3}\right)$, the symbols $\perp$ and $\parallel$ indicating the diffusion direction with respect to the axis of six-fold symmetry.

(5) f (divacancies) $< f$ (vacancies), see Table I.

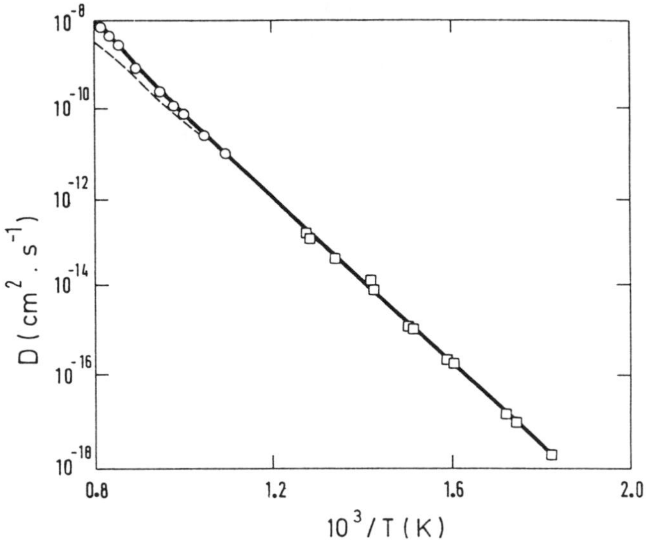

Fig. IV.7. — Self-diffusion in silver. Compilation of experimental results by Le Claire (1976).

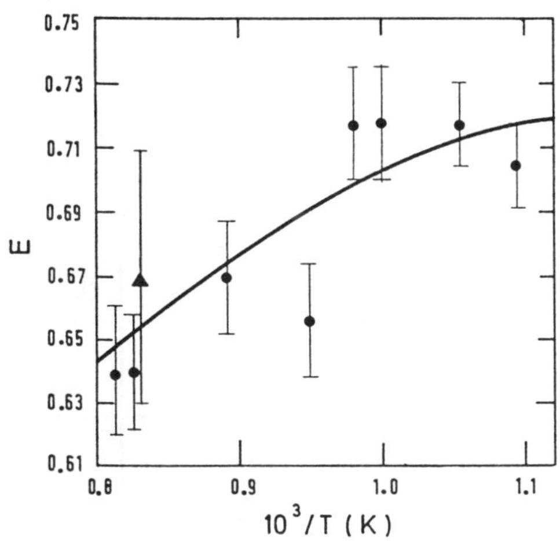

Fig. IV.8. — Isotope effect E as a function of $1/T$ for self-diffusion in silver. Compilation by Peterson (1975).

The jump-frequency ratio can be calculated from the values of $(c/a)^2\ (D_\perp/D_\parallel)$: $w_a/w_b > 1$ for Be, Mg, and Tl $\left(c/a > \sqrt{8/3}\right)$. The Arrhenius plot is linear for Zn and Cd, which have been studied over a wide range of temperature.

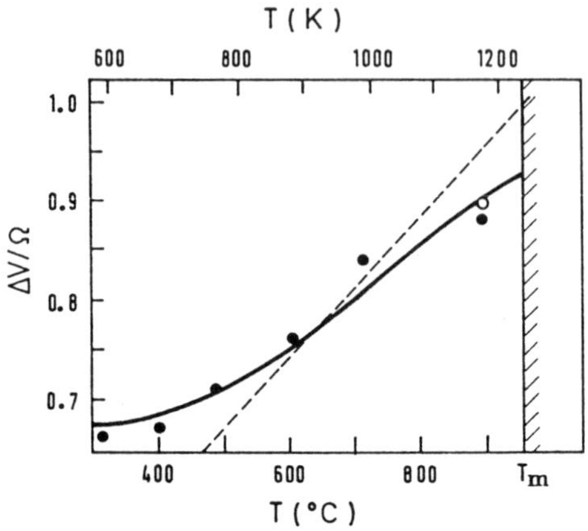

Fig. IV.9. — Volume of activation for self-diffusion in silver, compiled by Rein and Mehrer (1982). The solid line was calculated with $\Delta V = 0.67\Omega$ for single vacancies and $\Delta V = 1.1\Omega$ for divacancies.

3) For *BCC metals* different types of Arrhenius plots are observed (Table IV):

Table IV. — *Self-diffusion in some BCC metals with curved Arrhenius plots* [6] (After N. L. Peterson, (1978) and J. N. Mundy *et al.* (1981))

	D_{01} ($cm^2\ s^{-1}$)	Q_1 (kcal/mol)	D_{02} ($cm^2\ s^{-1}$)	Q_2 (kcal/mol)	$0.034\ T_m$ (kcal/mol)
Li	0.038	12	9.5	15.9	15.4
Na	0.004	8.4	0.29	10.6	12.6
V	0.014	67.5	7.5	85.7	74.4
Nb	0.008	83.4	3.7	104.6	91.6
Ta	0.0013	92.2	10	123.3	111.1
Mo	0.13	104	139	130.6	98.0
W	0.04	125	46	158	125.7
ß – Ti	3.58×10^{-4}	31.1	1.09	59.9	65.9
ß – Zr	8.5×10^{-5}	27.6	1.34	65.2	72.4

(6) The latest study on vanadium (George, Janot, Ablitzer, and Chabre 1981) proposes three mechanisms: divacancies, free single vacancies, and interstitial impurity-vacancy complexes.

– Linear: $\alpha + \delta$ Fe, Cr, ...

– Slightly curved: Alkali metals, elements of groups VB and VIB. The mechanism is especially controversial for sodium, the metal for which the most data are available: D^* by tracers and NMR, the isotope effect, and the volume of activation, all as a function of temperature. The interpretation in terms of vacancies and divacancies leads to a rather small $\Delta V = 0.33\Omega$, $\Delta K = 0.6$, and is not universally accepted (see below, Köhler and Herzig, (1988)).

– Abnormally high activation energy: Cr with $Q = 0.049\ T_m$,

– Strong curvature, accompanied by values of D_0 and Q which are smaller than normal and lead to higher than normal values of D when extrapolated to T_m.

There are about ten of these anomalous metals:

– The high-temperature β phases of group IVB metals (which have a HCP structure at low temperature): Ti, Zr, Hf (Fig. IV.10);

– The lanthanides: γLa, δCe, βPr, γYb, βGd;

– The actinides: γU, εPu.

Only βTi and βZr have been studied in detail and over a wide temperature range. For βZr, Q/T_m varies from 0.0104 to 0.022 as the temperature increases. The isotope effect increases significantly with temperature, leveling off at $T \geq$ 1800 K.

All the mechanisms listed in § IV above have been advanced to explain this anomalous behavior. Special variations proposed for these metals have been the effect of dislocations introduced by heating through the $\alpha - \beta$ phase change, or a region of extrinsic diffusion due to interstitial impurities which are very soluble in βTi and βZr.

A completely different explanation is based on a vacancy mechanism with ΔH^m decreasing with decreasing temperature, due to a phonon-assisted jump process particular to the BCC structure. In this structure, the complex formed by the jumping atom at the saddle point and its three neighbors has a structure identical to that of the hexagonal ω phase which appears in these metals either at low temperature because of alloying, or under hydrostatic pressure. The saddle-point configuration, as the embryo of the ω phase, would result from a local "condensation" of a soft phonon mode; the longitudinal phonon dispersion curves for βZr show a sharp minimum for $\mathbf{q} = 2/3 < 111 >$; this reduces the restoring forces that oppose the atom movement along the <111> direction. The same type of process can account for the lesser curvature observed for the other BCC metals (Köhler and Herzig, 1988), where the mode softening is less marked. The mode softening is also observed in BCC metals that do not form ω phase. Because of these structural fluctuations, ΔH^m should decrease with decreasing temperature.

4) *Ferromagnetic metals.* — Near the Curie point (770° C), the Arrhenius plot for αFe shows a strong anomaly (Fig. IV.11): a large reduction in the value of the self-diffusion coefficient which begins already in the paramagnetic

region and extends over the paramagnetic-ferromagnetic transition. This effect appears to be connected to a strong anomaly in the elastic constants in the same temperature region. The anomaly is weaker for cobalt. The Curie point of nickel is too low (360° C) to allow diffusion measurements at the magnetic transition.

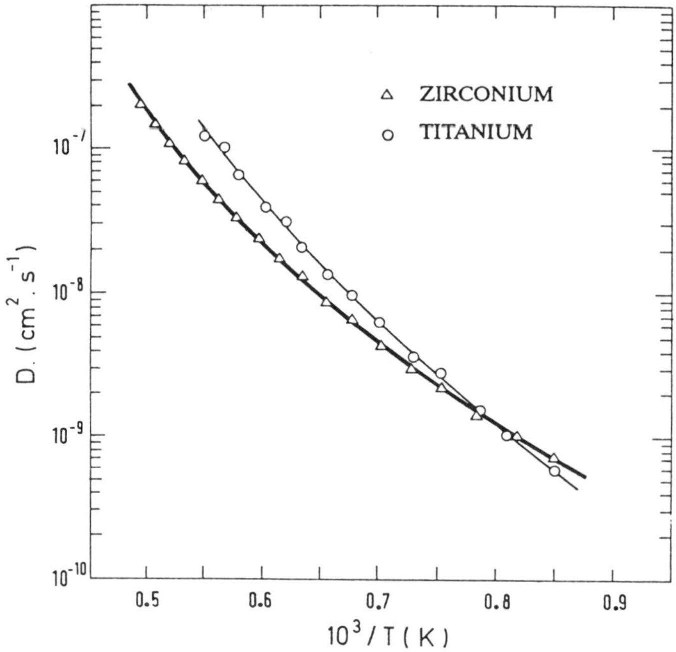

Fig. IV.10. — Arrhenius plots for self-diffusion in titanium and zirconium. Results compiled by Le Claire (1976).

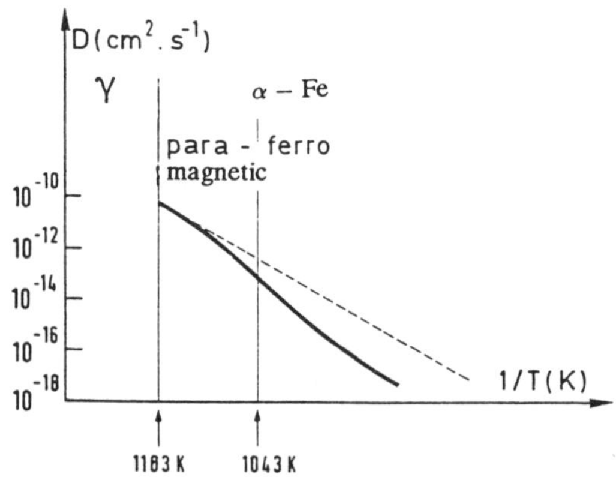

Fig. IV. 11. — Anomaly in the self-diffusion in αFe at the Curie point. After Mehrer (1983).

IX. Self-diffusion in semiconductors.

The diffusivity at the melting point in elemental semiconductors is 10^4 times smaller that in metals i.e. $D(T_m) \sim 10^{-13}$-10^{-12} cm^2 s^{-1} ; this smallness limits the range of temperature over which the diffusion coefficients are measurable. There is an additional difficulty in self-diffusion measurements in silicon: it has no convenient radioisotope (the half-life of ^{31}Si is 2.6 h). Thus one must use either the stable isotope ^{30}Si as a tracer, or be content with a less direct method than the tracer-sectioning technique. The formation of surface oxides SiO_2 or SiO further impedes the diffusion measurements, and affects the defect equilibria. An additional experimental difficulty can be caused by the equilibration of the silicon vacancies, V_{Si}, and the self-interstitials, Si_I, because silicon crystals usually contain far fewer dislocations than do crystals of metals.

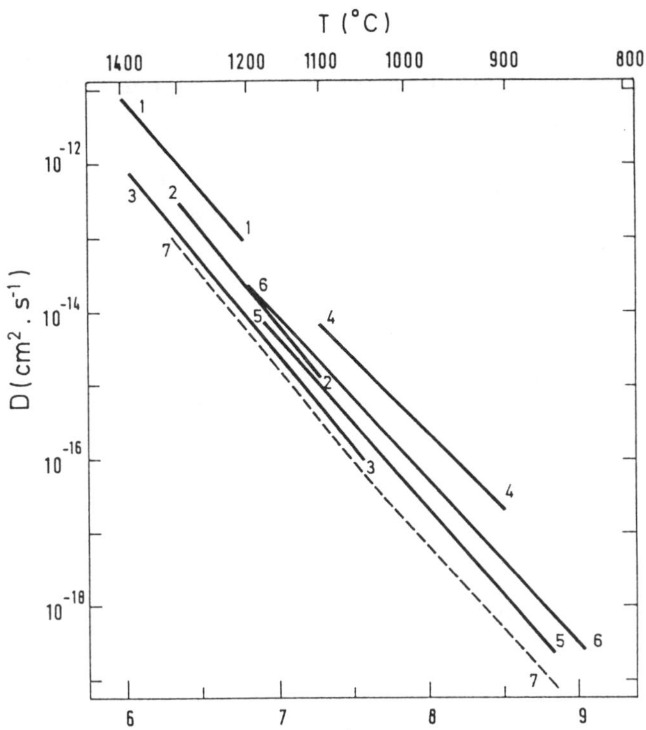

Fig. IV.12. — Self-diffusion in intrinsic silicon. Results from different authors compiled by Demond *et al.* (1983). Methods: 1) ^{31}Si, mechanical sectioning; 2) ^{31}Si, electrochemical sectioning; 3) ^{31}Si, ion milling; 4)implantation of ^{30}Si and nuclear reaction analysis by the resonant (p, α) reaction; 5) SIMS with ^{30}Si; 6) implantation of ^{30}Si and (p, α) nuclear reaction; 7) solute diffusion of germanium.

In a crystal free of dislocations, the only sources and sinks for point defects are the surfaces of the sample, where $[V_{Si}] \times [Si_I]$ = constant. Equilibration in the volume requires quite long annealing times.

In germanium, the Arrhenius law is followed over quite a large domain of temperature, down to $T/T_m = 0.68$, and the mechanism is very likely to be diffusion by single vacancies ($\Delta H = 3.23$ eV, $D_0 = 30$ cm^{2}s^{-1}, Werner and Mehrer 1985).

The results for silicon are quite scattered (Fig. IV.12), with $\Delta H \approx 4.7$ to 5.1 eV. The relation $\Delta H = \Delta H^f + \Delta H^m$ does not hold if the vacancy enthalpies are determined after quenching. However, a model based on two mechanisms, vacancies and self-interstitials, with the latter playing the dominant role, appears to explain the experimental results from 850 to 1150° C quite well.

For both Ge and Si, the preexponential factors are high, leading to large values of the entropy of activation, $\Delta S \approx 10k$. This result is interpreted in terms of a strong relaxation around the point defects (cf. the small volume of activation for self-diffusion in germanium, $0.3 - 0.4\Omega$). Seeger and Chik (1968) introduced the idea of the "extended" point defect; the vacancy would be comparable to a microvolume ($\approx$ 10 atoms) of amorphous silicon (which has a density 90–95% that of crystalline Si), and the interstitial would be comparable to a microvolume of liquid silicon (the coordination number of liquid silicon is 6 compared to 4 for the crystal). We mention also the "concerted exchange" mechanism proposed by Pandey (1986) on the basis of theoretical calculations.

IX.1 IONIZATION OF THE POINT DEFECTS. — Defects in semiconductors can be neutral or ionized, depending on the position of the Fermi level (*vide infra*). For example, in the case of a vacancy mechanism with neutral and negatively ionized vacancies (if the vacancies behave as acceptors), the diffusion coefficient is given by:

$$D^* = \frac{1}{8}\, a^2\, f \left\{ [V^X]\, w^X + [V']\, w' \right\}, \qquad \text{(IV.43)}$$

where $[V^\times]$ and $[V']$ mean the mole fractions of neutral and singly ionized vacancies, and $w^\times$ and w' their respective jump frequencies.

The change of the effective charge state of the vacancy can, in effect, induce a change in the migration enthalpy (Fig. IV.13a) and therefore in the jump frequencies. In n-type Si, for example, where negatively charged vacancies are the dominant species, $\Delta H^m = 0.18$ eV, whereas in p-type Si, where neutral vacancies are dominant, $\Delta H^m = 0.33$ eV, as measured by low-temperature anneals. The jump frequency is thus increased by ionization.

Bourgoin and Corbett have proposed a mechanism of *athermal migration*, which would be made possible by alternate changes of the charge state of a defect; in this case the energy necessary for a jump is furnished by the charge carriers, because of the modification of the electronic configuration of the defect (Fig. IV.13b).

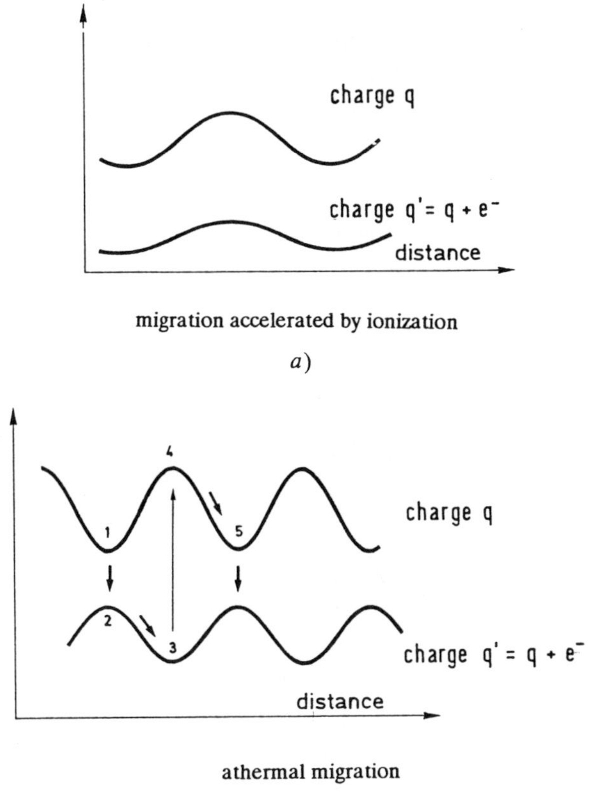

Fig. IV.13. — *a*) Change of the enthalpy of migration of a defect as a function of its charge in a semiconductor. *b*) Bourgoin and Corbett's mechanism of athermal migration. After Bourgoin and Lannoo (1981).

Diffusion in semiconductors can be *accelerated* by *light irradiation*, which essentially affects the migration process, either by an ionization effect, as described above, or by a local production of phonons. When a charge carrier is trapped at a defect site, an energy equal to the difference of the energy of the conduction or valence band (depending on the nature of the defect) and the energy level of the defect is dissipated by one of the following three mechanisms: 1) radiative emission, 2) transmission to other carriers (the Auger effect), 3) production of phonons, either by a cascade of successive one-phonon transitions, or by a single multiphonon transition. The local production of phonons can lower the energy barrier for migration of the defect and cause migration at low temperature (see Lanoo and Bourgoin, *op. cit.* vol. **2**, ch.6).

Effect of impurities. — If the impurities act as dopants, i.e., as acceptors or donors, they lower or raise the Fermi level. This leads to a change of the charge state of the point defects, which in turn means a change in the defect population, and perhaps a change in the migration energy.

Consider a defect with two charge states, e.g., a vacancy which is either neutral or negatively charged (7):

$$\mathrm{V}' \rightleftharpoons \mathrm{V}^{\times} + \mathrm{e}^- .$$

This defect (the vacancy above is an acceptor) introduces a localized energy level E_{A} in the band gap. If the Fermi level is located between the valence band and E_{A} (p-type), the neutral state $\mathrm{V}^{\times}$ is favored, but if E_{F} is situated between E_{A} and the conduction band, the V' state is favored.

The ratio of the concentrations (8) is obtained from statistical thermodynamics (9):

$$\frac{[\mathrm{V}']}{[\mathrm{V}^{\times}]} = \exp\left[-\frac{E_{\mathrm{A}} - E_{\mathrm{F}}}{kT}\right], \tag{IV.44}$$

with:

$$[\mathrm{V}^{\times}] = \exp\left(\frac{\Delta S^{\mathrm{f}}_{\mathrm{V}^{\times}}}{k}\right) \exp\left(-\frac{\Delta H^{\mathrm{f}}_{\mathrm{V}^{\times}}}{kT}\right).$$

Comparing the doped and intrinsic states (the latter denoted by i) (10):

$$\begin{aligned}\frac{[\mathrm{V}']}{[\mathrm{V}^{\times}]} &= \exp\left(-\frac{E_{\mathrm{A}} - E^{\mathrm{i}}_{\mathrm{F}}}{kT}\right) \exp\left(-\frac{E^{\mathrm{i}}_{\mathrm{F}} - E_{\mathrm{F}}}{kT}\right) \\ &= \frac{[\mathrm{V}']_{\mathrm{i}}}{[\mathrm{V}^{\times}]} \cdot \frac{n}{n_{\mathrm{i}}},\end{aligned} \tag{IV.45}$$

where n and n_{i} are the electron concentrations in the doped and intrinsic materials, respectively. Doping with donor elements (n-type) increases the concentration of negatively charged vacancies.

Since $[\mathrm{V}^{\times}]$ is independent of doping, the self-diffusion coefficient varies

(7) See chapter VI, Appendix II, for the designation of charge according to the Kröger-Vink notation.

(8) The mole fraction of vacancies is denoted by N_{V} or [V].

(9) See Lannoo and Bourgoin, *op. cit.* vol **1**, p. 200.

(10) Recall that $n_{\mathrm{i}}/n = \exp\{(\mu_{\mathrm{i}} - \mu)/kT\}$, with $\mu = E_{\mathrm{F}}$ and $\mu_{\mathrm{i}} = E^{\mathrm{i}}_{\mathrm{F}}$.

linearly with the ratio n/n_i (Eq. (IV.45)), increasing with n-type doping, decreasing with p-type, a result verified experimentally for the case of diffusion in germanium (Werner and Mehrer, 1985).

In silicon, the vacancies introduce two new acceptor levels, E_{A1} and E_{A2}, and two new donor levels, E_{D1} and E_{D2}. The concentration of doubly charged vacancies V″ is given by:

$$\frac{[V'']}{[V^{\times}]} = \left(\frac{n}{n_i}\right)^2 \cdot \exp\left[-\frac{E_{A1}+E_{A2} - 2\,E_F^i}{kT}\right], \tag{IV.46}$$

obtained by applying equation (IV.45) for $[V']$ with $E_A = E_{A1}$. In the same way, for donor levels:

$$\frac{[V^{\bullet\bullet}]}{[V^{\times}]} = \left(\frac{n_i}{n}\right)^2 \cdot \exp\left[-\frac{E_{D1}+E_{D2} - 2\,E_F^i}{kT}\right]. \tag{IV.47}$$

IX.2 COMPOUND SEMICONDUCTORS. — Self-diffusion of the constituents of the III-V or II-VI compounds has been measured only infrequently, and the experimental results are for the most part not to be trusted too far. There are three reasons for this situation: 1) the diffusion coefficients are very small ($< 10^{-14}$ cm^2s^{-1}), 2) certain elements (groups V and VI) are so volatile that thermal etching is difficult to avoid, 3) the partial pressures of the components have to be carefully controlled during the diffusion anneal in order to keep the compound in a thermodynamically well defined state. The technique proposed by Mathiot and Edelin (1980) allows the experiments to be carried out for compositions at the boundaries of stability of the parent phase, i.e., saturated in A or in B, by annealing in a sealed ampule containing a saturating source. Note that the deviation from stoichiometry in these compounds is small (10^{-4}). Diffusion in GaSb has been reported by Weiler and Mehrer (1984). In recent years, much work has been done on diffusion in GaAs and AlGaAs quantum well heterostructures (Tan and Gösele, 1988, Deppe and Holonyak, 1988) which has thrown some light on diffusion mechanisms in GaAs (see also Tuck, 1988).

The possible mechanisms of diffusion in these compounds are the same as those proposed for ordered alloys: 1) independent vacancy mechanisms on each sublattice, 2) a ring mechanism which does not destroy the overall order, 3) diffusion by vacancy pairs V_AV_B and/or 4) diffusion by so-called triple defects $2V_A + A_B$, which involves an antisite defect (see Ch. V, § VIII.2).

X. Self-diffusion in ionic crystals.

Ionic crystals are very interesting from the point of view of their diffusional properties because measurement of the electrical conductivity furnishes an additional access to the diffusion parameters. From this point of view, the alkali halides, which have the FCC NaCl structure, are a textbook example that will be the central subject of this paragraph (Fig. IV.14). Several oxides are similar to this case, but they do not present the same advantages from the point of view of the analysis, be they insulators like MgO, or semiconductors like NiO, CoO, MnO, and FeO. Fluorite has also been the subject of much study, both for its own interest and as a model for UO_2.

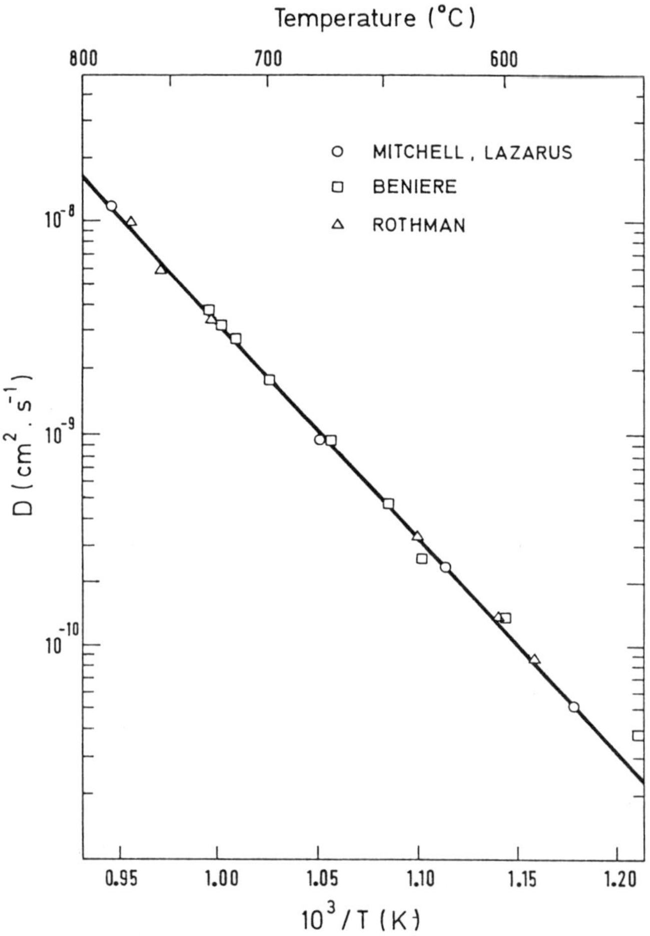

Fig. IV.14. — Arrhenius plot for the self-diffusion of ^{22}Na in NaCl single crystals. Comparison of the results of three different groups by Mitchell and Lazarus (1975).

The diffusion coefficients of the two ionic species can be of the same order of magnitude or very different, depending on the material. The cation and anion *transport numbers* t^+ and t^- (defined as the fraction of the current carried by the cations and anions respectively) are given to a good approximation by the expression:

$$t^{+(-)} = \frac{D^{+(-)}}{D^+ + D^-}. \qquad \text{(IV.48)}$$

In silver halides, $D^+ >> D^-$ (and $t^+/t^- \approx 10^4$ to 10^6). On the other hand, for the alkali halides, $D^+ \gtrsim D^-$; $t^- \approx 5\%$ in NaI, $\approx 30\%$ in NaCl and KCl, and even 50% in CsCl. The situation is reversed in alkali-earth halides: $D^+ << D^-$. Both cases are found in binary oxides.

The predominant defects are:

– Schottky defects in alkali halides (FCC, NaCl type),

– Frenkel defects on the cation sublattice in silver halides (FCC, NaCl type),

– Frenkel defects on the anion sublattice in the alkali-earth halides (cubic, CaF_2 type), as well as chalcogenides and oxides (UO_2, ZrO_2) with the fluorite structure.

X.1 Alkali halides. — Schottky defects are predominant; the motion of cations and anions takes place by site exchange with a vacancy on the same sub-lattice. The vacancy jump frequencies are given by:

$$w_{+,-} = \tilde{\nu} \exp\left(\frac{\Delta S_m^{+,-}}{k}\right) \exp\left(-\frac{\Delta H_m^{+,-}}{kT}\right). \qquad \text{(IV.49)}$$

The anion-cation separation a is usually taken as the reference length. The lattice parameter is then $2a$, hence for the self-diffusion coefficient by the vacancy mechanism ([11]):

$$\begin{cases} D_v^+ = 4\,a^2\,[V']\,f\,\tilde{\nu}\exp\left(\dfrac{\Delta S_m^+}{k}\right)\cdot\exp\left(-\dfrac{\Delta H_m^+}{kT}\right) = 4\,a^2\,f\,[V']\,w^+, \\ D_v^- = 4\,a^2\,[V^\bullet]\,f\,\tilde{\nu}\exp\left(\dfrac{\Delta S_m^-}{k}\right)\cdot\exp\left(-\dfrac{\Delta H_m^-}{kT}\right) = 4\,a^2\,f\,[V^\bullet]\,w^-, \end{cases} \qquad \text{(IV.50)}$$

where f_0 is the correlation factor (0.78 for the FCC lattice). V' indicates a cation vacancy (charge negative with respect to the lattice), and $V^\bullet$ the anion vacancy (positive charge with respect to the lattice). For the notation, see chapter VI, Appendix II.

The electrical conductivity is due to the migration of cation and anion

([11]) The mole fraction of vacancies is here denoted by [V].

vacancies. Following the Einstein relation:

$$\sigma = 4\,a^2\,\frac{ne^2}{kT}\,\left\{[\mathrm{V}']\,w^+ + [\mathrm{V}^\bullet]\,w^-\right\}, \tag{IV.51}$$

which is often expressed as a "conductivity diffusion coefficient":

$$\begin{aligned} D_\sigma &= \frac{kT}{ne^2}\,\sigma \\ &= 4\,a^2\,\left\{[\mathrm{V}']\,w^+ + [\mathrm{V}^\bullet]\,w^-\right\} \end{aligned} \tag{IV.52}$$

in order to allow comparison to D^* measured with tracers (here n is the number of cations (or anions) per unit volume). In the ***intrinsic region,*** dominated by Schottky defects at thermal equilibrium, the equation describing the defect equilibrium is:

$$[\mathrm{V}']\ [\mathrm{V}^\bullet] = \exp\left(-\frac{\Delta G_{\mathrm{S}}}{kT}\right) = K_{\mathrm{S}}, \tag{IV.53}$$

where ΔG_{S} is the Gibbs free energy for the formation of a Schottky pair. The condition of electrical neutrality is:

$$[\mathrm{V}'] = [\mathrm{V}^\bullet]. \tag{IV.54}$$

The activation enthalpy for self-diffusion is then:

$$Q^+ = \frac{\Delta H_{\mathrm{S}}}{2} + \Delta H_{\mathrm{m}}^+, \tag{IV.55a}$$

for the cations, and

$$Q^- = \frac{\Delta H_{\mathrm{S}}}{2} + \Delta H_{\mathrm{m}}^-. \tag{IV.55b}$$

for the anions. The electrical conductivity, on the other hand, follows the Arrhenius form (cf. Eq. (IV.51)) only if the jump frequency of one species is negligible, or in other words, if the self-diffusion coefficient of that species is negligible ([12]). If $D^+ >> D^-$, then $D^+ = fD_\sigma$. Otherwise, the cation transport number t^+ has to be evaluated, and the comparison is made between D_σ^+ and D^+, via the equations $D_\sigma^+ = t^+ D_\sigma$ and $t^+ = D^+/\ (D^+ + D^-)$, following equations (IV.50) and (IV.52).

These equations do not describe the experimental results correctly, as $D^+/D_\sigma \approx 1.1$, instead of $f < 1$. The disagreement is due to the existence of *neutral vacancy pairs*, which contribute to self-diffusion, but not to the conductivity; the proof of their role is furnished by experiments on the effect dopants (see below).

([12]) Strictly it is σT rather than σ that is expected to follow the Arrhenius form.

The diffusion coefficients are then given by:

$$D^+ = D_V^+ + D_p^+ \tag{IV.56a}$$

$$D^- = D_V^- + D_p^-, \tag{IV.56b}$$

where the subscripts V and p indicate the self-diffusion coefficients by single vacancies (Eq. (IV.50)) and neutral pairs, respectively. For the latter, reasoning similar to that of § II.2 leads to([13]):

$$\begin{cases} D_p^+ = 8\,a^2 f'_+\,\tilde{\nu} \exp\left(\dfrac{\Delta S_S + \Delta S_m'^+ + \Delta S_p}{k}\right) \exp\left(-\dfrac{\Delta H_S + \Delta H_m'^+ + \Delta H_p}{kT}\right) \\ D_p^- = 8\,a^2 f'_-\,\tilde{\nu} \exp\left(\dfrac{\Delta S_S + \Delta S_m'^- + \Delta S_p}{k}\right) \exp\left(-\dfrac{\Delta H_S + \Delta H_m'^- + \Delta H_p}{kT}\right), \end{cases} \tag{IV.57}$$

where $\Delta S'_m$ and $\Delta H'_m$ are the entropies and enthalpies of migration for an exchange with a vacancy pair, different for anions and cations, $\Delta G_p = \Delta H_p - T\,\Delta S_p$, the Gibbs free energy for the formation of a pair from a cation and an anion vacancy. Note that the correlation factors f'_+ and f'_- are different for anions and cations. In effect, if the lifetime of the vacancy pair is long compared to the time between atomic jumps, its motion involves the jumps of both the cation and the anion species, with frequencies w'_+ and w'_-. The values of $f'_{+,-}$ are thus functions of the ratio w'_+/w'_-, which have been tabulated by Compaan and Haven and by Howard.

It follows that:

$$\frac{D_p^+}{D_p^-} = \frac{f'_+}{f'_-}\frac{w'_+}{w'_-}. \tag{IV.58}$$

The expressions for the electrical conductivity and D_σ (Eqs. (IV.51) and (IV.52)) are not changed.

Fitting of the Arrhenius plot is then no longer possible because the vacancy pairs introduce too many new parameters. Also, the correlation factor cannot be calculated directly from D^+, D^-, and D_σ because:

$$f = \frac{D_V^+ + D_V^-}{D_\sigma} = \frac{D^+ - D_p^+ + D^- - D_p^-}{D_\sigma}.$$

The isotope effect for the cation is given by:

$$E^+ = \alpha f\ \Delta K + (1-\alpha)\, f'_+\ \Delta K',$$

where $\alpha = D_V^+/D^+$ and $(1-\alpha) = D_p^+/D^+$.

([13]) The concentration of paired vacancies of each sign is here equal to the concentration of pairs. Equation (IV.13) then becomes $\Gamma_s = 1/3\ N_p w_p$, and since the lattice parameter is $2a$, one obtains equation (IV.57) instead of equation (IV.15).

Effect of dopants, extrinsic region. — The experimental results always show a break in the Arrhenius plot; below a certain critical temperature, the slope is always smaller. This behavior is attributed to the presence of heterovalent impurities. Doping with heterovalent cationic impurities raises the lower part of this line without changing its slope, the transition temperature increases with doping (Fig. IV.15).

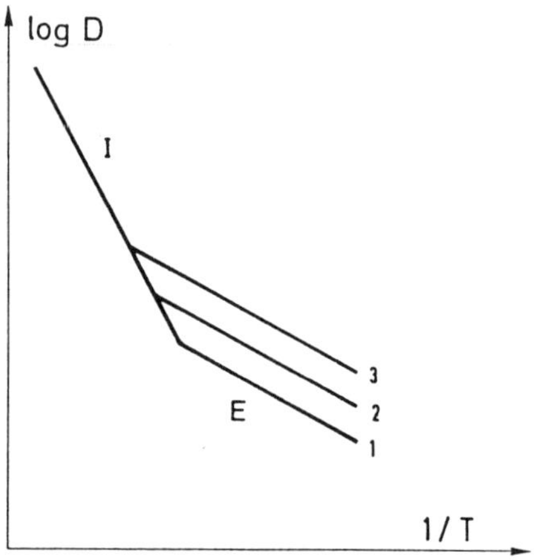

Fig. IV.15. — Self-diffusion of the cation in an alkali halide. The intrinsic region (I), and the extrinsic region (E) for doping levels increasing in the order 1, 2, 3.

In the extrinsic region, the concentration of cation vacancies is fixed by the concentration of heterovalent dopant, according to the law of electrical neutrality. If $F^{\alpha\bullet}$ is the dopant of excess positive charge $+\alpha\,|e|$:

$$[V'] - [V^\bullet] = \alpha\,[F^{\alpha\bullet}]\,. \tag{IV.59}$$

For a real crystal, $F^{\alpha\bullet}$ should be replaced by $F^{\alpha\bullet} - F^{\alpha\prime}$, where the F′s indicate all the cationic and anionic impurities. In practice, doping generally means doping with cations because it is easier experimentally (greater solubility). If the content of dopant is large enough:

$$\begin{aligned} [V'] &\simeq \alpha\,[F^{\alpha\bullet}] = \text{const.} \\ [V^\bullet] &= K_S/\alpha\,[F^{\alpha\bullet}]\,. \end{aligned} \tag{IV.60}$$

The activation energies in the extrinsic region are thus:

$$Q^+ = \Delta H_m^+ \tag{IV.61a}$$

$$Q^- = \Delta H_S + \Delta H_m^-\,. \tag{IV.61b}$$

Thus one can deduce the values of $\Delta H_{\mathrm{m}}^{+}$ and ΔH_{S} or of $\Delta H_{\mathrm{m}}^{-}$ and ΔH_{S} from a comparison of the intrinsic and extrinsic domains, but only if the contribution of the vacancy pairs is negligible (which is usually not the case, as it can reach 50%).

It is important to note that the *vacancy-pair contribution is independent of the dopant concentration.* The quasi-chemical equation for the formation pairs is:

$$\mathrm{V}' + \mathrm{V}^{\bullet} \rightleftharpoons \mathrm{p},$$

thus

$$\frac{\mathrm{p}}{[\mathrm{V}'] \ [\mathrm{V}^{\bullet}]} = 6 \exp\left(-\frac{\Delta G_{\mathrm{p}}}{kT}\right), \qquad \text{(IV.62)}$$

which depends only on the temperature. This is the origin of the idea to separate the vacancy pair contribution by doping:
- by doping with cations $[\mathrm{V}'] = [\mathrm{F}^{\bullet}]$ and $[\mathrm{V}^{\bullet}] \to 0$, so that $D^{-} \approx D_{\mathrm{p}}^{-}$,
- by doping with anions $[\mathrm{V}^{\bullet}] = [\mathrm{F}']$ and $[\mathrm{V}'] \to 0$, so that $D^{+} \approx D_{\mathrm{p}}^{+}$.

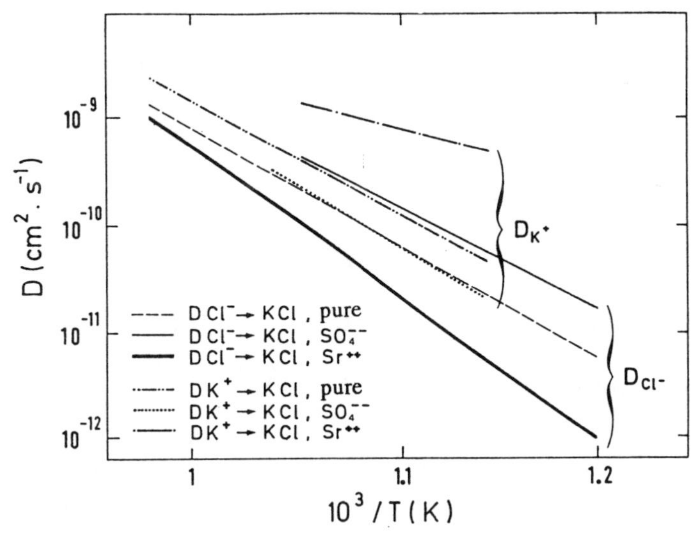

Fig. IV.16. — Self-diffusion of K^{+} and Cl^{-} in pure and doped KCl crystals. The addition of Sr^{2+} increases $D\left(K^{+}\right)$ and decreases $D\left(Cl^{-}\right)$; the addition of SO_4^{2-} increases $D\left(Cl^{-}\right)$ and reduces $D\left(K^{+}\right)$. After Bénière *et al.* (1976).

These predictions are borne out by experiment (Fig. IV.16). In fact, the limits given above are used as a first approximation, and all the parameters are then iteratively adjusted to the entire ensemble of data. The parameters and the data are given in the following tables, suggested by Bénière *et al.* (1976):

Table V. — *Possible (desirable) measurements*

1) Self-diffusion coefficient of the cation in pure salt		
2) " " " " " "	"	salt doped −
3) " " " " " "	"	salt doped +
4) " " " " " anion	"	pure salt
5) " " " " " "	"	salt doped −
6) " " " " " "	"	salt doped +
7) Electrical conductivity	"	pure salt
8) " "	"	salt doped −
9) " "	"	salt doped +
10) Isotope Effect E	"	pure salt

Table VI. — *Physical parameters*

Schottky Defects	ΔH_s	ΔS_s
Cation migration	ΔH_m^+	ΔS_m^+
Anion migration	ΔH_m^-	ΔS_m^-
Vacancy – dopant association	ΔH_a	ΔS_a
Cation diffusion by pairs	$D_0'^+$	H^+
Anion diffusion by pairs	$D_0'^-$	H^-
Correlation factor (vacancies)	f	

The measurements listed in table V represent a considerable effort. Doping with anions is not easy, as the solubility of anionic impurities is generally low. Several other effects must also be taken into account and corrected for:

1) *Dopant-vacancy association* which reduces the number of free vacancies. For divalent cationic dopants, the dopant concentration is given by:

$$c_F = [F^\bullet] + [(V'\ F^\bullet)]^x \tag{IV.63}$$

where $(V'F^\bullet)^x$ represents a neutral dopant ion-cation vacancy complex. The association reaction:

$$\mathrm{F}^{\bullet} + \mathrm{V}' \rightleftharpoons (\mathrm{V}'\ \mathrm{F}^{\bullet})^{\mathrm{x}} \tag{IV.64}$$

has associated with it an equation of chemical equilibrium:

$$\frac{[(\mathrm{V}'\ \mathrm{F}^{\bullet})^{\mathrm{x}}]}{[\mathrm{F}^{\bullet}]\ [\mathrm{V}']} = K_{\mathrm{a}}. \tag{IV.65}$$

The equation of charge balance is:

$$[\mathrm{V}'] = [\mathrm{F}^{\bullet}] + [\mathrm{V}^{\bullet}]. \tag{IV.66}$$

There are now 4 variables: the concentration of free cation vacancies, $[\mathrm{V}']$, of free anion vacancies $[\mathrm{V}^{\bullet}]$, of cation vacancy-dopant complexes $[(\mathrm{V}'\mathrm{F}^{\bullet})^{\mathrm{x}}]$, and of free dopant ions $[\mathrm{F}^{\bullet}]$, and these can be calculated from the three equations (IV.63), (IV.66), and (IV.65), plus the equation for Schottky equilibrium (IV.53). The concentration of free cation vacancies is given by the cubic equation:

$$K_{\mathrm{a}}[\mathrm{V}']^3 + [\mathrm{V}']^2 - (c_{\mathrm{F}} + K_{\mathrm{S}}K_{\mathrm{a}})[\mathrm{V}'] - K_{\mathrm{S}} = 0. \tag{IV.67}$$

The equilibrium constant for association K_{a} corresponds to the entropy and enthalpy of association ΔS_{a} and $\Delta \mathrm{H}_{\mathrm{a}}$, in table VI. Association is only considered for cationic impurities, as the solubility of anionic dopants is too small for the association of anionic vacancies to be significant.

Equation (IV.67) encompasses three temperature regions where it can be simplified. These are, in order of decreasing temperature:

- The intrinsic region: $[\mathrm{V}'] \approx K_{\mathrm{S}}^{1/2}$
- The extrinsic region: $[\mathrm{V}'] \approx c_{\mathrm{F}}$
- The association region: $[\mathrm{V}'] \approx (c_{\mathrm{F}}/K_{\mathrm{a}})^{1/2}$.

Finally, at even lower temperatures, a fourth region can exist where $[\mathrm{V}']$ and c_{F} are controlled by the precipitation of dopant, its solubility having been exceeded. The Arrhenius plot of $D_{\mathrm{V}'}$ and of D^{+} can thus show four different regions with four different slopes and four different activation energies.

2) *Deviation from ideality.* Rigorously, equations (IV.53) and (IV.65) should be written for the activities and not the concentrations; activity coefficients should be introduced for each type of defect. The deviation from ideality is due to the long-range electrostatic interactions between defects of opposite sign. Lidiard has shown that the model of Debye-Hückel atmospheres, originally developed for liquid electrolyte solutions, can be applied to alkali halides and leads to an activity coefficient:

$$\gamma = \exp\left[-\frac{\chi \mathrm{e}^2}{2\,\varepsilon k T\,(1+\chi R)}\right], \tag{IV.68}$$

where R is the distance of closest approach of the defects, ε the dielectric constant, and χ the Debye-Hückel screening constant. These long-range interactions also affect the mobility of the defects, and correction terms can be calculated for σ, D^+, and D^-.

3) *Isotope effects.* — The jump frequencies in electrical conductivity and in diffusion are slightly different, since in the former case, the jumping ions have the natural atomic mass, whereas the second case concerns isotopes. The correction is made using equation (IV.34), knowing the isotope effect E.

A very complete analysis of this type has been carried out for KCl and NaCl by Bénière *et al.* (1976), and it confirmed closely the model of diffusion by single vacancies and vacancy pairs. The contribution of the latter is surprisingly high, the concentration of pairs and free vacancies being nearly equal at high temperatures. In both mechanisms, the cation is more mobile than the anion; the exchange frequency with the divacancy is smaller than that with the free vacancy. The entropy terms are rather high: $\Delta S_S \approx 9-10\ k$, $\Delta S_m \approx 1.5-3\ k$.

X.2 Silver halides. — The silver halides AgCl and AgBr have been extensively investigated because of their practical importance in the photographic process. Notwithstanding the fact that their structure is the same as that of NaCl, their behavior is very different, the point defects being essentially Frenkel disorder on the cation sublattice. This means that:

- Self-diffusion of the cation is due to the motion of silver interstitials (and, to a lesser extent, silver vacancies).
- The self-diffusion of the cation is much faster than that of the anion, $D_+/D_- \approx 10^3 - 10^4$, whereas in the alkali halides $D_+ \approx D_-$ near the melting point. An extreme case is presented by Agl, where this ratio reaches 10^6 (see Ch. V, § VIII)
- In the intrinsic region, the dominant diffusion mechanism is the interstitialcy, with several different types of jump, collinear or non collinear (dogleg), characterized by different activation energies.
- The activation energy for interstitial migration is very low, 0.1-0.04 eV, a result probably due to a quadrupolar deformation along $\langle 110 \rangle$ on the silver ion.
- The Haven ratio $H = (D^*/\sigma) / (kT/ne^2)$ is not equal to the correlation factor (several types of defects, several types of jumps, jump distance of defect different from that of tracer), and depends on the temperature.
- At low temperature, the diffusion of the cation takes place via extrinsic vacancies.
- The diffusion coefficient of the anion exhibits both an intrinsic and an extrinsic region.

The cation self-diffusion coefficient can be expressed as a function of the diffusion coefficients of the vacancies and the interstitials:

$$D_+^* = N_v f_v D_v + N_i H_i D_i, \qquad \text{(IV.69)}$$

with H_i denoting the Haven ratio for the interstitials (see Ch. III, § VI and below).

For the conductivity, with negligible anion contribution:

$$\sigma = n_v e u_v + n_i e u_i = n_F e \left(u_v + u_i\right), \qquad \text{(IV.70)}$$

where n_F is the number of Frenkel defects per unit volume, and u_v and u_i are the defect mobilities. The Einstein relation for vacancies or interstitials is written:

$$u_{v,i} = (e/kT)\, D_{v,i}. \qquad \text{(IV.71)}$$

Whence:

$$H = D^*/D_\sigma = \left(f_v u_v + H_i u_i\right)/\left(u_v + u_i\right) = \left(f_v + \psi H_i\right)/\left(1 + \psi\right), \qquad \text{(IV.72)}$$

with $D_\sigma = \sigma\left(kT/ne^2\right)$ and ψ the mobility ratio u_i/u_v.. The value of H_i can be obtained from experiment; however, even though f_v is constant (0.78), ψ depends on the temperature, as does H_i if the interstitials execute several different kinds of jumps. The values of ψ can be obtained from the electrical conductivity of pure and doped crystals after applying a Debye-Hückel correction. The determination of $H(T)$ then allows the calculation of $H_i(T)$ from equation (IV.72).

Recall why H_i is different from the correlation factor f_i. From the theory of random motion, we can write for the tracer:

$$D^* = \frac{1}{6} f_i \zeta \Gamma_s \lambda^2 \qquad \text{(IV.73)}$$

where ζ is the number of the nearest neighbor sites, and λ the jump distance. For the interstitialcy mechanism, the total number of ionic jumps is equal to twice the total number of defect or tracer jumps since the elementary mechanism brings two jumps into play (two ions have to jump):

$$n\Gamma_s = 2 n_i\, w_i. \qquad \text{(IV.74)}$$

For conductivity via the same defects:

$$\sigma_i = n_i\, e u_i = \frac{n_i e^2}{kT}\, D_i, \qquad \text{(IV.75)}$$

with

$$D_i = \frac{1}{6}\, \zeta w_i\, \lambda_i^2,$$

since the jumps of the defects are not correlated. Note also that $\lambda_i \neq \lambda$, because if the tracer makes an $i \rightarrow L$ or an $L \rightarrow i$, jump, the defect (i.e., the electric charge) is displaced from one interstitial site to another interstitial site (Fig. IV.17).

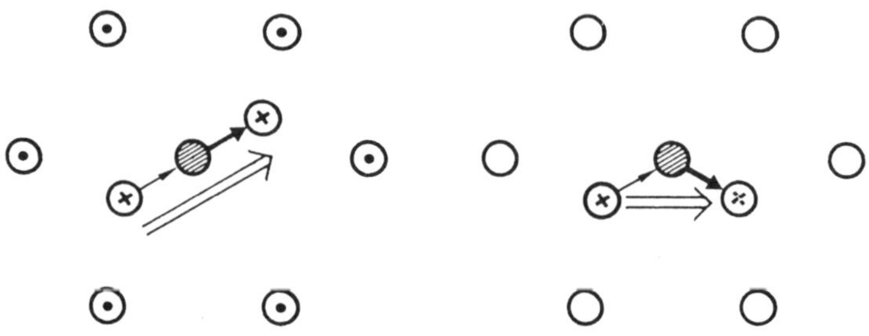

Fig. IV.17. — Collinear and dogleg interstitialcy jumps. The tracer jump λ_q is denoted by the single arrow ($\rightarrow$), the jump of the interstitial defect λ_{iq} by the double arrow ($\Longrightarrow$).

The Haven ratio compares D^* and $D_\sigma = \sigma kT/\left(ne^2\right)$:

$$H_i = \frac{D^*}{D_\sigma} = \frac{D^*}{(n_i/n)\ D_i} = 2\left(\frac{\lambda}{\lambda_i}\right)^2 f_i, \qquad \text{(IV.76)}$$

with $f_i = (1 + \langle\cos\Theta\rangle)$, where Θ is the angle between the two jumps $i \rightarrow L$ and $L \rightarrow i$ of the tracer. In the AgCl structure, the interstitial can carry out the three types of jump, one collinear and two non-collinear (Fig. IV.18) characterized by their frequencies w_1, w_2, and w_3:

Type of jump	Θ	f_i	$(\lambda/\lambda_i)^2$	H_i
w_1	0	2/3	1/4	1/3
w_2	arc cos (1/3)	32/33	3/8	0.7273
w_3	arc cos (−1/3)	0.9643	3/4	1.4464

The average Haven ratio for these three types of jump is then:

$$\bar{H}_i = 2\frac{\sum_q f_q\ \zeta_q w_q\ \lambda_q^2}{\sum_q \zeta_q\ w_q\ \lambda_{iq}^2}, \quad \lambda_q \neq \lambda_{iq}, \qquad \text{(IV.77)}$$

with q as the index for the type of jump. The experimental value of H_i for AgCl and AgBr is always between 0.727 and 0.33, which leads to an interpretation (neglecting vacancies) based on two types of interstitialcy jumps: collinear w_1 and non-collinear w_2. The value of w_1/w_2 can be calculated from the variation of H_i with T if the f_q and λ_q are known; it varies from 2 to 0.5 with increasing temperature.

This interpretation is consistent with the measurements of the isotope effect. In AgBr doped with Cd, $E = 0.51$; this represents the vacancy mechanism

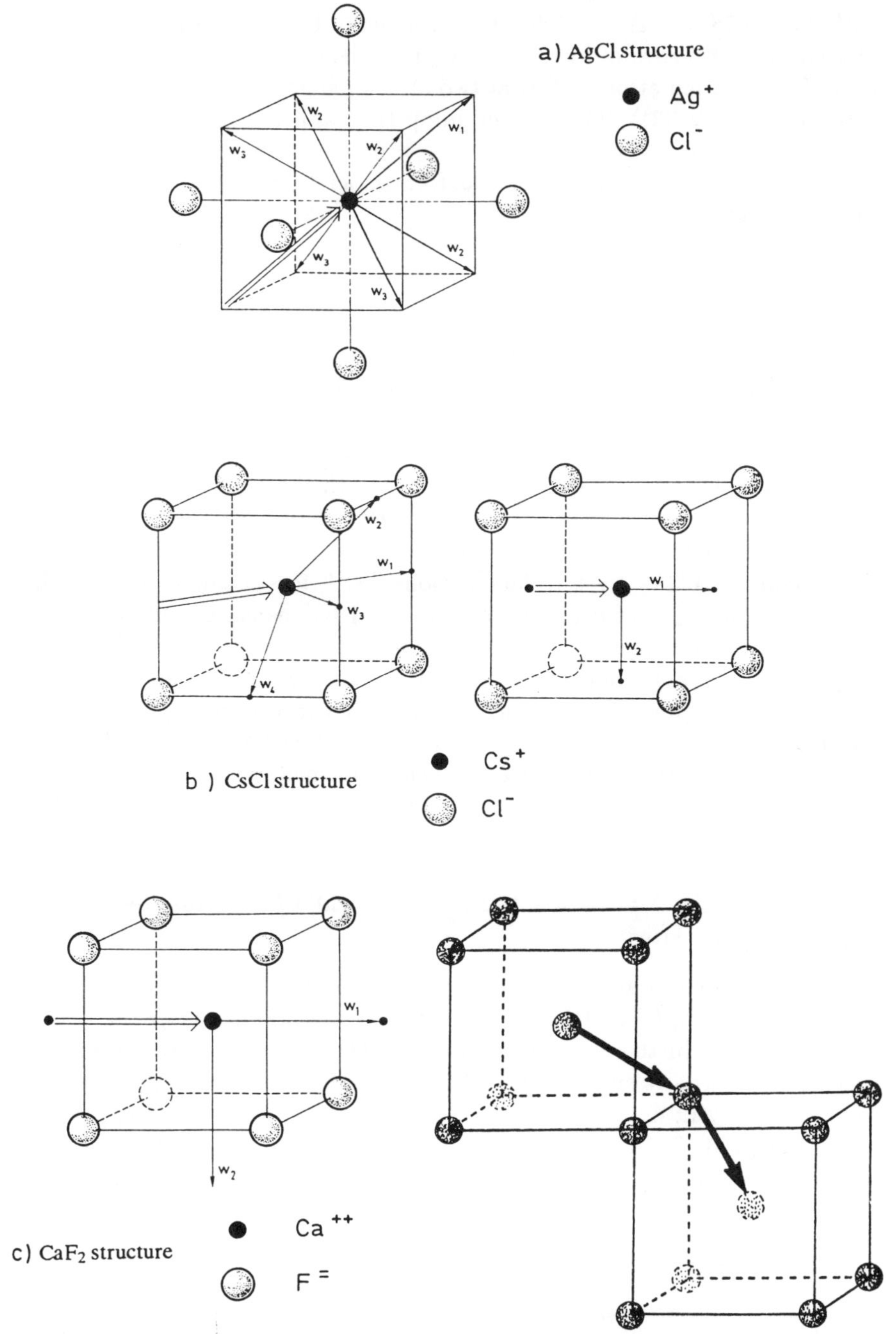

Fig. IV.18. — The interstitialcy mechanism in various ionic crystal structures. The small black circles indicate the various interstitial sites possible for the ions denoted by the large black circles. After Compaan and Haven (1958).

with $f_v = 0.78$ and $\Delta K = 0.66$. For pure crystals, interpretation of the results in terms of two types of interstitialcy jumps leads to $\Delta K_1 = 1$ and $\Delta K_2 = 0.1$ (note for the calculation of E that two atoms participate in the elementary process, cf. Eq. (IV.33)). The dog-leg jump thus appears to be quite delocalized. The interpretation of conductivity measurements is based on the same model, but Debye-Hückel type effects and defect association in doped crystals must be taken into account.

X.3 The fluorite structure (Fig. IV.18c). — The structure of fluorite, CaF_2, consists of two sublattices: the simple cubic anion sublattice, and the FCC cation sublattice, which can also be considered as simple cubic, identical to the anion sublattice but with one cation of every two missing. Recall that the dominant defects are anion Frenkel defects.

In $SrCl_2$, radiotracer measurements for both chemical species yield $D^+ << D^-$, in agreement with electrical measurements which yield a value of 1 for the anion transport number t^- in several fluorides: CaF_2, BaF_2,...

Doping of CaF_2 by trivalent cations (Y^{3+} for example) entails the formation of anion interstitials F_i, as indicated by measurements of the lattice parameter and the density. The Y^{3+} ions go into solution substitutionally as $Y^{\bullet}_{Ca}$ and the excess anion (YF_3 instead of CaF_2) goes into an interstitial site as F'_i. At low temperature, the formation of neutral pairs $(Y^{\bullet}_{Ca}F'_i)^{\times}$ on first or second nearest-neighbor sites can be expected, as well as clusters. The simplest example of the latter is the "2:2:2" cluster, formed from a dimer made up of two neutral pairs in a plane. Such clusters are made up of 2 substitutional cations $Y^{\bullet}_{Ca}$, 2 interstitials F'_i, and two vacancies $V^{\bullet}_F$, or two relaxed anions.

The alkali metals also dissolve substitutionally as M'_{Ca} with the formation of anion vacancies $V^{\bullet}_F$. At low temperatures, $(M'V^{\bullet}_F)^{\times}$ complexes can form on nearest-neighbor sites.

The diffusion of anions takes place essentially by the non-collinear interstitialcy mechanism, (Fig. IV.18c) as shown by measurements of the correlation factor in $SrCl_2$. In the extrinsic region, $D^-/D_\sigma > 1$, which confirms the presence of neutral pairs such as $(F'_i\ Y^{\bullet}_{Ca})^{\times}$. Further evidence for these is obtained from the study of depolarization thermocurrents (ITC) (Ch. IX, pt. 2, § I.3) and NMR (ibid. § II.2). At high temperatures, the anion mobility can increase dramatically, leading to the so-called "ionic superconductivity" (See Ch. V, § VIII).

X.4 Oxides. — A great diversity of structures and chemical bonding exists among the oxides. We restrict ourselves here to simple oxides of the type M_aO_b. Diffusion on the two sublattices of oxides is generally independent (there are no antistructure defects) but coupling is possible, e.g., the oxygen interstitial with the metal vacancy ("counter vacancy" mechanism).

The simplest case is that of stoichiometric oxides, MgO, Al_2O_3... Magnesia (MgO) is analogous to NaCl except for its ions being divalent. However, its behavior is almost always extrinsic; in view of the very high enthalpy of formation of a Schottky pair (theoretical estimate $\approx$ 8 eV), the concentration of intrinsic defects is very small, and a few ppm of impurities are enough to bring it into the extrinsic region, with $[V_M] >> [V_O]$ so that $D^*_M >> D^*_O$.

The oxides of transition metals with NaCl structure, such as NiO, CoO, MnO, and FeO, deviate from stoichiometry by being metal deficient: $M_{1-\delta}O$. The concentration of cation vacancies responsible for this deviation can be high: $\delta \approx 10^{-4}$ in NiO to more than 0.1 in FeO, so that the behavior of these oxides is normally intrinsic. The deviation from stoichiometry depends on the temperature and the partial pressure of oxygen.

The self-diffusion coefficient of the cation (or the anion) is written:

$$D^* = \sum_d f_d N_d D_d, \tag{IV.78}$$

where the sum is over all the types of point defects present on the sublattice under consideration, and D_d is the diffusion coefficient of a defect:

$$D_d = \beta a^2 \, \tilde{\nu} \, \exp(-\Delta G_d^{\mathrm{m}}/kT) \,. \tag{IV.79}$$

The defects are formed according to an oxidation reaction; the laws of mass action and charge balance lead to a defect concentration:

$$N_d = A_d \, (P_{O_2})^m \, \exp\left(-\widetilde{\Delta G}_d/kT\right), \tag{IV.80}$$

where $\widetilde{\Delta G}_d$ is the effective Gibbs free energy of formation ([14]). The exponent m is characteristic of the defect and its charge ([15]). The self-diffusion coefficient is then:

$$D^* = \beta a^2 \, \tilde{\nu} f_d \, A_d \, (P_{O_2})^m \, \exp\left(\frac{\Delta S_d^{\mathrm{m}} + \widetilde{\Delta S}_d}{k}\right) \exp\left(-\frac{\Delta H_d^{\mathrm{m}} + \widetilde{\Delta H}_d}{kT}\right). \tag{IV.81}$$

Example: Consider the oxide MO. If the majority cation defect is the neutral vacancy, V_M^x, the formation reaction is (see Ch. VI, App. II for the notation):

$$\frac{1}{2} \, O_2 \rightleftharpoons V_M^x + O_O, \qquad ...\Delta G^f_{V_M},$$

([14]) Called "effective" because it is different from the Gibbs free energy of the oxidation reaction expressing the formation of the defect (cf. Eq. (IV.82)).

([15]) In this equation, the units of P_{O_2} are atmospheres. In other units, $(P_{O_2}/P^*_{O_2})$ must be written, where $P^*_{O_2}$ is a reference pressure.

whence

$$[V_M^x] = K^f_{V_M^x}\,(P_{O_2})^{1/2}\,,$$

with

$$K^f_{V_M} = \exp\left(-\Delta G^f_{V_M}/kT\right).$$

In general, the defects are charged. For a negatively charged defect, for example:

$$\frac{1}{2}\,O_2 \rightleftharpoons V'_M + O_O + h^\bullet, \qquad ...\Delta G^f_{V'_M}$$

together with the equation for charge neutrality:

$$[V'_M] = [h^\bullet]$$

leads to:

$$[V'_M] = K^{1/2}_{V'_M}\,(P_{O_2})^{1/4}\,,$$

and thus:

$$m = 1/4 \text{ and } \widetilde{\Delta G} = \frac{1}{2}\,\Delta G^f_{V'_M}\,. \qquad \text{(IV.82)}$$

The table below summarizes the values of m obtained for the different types of majority defects in the oxide MO.

Table VII. — *Exponent of P_{O_2} for different defects*

Defect	V_M^x	V'_M	V''_M	M_i^x	$M_i^\bullet$	$M_i^{\bullet\bullet}$
m	1/2	1/4	1/6	−1/2	−1/4	−1/6
Defect	V_O^x	$V_O^\bullet$	$V_O^{\bullet\bullet}$	O_i^x	O'_i	O''_i
m	−1/2	−1/4	−1/6	1/2	1/4	1/6

In most cases, the same defect can exist in several different charge states. It is often assumed that the Gibbs free energy of migration is the same no matter what the charge on the defect. This simplification is based on chemical diffusion experiments (see Ch. VI), which directly yield the diffusion coefficient

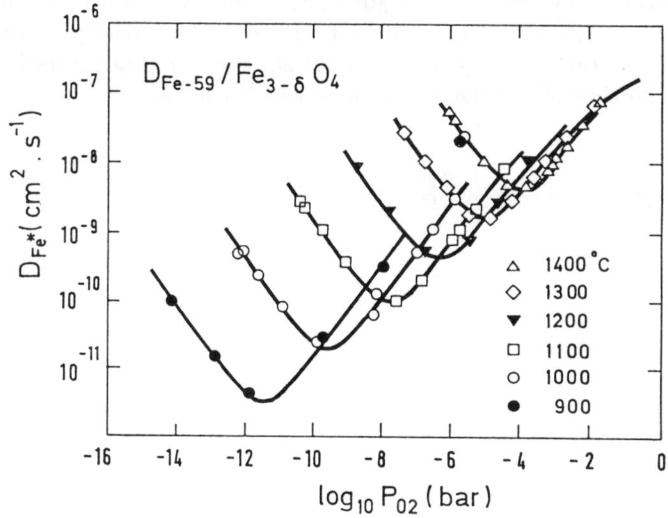

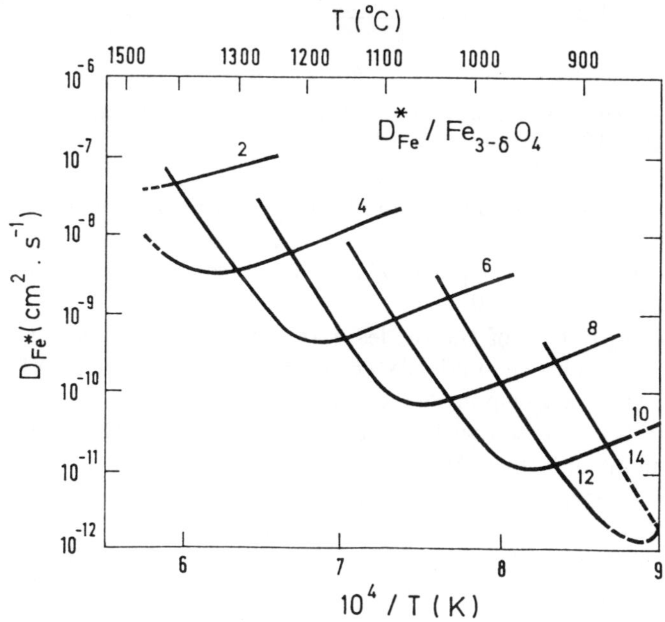

Fig. IV.19. — Self-diffusion of Fe in magnetite $Fe_{3-\delta}O_4$. Effect of temperature and oxygen partial pressure (the values of $-\log_{10} P_{O_2}$ (in bars) are noted on each curve). After Dieckmann and Schmalzried (1977).

D_d of the majority defects ([16]). Figures IV.19 and IV.20 show the variation of the cation and anion diffusion coefficient with oxygen partial pressure. Several sets of experimental results, e.g., figure IV.20, suggest that interstitials play a role in the diffusion of oxygen as D_O^* increases with P_{O_2}.

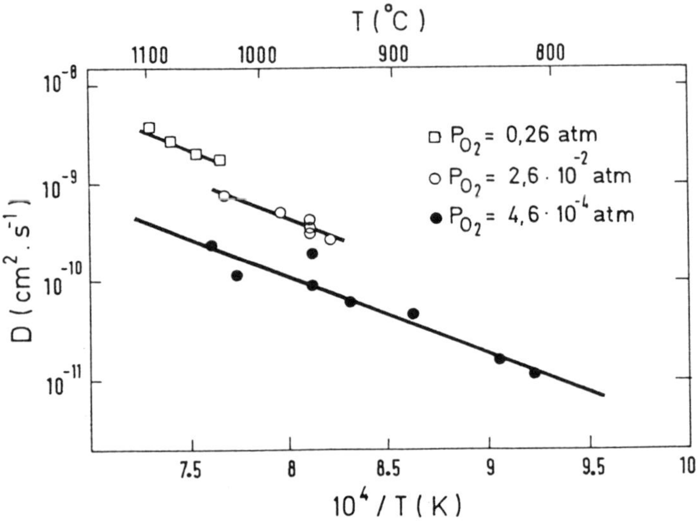

Fig. IV.20. — Self-diffusion of oxygen in $Cu_{2-\delta}O$. Effect of temperature and oxygen partial pressure. After Périnet *et al.* (1982).

The above reasoning and calculations can easily be generalized to oxides with the formula M_aO_b. In the series NiO,... FeO, the majority defects are cation vacancies, so that $D_M^* >> D_O^*$. In fact, these defects are so much in the majority that $D_M^*/D_O^* \approx 10^4 - 10^5$.

The reverse is true of the oxides with a fluorite structure: CeO_2, ZrO_2, ThO_2, and UO_2, where the majority defects are oxygen vacancies $V_O^{\bullet\bullet}$ (or cation interstitials) which are responsible for the hypostoichiometry ($MO_{2-\delta}$). Hyper-stoichiometry is connected with anion interstitials. Since the majority defects are on the anion sublattice, $D_M^* < D_O^*$. The same is the case for cubic zirconia, $ZrO_{2-\delta}$, where oxygen vacancies are the majority defect, introduced as an extrinsic defect, the cubic structure being stabilized by a high concentration of bi- or trivalent dopant. CeO_2 would be the model for this type, as it does not undergo an allotropic transformation.

As soon as the deviation from stoichiometry becomes large ($> 10^{-3}$), a description in terms of simple point defects becomes unrealistic; the lower the temperature, the more unrealistic this simple treatment becomes. Neutron and X-ray diffraction studies have shown the presence of "complexes" of vacancies and interstitials (the defects proposed by Koch and Cohen, or possibly larger

([16]) The majority defects are those responsible for the departure from stoichiometry. In most cases, they involve only one sublattice.

clusters in FeO, Willis's complexes in UO_2,...). The diffusion coefficient then no longer follows simple laws; e.g., in FeO, it decreases with increasing deviation from stoichiometry δ if the temperature is below 800° C.

The effect of doping in oxides can be described in the same way as in ionic crystals, as has already been suggested concerning the extrinsic behavior of MgO. The change of the concentrations of the majority and minority defects with doping can be predicted from the formation reactions for the defects and from the charge-balance equation; this is interesting for several reasons:

1) to establish the concentration of the majority defects. This is the case of zirconia stabilized in the cubic phase by doping with 10 – 15% CaO or Y_2O_3, which fixes the concentration of the majority defects. In this material, a high ionic conductivity associated with the oxygen vacancies is responsible for numerous applications employing zirconia electrolyte in fuel cells, oxygen sensors, etc. (see Ch. V, § VIII);

2) to study the minority defects. In a $M_{1-\delta}O$ oxide, monovalent lithium dopant occurs in the form Li'_M; it lowers the concentration of the majority defects V'_M and increases the concentration of the minority defect $V^{\bullet}_O$. At constant T and $P(O_2)$, an increase in the lithium content leads to corresponding variations in the self-diffusion coefficients, the inverse effect to doping with Al. These results support the hypotheses proposed on the nature of the minority defects.

For an anlysis of all the different types of oxides and a review of the experimental data, the reader can consult Kofstad's book (1972), the Nostoco series (the Alenya colloquium (1984), Simkovich and Stubican (1985), Catlow and Mackrodt (1987)), the Confolant Summer School (1977), or Matzke's review articles (1981, 1986).

XI. Molecular crystals.

Molecular crystals are characterized by a cohesive energy that is low (0.4 – 1 eV/molecule) compared to other solids, because the bonds that hold the crystals together are due to van der Waals forces, or eventually, to hydrogen bonding. This class of crystals contains the rare-gas solids, as well as crystals made up of polyatomic molecules, inorganic, or, more often, organic. Inside the crystal, the molecule remains an individual entity, the arrangement of which defines the symmetry of the crystal. The intramolecular energies (atomic bonds) are at least an order of magnitude higher than the intermolecular cohesion energies. The point defects are *molecular defects*, for example, the vacancy is the absence of a molecule from a lattice site.

Self-diffusion experiments can be carried out with radiotracers, or, especially for organic crystals, by proton NMR. Molecules can be labeled with ^{14}C or ^{3}H (tritium); they can also be deuterated, which facilitates the measurement of the isotope effect by increasing the mass difference.

In the *rare-gas solids*, the activation energy for self-diffusion is very close to $2L_s$ (L_s = enthalpy of sublimation), a result in good agreement with theoretical estimates for vacancies: $\Delta H_v^f \approx \Delta H_v^m \approx L_s$. The estimates are based on static and dynamic calculations; for the rare gases, some good central-force potentials are available. However, this interpretation is controversial, and the participation of divacancies is not excluded.

The situation is more varied in *organic crystals*. These are divided into two classes: 1) crystals called "plastic", which for certain substances correspond to a high-temperature phase (also called a "rotator" phase) characterized by a high orientational disorder (molecular reorientation frequency $\approx 10^8 - 10^{11}$ s^{-1}). This disorder manifests itself as a high entropy of phase change, and a low entropy of fusion ($< 2.5\ R$). These materials have a high crystal symmetry (cubic or hexagonal). 2) non-plastic crystals (naphtalene, anthracene) with a normal entropy of fusion.

In the latter crystals, the diffusivities are very low; at the melting point T_m, $D \approx 10^{-12} - 10^{-11}$ cm^2s^{-1}. On the other hand, in the plastic crystals, values more typical of metals are found: $D(T_m) \approx 10^{-8} - 10^{-9}$ cm^2s^{-1}. The activation energies are higher than $2L_s$ in the first case ($2.3 - 2.4\ L_s$), equal to $2L_s$ or considerably less in the second case, a fact which suggests a strongly relaxed vacancy. These values can be correlated with the entropy of fusion, i.e., with the degree of orientational disorder, as shown in the table below:

Table VIII. — *Self-diffusion in molecular crystals*

Structure	Molecule	$\Delta S^m/R$ (melting)	Q/L_s (self-diffusion)
FCC	Adamantane $C_{10}H_{16}$	2.5	2.1-2.3
	Cyclohexane C_6H_6	1.1	≈1
	Pivalic acid $(CH_3)_3$-C-COOH	0.8	≈1
BCC	Hexamethylethane $(CH_3)_3$-C-C$(CH_3)_3$	2.4	≈2
	Succinonitrile NH_2-CH_2-CH_2-NH_2	1.35	≈1.2

The relation $Q = \Delta H_v^f + \Delta H_v^m$ cannot be verified as the vacancy enthalpies are known neither experimentally nor theoretically, except for a simulation on adamantane. The results are compatible with a vacancy mechanism, which could be expected in the FCC and BCC structures. The vacancies are more relaxed the higher the orientational disorder (translational and rotational relaxation near the defect), and hence the lower values of ΔH_v^{f} and ΔH_v^{m}. This hypothesis is supported by some measurements of the isotope effect: in adamantane, $f\ \Delta K = 0.8$, but in pivalic acid, $f\ \Delta K$ is very small (< 0.2). The low value could be attributed to relaxation on the one hand, and to a correlation factor lower than 0.8 on the other, the latter caused by local order (this solid is in fact made up of dimers). But the role of divacancies is not excluded, a curved Arrhenius plot is observed for hexamethylethane.

Bibliography

I. General references

Bocquet J.L., Brebec G., and Limoge Y., Diffusion in Metals and Alloys, Ch. 8 in *Physical Metallurgy*, R.W. Cahn and P. Haasen, eds. 3rd ed. (North Holland) 1984, p. 385-474.

Brebec G., Diffusion in Metals, *Mass Transport in Solids, op. cit.* Ch. 10, p. 251-281.

Casey H.C. and Pearson G.L., Diffusion in Semiconductors, in *Point Defects in Solids*, v.**2**, *op. cit.* (1975) p. 163-255.

Kofstad P., *Non-stoichiometry, Diffusion, and Electrical Conductivity in Binary Metal Oxides* (Wiley Interscience, New York) 1972.

Chadwick A.V., and Sherwood J.N., Point Defects in Molecular Crystals, in *Point Defects in Solids*, v**2**, *op. cit.* (1975) p. 441-475.

Défauts Ponctuels dans les Solides. – Ecole d'Eté de Confolant (1977), (Les Editions de Physique, Les Ulis)

Willoughby A.F.W., in *Defects in Semiconductors* II, S. Mahajan, J. W. Corbett, eds. (North Holland, 1983), p. 237.

Transport in Non-Stoichiometric Compounds, Workshop Alenya (1982), *Solid State Ionics,* **12** (1984).

Nonstoichiometric Compounds, Workshop, University of Keele (1986) *Advances in Ceramics* V. **23**, Eds. C. R. A. Catlow and W. C. Macrodt (American Ceramic Society, 1987).

Transport in Nonstoichiometric Compounds, Workshop, Pennsylvania State University (1984), Eds. G. Simkovich and V. S. Stubican (Plenum, 1985).

Frank W., Gösele U., Mehrer H. and Seeger A., Diffusion in Silicon and Germanium, in *Diffusion in Crystalline Solids, op. cit.* (1984) 63-142.

II. Specific references

Bénière M., Chemla M., and Bénière F., *J. Phys. Chem. Solids,* **37** (1976) 525-538.

Brown A.M. and Ashby M.F., *Acta Metall.* **28** (1980) 1085-1101.

COMPAAN K. and HAVEN Y., *Trans. Faraday Soc.* **54** (1958) 1498-1508.
DEMOND F.J., KALBITZER S., MANNSPERGER H., and DAMJANTSCHITSCH H., *Phys. Lett.* **93A** (1983) 503-506.
DEPPE D. G. and HOLONYAK jr., N., *J. Appl. Phys.* **64** (1988) R93-R113.
DIECKMANN R., and SCHMALZRIED H., *Ber. Bunsen. Phys. Chem.* **81** (1977) 344-377 and 414-419.
GEORGE B., JANOT C., ABLITZER D., and CHABRE Y., *Philos. Mag.* **44a** (1981) 763-778.
ISHIOKA S., NAKAJIMA H., KOIWA M., *Philos. Mag. A* **55** (1987) 359-374.
KÖHLER U. and HERZIG C., *Philos. Mag.* **58a** (1988) 769-786.
LANNOO M., and BOURGOIN J. *Point Defects in Semiconductors, op. cit., vol.* **1** 234-236.
LAZARUS D., Dimeta-82, *op. cit.* (1983) 134-144.
LE CLAIRE A.D., Colloque Saclay (1976) vol. **1** 17-53.
MARCHESE M., DE LORENZI G., JACUCCI G., and FLYNN, C.P. *Phys. Rev. Lett.* **57** (1986) 3280-3283.
MARCHESE M. and FLYNN C. P., *Phys. Rev.* **B38** (1988) 12200-12207.
MATHIOT D. and EDELIN G., *Philos Mag.* **A41** (1980) 447-458.
MATZKE Hj., Diffusion in Non-Stoichiometric Oxides, in *Non-Stoichiometric Oxides*, Ed. O. T. Sorensen (Academic Press, 1981) p. 156-232.
MATZKE Hj., Diffusion in Ceramic Oxide Systems, in *Fission Product Behavior, Advances in Ceramics* V. **17**, Ed. I. J. Hastings (American Ceramic Society, 1986) p. 1-54.
MEHRER, H., Dimeta-82, *op. cit.* (1983) 633-635.
MEYER M., *M.E.S. Rev. Metall.* **9** (1984) 472.
MITCHELL J., and LAZARUS D., *Phys. Rev.* **B12** (1975) 734-752.
MULLEN J.G., *Phys. Rev.* **124** (1961) 1723-1730.
MUNDY J.N., HOFF H.A., PELLEG J., ROTHMAN S.J., NOWICKI L.J., and SCHMIDT F.A., *Phys. Rev.* **B24** (1981) 658-665.
MURCH G.E., in *Diffusion in Crystalline Solids, op. cit.* (1984) Ch. 7.
PANDEY K. C., *Phys. Rev. Lett.* **57** (1986) 2287.
PERINET F., BARBEZAT S., and PHILIBERT J., in *Reactivity of Solids.* K. Dyrek, J. Haber, and J. Nowotny, eds. Materials Science Monograph # 10 (Elsevier, Amsterdam) 1982, p. 234-239.
PETERSON N.L., *Diffusion in Solids, Recent Developments* (1975) 115-170.
PETERSON N.L. *J. Nucl. Mater.* **69-70** (1978) 3-37.
REIN G. and MEHRER H., *Philos. Mag.* **A45** (1982) 467-492.
SEEGER A. and CHIK K.P., *Phys. Status Solidi,* **29** (1968) 455-542.
TAN T. Y. and GÖSELE U., *Mater. Sci. Eng.* **B1** (1988) 47-65.
TUCK B.; *Atomic Diffusion in III-V Semiconductors* (Adam Hilger, 1988).
VAROTSOS P. and ALEXOPOULOS K., *Thermodynamics of Point Defects and their Relation with Bulk Properties* (North Holland, 1986).
WEILER D. and MEHRER H., *Philos. Mag.* **A49** (1984) 309-325.
WERNER M. and MEHRER H., *Dimeta-82, op. cit.* (1983) 393-397.

WERNER M. and MEHRER H., *J. Electronic Materials* **14A** (1985) 301-307.
WERNER M. and MEHRER H., *Phys. Rev.* **B32** (1985) 3930.
ZENER C., *J. Appl. Phys.* **22** (1951) 372-375.

CHAPTER V

Solute diffusion in pure materials
Diffusion in alloys

I. Introduction.

From the point of view of atomic diffusion theory, the complexity of diffusion phenomena increases in the following order:

1. *Self-diffusion*, i.e., the motion of A atoms in a lattice of A atoms (or a sublattice of A atoms if there are no antistructure defects): $D_{\mathrm{A}}^{\mathrm{A}}$.

2. *Diffusion of a tracer* A* in pure A, which is usually called self-diffusion. This nomenclature is not strictly correct, as $D_{\mathrm{A}^*}^{\mathrm{A}} \neq D_{\mathrm{A}}^{\mathrm{A}}$, for the following reasons:

– correlation: in the case of a defect mechanism, the jumps of the tracer are correlated because it is a tagged atom:

$$D_{\mathrm{A}^*}^{\mathrm{A}} = f D_{\mathrm{A}}^{\mathrm{A}}, \quad f \leq 1.$$

– the isotope effect: the difference between the masses of the tracer isotopes and the natural isotopes of the material manifests itself as a difference in the vibration frequency and hence in the preexponential factor D_0.

3. *Solute diffusion at infinite dilution*, i.e., the diffusion of one atom of an impurity B in pure A: $D_{\mathrm{B}}^{\mathrm{A}}$, often called *impurity diffusion*, especially in the American literature. In practice, of course, the dilution is not infinite, but the sensivity for the detection of B, especially with radioisotopes, makes possible the measurement of $D_{\mathrm{B}}^{\mathrm{A}}$ at very high dilution (1). Since the B atoms are tagged, their jumps are correlated, but the correlation factor is no longer a simple number, independent of temperature.

4. *Self-diffusion of* A *and solute diffusion of* B *in dilute alloys of* B *in* A. The motion of the A and B atoms in such alloys can still be analysed as long as the concentration of B atoms is not too high. $D_{\mathrm{A}^*}^{\mathrm{AB}}$ and $D_{\mathrm{B}^*}^{\mathrm{AB}}$ are studied as functions of N_{B}, the mole fraction of the solute B, both experimentally and theoretically.

---

(1) That is, in the absence of very strong interactions, isolated solute atoms predominate, since the concentrations of solute atom pairs, triplets, etc., which are proportional to c^2, c^3, etc., respectively, are negligible when the solute concentration is <1% .

5. *Diffusion of* A^* *and* B^* *in concentrated* AB *alloys.* The theoretical analyses must be approximate to avoid intractable calculations. Results can be compared to Monte Carlo simulations.

This chapter essentially covers cases 3, 4, and 5.

II. Solute diffusion at infinite dilution.

The theory of random walk developed previously can be applied just as well to a B atom as to an A atom. The expression for D_B^A differs from that for D_A^A for two reasons:

– the probability of a vacancy occupying one of the nearest-neighbor sites to a B atom is not equal to the equilibrium concentration $\bar{N}_v = \exp(-\Delta G_v^f/kT)$ due to the solute-vacancy interaction (attractive or repulsive).

– the correlation factor is no longer a purely geometric factor but depends on the various local jump frequencies: $f(w_i)$.

The formula $D = \beta f \Gamma_s a^2$ established for a cubic crystal is thus written:

$$D_B^A = \beta a^2 \frac{p}{\zeta} f_2(w_i) \cdot w_2 = \beta a^2 N_v f_2(w_i) w_2. \tag{V.1}$$

$N_v \neq \bar{N}_v$ is sometimes referred to as "vacancy availability". One often sets $p = N_v \zeta$, where p is the degree of association, or the fraction of B atoms that has a vacancy on a nearest neighbor site (since there are ζ nearest neighbor sites, the probability of a vacancy being on one of them is p/ζ), and w_2 is the frequency of vacancy-solute exchange. The geometric factor β equals 1, except for the case of the diamond cubic structure, where it is 1/8.

We shall analyze the classic case of the FCC structure. The analysis for the other structures is similar, but generally more complicated.

II.1 The five frequency model (FCC). — The schematic diagram of figure V.1 shows that if only nearest-neighbor vacancy-solute interactions are considered, there are four frequencies to be considered in addition to the unperturbed host atom-vacancy exchange frequency w_0:

w_2, the frequency of B atom-vacancy exchanges;

w_1, the frequency of A atom-vacancy exchanges that do not dissociate the vacancy-solute pair (there are 4 such jumps);

w_3, the frequency of an A atom-vacancy exchange that dissociates the solute-vacancy pair, without distinguishing whether the vacancy moves to a 2nd or 3rd nearest-neighbor position (there are 7 such jumps);

w_4, the frequency of an atom-vacancy exchange that brings the vacancy into a nearest-neighbor position to the impurity.

From the above description, w_4 is the frequency of association jumps, and w_3 the frequency of dissociation jumps. At steady state, the numbers of association and dissociation jumps are equal:

$$(p/12)\ w_3 = \bar{N}_v\ w_4$$

or

$$p = 12\ \bar{N}_v (w_4/w_3). \tag{V.2}$$

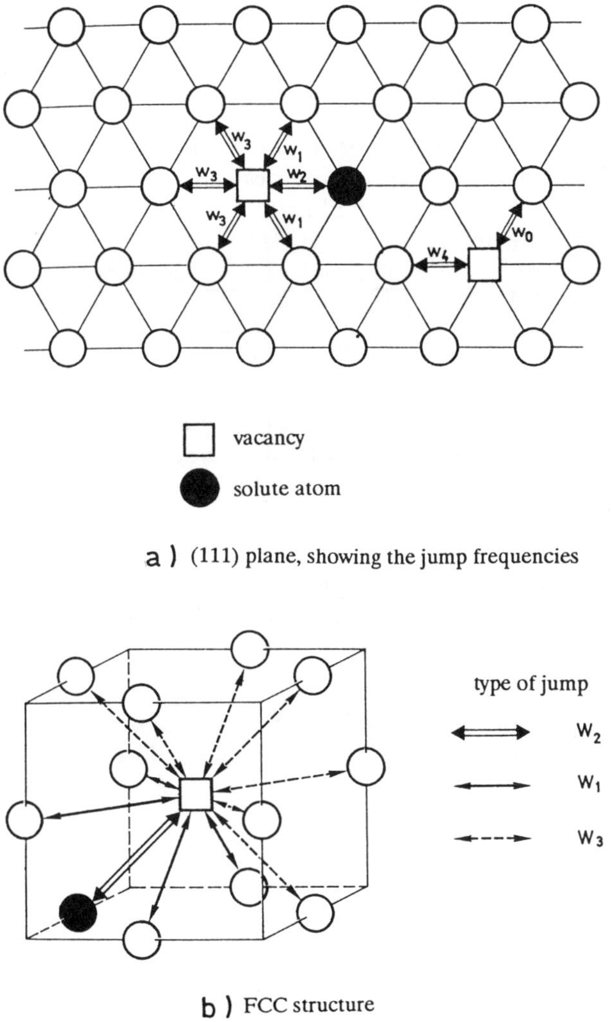

Fig. V.1. — Solute diffusion in a FCC lattice. Five-frequency model for vacancy (□) jumps in the presence of a foreign atom (•).

The vacancy concentration in pure A at thermal equilibrium is:

$$\bar{N}_v = \exp(-\Delta G_v^{\mathrm{f}}/kT),$$

whence

$$p = 12 \exp\left(-\frac{\Delta G_v^{\mathrm{f}} + \delta g^{\mathrm{f}}}{kT}\right),$$

and

$$w_4/w_3 = \exp\left(-\delta g^{\mathrm{f}}/kT\right), \tag{V.2a}$$

where δg^{f} is the Gibbs free energy for vacancy-impurity association. The Gibbs free energy for forming a vacancy on a nearest-neighbor site of a solute atom is $\Delta G_v^{\mathrm{f}} + \delta g^{\mathrm{f}}$, with attraction corresponding to $\delta g^{\mathrm{f}} < 0$, and repulsion to $\delta g^{\mathrm{f}} > 0$.

The correlation factor f_2 is then written:

$$\boxed{f_2 = \frac{w_1 + (7/2)\ F w_3}{w_1 + w_2 + (7/2)\ F w_3}}, \tag{V.3}$$

where F is a function of the ratio w_0/w_4 only and has been calculated by Manning. The derivation of equation (V.3) is beyond the scope of this book (see Manning, 1968). If $w_0 = w_4$, $7F = 5.15$; for $w_4 >> w_0$, $7F = 2$; and for $w_4 << w_0$, $F = 1$ (Fig. V.2). Combining equations (V. 1, 2, and 3), we obtain:

$$D_{\mathrm{B}}^{\mathrm{A}} = a^2\ \bar{N}_v\ \frac{w_4}{w_3} \cdot \frac{w_1 + (7/2)\ F w_3}{w_1 + w_2 + (7/2)\ F w_3} \cdot w_2. \tag{V.4}$$

The expression (V.3) for the correlation factor may appear complicated at first glance, but consideration of a few limiting cases should aid in getting a feel for it:

– no interaction: $w_1 = w_3 = w_4 = w_0$,

$$f_2 = \frac{w_0 \cdot (3.57)}{w_0 \cdot (3.57) + w_2},$$

and if, in addition to the above, $w_2 = w_0$, one obtains $f_2 = 0.781$, the geometric correlation factor for self-diffusion.

– tight binding of the vacancy to the solute: w_3 negligible (no dissociation):

$$f_2 = \frac{w_1}{w_1 + w_2}.$$

If in addition to w_3 being negligible, $w_1 > w_2$, $f_2 \approx 1$. In effect, since w_1 jumps are the more probable, the vacancy moves rapidly around the solute atom, and the vacancy-solute exchange can take place for any position of the vacancy. Thus the successive jumps of the solute are not correlated, $f_2 = 1$, and the diffusion rate is determined by the jump frequency w_2 (Eq. (V.4)).

– if $w_1 < w_2$, $f_2 \approx w_1/w_2$ is very small; the solute and the vacancy exchange continuously, and successive solute jumps are strongly correlated. If this value of f_2 is substituted in equation (V.4), w_2 disappears, and w_1 becomes the jump frequency that determines the rate of diffusion.

II.2 MODEL FOR THE BCC STRUCTURE. — The solute-vacancy interaction at the second nearest-neighbor position is not negligible for this structure, as the distances to first and second nearest neighbors differ by only 15% . Four different dissociative jump frequencies have to be defined, depending on whether the vacancy leaves a nearest-neighbor position (w_3, w_3', and w_3'') or a next-nearest-neighbor position (w_5) of the solute (Fig. V.1a). The reverse jumps are denoted by w_4, w_4', w_4'' and w_6, respectively. In the BCC structure, there is no jump corresponding to the w_1 jump in the FCC structure, because any exchange of the vacancy with a solvent dissociates the solute-vacancy pair. The solute-vacancy exchange frequency is as usual denoted by w_2. The condition of dynamic equilibrium leads to relations analogous to equation (V.2a):

$$\begin{aligned} w_4'/w_3' = w_4''/w_3'' &= \exp\left(-\delta g_1^{\mathrm{f}}/kT\right), \\ w_6/w_5 &= \exp\left(-\delta g_2^{\mathrm{f}}/kT\right), \end{aligned} \qquad \text{(V.5)}$$

and

$$\frac{w_6}{w_5}\,\frac{w_4}{w_3} = \frac{w_4'}{w_3'}, \qquad \text{(V.6)}$$

where δg_1^{f} and δg_2^{f} are the Gibbs free energies of vacancy-solute association for the first and second-nearest neighbor positions, respectively, with $\delta g^f < 0$ again indicating attraction.

The correlation factors are expressed in the framework of two simplified models:

Model I: Simplification of certain associative jumps:

$$w_4' = w_4'' = w_6 = w_0.$$

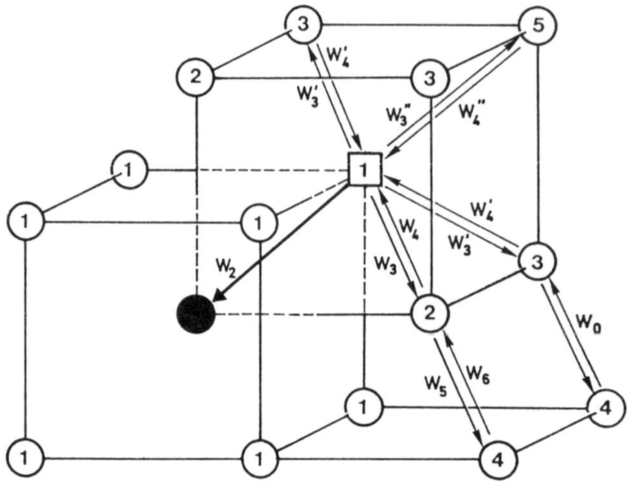

Fig. V.1a. — Solute diffusion in the BCC structure, showing the jump frequencies of the vacancy (□) for the case of an interaction with a solute atom (•) in the first or second nearest-neighbor position. The atoms labeled 1 are the nearest neighbors of the solute, atoms labeled 2 are the second nearest neighbors, etc.

It follows that $w_3/w_3' = w_4/w_5$; this is a five-frequency model:

$$f_2 = \frac{7\ Fw_3'}{2\ w_2 + 7\ Fw_3'}, \tag{V.7}$$

with $F = F(w_3/w_3')$ and $7F = 5.33$, 3, and ∞ for $w_3/w_3' = 1$, ∞, and 0, respectively.

According to (V.5):

$$w_3' = w_0 \exp\left(\delta g_1^{\mathrm{f}}/kT\right),$$
$$w_5 = w_0 \exp\left(\delta g_2^{\mathrm{f}}/kT\right).$$

Model II: Second nearest-neighbor interactions are neglected: $\delta g_2^{\mathrm{f}} = 0$. Then only the jump frequencies w_0, w_2, w_3, and w_4 remain, since:

$$\begin{array}{ccccc} w_5 & = & w_6 & = & w_0 \\ w_3 & = & w_3' & = & w_3'' \\ w_4 & = & w_4' & = & w_4''. \end{array}$$

The model is similar to the five-frequency model for FCCs (w_1 does not exist in BCCs):

$$f_2 = \frac{7\ Fw_3}{2\ w_2 + 7\ Fw_3}, \tag{V.8}$$

with $F = F(w_4/w_0)$ (Fig. V.2).
In the three models above, f_2 is always of the form:

$$f_2 = \frac{u}{2\ w_2 + u}, \tag{V.9}$$

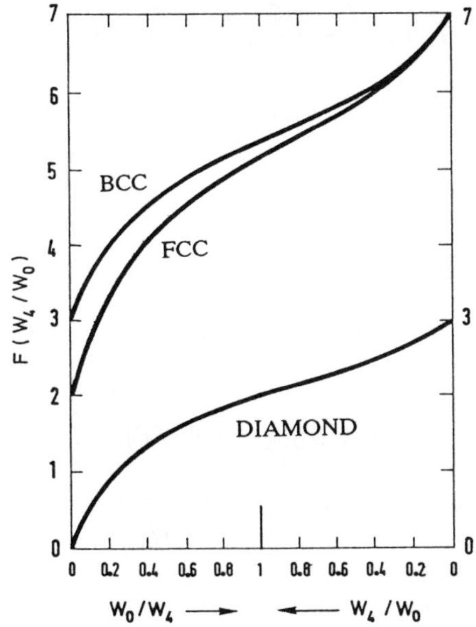

Fig. V.2. — The functions 7 $F(w_4/w_0)$ in the BCC and FCC structures and 3 $F(w_4/w_5)$ for the diamond structure. After Manning (1968).

where u is a function of the solvent jump frequencies. In the simplified reasoning at the end of § **II**.1, u is interpreted as the effective escape frequency of the vacancy, since its value relative to w_2 governs the migration of the vacancy after its first exchange with the solute atom, and with it the correlation between the first and second jumps of the solute atom.

The same form for f_2 is obtained for other cubic structures as well.

In the *diamond structure*:

$$f_2 = \frac{3\ Fw_3}{2\ w_2 + 3\ Fw_3}, \tag{V.10}$$

with F given by figure V.2.

In the *simple cubic structure:*

$$f_2 = \frac{5\ Fw_3}{2\ w_2 + 5\ Fw_3}. \tag{V.11}$$

II.3 Comparison of self- and solute diffusion. — It is interesting to compare the self-diffusion coefficient $D^{\mathrm{A}}_{\mathrm{A}^*}$ in a given solvent A with the diffusion coefficient $D^{\mathrm{A}}_{\mathrm{B}}$ of a solute B in the same solvent. In a cubic crystal:

$$\begin{aligned} D^{\mathrm{A}}_{\mathrm{B}} &= \beta a^2 (p/12)\ f_2\ w_2 \\ &= \beta a^2\ \tilde{\nu} f_2 \exp\left(-\ \frac{\Delta G^{f}_{v} + \delta g^{f}}{kT}\right) \exp\left(-\ \frac{\Delta G^{\mathrm{m}}_{2}}{kT}\right), \end{aligned} \qquad \text{(V.12)}$$

with the free energy of the jump having a value $\Delta G^{\mathrm{m}}_{2}$, different from $\Delta G^{\mathrm{m}}_{v}$ for self-diffusion, because $w_2 \neq w_0$.

The enthalpy of activation for solute diffusion is thus:

$$Q_2 = \Delta H^{\mathrm{f}}_{2} + \Delta H^{\mathrm{m}}_{2} - C, \qquad \text{(V.13)}$$

the three terms arising from the temperature dependence of p, w_2, and f_2, respectively:

$$C = k\ \partial \log f_2 / \partial(1/T). \qquad \text{(V.14)}$$

Although f_2 is a function of several jump frequencies, its behavior may still be nearly Arrhenius if $w_2 >> w_1$, w_3; since $f_2 = u/(w_2+u) \approx u/w_2$, f_2 follows an Arrhenius law. In the opposite case, $w_2 << u$, and $f_2 \approx 1$, so that C becomes negligible.

Comparing the expression for solute diffusion to that for self-diffusion, we find:

$$D^{\mathrm{A}}_{\mathrm{A}^*} = \beta a^2\ \tilde{\nu} f_0 \exp\left(-\ \frac{\Delta G^{\mathrm{f}}_{v} + \Delta G^{\mathrm{m}}_{v}}{kT}\right).$$

With

$$Q_1 = \Delta H^{\mathrm{f}}_{1} + \Delta H^{\mathrm{m}}_{1},$$

the difference between the activation enthalpies becomes:

$$\boxed{\Delta Q = Q_2 - Q_1 = \delta H^{\mathrm{f}} + \delta H^{\mathrm{m}} - C}, \qquad \text{(V.15)}$$

with:

$$\delta H^{\mathrm{f}} = \Delta H^{\mathrm{f}}_{2} - \Delta H^{\mathrm{f}}_{1},$$

where δH^{f} is just the enthalpy of solute-vacancy interaction for nearest-neighbor sites, and:

$$\delta H^{\mathrm{m}} = \Delta H^{\mathrm{m}}_{2} - \Delta H^{\mathrm{m}}_{1}$$

is the difference of the enthalpies of migration, which depends on the vacancy-solute interaction at the saddle point.

II.4 Application to metals.

a) *Experimental results.* — In a number of solvents, ΔQ varies monotonically with $Z_2 - Z_0$, the difference between the valences of the solute and solvent. For fast diffusers, $\Delta Q < 0$ and $Z_2 > Z_0$, and the reverse is true for slow diffusers, $\Delta Q > 0$ and $Z_2 < Z_0$ (Fig. V.3). The solute diffusion coefficient D_2 generally increases with $Z_2 - Z_0$. This type of behavior is observed for noble metals and for the group IIB metals Zn and Cd. It is not the case for Al and Mg, Pb, the transition metals, or the alkali metals. In Al, for example, ΔQ is always quite small ($|\Delta Q| \leq 0.13$ eV, and independent of the size and sign of the valence difference), except for transition elements, for which $\Delta Q \sim +0.5$ to 1.3 eV.

b) *Theoretical models.* — The value of ΔQ for a vacancy mechanism in simple metals can be calculated from a theory based on the free electron model. Following Lazarus, the vacancy in a solvent of valence Z is considered to have a charge $-Ze$, and the solute atom of excess charge ze is considered to behave as a point charge $(Z + z)e$. The excess charge of the solute gives rise to a perturbing potential $V(r)$. The quantity δH^{f} is equal to the electrostatic energy of a charge $-Ze$ (the vacancy) located at a distance equal to the interatomic distance d from the center of the potential $V(r)$:

$$\delta H^{\mathrm{f}} = -ZeV(d). \tag{V.16}$$

The calculation of δH^{m} is more subtle. Following Le Claire (1962), the saddle-point configuration is described as follows: the excess charge of the solute ze is located halfway between two "half vacancies" located on two neighboring sites (Fig. V.4). Each half vacancy carries a charge of $-Ze/2$ located at a distance of 11/16 d from the saddle point. δH^{m} is then given by the difference of the electrostatic energies corresponding to this configuration, and the equilibrium configuration corresponding to the two charges $-Ze$ and ze at a distance d:

$$\delta H^{\mathrm{m}} = -2 \cdot \frac{Ze}{2} V\left(\frac{11}{16}d\right) + Ze \cdot V(d), \tag{V.17}$$

whence:

$$\delta H^{\mathrm{f}} + \delta H^{\mathrm{m}} = -Ze \cdot V\left(\frac{11}{16}d\right). \tag{V.18}$$

The term C can not be completely calculated on the basis of these assumptions because even though the activation energies for the w_1, w_2, and w_3 jumps can be calculated on this model, the preexponential factors cannot.

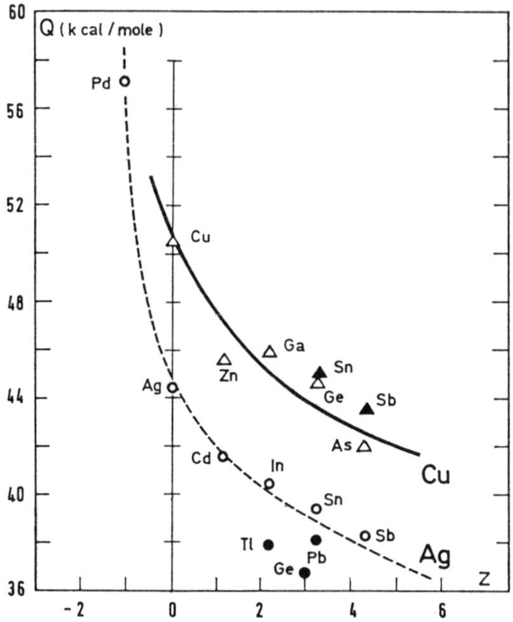

Fig. V.3. — Solute diffusion at infinite dilution in silver and copper: variation of the activation energy with the difference in valence between solute and solvent. Open symbols (△, o): elements on the same row of the periodic table as the solvent, full symbols (▲, •): elements of a different row.

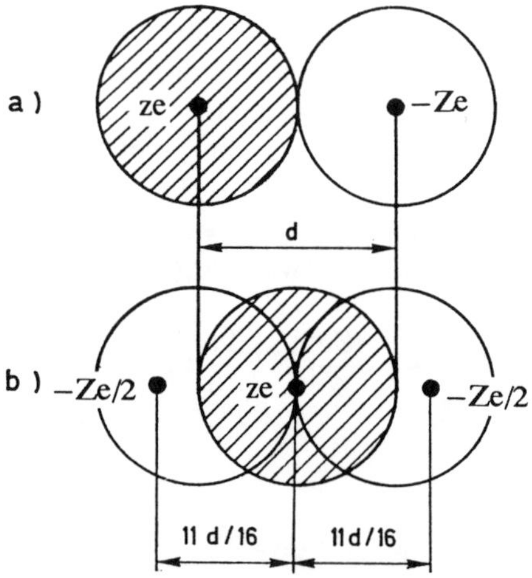

Fig. V.4. — Le Claire's model for the electrostatic interaction between a vacancy of relative charge $-Z\,|e|$ and a solute atom of excess charge $z\,|e|$ in the stable position (a) and at the saddle point (b).

To overcome this difficulty, Le Claire used the preexponential D_0 factors for self-and solute diffusion. Thus for any of the w_i:

$$w_i = \nu_i \exp\left(-\Delta H_i^{\mathrm{m}}/kT\right) = \nu_i \exp\left(-\frac{\Delta H_i^{\mathrm{m}} + \delta H_i^{\mathrm{m}}}{kT}\right).$$

For jumps of the solvent A atoms, $\nu_1 = \nu_3 = \nu_0$. The ratio ν_2/ν_0 is evaluated from:

$$\frac{D_{\mathrm{B}}^{\mathrm{A}}}{D_{\mathrm{A}^*}^{\mathrm{A}}} = \frac{f_2}{f_0}\frac{\nu_2}{\nu_0} \exp\left(-\frac{\delta H^{\mathrm{f}} + \delta H_2^{\mathrm{m}}}{kT}\right) = \frac{D_0^{\mathrm{B}}}{D_0^{\mathrm{A}}} \exp\left(-\frac{\Delta Q}{kT}\right),$$

whence:

$$\frac{\nu_2}{\nu_0} = \frac{D_0^{\mathrm{B}}}{D_0^{\mathrm{A}}}\frac{f_0}{f_2} \exp\left[-\frac{\Delta Q - (\delta H^{\mathrm{f}} + \delta H_2^{\mathrm{m}})}{kT}\right]$$

$$\frac{\nu_2}{\nu_0} = \frac{D_0^{\mathrm{B}}}{D_0^{\mathrm{A}}}\frac{f_0}{f_2} \exp\left(C/kT\right).$$

The following implicit expression then allows the evaluation of C (with F of equation (V.3) constant):

$$C = \frac{D_0^{\mathrm{B}}}{D_0^{\mathrm{A}}} f_0 \exp\left(\frac{C}{kT}\right)\left(\frac{(\delta H_2^{\mathrm{m}} - \delta H_1^{\mathrm{m}}) \exp\left(-\frac{\delta H_2^{\mathrm{m}} + \delta H_1^{\mathrm{m}}}{kT}\right)}{\left[\exp(-\delta H_1^{\mathrm{m}}/kT) + \frac{7}{2}\exp(-\delta H_3^{\mathrm{m}}/kT)\right]^2} + \frac{\frac{7}{2}(\delta H_2^{\mathrm{m}} - \delta H_3^{\mathrm{m}}) \exp\left(-\frac{\delta H_2^{\mathrm{m}} + \delta H_3^{\mathrm{m}}}{kT}\right)}{\left[\exp(-\delta H_1^{\mathrm{m}}/kT) + \frac{7}{2}\exp(-\delta H_3^{\mathrm{m}}/kT)\right]^2}\right). \qquad \text{(V.19)}$$

The calculation of C is then carried out using the experimental values of the preexponential factors D_0^{A} and D_0^{B} and the three differences in the migration enthalpies δH_i^{m} $(i = 1,\ 2,\ 3)$ calculated from the half-vacancy model.

The remaining problem is to find a good expression for the perturbing potential $V(r)$. In the noble metals, the approximate Thomas-Fermi solution, expressed as a screened Coulomb potential, is usually used:

$$V(r) = \alpha\frac{ze}{r} \exp\left(\frac{r}{\rho}\right), \qquad \text{(V.20)}$$

where ρ is the screening radius:

$$\rho^2 = \frac{\hbar^2}{4\,me^2}\left(\frac{\pi}{3\,n_e}\right)^{1/3},$$

with m and e the mass and charge of the electron and n_e the number of electrons per unit volume. (The parameter α is described in Le Claire's paper.) To get an idea of the orders of magnitude involved, we calculate for a divalent solute in silver, $Z = 1$ and $z = 1$:

$$\begin{aligned} \delta H^{f} &= -0.028 \text{ eV} \\ \delta H_2^{m} &= -0.16 \text{ eV} \\ \delta H_1^{m} &= +0.013 \text{ eV} \\ \delta H_3^{m} &= +0.005 \text{ eV}. \end{aligned}$$

There is a vacancy-impurity attraction since δH^{f} is negative, but it is weak ($\approx 1/40$ eV). The jump frequency w_2 is increased ($\delta H_2^{m} < 0$). The jump frequencies of solvent atoms are somewhat reduced in the neighborhood of the solute. Since w_2 is larger than all other jump frequencies, the correlation effect is strong. A value of -0.1 eV is calculated for C at 800°C from the experimental values of D_0. Then:

$$\begin{aligned} \Delta Q &= -0.028 - 0.16 + 0.1 \\ &\simeq -0.09 \text{ eV}. \end{aligned}$$

The C term is thus far from negligible. The theoretical value is in excellent agreement with experimental results for the Ag-Cd, Ag-Zn, and Cu-Zn systems.

The theoretical model described above, based on the screened Coulomb potential, gives values of ΔQ which are in good agreement with experiment for solutes of the B groups in the periodic table dissolved in the noble metals: ΔQ is negative and its absolute value increases almost linearly with the difference in valence ([2]) (Fig. V.3). The same model, adapted to the HCP structure, gives equally good agreement with experiment, including the sign of the anisotropy, for diffusion in zinc.

In other systems where the vacancy mechanism is considered operative, e.g., the monovalents (alkali metals), divalents (Mg), trivalents (Al), the values of ΔQ calculated from the screened Coulomb potentials do not agree with experiment, even as to sign. Indeed, in Mg, Q increases with the valence of the solute, in contradiction to the trend predicted by theory. In aluminum, $\Delta Q \approx 0$ for a large number of elements.

([2]) The variation of ΔQ with impurity z is not completely linear in the theoretical model, because in the expression for the potential, equation (V.20), α is a function of z.

There are several possible reasons for this disagreement:

– the choice of the potential around the impurity. Instead of a screened Coulomb potential, a self-consistent potential should be chosen; for example one with a simpler asymptotic form, valid for distances greater than a fraction of d. This potential has an oscillatory form $\cos(2k_F r + \phi)/r^3$, where k_F is the Fermi wave vector; these oscillations explain how the sign of the potential can change depending on the value of r, and not always be the same, independent of r, as predicted by equation (V.20). Thus the vacancy-solute interaction can be attractive at the nearest-neighbor position but repulsive at the saddle point. Unfortunately for the calculations, the phase difference ϕ is generally not known;

– the model itself. In considering only the difference in valence between the solute and solvent, the model neglects the *effect of atomic size*; in fact, a correlation between D (or Q) and the size of the solute is observed. Finally, the description of the saddle point is rather schematic, in that relaxation effects are neglected. Recent calculations have also used pseudo-potentials for calculating the enthalpy of migration in normal metals.

In some cases, e.g., BCC metals, one cannot exclude a global alloying effect on the activation enthalpy, the effect being due to the change of the d-electron density on alloying, rather than to a localized vacancy-solute interaction.

Finally, there exists an empirical relation for the activation enthalpy:

$$Q_2 = Q_1 + k\Delta T_M, \tag{V.21}$$

where ΔT_M is the difference in melting temperatures, and also a relation for the preexponential factor:

$$\mathrm{Log}\frac{D_0^{(2)}}{D_0^{(1)}} = \mathrm{const}\,(Q_2 - Q_1)\,,$$

deduced from Wert's and Zener's relationship of D_0 to Q (Ch. IV, Eq. (IV.42)).

II.5 Ultra-fast diffusers. — The preceding analysis applies to so-called "normal" diffusion, characterized by a solute diffusion coefficient that does not differ by more than an order of magnitude or so from the self-diffusion coefficient, or, more precisely, diffusion such that:

$$0.1 < D_0^{(2)}/D_0^{(1)} < 10 \qquad \text{and} \qquad 0.75 < Q_2/Q_1 < 1.25,$$

values which suggest that the same mechanism is operative for both solute and self-diffusion.

On the other hand, the diffusivities of certain solutes exceed the self-diffusion coefficient of the solvent by many orders of magnitude: $D_2/D_1 \approx 10^2 - 10^5$ at T_M. The classic example is also one of the first published studies of diffusion in the solid state: gold in lead (Roberts-Austen, 1896). Here $D^{Pb}_{Au}/D^{Pb}_{Pb} \approx 10^5$ at 175°C; only recently has it been demonstrated that this phenomenon truly represents lattice diffusion and not diffusion along some kind of short-circuiting path such as grain boundaries or dislocations.

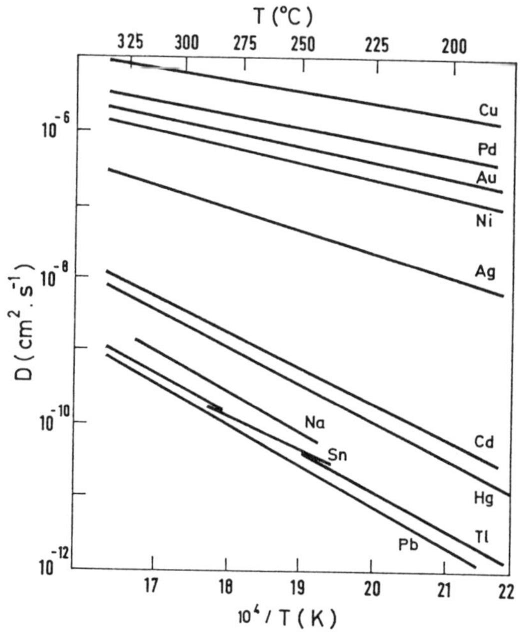

Fig. V.5. — Arrhenius plot for self- and solute diffusion in lead. After Warburton and Turnbull (1975).

This type of very rapid diffusion is observed in metals of Group IIIB (In, Tl), and IVB (Sn, Pb). Figure V.5 compares self-diffusion with the diffusion of several solutes in lead. The case of the noble metals Cu, Ag, Au, and of Pd and Ni are particularly spectacular; but even for Cd and Hg the diffusivity is quite large. Although the activation energy for self diffusion in Pb is about 1.1 eV, it is only 0.35 and 0.43 eV for the diffusion of copper and gold solutes, respectively. Another example is diffusion in thallium: the self-diffusion coefficient in the BCC phase is much higher than in the (lower-temperature) HCP phase, quite in accord with the general rule; but the solute diffusion of Ag in Tl is practically independent of the phase change. Rapid diffusion of certain impurities is also observed in some of the BCC anomalous diffusers (Ti, Zr, U) (see Ch. IV, VIII.3), as well as in their HCP modifications (α-Ti and Zr); the variety of systems is large enough so that no simple correlations appear readily. There is no simple rule for predicting whether solute diffusion will be anomalous.

Such high diffusivities are generally thought to result from the operation of an interstitial mechanism, even though the ratios of the atomic radii of solute to solvent lie in the interval 0.62-0.93, significantly above Hägg's criterion of 0.59 as the limit of an interstitial solid solution. The vacancy mechanism is excluded for several reasons:

1. The effect of additions of a fast-diffusing solute on self-diffusion of the solvent can be calculated from the theory (5-frequency model), yielding a minimum possible value of b_1 (*vide infra* § VI); the experimental value of b_1 is always below the calculated b_1 (min).

2. The activation volume is very small (0.28 Ω for Au and Zn in Pb, 0.15 Ω for Cu and Pd), unlike the values obtained for self-diffusion by vacancies (Ch. IV, Tab. III).

There are at least three mechanisms possible:

1. simple interstitial,

2. mixed dissociated interstitial (or mixed dumbbell), either by the translational + rotational process (see Fig. III.4), or by the "looping" mechanism, (dissociation of the mixed dumbbell, migration of the normal dumbbell around the solute atom, and recombination in a mixed dumbbell) - which gives rise to a low isotopic effect (Johnson and Lam),

3. vacancy-interstitial pair. In this last case (Fig. V.6a), the interstitial solute is a nearest neighbor of a vacancy, with strong binding between the two defects. Six jump frequencies are considered in the FCC structure: v_1 denotes creation of the pair and v_2 its annihilation (both according to the reaction: $B_A \rightleftharpoons B_i + V_A$ where the subscript i means interstitial); k_1 and k_2 denote frequencies of i $\rightarrow$ i jumps of the solute that dissociate or do not dissociate the pair, respectively, and w_1 and w_2 denote solvent jumps which dissociate or do not dissociate the pair, respectively.

The correlation factor for solute diffusion must be small, as the solvent jump frequencies w_1 and w_2 are probably small with respect to k_2 and especially v_2; if $k_2 >> w_2$, the interstitial circles around the vacancy and $f \approx 0$. The measured values for the isotope effect $f\Delta K$ are about 0.25-0.30 for Cu, Ag, Au, and Hg in Pb. Further, the solute atom cannot move over any distance unless the solvent jumps (w_2) allow reorientation of the pair, which leads to an effect of the solute on solvent self-diffusion, which would not be predicted on the basis of mechanisms 1 and 2.

It appears that both mechanisms 2 and 3 are operative in metals of groups IIIB and IVB, but with differing importance for the very fast diffusers (noble metals) and the fast diffusers such as Cd or Ag (cf. Fig. V.5). The same vacancy-interstitial pair mechanism appears to explain equally well the large values of the diffusion coefficients of certain transition metal solutes (Fe, Co, Ni) in the anomalous BCC metals: β-Ti, β-Zr, or in V and Nb and in HCP α-Ti and α-Zr as well. Among the fast diffusers in general, D increases as the solute radius decreases.

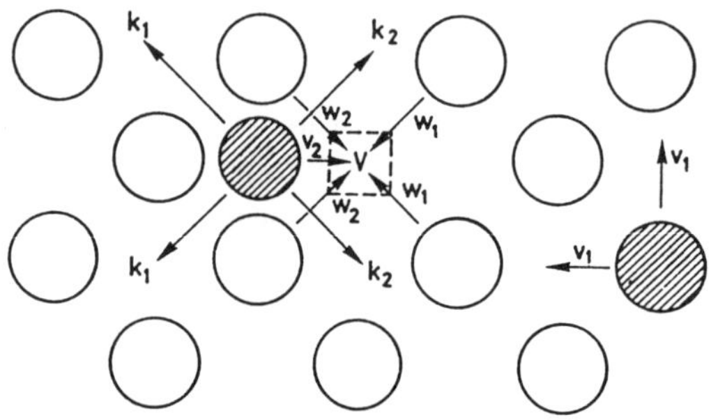

Fig. V.6a. — Miller's model for diffusion by interstitial-vacancy pairs. The solute atom is hatched.

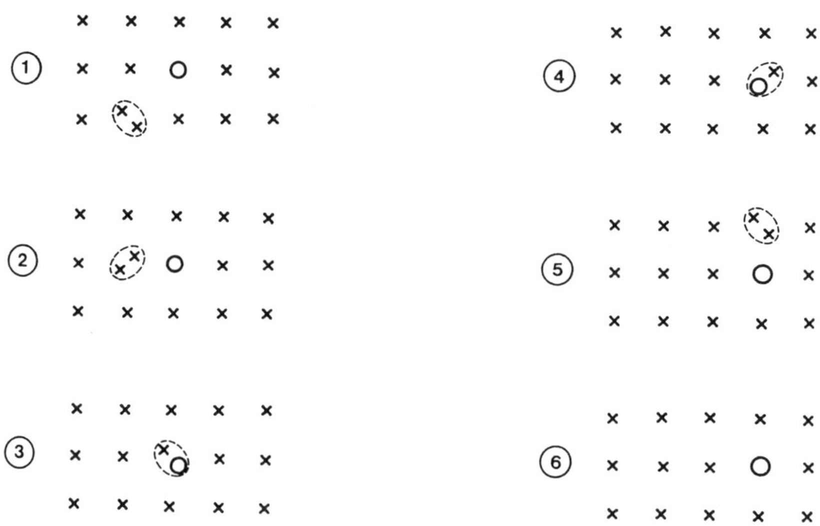

Fig. V.6b. — Migration of a substitutional impurity by the interstitialcy mechanism (mixed dumbell).

III. Interstitial solid solutions.

When a solute (H, B, C, N, O) is normally dissolved in interstitial sites in the base metal, its migration takes place by a purely interstitial process:

$$D = \beta w a^2 = \beta a^2/\tau, \tag{V.22}$$

where β is a geometric factor which depends on the structure and on the type of interstitial site (octahedral (O) or tetrahedral (T)) (see Ch. II, § III).

III.1 THE SOLUTES C, N, AND O. — There appear to be no empirical correlations that aid in systematizing the experimental results, except for the BCC metals, where in each group of the periodic table (IVA, VA,...) the activation energies are related to the size of the interstitial atom, and $D \approx 10^{-5} - 10^{-4}$ cm^2 s^{-1} at the melting point.

Considering the atom as a small harmonic oscillator of mass m, vibrating in a sinusoidal potential well of depth ΔH^{m}, Wert and Zener found:

$$\tilde{\nu} = \left(\frac{\Delta H^{\mathrm{m}}}{2\, md^2}\right)^{1/2}, \tag{V.23}$$

where d is the distance between two neighboring sites. This approximation is reasonable (error <15%) as long as $\Delta H^{\mathrm{m}} > 2kT$. The entropy of activation ΔS^{m} can be calculated from the preexponential factor $D_0 = \beta a^2\, \tilde{\nu} \exp(\Delta S^{\mathrm{m}}/k)$.

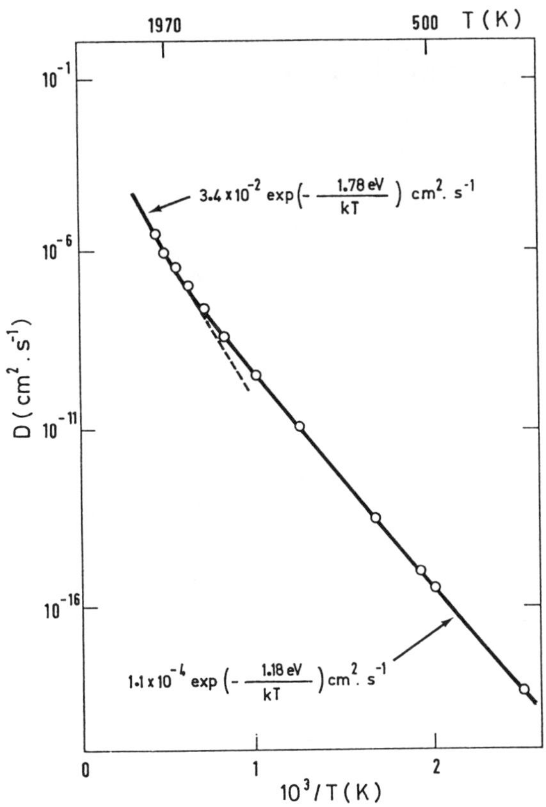

Fig. V.7. — Arrhenius plot for the diffusion of carbon in molybdenum. After Kimura and Yoshioka (1976).

The diffusion coefficient has often been measured over an exceptionally wide range of temperature, owing to the determinations of D at low temperature by magnetic or mechanical relaxation (internal friction, elastic after-effect). This has revealed curvature in the Arrhenius plot in several cases: C in α-Fe and Mo (Fig. V.7), although for N and O in Ta or N in α-Fe, the plot is linear over 16 orders of magnitude (Fig. V.8). Four mechanisms have been invoked to explain this curvature:

– formation of interstitial-vacancy pairs;

– formation of di-interstitials which migrate by an O-T-O mechanism at low temperature;

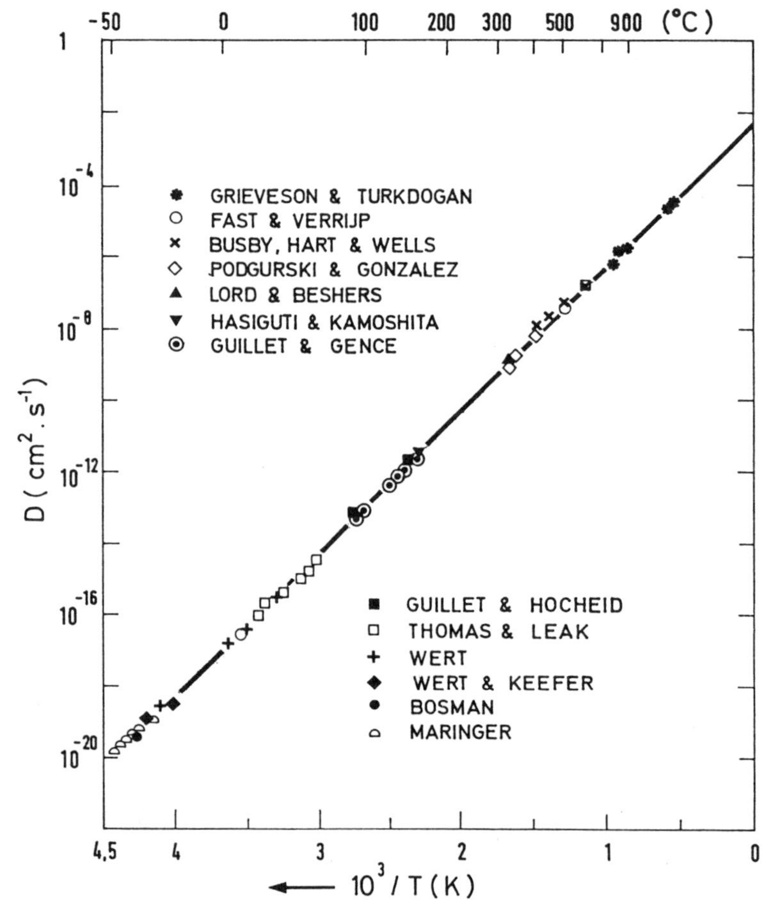

Fig. V.8. — Arrhenius plot for the diffusion of nitrogen in α-Fe. Compilation of 13 sets of results by Beshers (1973).

– temperature dependent $\Delta H^{\mathrm{m}}$and ΔS^{m} (m = migration) due to a non-linear temperature dependence of the elastic constants. This explanation is based on an old (1952) idea of Zener's, who assumed that ΔH^{m} is proportional to an elastic modulus μ; then it follows that (see Ch. IV, § IV and VII):

$$\Delta S^{\mathrm{m}} = \lambda\theta\ \Delta H^{\mathrm{m}}/T_{\mathrm{M}}, \qquad \text{with } \lambda = \text{ const.} \qquad \theta = -\frac{T_{\mathrm{M}}}{\mu_0}\frac{\mathrm{d}\mu}{\mathrm{d}T}; \qquad \text{(V.24)}$$

– two diffusion paths, direct: O→T→O, or complex: O → T → T → ... ...T→O. Measurement of the isotope effect $^{12}\mathrm{C}/^{13}\mathrm{C}$ in α Fe confirmed the direct interstitial mechanism with $E = 1.07$, i.e., $f \approx 1$, $\Delta K \approx 1$.

III.2 Hydrogen and its isotopes (D, T). — The diffusion of hydrogen in metals has been much studied for technological as well as fundamental reasons. Hydrogen dissolves interstitially in most metals; the enthalpy of solution can be negative (metals that form hydrides, especially metals of groups IVA and VA) or positive (metals of groups VIA, VIIA, and VIIIA, and the noble metals). In the first case, the hydrogen is on tetrahedral sites; its solubility can be very large (H/M $\approx$ 1), and can decrease with increasing temperature. In the second case, the solubility is small, specially in BCC metals, and decreases with the temperature; the hydrogen occupies tetrahedral sites, and perhaps, if the temperature is not too low (>80°C for α-Fe), octahedral sites as well. The principal results can be summarized as follows:

1. The diffusion of hydrogen is extremely rapid, for example 2×10^{12} jumps per second in vanadium at ambient temperatures, that is, 1/3 of the highest phonon frequency, with $D = 5 \times 10^{-5}$ cm^2 s^{-1}. It exceeds the diffusivities of the heavier interstitials (C, N, O) by 15 to 20 orders of magnitude (Fig. V.9). The activation energy is small (FCC) to very small (BCC):

	V	Fe	Ni	Pd
Q (eV)	0.05	0.08	0.4	0.23

2. The diffusion of hydrogen can be observed down to temperatures low enough that quantum effects are possible because of the low mass of the particle; in addition to thermally activated jumps, migration by tunneling is possible at low temperature ($T < 20$ K).

3. Three isotopes are available: hydrogen, deuterium, and tritium, yielding the largest mass ratios in the periodic table. Further, using the positive muon as a "light proton" (its mass is one ninth of the proton's), the mass ratio can be stretched to 27! The muon has a very short lifetime (2.2 μs); its concentration is thus always low, and its diffusion can truly be studied at infinite dilution.

4. Metal-hydrogen alloys can be prepared by charging the metal with hydrogen by heating in a hydrogen atmosphere, or electrolytically (cathodic

charging), or by implantation (Ch. IX, part 1, § I). Either very dilute or very concentrated alloys (up to 50-50) can be studied, the latter for a study of cooperative effects.

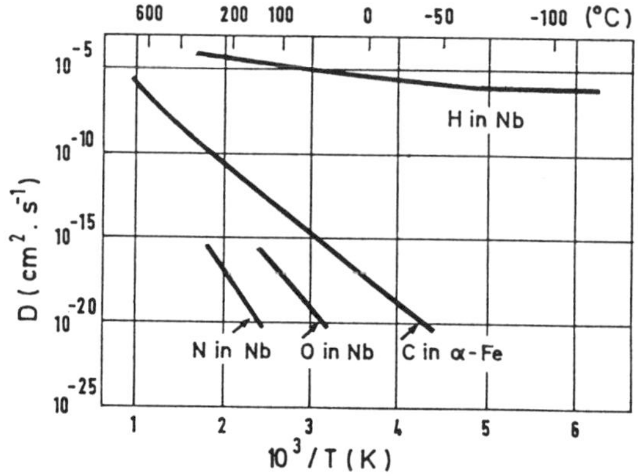

Fig. V.9. — Arrhenius plot for the diffusion of some interstitial solutes. Compilation by Völkl and Alefeld (1975).

5. In addition to the classical measurements such as permeation, radiotracers (tritium), and electrochemistry, there exist methods for measuring the jump frequency in a solid solution in equilibrium without parasitic effects connected with surfaces: mechanical or magnetic relaxation, Gorski effect, NMR, quasi-elastic neutron scattering, muon spectroscopy, etc.

In view of the wide range of temperature over which the diffusion of hydrogen has been studied, it is not surprising that Arrhenius behavior is not followed if the temperature range under consideration is wide enough; a curvature is indeed observed for diffusion in BCC metals such as Nb and Ta as well as in FCC metals such as Ni. In α-Fe, in spite of an enormous bibliography, the scatter of the experimental results is disheartening. ([3])

BCC metals are more prone to such anomalies, because the activation energy for hydrogen diffusion (0.05-0.1 eV) is smaller than the depth of most traps (a few tenths of an eV).

Hydrogen diffuses more rapidly than deuterium, and deuterium diffuses more rapidly than tritium, but the isotope effect does not follow the classical $m^{-1/2}$ law. In the BCC metals, the values of D_0 are practically the same,

([3]) This scatter has four possible origins: 1) because of the low solubility, methods not sensitive to the state of the surface, such as relaxation or resonnance, are not applicable; 2) trapping of hydrogen at impurities and crystal defects (dislocations, grain boundaries); 3) formation of immobile di-interstitials; 4) the formation of molecular hydrogen in various defects.

but the activation energies differ ($Q_H < Q_D$) so that D_H/D_D depends on the temperature, but a different type of behavior appears at low temperature (Fig. V.10). This ratio can even exceed 20 in metals of group V; such an effect can be due to a local distortion of the lattice ("polaron"). In FCC metals, the values of D_0 do indeed vary as $m^{-1/2}$, but owing to the relative values of the activation energies ($Q_H > Q_D > Q_T$ at high temperature), an inverse isotope effect is observed at low temperature.

At high temperature, the diffusion is normally thermally activated. Quantum effects can manifest themselves if $h\nu > kT$, where ν is the vibration fre-

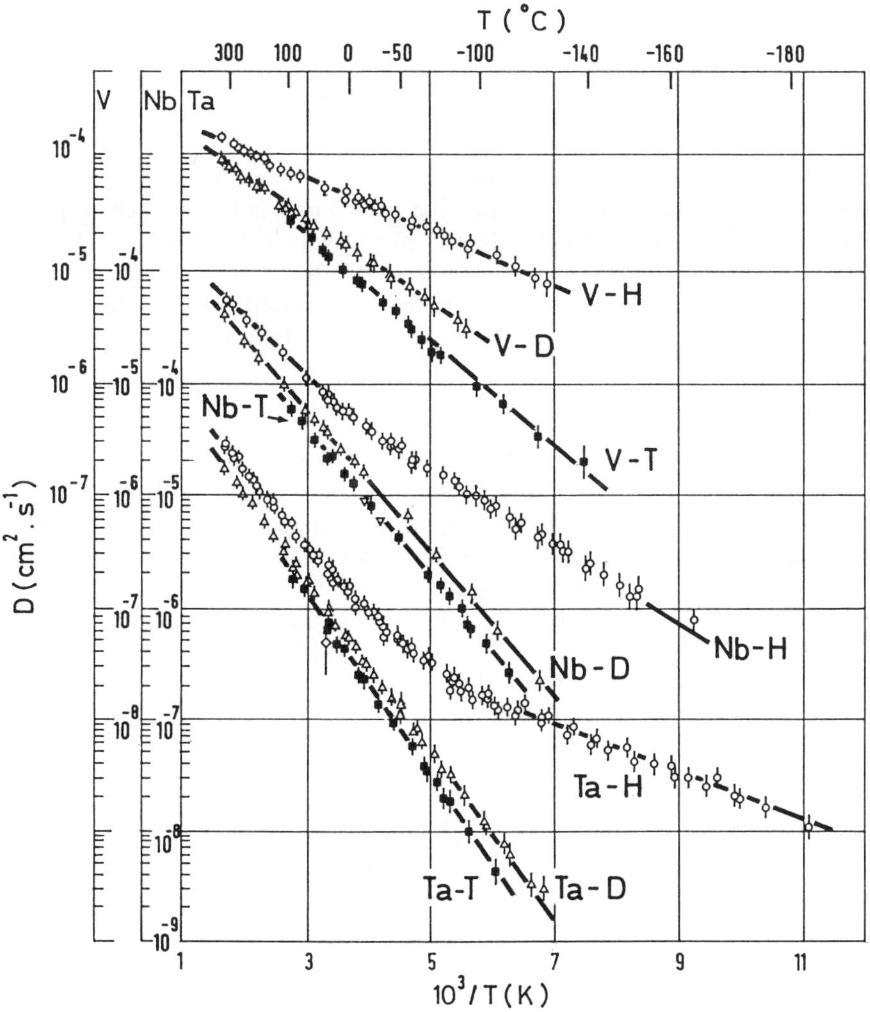

Fig. V.10. — Arrhenius plot for the diffusion of hydrogen and its isotopes deuterium (D) and tritium (T) in transition metals of group V. Compilation by Qi *et al.* (1983).

quency of the interstitial. According to equation (V.23), $\nu_T < \nu_D < \nu_H$; with decreasing temperature, quantum effects first show themselves for the proton. Below about 300 K, a progressive transition to one or more mechanisms with very low activation energy appears: transition between two states (i.e., two neighboring sites) by photon-assisted tunneling (Fig. V.11). As the transition appears at lower temperature for D, and does not take place at all for T because its mass is too great, anomalies of the isotope effect at low temperature can indeed be understood. (See also Exercise 37).

Coherent diffusion by tunneling (delocalized states, Bloch waves) is possible only at very low temperature. It has been observed for protons trapped by impurities (oxygen) in niobium; it is characterized by the complete absence of thermal activation (Fig. V.12). See Exercise 37.

The quantum effects are even more evident with the muon, owing to its very low mass. Below 50 K it can diffuse freely by tunneling, as an electron does. Such coherent diffusion is limited by thermal vibrations at higher temperatures; D thus decreases with increasing temperature (Fig. V.12). At still higher temperatures, the mean free path of the muons becomes equal to the distance between sites, and the transition between sites takes place by photon-assisted tunneling (the "small polaron" model which describes the distortion of the site by the muon). D then increases with temperature. Finally, at high temperature, all possible states are thermally excited and the muon jumps from site to site hopping over the potential barriers by thermal activation (Fig. V.13).

Trapping. — Hydrogen or deuterium can be trapped in certain sites. In general, the traps are, in order of increasing interaction energy, interstitials ($\lesssim$ 0.1 eV), dislocations ($\approx$ 0.2 eV), grain boundaries ($\approx$ 0.3 eV), interphase interfaces (precipitates, inclusions, $\approx$ 0.3 to 0.55 eV), surfaces. At 25°C, the solubility in iron can reach 1 cm^3 H_2 at NTP per cm^3, but the "ideal" solubility under 1 atm. would be about 5×10^{-5} cm^3/cm^3. On a well annealed, very high purity iron, D at 25°C is measured to be 3 to 7×10^{-5} $cm^2\ s^{-1}$, while on steels the measured values range from 10^{-5} to 10^{-12} $cm^2\ s^{-1}$. Trapping must be taken account of in analyzing the measurements. In Oriani's model (1970), an equilibrium is assumed between the free and the trapped hydrogen, the latter not participating in the macroscopic diffusion process. Then the measurement yields $D_{\text{apparent}} < D_{\text{pure metal}}$ (see Ch. I, § VI, Ex. 2.) so that trapping gives rise to an Arrhenius plot convex upwards. The agreement of this model with experiment is not quantitative (several types of traps? jump frequency modified by the traps?...See also Exercise 31).

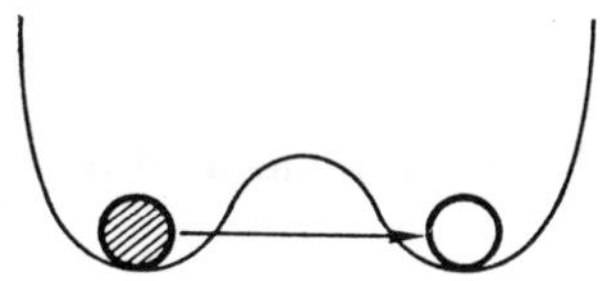

Fig. V.11. — Migration by tunneling.

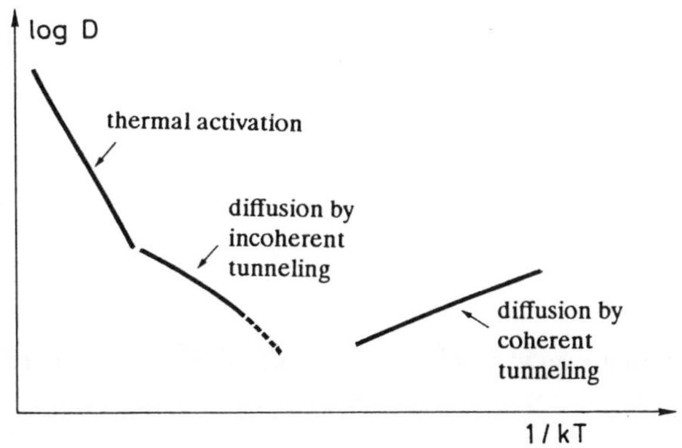

Fig. V.12. — Diffusion of a light interstitial at very low temperature.

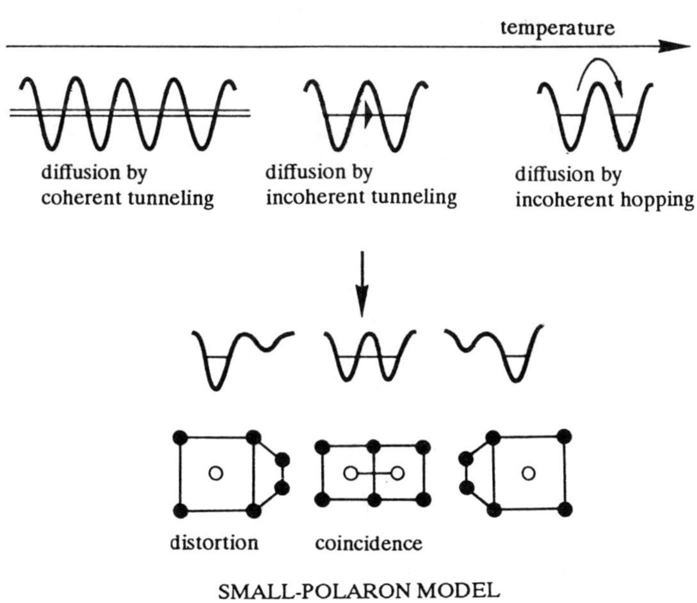

Fig. V.13. — Diffusion of a muon in a perfect crystal, according to Chappert (1979).

IV. Ionic crystals.

A distinction is made here between homo- and aliovalent solutes.

IV.1 DIFFUSION OF HOMOVALENT SOLUTES. — Few systematic measurements have been carried out, perhaps because there is no simple theory as in the case of normal metals. Among the correlations invoked to systematize the results, size effects and the polarizability of the ions should be mentioned.

The Arrhenius plot is similar to the one for self-diffusion, with an intrinsic and an extrinsic region. As for self-diffusion, the difference of the enthalpies of activation for these two regions is equal to the enthalpy of formation, if the degree of association is negligible. In effect, the solute diffusion coefficient is given by:

$$D_{\mathrm{B}^*} = \beta \cdot a^2 \, f_2 \, w_2 \, N_v .$$

In the intrinsic region, $N_v = \bar{N}_v \exp(-\delta H^{\mathrm{f}}/kT)$, where δH^{f} is the enthalpy for vacancy-impurity association. In the extrinsic region, $N_v = (1/\alpha) N_{\mathrm{D}} \exp(-\delta H^{\mathrm{f}}/kT)$, where N_{D} is the atomic fraction of cationic dopant of relative charge α. It follows that:

$$\begin{aligned} D_{\mathrm{B}}^{\mathrm{int}}/D_{\mathrm{B}}^{\mathrm{ext}} &= \bar{N}_v \, (N_{\mathrm{D}}/\alpha) \\ Q^{\mathrm{int}} - Q^{\mathrm{ext}} &= \Delta H_v^{\mathrm{f}} \end{aligned} \tag{V.25}$$

For an NaCl-type crystal, $\Delta H_v^{\mathrm{f}} = 1/2 \, \Delta H_{\mathrm{s}}^{\mathrm{f}}$, the formation enthalpy of a Schottky defect.

IV.2 DIFFUSION OF HETEROVALENT SOLUTES. — Owing to the association of a heterovalent cation with the defects introduced for charge compensation, a strong variation of the solute diffusion coefficient with solute concentration is expected. Figure V.14 shows this effect as a calculated by Howard and Lidiard. The strong variation of $D_{\mathrm{B}^*}^{\mathrm{A}}$ with c_{B} at low concentrations results from the fact that the impurity cannot diffuse unless a vacancy is on a nearest-neighbor site, from which the flux of impurity-vacancy pairs is given by:

$$J_{\mathrm{B}} = -D_{\mathrm{B}^*} \, \nabla c_{\mathrm{pairs}}. \tag{V.26}$$

It is only when the degree of association is equal to unity that $D_{\mathrm{B}} = (a^2/3) \, f_2 \, w_2$ (recall that the lattice parameter is $2a$, cf. Eq. (V.1)). To calculate D_{B}, the flux equations must be written with the coupling by the Nernst field, since the difference between the fluxes of B and of vacancies produces a space charge and a local electric field (see Ch. VI). One finds:

$$D_{\mathrm{B}^{\bullet}}^{\mathrm{A}} = \frac{2}{3}\, a^2\, w_2\, f_2\, \frac{p}{p+1},$$

where p is the degree of association.

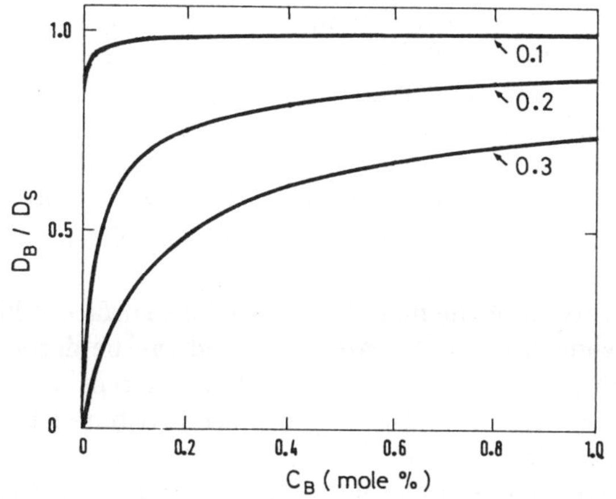

Fig. V.14. — Theoretical variation of the solute diffusion coefficient D_{B} with concentration for a divalent cation in a NaCl-type crystal, for 3 values of the temperature, (given as the enthalpy of association/kT). After Howard and Lidiard (1964).

The penetration profiles do not follow a simple analytical law because the diffusion coefficient varies with the concentration. The value of D at saturation ($p \to 1$), D_s, follows an Arrhenius law, the slope of which gives the enthalpy of migration because $D_s = (1/3)\, a^2\, f_2\, w_2$ (we neglect the temperature dependence of f_2). If $p < 1$, the Arrhenius plot is curved.

V. Semiconductors.

Our discussion is limited to the elemental semiconductors Si and Ge. Two kinds of specific effects are observed, due to 1) doping by the diffusing species itself, and 2) the very low density of point defect sources and sinks in the bulk, leading to the dominant effects of the surfaces. The reader should also see the review by Fahey *et al.* (1989). The technologically important problem of dopant diffusion in III-V compounds, especially GaAs, is more complicated because of the presence of two lattices and anti-site defects (Stolwijk, Perret nd Mehrer (1988), Deppe and Holonyak (1988), Tan and Gösele (1988)). See the reviews on III-V and II-VI compounds by Sharma (1989) and by Tuck (1988).

V.1 SUBSTITUTIONAL SOLUTES. — These are the elements of columns III and V of the periodic table (B, Al, Ga, and In from the former, P, As, and Sb from the latter) which have been thoroughly studied because of their role as dopants. Owing to the nature of their chemical bonding, it is natural to think that these atoms dissolve substitutionally, and diffuse by the same point-defect mechanisms as self-diffusion. These impurities diffuse faster than self-diffusion in Si and Ge, $D^{A}_{B^\bullet} > D^{A}_{A^\bullet}$ and $\Delta Q = Q_{B^\bullet} - Q_{A^\bullet} < 0$. In Si, $\Delta Q \approx -1$ eV. Diffusion in an intrinsic base material is in fact ambipolar diffusion (see Ch. VI, § III.2), since the diffusion of the dopant atom is accompanied by the diffusion of the charge carriers formed (which have much higher mobility); this gives rise to an internal (Nernst) electric field, which has as an effect the doubling of the diffusivity of the dopant (cf. Eq. (VI.51)).

If the concentration in the neighborhood of the surface is high enough, the diffusion profiles show a *characteristic break*, and can be decomposed into two independent depth profiles (Fig. V.15) with two different diffusion coefficients and activation energies. These two regimes are attributed to the *two charge states of the vacancy* (cf. Ch. IV, § IX). In effect, the solute behaves like a dopant that displaces the Fermi level; its position in the band gap with respect to the possible charge state of the vacancy varies with the depth of penetration x in such a way that the charge on the vacancies can be changed. This can in turn change the vacancy-impurity interaction energy, and eventually lead to different enthalpies of migration.

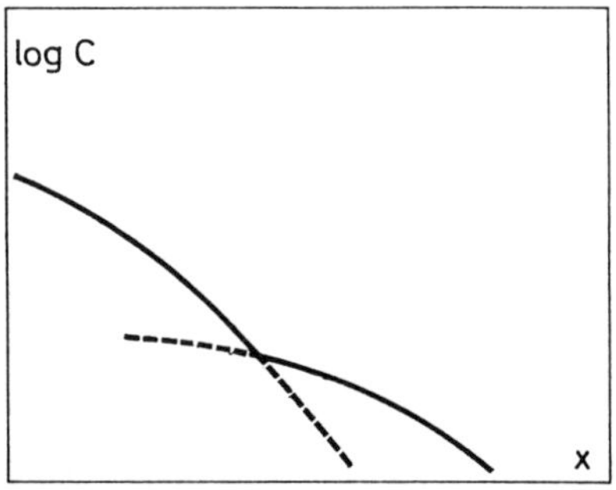

Fig. V.15. — Depth profile for a substitutional solute in silicon (schematic).

The depth profiles of phosphorous in silicon, for example, show a very strong *dependence on its surface concentration*. If this is less than the concentration of the intrinsic charge carriers, the profile is "normal". When it is increased, the profile displays an inflection point, followed by a long tail extending well into the interior of the sample. For still higher surface concentrations, a plateau appears near the surface. Analogous effects are observed for boron and arsenic, and are attributed to the predominance of doubly ionized vacancies in the near-surface (most highly doped) region, with singly ionized or neutral vacancies towards the center. The depth profiles are interpreted on the basis of three diffusion coefficients.

Other effects, occasionally quite curious, are observed, especially during the simultaneous diffusion of two dopants of opposite sign. During the preparation of n-p-n devices, the diffusion of phosphorus at high concentration to make up the emitter is accompanied by an advance of the base/collector junction which can then occur at a depth of several micrometers (the "emitter-push effect", Fig. V.16). These effects are a mark of a strong supersaturation of point defects, probably self-interstitials generated by the high concentration of phosphorus. This supersaturation would explain the enhanced diffusion of boron and arsenic, whereas the corresponding undersaturation of vacancies would cause a lower diffusivity of antimony (an element supposed to diffuse via vacancies).

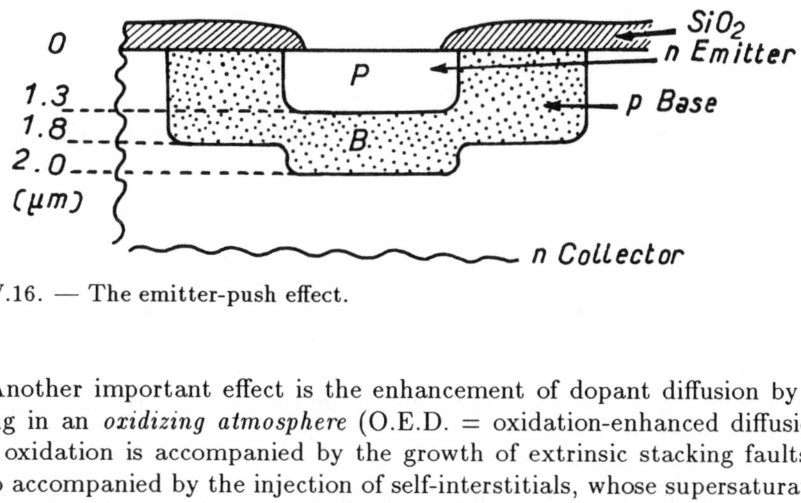

Fig. V.16. — The emitter-push effect.

Another important effect is the enhancement of dopant diffusion by annealing in an *oxidizing atmosphere* (O.E.D. = oxidation-enhanced diffusion). Since oxidation is accompanied by the growth of extrinsic stacking faults, it is also accompanied by the injection of self-interstitials, whose supersaturation can exceed a factor of ten. The self-interstitials exchange with the dopant atoms, creating interstitial dopant atoms, which diffuse very rapidly. The enhancement depends on the rate of oxidation, and hence on the orientation of the sample surface.

V.2 Interstitial impurities. — A number of elements diffuse in Si and Ge with very high diffusivities and low activation energies (0.4–1.2 eV). Among these are hydrogen, the alkali metals, the noble metals, Fe, Ni... Noble metals dissolve either as interstitials (e.g. Cu in Ge and Si) or partly interstitially, partly substitutionally (e.g., Au in Si is mainly substitutional).

The difference from self-diffusion can be spectacular: a factor of 10^9 in the case of Cu in Ge at 800°C. Gold and silver are fast diffusers in silicon ($D \approx 7 \times 10^{-5}$ cm^2 s^{-1} at 1200°C); on a single-crystal wafer of Si, coated on one face with Au, the depth profile of gold after diffusion is U-shaped (Fig. V.17), with an accumulation of gold near each face. The effect of crystalline perfection is here very important: accumulations of the diffusing element similar to those near the surface are found around dislocations.

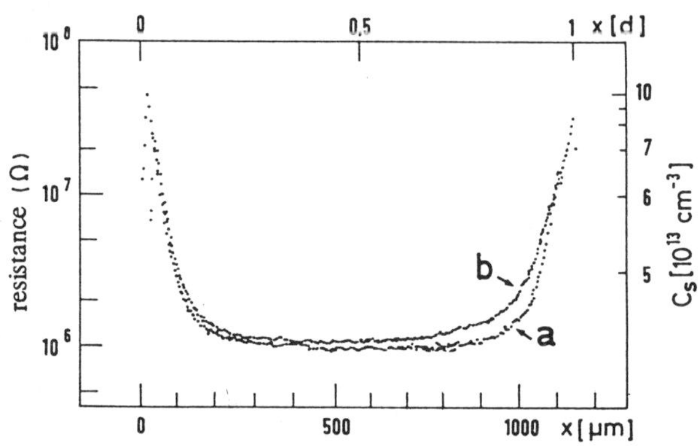

Fig. V.17. — Depth profile of (substitutional) gold in a wafer of silicon. Diffusion annealed 5 h at 1100 K. a) Au deposited initially on both faces of the wafer; b) deposited on the left face only. "d" denotes the thickness of the wafer. After Frank *et al.* (1981).

Diffusion this rapid must necessarily involve ***interstitials***. By this, we do not mean small atoms which dissolve interstitially and which diffuse by the direct interstitial mechanism, a hypothesis confirmed by measurements of the isotope effect ($E = 0.94$ for ^{6}Li/^{7}Li in Si). On the contrary, the mechanism for the noble metals or transition metals must be different. Frank and Turnbull (1956) have proposed the following model: the fast-diffusing solute B is dissolved in the solvent A interstitially ***and*** substitutionally, according to a *local* equilibrium reaction:

$$B_A \rightleftharpoons V_A + B_i \qquad \text{(I)}$$

between the substitutional solutes B_A, the interstitial solutes B_i, and the vacancies V_A. If the dominant defect in the solvent is not the vacancy but the self-interstitial, the "kick-out" reaction of Frank and Seeger is more likely to describe the equilibrium:

$$A_A + B_i \rightleftharpoons B_A + A_i. \qquad \text{(II)}$$

The analyses of these mechanisms are quite similar; they are based on the system of balance equations for the different types of defects, to which the local equilibrium reaction corresponding to equation I or II is added. In case I:

$$\frac{N_i\ N_v}{N_s} = K(T), \tag{V.27}$$

with $N_i < N_s$ and where N_i, N_v, and N_s indicate the mole fractions of interstitial solute, vacancies, and substitutional solute, respectively. If the crystal contains no dislocations, the B atoms dissolve at the surface interstitially, and penetrate rapidly in this form into the crystal, until they are trapped by vacancies, according to the reaction $\mathrm{B_i + V_A \longrightarrow B_A}$, and thus become much less mobile. Thus in the absence of internal sources, there is a reduction of the vacancy concentration at thermal equilibrium. The crystal "attempts" to remedy this situation by injection of vacancies from the surface. Owing to the draining away of the vacancies, the newly introduced interstitials penetrate very deeply into the materials, at least until they reach the neighborhood of a source of vacancies, a surface or a dislocation. Then they become substitutional and accumulate locally to the extent that vacancies are produced. This mechanism is responsible for the large concentrations of solute observed near surfaces (Fig. V.17) or dislocations after diffusion; it is easy to understand that the calculation of the concentration profiles can become very complicated.

If the point defects (e.g., vacancies) are at equilibrium everywhere, owing to a high enough density of sources, then $N_v = \bar{N}_v$ (the bar denotes thermal equilibrium) and equation (V.27) becomes for this situation:

$$\frac{\bar{N}_i}{\bar{N}_s} = \frac{K(T)}{\bar{N}_v}. \tag{V.28}$$

If, on the contrary, the dislocation density is small, then $N_v < \bar{N}_v$ and the concentration of vacancies decreases during the penetration of the interstitials. The under-saturation of vacancies leads to a high concentration of interstitial impurities:

$$\frac{N_i}{N_s} > \frac{\bar{N}_i}{\bar{N}_s}.$$

The measured diffusion coefficient is then practically that of the interstitials:

$$D_{\mathrm{meas}} \approx D_i. \tag{V.29}$$

If, however, the defects are everywhere in equilibrium, solution of the balance equations yields:

$$D_{\mathrm{meas}} = D_{\mathrm{eff}} = \frac{\bar{N}_i}{\bar{N}_i + \bar{N}_s}\, D_i. \tag{V.30}$$

The depth profiles are made up of two parts, which could be mistaken for a volume diffusion part followed by a "tail" due to short-circuit diffusion along grain boundaries or dislocations (see Ch. VII). The interpretation here is, however, different: the "tail" of the curve exists even in a perfect crystal. If the density of dislocations increases, the tail becomes more and more flat, finally resembling a plateau. Figure V.18 shows profiles calculated for a perfect crystal, assuming that $D_i >> D_v$.

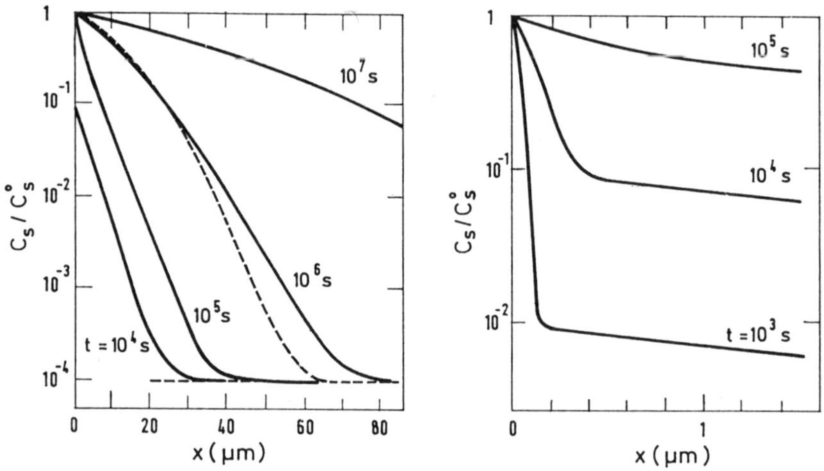

Fig. V.18. — Dissociative mechanism. Depth profiles calculated by Edelin (1980) with: $D_s = 10^{-14}$ cm^2 s^{-1}, $D_v = 10^{-8}$ cm^2 s^{-1}, $[B_A] = 10^{-2}$, $B_A \underset{k_2}{\overset{k_1}{\rightleftharpoons}} B_i + V_A$, $k_1 = 10^{-5}$ s^{-1}, $k_3 = 0$ (perfect crystal, cf. Eq. (V.33)). Left: Dislocation-free crystal. Dashed line: erfc(x). Right: blowed up near the surface.

The diffusion equations. — The appropriate system of equations to be solved for an analysis of the Frank-Turnbull mechanism is as follows. If k_1 and k_2 are the constants for local equilibrium:

$$B_A \underset{k_2}{\overset{k_1}{\rightleftharpoons}} B_i + V_A. \tag{I}$$

For substitutional solutes (⁴):

$$\frac{\partial N_s}{\partial t} = \frac{\partial}{\partial x}\left(D_s \frac{\partial N_s}{\partial x}\right) - k_1\, N_s + k_2\, N_i\, N_v. \tag{V.31}$$

(4) D_s is proportional to N_v and thus depends on the penetration depth,

For interstitial solutes:

$$\frac{\partial N_i}{\partial t} = D_i \frac{\partial^2 N_i}{\partial x^2} + k_1\ N_s - k_2\ N_i\ N_v. \tag{V.32}$$

Finally, for the vacancies:

$$\frac{\partial N_v}{\partial t} = D_v \frac{\partial^2 N_v}{\partial x^2} + k_1\ N_s - k_2\ N_i\ N_v + k_3 \left(\bar{N}_v - N_v\right), \tag{V.33}$$

where the last term on the right hand side is the production of vacancies by sources in the volume (dislocations).

If the dislocation density is very high, $N_v = \bar{N}_v$ and $\partial N_v / \partial t = 0$. Adding equations (V.31) and (V.32), we obtain:

$$\frac{\partial \left(\bar{N}_i + \bar{N}_s\right)}{\partial t} = D_i \frac{\partial^2 \bar{N}_i}{\partial x^2} + D_s \frac{\partial^2 \bar{N}_s}{\partial x^2}. \tag{V.34}$$

In general, $\bar{N}_s > \bar{N}_i$; the measured concentration profile thus represents atoms in substitutional sites, even though the transport takes place via interstitials.

According to the equation for local equilibrium (V.28), $\bar{N}_s = \alpha\ \bar{N}_i$ at constant temperature, where α is a constant. Then:

$$\frac{\partial \bar{N}_s}{\partial t} = D_{\text{eff}} \frac{\partial^2 \bar{N}_s}{\partial x^2}, \tag{V.35}$$

$$D_{\text{eff}} = \frac{\bar{N}_s}{\bar{N}_i + \bar{N}_s} D_s + \frac{\bar{N}_i}{\bar{N}_i + \bar{N}_s} D_i, \tag{V.36}$$

an expression that reduces to equation (V.30) if $\bar{N}_s\ D_s << \bar{N}_i\ D_i$.

VI. Dilute alloys.

VI.1 Effect of the solute concentration. — In a number of dilute alloys, the self-diffusion coefficient varies with the solute concentration according to the simple relation:

$$D_{\mathrm{A}^*}^{\mathrm{AB}} = D_{\mathrm{A}^*}^{\mathrm{A}} \cdot \exp\left(b N_{\mathrm{B}}\right), \tag{V.37}$$

which can be approximated by:

$$D_{\mathrm{A}^*}^{\mathrm{AB}} = D_{\mathrm{A}^*}^{\mathrm{A}} \left[1 + b_1\ N_{\mathrm{B}} + b_2\ N_{\mathrm{B}}^2 + \cdots\right], \tag{V.38}$$

and often for N_B not greater than 10^{-2},

$$\boxed{D_{A^*}^{AB} = D_{A^*}^{A} \, [1 + b_1 \, N_B] \, .} \tag{V.39}$$

The term $b_1 \, N_B$ expresses the change in the jump frequencies of the solvent atoms neighboring the solute atom B; the term $b_2 \, N_B^2$ corresponds to the effect of B atom pairs. An analogous expression is observed for the diffusion of the solute B atoms:

$$D_{B^*}^{AB} = D_{B^*}^{A} \left[1 + B_1 \, N_B + B_2 \, N_B^2 + \cdots\right] , \tag{V.40}$$

but the B_1 term expresses the effect of solute atom pairs. Thus in the approximation where $b_2 = 0$, also $B_1 = 0$. The addition of solute has a stronger effect on self-diffusion than on solute diffusion if B-B pairs are negligible. Conversely, if D_B^* varies with the addition of B, the solute atom pairs are not negligible and equation (V.39) is not sufficient to describe the effect of solute on self-diffusion: b_2 is not negligible. The experimental values of b_1 and B_1 usually have the same sign.

Certain solutes accelerate the diffusion of the solvent $(b_1 > 0)$, others slow it down $(b_1 < 0)$. One would predict that b_1 is related to the various jump frequencies in the neighborhood of the solute; it is possible to calculate b_1 within the framework of the five frequency model for the FCC lattice described above for the calculation of f_2 (Howard and Manning, 1967).

The calculation of the total number of jumps of a B atom in an FCC lattice leads to a verification of the linear formula (V.39) with:

$$\begin{aligned} b_1 &= -18 + \frac{w_4}{w_0}\left[4\frac{\chi_1}{f_0}\,\frac{w_1}{w_3} + 14\frac{\chi_2}{f_0}\right] \\ &= -18 + \left[4\frac{\chi_1}{f_0}\,\frac{w_1}{w_0} + 14\frac{\chi_2}{f_0}\,\frac{w_3}{w_0}\right] \exp\left(-\delta g^{\mathrm{f}}/kT\right) , \end{aligned} \tag{V.41}$$

since $w_4/w_3 = \exp\left(-\delta g^{\mathrm{f}}/kT\right)$. Here f_0 is the usual correlation factor for the w_0 jumps in the pure solvent, χ_1 and χ_2 are the partial correlation factors ([5]) which relate to w_1 jumps and $(w_3, \; w_4)$ jumps, respectively. These expressions are based on the assumption that the w_i are independent of concentration. Evidently the increase of the solvent diffusivity can stem from the increase in the vacancy concentration due to the presence of the solute (attraction, $w_4/w_0 > 1$) and/or the increase of the solvent jump frequencies near the solute $(w_1 > w_3)$. With the same approximations, the diffusivity of the solute is not affected by its concentration.

([5]) The factors χ_1 and χ_2 have been tabulated by Howard and Manning (1967).

With the aid of equations (V.3), (V.4), and (V.41), it is possible to express b_1 as a function of the ratio $D_B^A/D_{A^\bullet}^A$:

$$b_1 = -18 + \frac{D_B^A}{D_{A^\bullet}^A}\frac{4f_0}{1-f_2}\left[\frac{4\chi_1(w_1/w_3)+14\chi_2}{f_0(4w_1/w_3+14F)}\right], \tag{V.42}$$

a more useful expression for discussing the possible values of b_1. In effect, if $D_B^A > D_{A^\bullet}^A$, it is very likely that $b_1 > 0$; *fast diffusing solutes have a tendency to increase the diffusivity of the solvent*, an effect essentially due to the increase of the vacancy concentration. With $D_B^A/D_{A^\bullet}^A$ in the range 1 to 10 and $f_2 \approx 1/2$, b_1 lies between 0 and 100. For slowly diffusing solutes, b_1 tends to be negative, but never less than -18.

The calculations of b_2 and of B_1 are much more complicated, because of the much larger number of configurations involving solute pairs. Nevertheless, Bocquet (1973) has given a greatly simplified expression for FCC alloys:

$$B_1 = 18\left[\exp\left(-g^f/kT\right) - 1\right], \tag{V.43}$$

where g^f is the Gibbs free energy for vacancy-impurity binding. The factor B_1 has also been calculated for the enhancement of self-diffusion by interstitial solutes (Jones and Le Claire, 1972).

Analogous expressions, albeit less simple ones, have also been established for BCC solid solutions. With the notation of § II.2, these are:

BCC Model I:

$$b_1 = -20 + 14\frac{\mu_1}{f_0} + 6\frac{w_3}{w_0}\frac{\nu_1}{f_0}\exp\left(-\frac{\Delta g_1^f}{kT}\right),$$

with

$$w_0 = w_3^{'}\exp\left(-\Delta g_1^f/kT\right) = w_5\exp\left(-\Delta g_2^f/kT\right).$$

BCC Model II:

$$b_2 = -20 + 6\frac{\mu_2}{f_0} + 14\frac{w_3}{w_0}\frac{\nu_2}{f_0}\exp\left(-\frac{\Delta g_1^f}{kT}\right),$$

where $\mu_1,\ \nu_1\left(w_2/w_3^{'},\ w_3/w_3^{'}\right)$ and $\mu_2,\ \nu_2\left(w_2/w_3,\ w_4/w_0\right)$ are partial correlation factors which have been tabulated ([6]). As usual, f_0 is the correlation factor of the solvent, $= 0.727$.

([6]) The factors μ_1, μ_2, ν_1, and ν_2 have been tabulated by Jones and Le Claire (1972).

VI.2 Determination of the jump frequency ratios. — The various jump frequencies w_i enter into the expressions for both the correlation factor f_2 and the coefficient b_1. These two quantities can be determined experimentally, the first from the isotope effect for the diffusion of the solute, the second from the effect of solute on the solvent self-diffusion coefficient. Then for FCC metals:

$$f_2 = \frac{2\,w_1 + 7\,Fw_3}{2\,w_1 + 2\,w_2 + 7\,Fw_3}, \quad \text{with } F \equiv F(w_4/w_0) \tag{V.3}$$

$$b_1 = -18 + 4\frac{w_4}{w_0}\left[\frac{w_1}{w_3} + \frac{7}{2}\right] \tag{V.44}$$

or equation (V.41), which is more accurate than equation (V.44).

Further, in the pure solvent, according to (V.1) and (V.2):

$$\frac{D^{\mathrm{A}}_{\mathrm{B}\bullet}}{D^{\mathrm{A}}_{\mathrm{A}\bullet}} = \frac{w_4}{w_3}\,\frac{w_2}{w_0}\,\frac{f_2}{f_0}, \tag{V.45}$$

with f_0, the correlation factor for the pure solvent, $= 0.78$.

These three equations allow the calculation of the three jump-frequency ratios: w_3/w_1, w_2/w_1, and w_4/w_0 from three independent measurements: $D^{\mathrm{A}}_{\mathrm{B}\bullet}/D^{\mathrm{A}}_{\mathrm{A}\bullet}$ and f_2 in pure A, and b_1 in dilute AB alloys. The analysis is not completely rigourous, since the isotope effect measures $f\,\Delta K$ only if the correlation factor can be expressed in the form (V.9). Further, it must be assumed that $\Delta K_2 = \Delta K_0$, where for pure A, $E = f_0\Delta K_0$ and $f_0 = 0.78$ for an FCC structure. The latter hypothesis is valid only if the atomic masses of the solute and solvent do not differ too greatly. The difficulty can be avoided by using a fourth relation among the three jump frequency ratios and the ratio of the phenomenological coefficients $L_{\mathrm{AB}}/L_{\mathrm{BB}}$ which represents the "vacancy wind" effect. This effect manifests itself in several processes: 1) interdiffusion (Heumann, 1979); 2) flux of solute introduced by a supersaturation of vacancies (quenched or irradiated alloys (see Antony, 1975)); 3) displacement of inert markers during electromigration (Doan, 1972) (see Ch. VIII, § III.2, III.3, III.4).

Heumann's relation is particularly useful because no radiotracer is needed for the solute; the relation concerns experiments on interdiffusion. Heumann's relation is written:

$$\lim_{N_{\mathrm{B}}\to 0}\left(\frac{L_{\mathrm{AB}}}{L_{\mathrm{BB}}}\right) = \frac{D^{\mathrm{A}}_{\mathrm{A}\bullet}}{D^{\mathrm{A}}_{\mathrm{B}\bullet}}\left[\frac{1}{f_0} - \frac{\tilde{D}_{\mathrm{A}}(0)}{D^{\mathrm{A}}_{\mathrm{A}\bullet}}\,\frac{\bar{V}_{\mathrm{B}}}{\bar{V}_{\mathrm{A}}}\right] \tag{V.46}$$

where $\tilde{D}_{\mathrm{A}}(0)$ is the intrinsic diffusion coefficient of the solvent, extrapolated to pure A, and the $\bar{V}_i$ are the partial molar volumes. The phenomenological coefficients can be expressed as a function of the three jump-frequency ratios (see Ch. VIII, Eq. VIII.35).

A large number of FCC systems have been studied in detail (see, e.g., Bocquet *et al.* 1984). Some examples are given in table I. Since Zn diffuses more rapidly than the solvent, $w_2 > w_1$; and since Zn attracts vacancies, $w_4 > w_0$.

The self-consistency of the model can be verified because the correlation factor for the diffusion of the solute can be obtained from experimental data on dilute alloys:

$$f_2 = 1 - \frac{4f_0}{b_1 + 18} \cdot \frac{D_{B^\bullet}^{A}}{D_{A^\bullet}^{A}}, \tag{V.47}$$

an expression which can be deduced from equations (V.3), (V.41), and (V.45). (The value of $b_{\min}$ mentioned in the section on ultra-fast diffusers (§ II.5) is derived from this equation by setting $f_2 = 0$.) The correlation factor f_2 can also be calculated theoretically under the same assumptions as ΔQ (see § VI.1). Figure V.19 shows the variation of the correlation factor for the diffusion of Zn in Ag as a function of temperature, the experimental values having been determined from the isotope effect. The values calculated from the jump frequencies of table I and those calculated from the electronic theory based on Le Claire's model are in good agreement with experiment.

Unfortunately, the ratio w_4/w_3, from which the binding free energy can be obtained, cannot be extracted from the experimental results. According to Le Claire, w_4/w_3 refers to the jumps of a vacancy in opposite directions between two sites, whereas diffusion data can only yield jump frequency ratios for jumps leaving from the *same* site.

Table I. — *Jumps frequencies of a vacancy neighboring a* Zn *atom in* Ag *and* Cu.

(After Bocquet *et al.* (1984))

System	T(K)	$D_{B^\bullet}/D_{A^\bullet}$	f_B	b_1	b_T(*)	L_{AB}/L_{BB}	w_2/w_1	w_3/w_1	w_4/w_0
AgZn	1010	4.1	0.52	12.6			1.53	0.27	1.15
	1153	3.9	0.57	12.7			1.54	0.39	1.30
	1153	3.9		12.6	6		1.20	0.26	1.12
CuZn	1168	3.56	0.47	7.3			2.5	0.5	1.2
	1168	3.3		8		−0.22	3	0.5	1

(*) b_T measures the effect of solute on the rate of the displacement of the distribution of solvent tracers during self-electromigration.

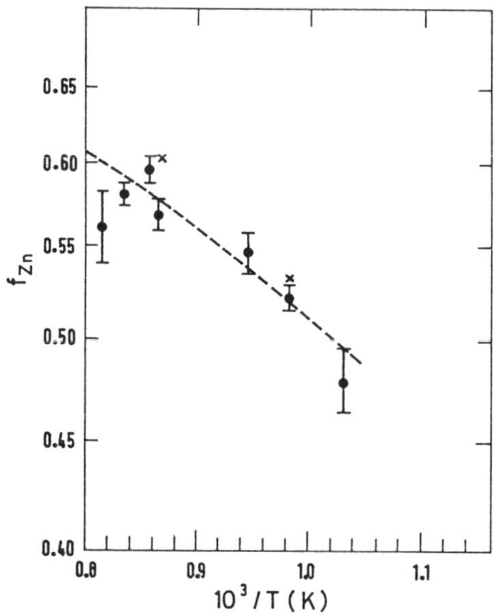

Fig. V.19. — Variation of the correlation factor for the diffusion of zinc in silver (solute diffusion at infinite dilution) from measurements of the isotope effect (ɸ) and calculated from experimental values of b according to equation (V.47) (X). After Rothman and Peterson (1967).

In the BCC structure, only two jump-frequency ratios are accessible:

w_3/w_3' and w_2/w_3' for model I,

w_2/w_3 and w_4/w_0 for model II,

(see above § II.3).

VI.3 The effect of substitutional impurities on the diffusion of interstitials. — This effect can be due to two causes :

– *site blocking* (excluded sites) when the impurities repel the interstitials. If the impurities are immobile, $D = D_{\text{pure}}\ (1 - \beta C_{\text{impurity}})$. For high enough concentrations of impurities, this becomes a percolation problem ;

– *trapping*, when there is a strong chemical attraction between the interstitial and the impurity. The measured diffusion coefficient is then an apparent diffusion coefficient smaller than the "true" one. (See Ch. I § VI, Eq. (I.29) and Exercise 32). The Arrhenius plot is curved (convex upwards).

VII. Diffusion in homogeneous concentrated alloys.

The experimental methods for studying self-diffusion in an elemental solid, such as the thin layer geometry, can equally well be applied to a concentrated alloy. Consider a binary alloy AB, into which the tracers A* and B* are diffused. In view of the small quantity of material in the tracers, the chemical composition of the system remains the same. Two tracer self-diffusion coefficients can be defined, $D_{A^*}^{AB}$ and $D_{B^*}^{AB}$, and the variation of these coefficients with composition can be determined by studying a series of alloys of varying composition. Figure V.20 shows a possible example of this, indicating the limiting values: on the side of pure A, $D_{A^*}^{AB}$ approaches $D_{A^*}^{A}$, the self diffusion coefficient of pure A, and $D_{B^*}^{AB}$ approaches $D_{B^*}^{A}$, the solute diffusion coefficient of B in pure A at infinite dilution, and similarly on the side of pure B. The curves $D_{A^*}(N_A)$ and $D_{B^*}(N_A)$ can take vastly different forms and can even cross each other.

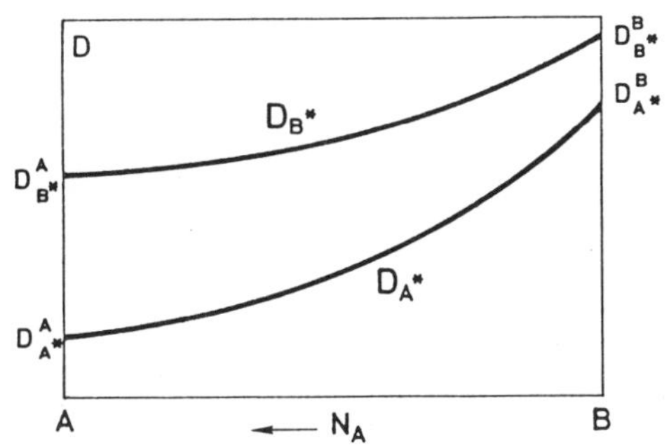

Fig. V.20 — Diffusion of tracers of the elements A and B in homogeneous binary AB alloys.

The mechanisms of diffusion via point defects also operate in alloys, but the theoretical analysis is much more complicated owing to the large number of possible jump frequencies and configurations to be considered, depending on the local order in the neighborhood of a vacancy. However, the tracer diffusion coefficients can always be expressed in the same simple form as in an elemental solid. For, say, A atoms (see Hehenkamp, 1983):

$$D_A = f_A \, p_A \, w_A \, a^2, \tag{V.48}$$

where w_A is an average jump frequency, p_A is a factor which measures the probability, averaged over all the possible configurations, of finding a vacancy as a nearest neighbor of an A atom (p_A is sometimes called the vacancy availability factor), and f_A is the correlation factor. At equilibrium, the degree of order of the solution directly determines p_A, and indirectly determines w_A. The correlation factors f_A and f_B can vary with temperature, and this variation can be very strong in the case of long-range order.

VII.1 Disordered alloys. — A simplified description for a random alloy with no vacancy binding to either chemical species has been given by Manning. The model is based on the vacancy's having only *two* jump frequencies, w_A and w_B, depending on whether the vacancy exchanges with an A or a B atom, and independent of the local configuration around the vacancy. Then the average jump frequency of the vacancy is given by:

$$w_v = N_A\, w_A + N_B\, w_B, \tag{V.49}$$

where N_A and N_B are the mole fractions of A and B, respectively.

Such a "mean field" model cannot be applied to dilute alloys, where the solute-vacancy interaction can be very strong. Since $w_A \neq w_B$, the motion of the vacancy is not perfectly random ("physical correlation").

With the effective vacancy jump frequency, w_v, the jump vector of the vacancy again becomes an axis of symmetry, so that the expression given in chapter III can be applied: the probability for an A atom to return to its initial site after a first exchange with a given vacancy is:

$$-t_A = \frac{w_A}{w_A + H}, \tag{V.50}$$

where H is the probability that the vacancy definitely escapes from A. Since (see Ch. III):

$$f_A = \frac{1 + t_A}{1 - t_A},$$

equation (V.50) becomes:

$$f_A = \frac{H}{2\, w_A + H}, \tag{V.51}$$

and an analogous expression for f_B. Manning has shown that:

$$\begin{aligned} H &= \frac{2}{1 - f_0}\,(w_A\, N_A\, f_A + w_B\, N_B\, f_B) \\ &= M_0\, w_v\, f_v, \end{aligned} \tag{V.52}$$

where f_0 is the correlation factor for diffusion in the pure metal, f_v the correlation factor for the diffusion of vacancies and:

$$M_0 = \frac{2\, f_0}{1 - f_0}.$$

Implicit equations for f_A and f_B can be deduced from equations (V.51) and (V.52); these can then be calculated numerically as functions of w_A/w_B.

The values of f_A and f_B can also be obtained from measured values of

D_{A^*} and D_{B^*}. Knowing that:

$$D_{A^*} = \frac{1}{6} \zeta \lambda^2 N_v w_A f_A,$$

equations (V.51) and (V.52) lead to:

$$f_A = \frac{(M_0 + 2)(N_A D_{A^*} + N_B D_{B^*}) - 2 D_{A^*}}{(M_0 + 2)(N_A D_{A^*} + N_B D_{B^*})}, \qquad (V.53)$$

with M_0 = constant (7.15 in FCC, 5.33 in BCC).

An identical expression with $-2 D_{B^*}$ substituted for $-2 D_{A^*}$ gives f_B. If $D_{A^*} = D_{B^*}$, then $f_A = f_B = f_0$ (pure metal). If $D_{A^*} > D_{B^*}$, then $f_A < f_0$ and $f_B > f_0$; because of correlation effects, the ratio of the self-diffusion coefficients is less than the ratio of the jump frequencies w_A/w_B:

$$\frac{D_{A^*}}{D_{B^*}} = \frac{f_A}{f_B} \frac{w_A}{w_B}. \qquad (V.54)$$

In a disordered β-CuZn solid solution, the ratio of the isotope effects $f_A \Delta K_A / f_B \Delta K_B$ is in very good agreement with the ratio of the f's calculated from D_{A^*} and D_{B^*} according to equation (V.53), which indicates that $\Delta K_A \approx \Delta K_B$. The agreement is equally good for α-CuZn with 30% Zn, but it is not good, as could be expected, for 4% Zn (Peterson and Rothman, 1967, 1978). Manning's model is in equally excellent agreement with simulations on disordered BCC and FCC lattices.

VII.2 ORDERED ALLOYS. — The lattice of a CsCl-*type ordered alloy (B_2 structure)* can be considered as made up of two simple cubic sublattices α and β: the nearest neighbors of an α site are β type and vice-versa, so that the random movement of a vacancy is not possible without destruction of local order, i.e., the creation of antisite defects A_B and B_A (7).

If the formation energy of these defects is too high, especially if the degree of order is close to unity, several mechanisms are possible:

– Jumps to second nearest-neighbor sites; the diffusion of A atoms is limited to the α sublattice, that of B atoms to the β sublattice. This is the case for ionic crystals and for simple oxides; the preceding analysis for self- and solute

(7) Deviations from stoichiometry are accomodated either by antistructure or antisite defects (e.g., β-CuZn, AgCd...), or by structural vacancies (their energy of formation being different for each sublattice), the departure from stoichiometry being given by $[V_A] - [V_B]$, or by a mixture of both types: A_B dominant in alloys with an excess of A, V_A in alloys deficient in A (e.g., NiAl). Here the name "antistructure" is used for a pair of defects $A_B + B_A$ in a stoichiometric alloy, whereas "antisite" is used to mean a single defect like A_B.

diffusion in an elemental substance can be applied directly to each sub-lattice, considered as an independent entity.

– Complex jump cycles, such as the Huntington-McCombie-Elcock cycle: 6 consecutive jumps of the vacancy between nearest neighbor sites, such that at the end of the cycle, the vacancy is displaced a given distance and order is reestablished (Fig. V.21). Monte Carlo simulations confirm this mechanism, with strong physical correlation effects (incomplete cycles, and/or returning backwards). Their contribution to the mean square displacement can be very large ($\leq 50\%$) for a completely ordered alloy.

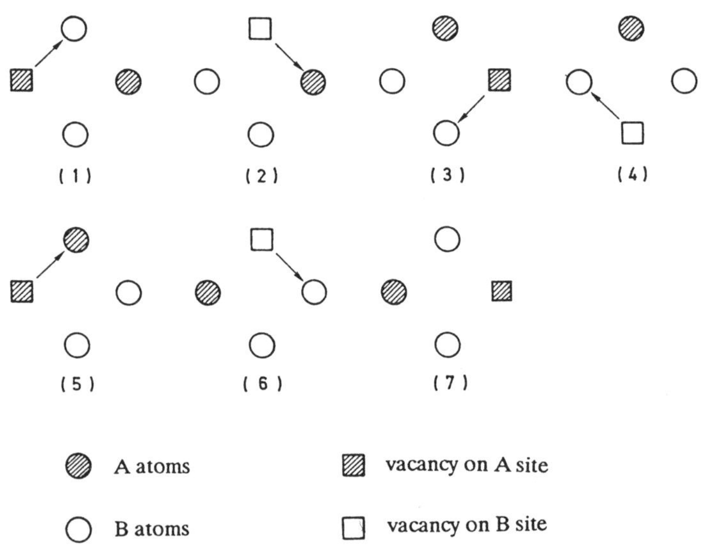

Fig. V.21. — Huntington-Elcock-McCombie cycle in a binary ordered alloy.

– *Vacancy pairs* $V_A - V_B$ (one vacancy on each sublattice) allow two successive correlated jumps, each one of length equal to the nearest-neighbor distance.

– Jumps due to *triple defects* $A_B + 2V_A$. At the stoichiometric composition, vacancies and anti-site defects can associate to form defects $A_B + 2V_A$. The migration of this complex involves vacancy jumps between nearest-neighbor sites for A atoms, and between second nearest-neighbor sites for B atoms (Fig. V.22). The isotope effects for the diffusion of A and of B should then vary in a different manner with composition.

In a general way, the diffusion coefficients of the two components can be expected not to differ greatly. For a six-atom ring in the B_2 structure, Elcock and McCombie (1958) obtained:

$$\frac{D_{A^\bullet}}{D_{B^\bullet}} = \frac{[V_\alpha]\, w_6^\alpha + 2\,[V_\beta]\, w_6^\beta}{2\,[V_\alpha]\, w_6^\alpha + [V_\beta]\, w_6^\beta} \tag{V.55}$$

where $[V_\alpha]$ and $[V_\beta]$ are the mole fractions of vacancies on the α and β sublattices, and $w_6^{(\alpha,\,\beta)}$ the frequencies of finishing a complete cycle beginning on an α or a β site, respectively. From equation (V.55) it is clear that $1/2 \le D_{A^\bullet}/D_{B^\bullet} \le 2$; there is no requirement that $D_{A^\bullet} = D_{B^\bullet}$.

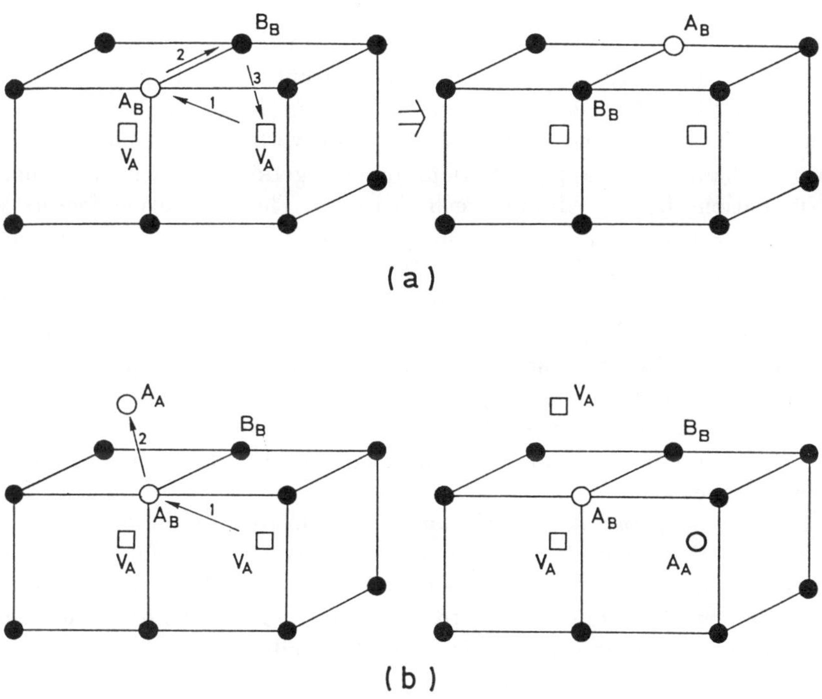

Fig. V.22. — Diffusion by triple defects 2 V_A+A_B in ordered B_2-type (CsCl) compounds. a) Migration of A and B (2nd-neighbor jumps). b) Migration of A only (nearest-neighbor jumps only).

The diffusion coefficients in the ordered state can also be expected to be smaller than those in the disordered state, and, at least for the ring type mechanisms, a smaller correlation factor is also expected. The first of these is convincingly shown in β-CuZn, where there is a break in the Arrhenius plot at $T = T_c$, with $Q_{dis} < Q_{ord}$. The self-diffusion coefficients also pass through a minimum at the stoichiometric composition. Measurements of the isotope effect in β-CuZn and FeCo have not yielded particularly low values ($f\ \Delta K = 0.325$ for β-CuZn), but as a matter of fact, the measurements were carried out on incompletely ordered alloys, which would perhaps not require the Huntington-Elcock-McCombie mechanism. The value of f is expected to remain approximately constant in the disordered alloy, and to decrease below T_c; the higher the ordering energy, h, [8] the greater the decrease in f. The difficulty of verifying this experimentally stems from the fact that the diffusivities are very small at the temperatures where the degree of order is high. Note also that recent Monte-Carlo simulations (Zhang *et al.* 1989) indicate that the isotope effect $\neq f\Delta K$ for diffusion in ordered alloys.

Notwithstanding these results, other mechanisms, especially second-nearest neighbor jumps, remain probable. In principle, what distinguishes an intermetallic compound from an ordered alloy is the independence of the sub-lattices. But this distinction is much more conceptual than operational.

Manning's model has been extended (Bakker (1979) and Stolwijk (1981)) to long- and short-range order, in the framework of the nearest-neighbor (Bragg-Williams) pair approximation, to give good agreement with numerical simulations by the Monte Carlo method. The correlation factors were calculated as a function of composition, reduced temperature h/kT and an interaction factor $U = (\Phi_{BB} - \Phi_{AA})/2h$.

In *compounds of the* $L1_2$ *structure*, the situation is simpler, at least for the A atoms. With the formula A_3B, each A atom has 8 A atoms (on a BCC lattice) and 4 B atoms as nearest neighbors; A atoms can thus diffuse by nearest neighbor jumps without creating antistructure defects, e.g., Ni in Ni_3Al. This is not the case for the diffusion of B atoms, which have 12 A nearest neighbors; $D_{B^*} << D_{A^*}$ is expected (known as the Cu_3Au rule), but so far no experiment has been carried out.

The DO_3 *structure* is somewhat more complicated, as there are three sub-lattices. However, departures from stoichiometry are frequent in these compounds and diffusion of the A atoms (the formula is A_3B) by nearest neighbor vacancy jumps is possible via constitutional vacancies. The activation energy for diffusion then reduces to that for migration only.

---

[8] The ordering energy, h, is given in a nearest-neighbor interaction model by:

$$h = \Phi_{AB} - \frac{1}{2}(\Phi_{AA} + \Phi_{BB}),$$

where the Φ_{ij} are the interatomic pair potentials.

VIII. Superionic conductors.

Superionic conductors (improperly called ionic superconductors) are compounds which have an ionic conductivity 0.1 to 1 $(\Omega \cdot \mathrm{cm})^{-1}$ in the solid state ([9]). The activation energy for conductivity and diffusion can be very low, as low as 0.1 eV.

Among these materials, certain ones display a first-order transition between a superionic phase at high temperature and a normally conducting phase at low temperature (Fig. V.23), whereas others are superionic at all temperatures. The "super" conductivity can be, depending on the material, either cationic or anionic in character. As a general rule, these phases have an *open crystal structure* which allows the rapid migration of "interstitial" ions.

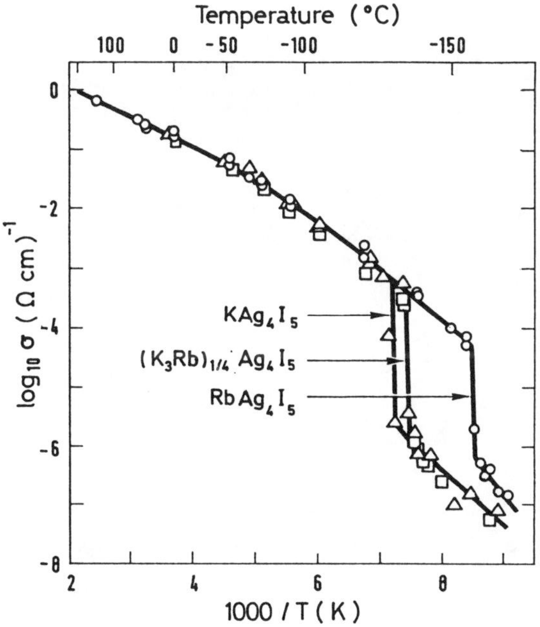

Fig. V.23. — Variation of the electrical conductivity with T^{-1}, showing the phase transition in compounds derived from AgI. After Lazarus (1976).

([9]) Points of comparison for the conductivity:
Cu at 300 K: 6×10^5 $(\Omega \cdot \mathrm{cm})^{-1}$;
typical ionic crystal near T_M : 10^{-4} $(\Omega \cdot \mathrm{cm})^{-1}$;
0.1 N aqueous solution of NaCl: 10^{-2} $(\Omega \cdot \mathrm{cm})^{-1}$.

Three classes of these materials can be distinguished:

Class I: compounds with a fixed anion lattice, three-dimensionally disordered cation lattice, ***phase transition***. Example of the *AgI* type: at low temperature, the β phase (wurtzite structure) with low diffusivity; above 146° C, the α phase with a BCC sublattice of I^- ions, each unit cell of which presents 42 interstitial sites of three different types over which the Ag^+ ions can distribute themselves (Fig. V.24.) Thermal activation is thus not needed for the creation of point defects, these being present in large numbers due to the structure. Some examples derived from α-AgI are MAg_4I_5, with M = Rb, K, Cs, NH_4. These compounds have the β-Mn structure and a first order-phase transition at - 150° C ($\Delta\sigma \approx 3$ orders of magnitude); they are thus "superionic" at ambient temperatures. The fast diffusing cation is generally Ag^+, less often Cu^+.

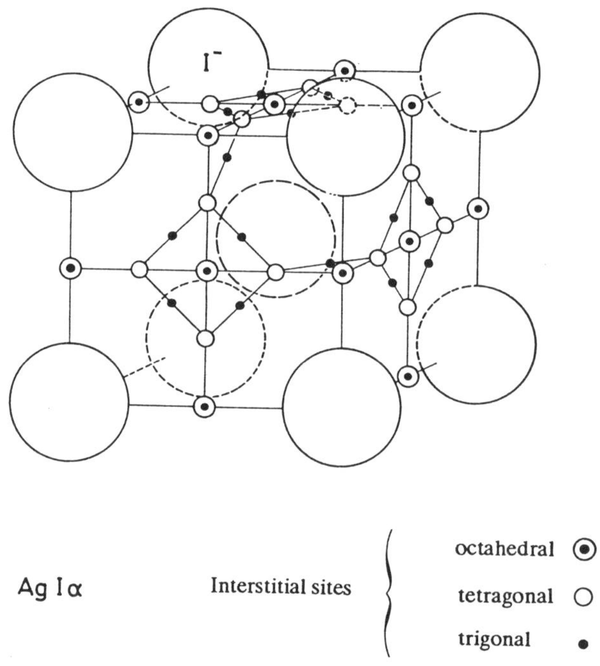

Fig. V.24. — Crystal structure of α-AgI.

Class II: *compounds with a hexagonal structure* in which small cations are mobile only along channels. An example of two-dimensional diffusion: the β *aluminas*, which are in fact, aluminates, corresponding to a general formula M_2O-n Al_2O_3 with M = Na, Li, Ag, ... and $n = 5$ to 11. The structure, made up of layers of spinel-type blocks, determines intermediate planes of M^+ and $O^=$ in which the number of possible cation sites is greater than the number of cations. There are three possible types of sites for the M^+ ions; at high temperature they are energetically more or less equivalent, so that the cations move following zig-zag paths that wind around the "columns" of oxygen ions (Fig. V.25). The best known of these compounds is $(1+x)$ Na_2O, 11 Al_2O_3, with $x < 1$.

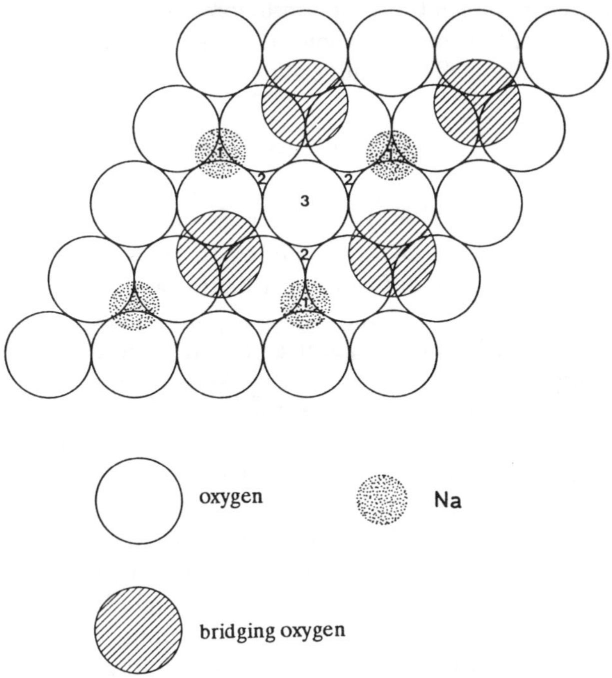

Fig. V.25. — Structure of β-Al_2O_3, parallel to the conducting plane (Na^+ ions). The large open circles indicate oxygen ions in the planes above and below the conduction plane. The large hatched circles correspond to "bridging" oxygen ions situated in the conduction plane. In this plane, the Na^+ ions can migrate following the zig-zag trajectory 1-2-3-2-1; these three types of sites form a plane honeycomb lattice. Site 1 is called a Beevers-Roth site, site 2 an anti-Beevers-Roth site, and site 3 a mid-oxygen site.

Another example of two-dimensional diffusion is the intercalation compounds, such as those based on TiS_2: Li_xTiS_2. An example of one-dimensional diffusion is β-eucryptite, $LiAlSiO_4$, in which the Li cations are located in channels parallel to the c -axis.

Class III: *compounds with the fluorite structure*, such as certain halides ($SrCl_2$, CaF_2, PbF_2, but not $PbCl_2$), oxides based on ThO_2, CeO_2, and even ZrO_2 stabilized in the cubic structure by the addition of CaO, Y_2O_3... (E.g., ZrO_2 xCaO with $x \approx 15\%$).

The transition to the superionic state is gradual without a change of phase, as opposed to class I; it is nevertheless marked by a peak in the specific heat. The conductivity is high ($\approx 1\,(\Omega \cdot \mathrm{cm})^{-1}$) only at very high temperatures ($T > 500°\mathrm{C}$ or $T/T_{\mathrm{M}} > 0.75$ for PbF_2 and $T \geq 1200°\mathrm{C}$ or $T/T_{\mathrm{M}} > 0.6$ for zirconia). The conductivity is ionic because of the large concentration of anion vacancies. The large concentration of heterovalent dopants, by introducing extrinsic vacancies, enlarges the superionic domain.

In a general way, the concentration, n, of mobile ions is fixed by the structure and the composition. The conductivity:

$$\sigma = nqu = nq \cdot \frac{D_\sigma\, q}{kT} \tag{V.56}$$

varies as D_σ/T, with $D \approx (1/6)\,\Gamma\lambda^2$ if there is only one type of jump. With $\lambda \approx 3$ Å, $\sigma \approx 1\ (\Omega \cdot \mathrm{cm})^{-1}$, D_σ is calculated to be $\approx 10^{-5}\ \mathrm{cm}^2\ \mathrm{s}^{-1}$, i.e., the same order of magnitude as diffusion in a liquid, and $\Gamma \approx 5 \times 10^{10}\ \mathrm{s}^{-1}$, the same order of magnitude as the jump frequency of a point defect: *every mobile ion in a superionic conductor is a point defect.* In compounds of classes I and II, these are interstitials. ([10])

Because of their high conductivity, these materials are considered "solid electrolytes" with many applications, both realized (zirconia cell for the measurement of oxygen activity) and potential (light-weight storage batteries, fuel cells).

THEORETICAL MODELS. — The diffusion of the rapidly moving species in these materials has given rise to much speculation about the mechanisms of migration of these ions. The simple models of correlation cannot be applied (see Ch. III, § VI). The experimental results, including neutron scattering and molecular dynamics simulations, favor well-defined jumps between stable sites (for example, the tetragonal sites in α-AgI).

An interesting example is the diffusion of sodium in β-alumina, which was modeled by Sato and Kikuchi (1971) as follows. The Na cations are located on a plane honeycomb lattice, which can be described as two interpenetrating sublattices A and B. The A sites are energetically more favorable than the B sites; the energy difference $W > 0$ in β-Al_2O_3 and $W = 0$ in β'' Al_2O_3 (i.e., the A and B sites are equivalent). An interaction energy Φ between Na ions located on nearest-neighbor sites is introduced ($\Phi < 0$ for repulsion). In addition to W and Φ, a third parameter, ρ, the concentration of sodium on the A and B sites, is introduced; for the stoichiometric composition $\rho = 1$ for β''-Al_2O_3 and $\rho = 1/2$ for β-Al_2O_3.

([10]) Note that in a compound, all sites not normally occupied in the base structure are called interstitial sites. (Constitutional vacancies in certain compounds can be considered interstitial sites).

Using the "path probably method" (see Sato 1984), Sato and Kikuchi (1971) calculated the diffusional flux at steady state and deduced from it the correlation factor, which expresses the efficiency of the participation of one jump of the diffusing species to the corresponding flux. Figure V.26a shows the variation of the correlation factor with the temperature (measured in units of $|\Phi|/k$) for β-alumina. It is seen that f passes through a sharp minimum at $\rho = 0.5$; this is the maximum of the effect of the non-equivalence of the neighboring sites. If an ion exchanges with a vacancy on a nearest-neighbor site, it passes onto the "bad" sublattice, and therefore has a high probability of returning to the initial site. Figure V.26b shows the physical correlation factor; note that $f_I = 1$ for β''-Al_2O_3, but if $W \neq 0$, as in the case of β- Al_2O_3, $f_I \neq 1$ and passes through a minimum for the stoichiometric composition $\rho = 1/2$ because of a strong ordering effect favoring backward jumps to restore the order.

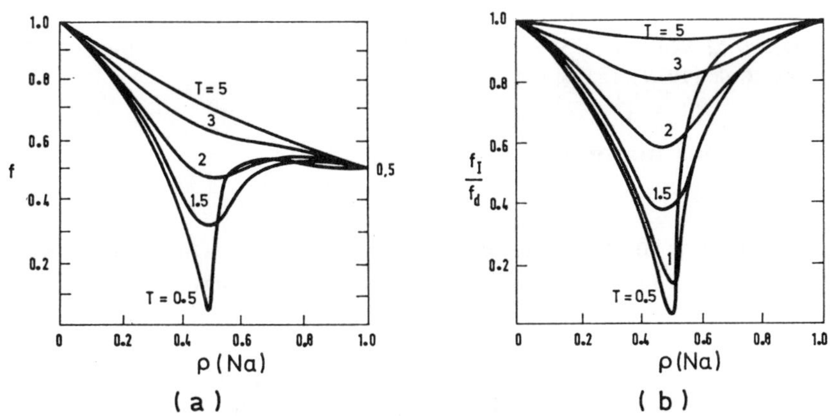

Fig. V.26. — a) Correlation factor f and b) physical correlation factor for the diffusion of Na in β-Al_2O_3. Calculation of Sato and Kikuchi (1971).

Figure V.27 shows the variation of the experimental Haven ratio, defined by equation (III.11), as a function of temperature for β-alumina. The theoretical models explain the anomaly at 200°C quite well; it is connected with interaction between the ions (see Murch 1980).

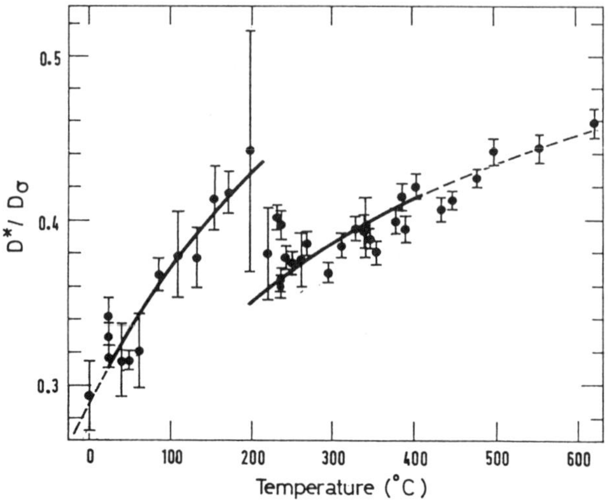

Fig. V.27. — Measurement of the Haven ratio D^*/D_σ in β-Al_2O_3 doped with Na. After Kim *et al.* (1979).

IX. Amorphous materials.

IX.1 AMORPHOUS METALS (OR METALLIC GLASSES). — The measurement of diffusion coefficients in these materials is fraught with practical difficulties (Fig. V.28).

1) The accessible temperature region is limited by the crystallization temperature – hence, very small diffusion coefficients $\left(< 10^{-13}\ \mathrm{cm}^2\ \mathrm{s}^{-1}\right)$.

2) The state of the material can be badly defined because its structure can depend on the preparation conditions (e.g., the quenching rate). Before a diffusion measurement, it is necessary to carry out *a relaxation anneal* in order to arrive at a well-defined structural state and an irreversible local "topological order" ([11]) ("isoconfigurational" measurements).

3) It is not evident that different experimental methods measure the same thing, and the exact nature of the diffusion coefficient being measured is often uncertain, since there is no good model of either the structure or the atomic jumps. In addition to the classical methods (determination of concentration profiles by sectioning-counting, etc... see Ch. IX), we mention:

– *the kinetics of crystallization* (kinetics of crystal growth, expressed in $\sqrt{Dt}$, see Ch. X, § IV.1) measures an interdiffusion coefficient;

([11]) The "topological" local order evolves towards that of a liquid (extrapolated), the stablest of the metastable states (Fig. V.28). The "chemical" local order is reversible; it involves diffusion over short distances.

– *interdiffusion in multilayer samples* with modulated concentration (see Ch. IX, § III.5) which allows measurements as low as 10^{-23} cm^2 s^{-1}. Extrapolation to $\lambda = \infty$ (λ is the wavelength of the modulation) gives the interdiffusion coefficient;

– *formation of the amorphous alloy* by the interdiffusion of its two constituents, originally in the crystalline state (Schwarz and Johnson, 1983);

– *the viscosity* measured by creep experiments, such as the deformation rate at steady state, proportional to the applied stress: $\dot{\varepsilon} = \sigma/\eta$, where η follows a Herring-Nabarro type law $\eta = kTd^2/16\Omega D$, with a structural parameter d of the order of a few nm;

– the local methods measure the ***structural relaxation*** by the variation of a physical property (resistivity, density, viscosity, modulus... see Fig. V.28, vertical arrows) as a function of time, but they do not involve diffusion over long distances. To go from a relaxation time to a diffusion coefficient requires a precise atomic model, as for crystalline media (see Ch. IX, part 2, §I).

The experimental results are difficult to rationalize; the reader should consult the review articles by Limoge *et al.* (1983), Cantor and Cahn (1984), Horvàth *et al.* (1987), or Hahn *et al.* (1987). The comparison of self-diffusion in the same alloy in the amorphous and crystalline states has not been carried out often, and the results depend on the system. The state of relaxation of the material is quite important; measurements on amorphous metals prepared by rapid quenching (Horvàth *et al.* 1987) have shown that the diffusion coefficient decreases by an order of magnitude when the material relaxes. Experiments on the time dependence of diffusion on samples prepared by the simultaneous

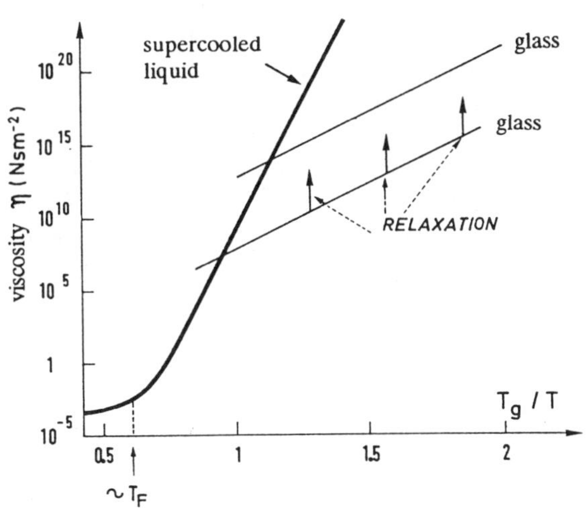

Fig. V.28. — Variation of the viscosity of a glass with the inverse reduced temperature T_g/T (T_g is the glass transition temperature).

evaporation of Ni and Zr have shown that samples prepared in this manner were completely relaxed (Hahn *et al.* 1987).

The similarity of the activation energies determined for isoconfigurational states by diffusion over a long distance and by relaxation (local order) or by viscosity suggests that the same mechanism could be operating in all these cases. A model of diffusion in terms of point defects is suggested by several experimental results: (1) experiments on magnetic anisotropy allow the determination of a formation energy and a migration energy, the first being much smaller than in a crystal ($\approx 1/5$), the second being about the same order; (2) definite stages are found on a relaxation anneal after quenching or irradiation; (3) the measurement of a volume of activation for crystallization under hydrostatic pressure; (4) the presence in irradiated alloys of cavities with volumes $\leq$ an atomic volume, as detected by positron annihilation; (5) the flux dependence of radiation-enhanced diffusion. However, there is so far no good theoretical model that applies in all amorphous materials. Actually, the values of D_0 for tracer diffusion in relaxed materials (Horvàth 1987) vary over 13 orders of magnitude; this may indicate some gross structural differences among these materials.

IX.2 Oxide glasses. — We consider separately the oxides which form a three-dimensional network ("continuous random network" model) such as SiO_2, GeO_2, B_2O_3, P_2O_5, Al_2O_3, V_2O_5, Ta_2O_5... The simplest and technologically most important case is silica, SiO_2. Its amorphous state is characterized by short-range order, both topological and chemical, as the structural unit consists of SiO_4 tetrahedra, bonded at their apices (the O sites). The Si-O-Si bond angle varies from 120 to 180°, a range of values to be compared to 152° observed in cubic cristobalite, the crystalline phase of silica most like the amorphous (their densities are equal to better than 0.5%).

Unfortunately, no data exist for the comparison of the diffusivities in amorphous silica with that in cristobalite. In amorphous SiO_2, oxygen diffuses much faster than silicon; at 1100°C, $D_{Si} \approx 10^{-19}$ cm^2 s^{-1} and $D_{Ox} \approx 10^{-13}$ cm^2 s^{-1}.

The latter value was determined from permeation experiments and was found to be proportional to p_{O_2}, a result that suggests that the diffusing species is the O_2 molecule.

In principle, at least two mechanisms could allow oxygen diffusion in a-SiO_2:

– interstitial diffusion of oxygen molecules,

– atomic diffusion involving some point defects (oxygen vacancy, or dumbbell).

In the former mechanism, oxygen diffuses without interacting with the silica network, whereas in the second, the in-diffusing oxygen (e.g., ^{18}O) exchanges with network oxygen. The expression for $D^*_{O_2}$ given above refers to the first mechanism; the second would lead to diffusivities lower by orders of magnitude.

Oxygen diffusion experiments in silica are frequently unreliable because of the presence of traces of water vapor. It is well known that H_2O molecules exchange very easily with the network oxygen, thus modifying the structure and allowing another path for oxygen migration. Because of the strong binding energy of Si-O bonds, very high activation energy is expected for Si, with $D_{Si} << D_{ox}$.

Noble gases: The activation energy for the diffusion of noble gases in a-SiO_2 increases with the square of the atomic radius.

Molecular gases (H_2, H_2O): The diffusivities of these gases follow the same correlation with molecular radius as noble gases, as long as the temperature is not too high; otherwise, they react chemically with the SiO_2 network. In the case of water:

$$H_2O + \;\; \begin{array}{c} | \\ -Si \\ | \end{array} -O \begin{array}{c} | \\ -Si \\ | \end{array} \rightleftharpoons 2 \begin{array}{c} | \\ -Si \\ | \end{array} -OH.$$

Interstitial diffusion of water is a complex process, involving rapid exchanges with SiO_2 according to the above reaction, the reaction going either from left to right or vice versa, a process akin to a rapid succession of trapping and detrapping steps.

The data on the diffusion of oxygen in silicate glasses, are highly scattered. There are no data on the diffusion of Si. Small network modifiers such as Na^+ are very mobile ($D \approx 10^{-11}$ cm^2 s^{-1} at 200°C). The reader should consult Frischat's review article (1975).

Bibliography

I. General references

Bocquet J. L., Brebec G. and Limoge Y., Diffusion in metals and alloys, Ch. 8, in *Physical Metallurgy*, R. W. Cahn, P. Haasen, eds. (North Holland) 1984, p. 385-474.

Larikov L. N., Geichenko V. V. and Falchenko V. M., *Diffusion Processes in Ordered Alloys*, Kiev, 1975. Translation from: U. S. Department of Commerce, National Technical Information Service, Springfield, Va. USA.

Le Claire A. D., Solute Diffusion in Dilute Alloys, *J. Nucl. Mater.* **69-70** (1978) 70-96.

MANNING J., *op. cit.* (1968) Ch. 3.
NEUMANN G., Impurity Diffusion in Pure Metals, *Dimeta-82, op. cit.* p. 3-22.
PETERSON N. L., Isotope Effects in Diffusion, in *Diffusion in Solids, Recent Developments, op. cit.* (1975) 115.
VÖLKL J. and ALEFELD G., *Hydrogen in Metals*: Part I, Basic Properties, Part II, Applications-Oriented Properties, *Topics In Applied Physics*, vols. **28** & **29** (Springer-Verlag, Berlin) 1978.
WARBURTON W. K. and TURNBULL D., Fast Diffusion in Metals, in *Diffusion in Solids, Recent Developments, op. cit.* (1975) 171.
Diffusion in BCC Metals, A. S. M. Cleveland (1965).
Vacancies and Interstitials in Metals and Alloys, op. cit., Trans. Tech. Publications, Aedermannsdorf (1987).

II. SPECIFIC REFERENCES

ANTHONY T. R., *Diffusion in Solids, op. cit.*, Ch. 7 (1975) 353-379.
BAKKER H., *Philos. Mag. A* **40** (1979) 525-540.
BESHERS D. N., in *Diffusion*, A. S. M. *op. cit.* (1973) 209-240.
BOCQUET J. -L., Thesis, Université Paris-Sud, Orsay (1973).
BREBEC. G., SEGUIN R., SELLA C., BEVENOT J. and MARTIN J. C., *Acta Metall.* **28** (1980) 327-333.
CANTOR B. and CAHN R. W., in *Amorphous Metallic Alloys*, F. E. Luborsky, ed. (Butterworths, London) 1983, p. 487-505.
CHAPPERT J., Courrier du C. N. R. S., *Images de la Physique* (1979) 22-26.
DEPPE D. G. and HOLONYAK Jr., N., *J. Appl. Phys.* **64** (1988) R93-R113.
DOAN N. V., *J. Phys., Chem. Solids* **33** (1972) 2161-2166.
EDELIN G., *Phys. Status Solidi* **b 98** (1980) 699-708.
FAHEY P. M., GRIFFIN P. B. and PLUMMER J. D., *Rev. Mod. Phys.* **61** (1989) 289-384.
FAIR R. B., *Diffusion and Defect Data* **37** (1984) 1-24.
FRANK W., SEEGER A. and GÖSELE U., in *Defects in Semiconductors*, J. Narayan and T. Y. Tan, eds., Materials Research Soc., vol. **2** (North Holland) 1981, p. 31-54.
FRANK W., GÖSELE U., MEHRER H. and SEEGER A. in *Diffusion in Crystalline Solids, op. cit.* (1984) p. 63-142.
FRISCHAT G. H., *Ionic Diffusion in Oxide Glasses*, Trans. Tech. Pub. Aedermannsdorf (1975) Diff. and Defect Monograph Series, # 3/4.
HAHN H., AVERBACK R. S., DING F. -R., LOXTON C. and BAKER J., *Vacancies and Interstitials in Metals and Alloys, op.cit.* (1987) 511-516.
HEHENKAMP Th., *Dimeta-82, op. cit.* (1983) 100-116.
HEUMANN Th., *J. Phys. F, Metal Physics* **9** (1979) 1997-2010.
HORVATH J., PFAHLER K., ULFER W., FRANK W. and KRÖNMÜLLER H. *Vacancies and Interstitials in Metals and Alloys, op. cit.* (1987) 523-528.

HOWARD R. E. and MANNING J. R., *Phys. Rev.* **154** (1967) 561-568.
JONES M. J. and LE CLAIRE A. D., *Philos. Mag.* **26** (1972) 1191-1204.
KIM K. K., MUNDY J. N. and CHEN W. K., *J. Phys. Chem. Solids* **40** (1979) 743-755.
KIMURA H. and YOSHIOKA K., *Mater. Sci. Eng.* **24** (1976) 171-177.
LAZARUS D., Colloque Saclay, *op. cit.* (1976) 865-878.
LE CLAIRE A. D., *Philos Mag.* **7** (1962) 141-167.
LIMOGE Y., BREBEC G. and ADDA Y., *Dimeta-82, op. cit.* (1983) 285-302.
MURCH G. E., *Atomic Diffusion Theory in Highly Defective Solids,* Diff. and Def. Monograph Series # 6 (Trans. Tech. Pub., Aedermannsdorf, Switzerland) 1980.
ORIANI R. A., *Acta Metall.*, **18** (1970) 147-157.
ROBERTS-AUSTEN W. C., *Phil. Trans. Roy. Soc.* **A 187** (1896) 383-415.
SATO H. and KIKUCHI R., *J. Chem. Phys.* **55** (1971) 677-702, and 702-715.
SATO H., in *Non-traditional Methods in Diffusion, op. cit.* (1984) 203-235.
SCHWARZ R. B. and JOHNSON W. L., *Phys. Rev. Lett.* **51** (1983) 415-418.
SHARMA B. L., *Defect and Diffusion Forum* **64-65** (1989) 1-76 and 77-118.
STOLWIJK N. A., *Phys. Stat Solidi B* **105** (1981) 223-232.
STOLWIJK N. A., PERRET M. and MEHRER H., in *Diffusion Processes in High Technology Materials, op. cit.* (1988) 79-98.
TAN T. Y. and GÖSELE V., *Mater. Sci. and Eng.* **B1** (1988) 47-65.
TUCK B., *Atomic Diffusion in III-V Semiconductors* (Adam Hilger, 1988).
VÖLKL J. and ALEFELD G., in *Diffusion in Solids, op. cit.* Ch. 5 (1975) 232-302.
ZENER C., *J. Appl. Phys.* **22** (1951) 372- 375.
ZHANG LING, OATES W. A. and MURCH G. E., *Philos. Mag. A* (1989).
ZH QI, VÖLKL J., LÄSSER R. and WENZL H., *J. Phys. F, Metal Physics* **13** (1983) 2053-2062.

CHAPTER VI

Diffusion and drift in alloys and compounds

Whatever the experimental conditions, it is always possible to define a diffusion coefficient for a given species (chemical species, point defect) by the ratio of a flux to a concentration gradient, i.e., by the application of Fick's law. In the preceding chapters, the discussion has been restricted to homogeneous systems with a constant chemical composition, by virtue of considering the diffusion of a "tracer", i.e., a diffusing species the concentration of which is so small that it effectively does not change the chemical composition, be it chemically identical to (self-diffusion) or different from the base material (solute diffusion).

In this chapter, two types of complications to this simple case will be considered:

– chemically inhomogeneous systems; in addition to the term for diffusion in a concentration gradient, a term has to be introduced to account for the mass flow under the gradient of thermodynamic activity (the non-ideal part mentioned in Ch. I), which in fact amounts to consideration of flow under a gradient of chemical potential (see Ch. VIII). *Intrinsic diffusion coefficients* are introduced here;

– systems in which the fluxes of the different species are not independent but are "coupled" for various reasons such as constant composition, constant volume, electroneutrality... Because of the coupling, *"effective" diffusion coefficients* have to be introduced, but the underlying reasoning is similar, no matter what the physical origin of the coupling.

These diffusion coefficients are of great practical importance, and in their application and associated calculations, care has to be taken to be sure that the diffusion coefficient or its numerical value as taken from tables truly corresponds to the problem under consideration.

Except where indicated, this chapter will treat only binary alloys AB or binary compounds $A_\alpha B_\beta$. Ternary and higher systems are treated in chapter VIII.

I. Intrinsic diffusion coefficients.

The "intrinsic diffusion coefficient" ("intrinsic" is not a particularly appropriate modifier) is the total diffusion coefficient, which, in a solid solution, is used to group the term for proper diffusion and the term for drift or mass flow into a single term. A separation of the diffusion coefficient into two such terms is possible whenever the driving force for drift is proportional to the concentration gradient. For species A, following the Einstein equation (I.20):

$$\begin{aligned} J_{\mathrm{A}} &= -D_{\mathrm{A}} \frac{\partial c_{\mathrm{A}}}{\partial x} + \langle v_{\mathrm{A}} \rangle \; c_{\mathrm{A}} \\ &= -D_{\mathrm{A}} \frac{\partial c_{\mathrm{A}}}{\partial x} + \frac{F_{\mathrm{A}} D_{\mathrm{A}}}{kT} \; c_{\mathrm{A}}, \end{aligned} \tag{VI.1}$$

where D_{A} is the self-diffusion coefficient of A in the alloy. Then one can write:

$$J_{\mathrm{A}} = -\bar{D}_{\mathrm{A}} \frac{\partial c_{\mathrm{A}}}{\partial x}, \tag{VI.2}$$

with the *intrinsic diffusion coefficient* $\bar{D}_{\mathrm{A}}$ given by:

$$\bar{D}_{\mathrm{A}} = D_{\mathrm{A}} \left[1 - \frac{c_{\mathrm{A}}}{kT} \frac{F_{\mathrm{A}}}{\dfrac{\partial c_{\mathrm{A}}}{\partial x}} \right]. \tag{VI.3}$$

If there is no force F_{A}, $\bar{D}_{\mathrm{A}} = D_{\mathrm{A}^*}$, the self-diffusion coefficient in the alloy of composition c_{A} (neglecting correlation effects). Two illustrative examples are given.

I.1 Interdiffusion of two metals A/B. — The thermodynamic force is given by (Ch. I, Tab. II):

$$F_{\mathrm{A}} = -kT \frac{\partial \operatorname{Log} \gamma_{\mathrm{A}}}{\partial x}. \tag{VI.4}$$

This can be written:

$$\begin{aligned} F_{\mathrm{A}} &= -kT \frac{\partial \operatorname{Log} \gamma_{\mathrm{A}}}{\partial c_{\mathrm{A}}} \frac{\partial c_{\mathrm{A}}}{\partial x} \\ &= -kT \left(\frac{\partial \operatorname{Log} \gamma}{\partial \operatorname{Log} N} \right) \frac{1}{c_{\mathrm{A}}} \frac{\partial c_{\mathrm{A}}}{\partial x}, \end{aligned}$$

omitting the subscript from the term in the parentheses, since according to the Gibbs-Duhem equation, it is the same for A and B. N is the mole fraction. Then with $D_{\mathrm{A}} = D_{\mathrm{A}^*}$ (see the above remark on correlation) the Darken-Dehlinger relation is obtained:

$$\boxed{\bar{D}_{\mathrm{A}} = D_{\mathrm{A}^*} \left(1 + \frac{\partial \operatorname{Log} \gamma}{\partial \operatorname{Log} N} \right) = D_{\mathrm{A}^*} \Phi}. \tag{VI.5}$$

Note that $\bar{D}_A$, D_{A^*}, and Φ (Φ is the thermodynamic factor) are functions of composition.

The above derivation comes back to the fact that the diffusional flux is proportional to the gradient of chemical potential:

$$J_A = -(D_{A^*}\, n_A/kT)(\partial\mu_A/\partial x)\,, \tag{VI.1a}$$

i.e., to the gradient of Log ($\gamma_A\, N_A$); from this, the first term of the flux is proportional to the concentration gradient and the second to the gradient of activity coefficient, but the latter is precisely the driving force for drift (Eq. (VI.4)).

The Darken-Dehlinger relation (Eq. (VI.5)) is only approximate. Coupling between fluxes, which would lead to the inclusion of correlation effects ($D_A \neq D_{A^*}$), has been neglected, as has the "vacancy wind", the analysis of which is beyond the scope of this chapter ([1]). Because of it, a factor R, containing D_{A^*} and D_{B^*}, the concentration gradients, and a correlation term, has to be introduced into equation (VI.5). The intrinsic diffusion coefficients have been calculated by Manning for the model of the disordered alloy with two jump frequencies (see Ch. V, § VII). If $D_{A^*} > D_{B^*}$,

$$\begin{cases} \bar{D}_A = D_{A^*}\Phi\left[1+\left(\dfrac{1}{f_0}-1\right)\left(1-\dfrac{D_{B^*}}{N_A\, D_{A^*}+N_B D_{B^*}}\right)\right] \\ \bar{D}_B = D_{B^*}\Phi\left[1-\left(\dfrac{1}{f_0}-1\right)\left(\dfrac{D_{A^*}}{N_A\, D_{A^*}+N_B D_{B^*}}-1\right)\right], \end{cases} \tag{VI.6}$$

where Φ is the thermodynamic factor. The effect of the vacancy wind is given by the term $(1/f_0-1)\times(...)$; here f_0 is the correlation factor for self-diffusion in the pure substance with the same crystal structure. $1/f_0 - 1 = 0.28$ for FCCs, 0.375 for BCCs, and 1 for diamond cubic structures. In a binary alloy, this term is in the same direction as the flux of the more slowly moving species (B in our example). As the effect of the vacancy wind is opposite to the vacancy flux, it increases the flux of the faster moving species A and enhances $\bar{D}_A$ and reduces the flux of the more slowly moving species B and $\bar{D}_B$. Equations (VI.6) show that on extrapolating to pure A ($N_A \to 1$), lim $\bar{D}_A \neq D_{A^*}$ (the intrinsic diffusion coefficient of A does not approach the self diffusion coefficient of pure A). On the other hand, for $N_A \to 1$, lim $\bar{D}_B = D_{B^*}$ (the solute diffusion coefficient of B at infinite dilution).

I.2 INTERDIFFUSION OF TWO IONIC CRYSTALS AX/BX. — The drift term stems partly from the nonideality of the solution, as illustrated above, but even more

([1]) In certain unidirectional systems, an atom of the diffusing species sees vacancies arriving more often from one side than from the other, which biases the probability of jumps to the right and left. This situation arises every time that the diffusion process gives rise to a net flux of vacancies, which is especially the case in interdiffusion (see the following § and Ch. VIII, § III).

from the electrical diffusion potentials. In this section the first effect is neglected in order to analyze the second more clearly. Assume that only the species A and B are mobile, and that they diffuse on a common sublattice, the X sublattice being fixed and independent (no antisite defects). In the absence of a driving force:

$$\begin{cases} J_{\mathrm{A}} = -D_{\mathrm{A}^*} \dfrac{\partial c_{\mathrm{A}}}{\partial x} \\[2ex] J_{\mathrm{B}} = -D_{\mathrm{B}^*} \dfrac{\partial c_{\mathrm{B}}}{\partial x}, \end{cases} \tag{VI.7}$$

with:

$$\frac{\partial c_{\mathrm{A}}}{\partial x} = -\frac{\partial c_{\mathrm{B}}}{\partial x},$$

and neglecting correlation effects. Since $D_{\mathrm{A}^*} \neq D_{\mathrm{B}^*}$, $J_{\mathrm{A}} \neq J_{\mathrm{B}}$, and a net flux of ions $(D_{\mathrm{A}^*} - D_{\mathrm{B}^*})(\partial c_{\mathrm{A}}/\partial x) \neq 0$, and thus also a flux of electric charge, ensues. But the condition of electrical neutrality requires that the latter be zero. In fact, any difference at all in the fluxes creates a space charge, and therefore an electrical potential (diffusion potential or Nernst potential), which gives rise to an electric field E_{d} which modifies the flux. Thus, the flux equations should be written:

$$\begin{cases} J_{\mathrm{A}} = -D_{\mathrm{A}^*} \dfrac{\partial c_{\mathrm{A}}}{\partial x} + qE_{\mathrm{d}} \dfrac{D_{\mathrm{A}^*}}{kT} c_{\mathrm{A}} \\[2ex] J_{\mathrm{B}} = -D_{\mathrm{B}^*} \dfrac{\partial c_{\mathrm{B}}}{\partial x} + qE_{\mathrm{d}} \dfrac{D_{\mathrm{B}^*}}{kT} c_{\mathrm{B}}, \end{cases} \tag{VI.8}$$

introducing the drift term from the Nernst-Einstein equation. Here q denotes the (identical) charge of ions A and B.

The condition of electrical neutrality requires:

$$J_{\mathrm{A}} + J_{\mathrm{B}} = 0. \tag{VI.9}$$

It follows that:

$$E_{\mathrm{d}} = \frac{kT}{q} \frac{D_{\mathrm{A}^*} - D_{\mathrm{B}^*}}{c_{\mathrm{A}} D_{\mathrm{A}^*} + c_{\mathrm{B}} D_{\mathrm{B}^*}} \frac{\partial c_{\mathrm{A}}}{\partial x}. \tag{VI.10}$$

The driving force is proportional to the concentration gradient. Introducing the intrinsic diffusion coefficients and using equations (VI.8) and (VI.10):

$$\begin{cases} J_{\mathrm{A}} = -\bar{D}_{\mathrm{A}} \dfrac{\partial c_{\mathrm{A}}}{\partial x} \\[2ex] J_{\mathrm{B}} = -\bar{D}_{\mathrm{B}} \dfrac{\partial c_{\mathrm{B}}}{\partial x}, \end{cases} \tag{VI.11}$$

with:

$$\boxed{\bar{D}_{\mathrm{A}} = \bar{D}_{\mathrm{B}} = \frac{D_{\mathrm{A}^*}\, D_{\mathrm{B}^*}}{N_{\mathrm{A}}\, D_{\mathrm{A}^*} + N_{\mathrm{B}}\, D_{\mathrm{B}^*}}}. \tag{VI.12}$$

This is the Nernst-Planck equation. It can also be written:

$$\frac{1}{\bar{D}_{\mathrm{A}}} = \frac{1}{\bar{D}_{\mathrm{B}}} = \frac{N_{\mathrm{A}}}{D_{\mathrm{B}^*}} + \frac{N_{\mathrm{B}}}{D_{\mathrm{A}^*}}. \tag{VI.12a}$$

It should not be confused with the expression for the interdiffusion coefficient (Darken), nor with the effective or "molecular" diffusion coefficient to be derived later.

Note that:

1) For a non-ideal solution, the above expressions for $\bar{D}_{\mathrm{A,B}}$ are multiplied by the thermodynamic factor $(1 + \partial \text{ Log } \gamma / \partial \text{ Log } N)$.

2) The condition $J_{\mathrm{A}} \neq J_{\mathrm{B}}$ can also be obtained if the flux $J_{\mathrm{X}} \neq 0$. This comes back to the problem of the interdiffusion of three elements, with its complications, notably the Kirkendall effect.

II. The interdiffusion coefficient.

II.1 DARKEN'S EQUATIONS. — The interdiffusion coefficient measures in some way the rate of mixing during interdiffusion in a couple A/B. In a metallic A/B couple, $J_{\mathrm{A}} \neq J_{\mathrm{B}}$ in general, the fluxes being proportional to the intrinsic diffusion coefficients, which in turn are not necessarily equal, i.e., $\bar{D}_{\mathrm{A}} \neq \bar{D}_{\mathrm{B}}$. Thence arises a net flux of material $\Delta J = |J_{\mathrm{A}}| - |J_{\mathrm{B}}|$ in the direction A $\longrightarrow$ B. Note that ΔJ varies with x, and goes to zero at the ends of the sample; thus it can be expected that ΔJ, in the simplest case, passes through a maximum (Fig. VI.1). This is possible only for a mechanism of diffusion involving point defects. If the defects are vacancies, for example, the ***conservation of the number of sites per unit volume*** imposes a ***flux of vacancies*** J_v such that at every point:

$$J_v + J_{\mathrm{A}} + J_{\mathrm{B}} = 0.$$

The flux $|J_v| = |J_{\mathrm{A}}| - |J_{\mathrm{B}}| = \Delta J$ is ***not conserved***, i.e., it varies along the abscissa. Point defects must thus be created in certain parts of the sample (formation of new atomic planes in the vacancy-source region) and annihilated in others (destruction of atomic planes at vacancy sinks) (Fig. VI.2) ([2]).

([2]) The transverse dimensions of the sample are assumed to be large enough, so the effect of the lateral surface as a source or sink for defects can be neglected and only volume effects need be considered.

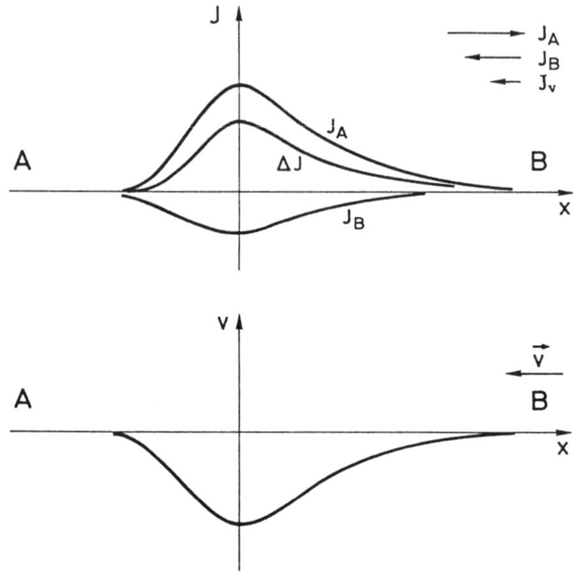

Fig. VI.1. — Interdiffusion of A and B, assuming $\bar{D}_A > \bar{D}_B$ everywhere.

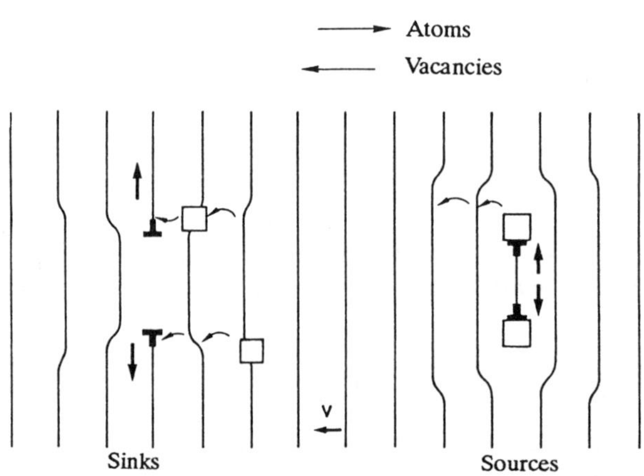

Fig VI.2 — Kirkendall effect: displacement of atomic planes resulting from a flux of vacancies. Vacancy sources are on the right, sinks on the left. The vertical arrows indicate dislocation climb.

Because of these creations and destructions, the atomic planes normal to the diffusion direction are displaced relative to the ends of the sample. This displacement is characterized by a non-uniform velocity **v** which varies as does ΔJ (Fig. VI.1). Two reference frames must be considered: a mobile one, tied to the crystal lattice, in which the equations for the fluxes J_A and J_B are written, and a fixed frame of reference, fixed in one end of the sample. It is in the fixed frame of reference, also called the laboratory frame of reference, that the concentration profiles $c_A(x)$ and $c_B(x)$ are determined experimentally, so that $c_A + c_B$ = constant. These two profiles are thus identical (but reversed). The unique coefficient of interdiffusion henceforth denoted by $\tilde{D}$ (Eq. (I.16)) is obtained, via a solution of Fick's equation, from this unique concentration profile. Thus in the fixed frame of reference:

$$\begin{cases} J_A^0 = -\tilde{D}\,\dfrac{\partial c_A}{\partial x} \\ J_B^0 = -\tilde{D}\,\dfrac{\partial c_B}{\partial x}, \end{cases} \tag{VI.13}$$

with:

$$J_A^0 + J_B^0 = 0 \tag{VI.14}$$

(conservation of matter: $c_A + c_B$ = const. in the fixed frame of reference).

The transformation between frames of reference is classically written:

$$\begin{cases} J_A^0 = J_A + c_A v \\ J_B^0 = J_B + c_B v. \end{cases} \tag{VI.15}$$

With the condition (VI.14) and the definition (VI.2), this becomes:

$$\boxed{v = (\bar{D}_A - \bar{D}_B)\,\frac{\partial N_A}{\partial x}}, \tag{VI.16}$$

with $N_A = c_A/(c_A + c_B)$ the mole fraction of A. Substituting in equation (VI.15):

$$\begin{aligned} J_A^0 &= -\bar{D}_A\,\frac{\partial c_A}{\partial x} + \frac{c_A}{c_A + c_B}\,(\bar{D}_A - \bar{D}_B)\,\frac{\partial c_A}{\partial x} \\ &= -\tilde{D}\,\frac{\partial c_A}{\partial x}, \end{aligned}$$

with:

$$\boxed{\tilde{D} = N_B\,\bar{D}_A + N_A\,\bar{D}_B}. \tag{VI.17}$$

This equation can again be written, following (VI.6):

$$\tilde{D} = (N_B\ D_{A^*} + N_A\ D_{B^*})\,\Phi\ R, \qquad \text{(VI.18)}$$

with $\Phi = 1 + (\partial \text{ Log } \gamma / \partial \text{ Log } N)$ the thermodynamic factor and R the vacancy-wind factor (see above).

The system of equations (VI.16), (VI.17) makes up *Darken's equations* (1948). According to Manning, if $D_{A^*}/D_{B^*} < 3$, $R < 1.07$ in the FCC structure. The thermodynamic factor is in general much more important, as values larger than 2 or 3, or smaller than 1/2, are not rare.

Note that:

1) The thermodynamic factor is >1 for solutions with a negative deviation from ideality (negative heat of mixing), and <1 in the opposite case.

2) The relations derived above assume a constant number of sites per unit volume, a condition fulfilled only if the molar volume is independent of composition. Further, if the partial molar volumes $\bar{V}_A$ and $\bar{V}_B$ differ greatly, interdiffusion brings about a displacement of the markers toward the side of the constituent with the larger molar volume; this behavior has nothing to do with the Kirkendall effect.

The complications stem from the variation of $\bar{V}_A$ and $\bar{V}_B$ with composition; it is no longer possible to exploit directly the measurements carried out in the laboratory frame of reference (fixed in the end of the sample). *Other frames of reference* are possible, but they are useful only if the diffusion equation and the calculation of an interdiffusion coefficient by the Matano method are applicable, and the counterparts of equations (VI.16) and (VI.17) are known (see App. I).

3) The thermodynamic factor can sometimes be negative; the diffusion then takes place up the concentration gradient ("uphill" diffusion). This situation is often encountered in ternary alloys, but it can also occur in binary systems, for solid solutions with compositions unstable vis-à-vis unmixing. Introducing the chemical potential μ in the expression for the thermodynamic factor, we obtain:

$$\Phi = 1 + \frac{\partial \text{ Log } \gamma}{\partial \text{ Log } N} = \frac{N}{kT}\,\frac{\partial \mu}{\partial N}.$$

Following a classical result for the thermodynamics of alloys, the chemical potentials are the partial Gibbs free energies which are determined by the intersection of the tangent to the free energy curve, $g(N_A)$, of the solution, with the vertical axes $N_A = 0$ and $N_A = 1$ (Fig. VI.3):

$$\mu_A = g + (1 - N_A)\,(\mathrm{d}g/\mathrm{d}N_A)_T\,.$$

Defining

$$g'' = \mathrm{d}^2 g/\mathrm{d}N_A^2,$$

we obtain:

$$\Phi = \frac{N_{\mathrm{A}}\,(1 - N_{\mathrm{A}})}{kT}\, g''. \tag{VI.19}$$

It is known that g'' is negative within the spinodal (Fig. VI.3). Any fluctuation in the composition will tend to grow, because the concentration gradient thus produced gives rise to a flux in the same direction, which in turn increases the gradient (coherent unmixing). (See Exercise 45.)

At the maximum of the unmixing curve, g'' goes to zero. Equation (VI.19) then predicts that $\tilde{D}$ goes to zero at the critical temperature, a result known as "critical slowing down". For a regular solution, $\Phi = 1 - \frac{T_s}{T}$, where T_s is the spinodal temperature. Critical slowing down was observed in the Nb-H system; near T_c, $\tilde{D}$ fell by several orders of magnitude, although the tracer diffusion coefficient followed a normal variation (Völkl and Alefeld 1978, reference cited in Ch. V).

4) It should be noted that Darken's equation is valid only if the system is in thermodynamic equilibrium at every point; if not, the velocity v cannot be calculated. To study this problem more precisely, a more detailed treatment is necessary, which explicitly takes into account the 3 components of the system: A atoms, B atoms, and vacancies, the last component not being conserved.

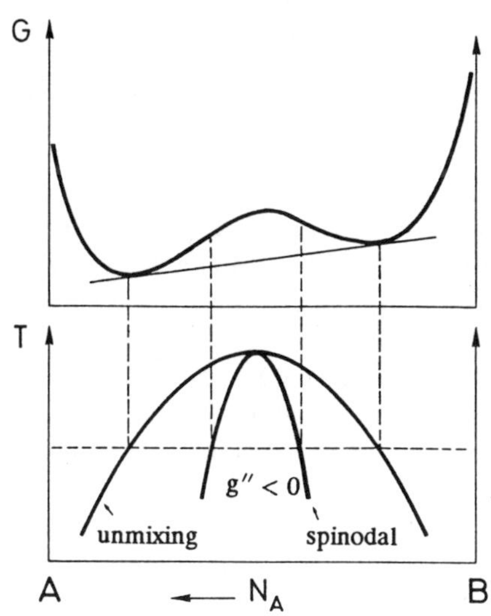

Fig. VI.3. — Spinodal decomposition.

II.2 Experimental verification and the Kirkendall effect. — The preceding analysis explains the origin of the *Kirkendall effect,* the motion of inert markers introduced at the initial weld plane of the A/B couple, motion in the direction of the "faster moving" component (Fig. VI.2). However, a quantitative verification of Darken's equations and of Manning's corrections requires an ensemble of experimental data on the same system which has rarely been obtained with the necessary precision.

1. *Measurement of the Kirkendall effect.* — The velocity, v, is generally determined for only one composition, that of the Kirkendall interface (the plane of the inert markers). This is a plane of constant composition (see below, § II.3) which therefore migrates with a velocity proportional to $\sqrt{t}$, which is obtained from the marker displacement ξ_K: $v_K = \xi_K/(2t)$. The measurement is carried out on a longitudinal section of the couple (Fig. VI.4a and b). The markers generally consist of very fine wires of a material insoluble in the components of the diffusion couple, or of very small particles of radioactive oxide, which is eventually detected during sectioning; a typical oxide for this purpose is $^{181}HfO_2$. Markers can equally well be introduced by ion implantation. The marker material must have vacancy formation and motion energies high enough to make it impermeable to a flux of vacancies. Under these conditions, the markers faithfully "follow" the atomic planes, whatever the nature of their interface with the matrix (coherent or incoherent). It is sometimes difficult to recognize the initial position of the markers. The Matano interface can be taken, or one can use two interfaces A/B/A, and measure 2 ξ_K. The annealing time is varied to check that ξ_K is indeed proportional to $\sqrt{t}$.

Knowing v_K, $\tilde{D}$, and dN_A/dx for the same composition in the concentration profile $N(x)$, $\bar{D}_A$ and $\bar{D}_B$ can be calculated for the marker plane via equations (VI.16) and (VI.17). To go farther, it is necessary to know the values of D_{A^*} and D_{B^*} in a homogeneous alloy of the same composition, as well as the corresponding thermodynamic factor. The uncertainty in the experimental data, especially in Φ and in the second derivative of the Gibbs free energy (Eq. (VI.19)), makes difficult the evaluation of equation (VI.18) with enough precision to verify the presence of the vacancy-wind term; hence this type of verification has been carried out only for a very small number of cases (Meyer, 1969). The experimental values of $\bar{D}_A$ and $\bar{D}_B$ are generally larger than the calculated values, and the same is true for the marker motion, even though Manning's vacancy-wind correction goes the right way (see Ch. VIII, Eq. (VIII.31)). To avoid the uncertainty in Φ, it is better to compare the measured ratio $\bar{D}_A/\bar{D}_B$ with the ratio containing D_{A^*} and D_{B^*} (Eq. (VI.6)). The disagreement (e.g., for CuBi or CuAu alloys) is not well understood (see Heumann, 1983).

2. *Heumann's Method.* — Another method for determining the intrinsic diffusion coefficients (Fig. VI.5) is based on the calculation of the flux across the Kirkendall interface taken as the reference plane. In this plane, the concentration c_A, and therefore $\bar{D}_A$, are constant; Fick's equation can thus be integrated

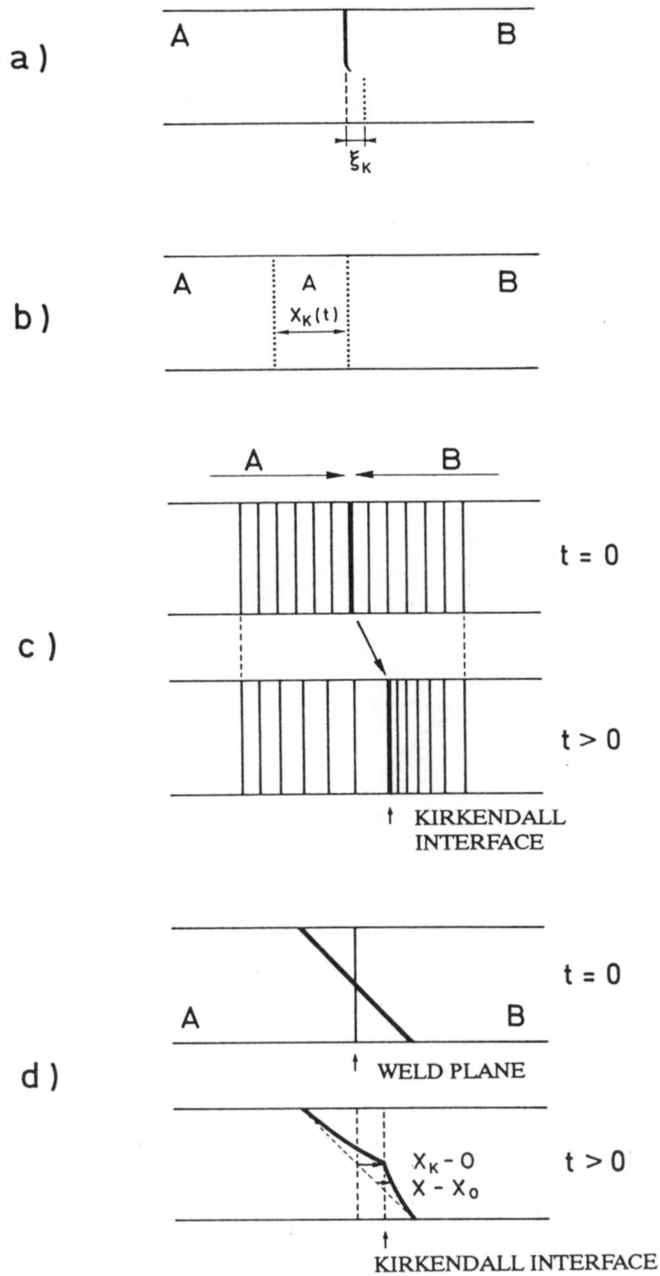

Fig. VI.4. — Measurement of the Kirkendall effect: a) a foil is placed in the upper part of the weld plane and wires in the lower part to mark the motion of this plane; b) measurement of the displacement with respect to an immobile plane; c) the method of many foils for measurement of marker motion as a function of distance; d) oblique interface.

over time from 0 to t:

$$\mathcal{N}_{\mathrm{A}} = \int_0^t J_{\mathrm{A}}\ \mathrm{d}t = -\bar{D}_{\mathrm{A}} \quad \int_0^t \left(\frac{\mathrm{d}c_{\mathrm{A}}}{\mathrm{d}x}\right)_{\mathrm{K}} \ \mathrm{d}t. \tag{VI.20}$$

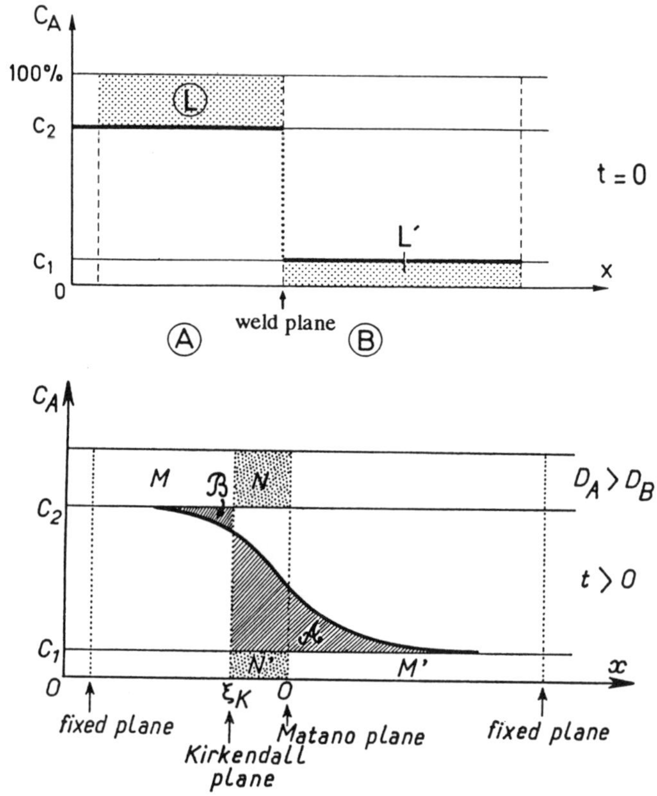

Fig. VI.5. — Heumann's Method

Side A $\left.\begin{array}{l}\mathrm{t} = 0 \ \text{ area } L \\ \mathrm{t} > 0 \ \text{ area } \mathcal{B} + M\end{array}\right\}$ difference $\mathcal{B} - N$

Side B $\left.\begin{array}{l}\mathrm{t} = 0 \text{ area } L' \\ \mathrm{t} > 0 \text{ area } \mathcal{A} + M' + N'\end{array}\right\}$ difference $\mathcal{A} + N'$, from which $\dfrac{\bar{D}_{\mathrm{A}}}{\bar{D}_{\mathrm{B}}} = \dfrac{\mathcal{A} + N'}{\mathcal{B} - N}$.

$\mathcal{N}_{\mathrm{A}}$ measures the quantity of A atoms which has traversed across the marker plane from left to right during the time t. As the variable, we choose $\lambda = x/\sqrt{t}$. Since c_{A} is a function of λ only, $\partial c_{\mathrm{A}}/\partial\lambda$ is a constant at all constant concentration planes, especially the Kirkendall marker plane:

$$\left(\frac{\mathrm{d}c_{\mathrm{A}}}{\mathrm{d}x}\right)_{\mathrm{K}} = \left(\frac{\mathrm{d}c_{\mathrm{A}}}{\mathrm{d}\lambda}\right)_{\mathrm{K}} \cdot \frac{\mathrm{d}\lambda}{\mathrm{d}x} = \left(\frac{\mathrm{d}c_{\mathrm{A}}}{\mathrm{d}\lambda}\right)_{\mathrm{K}} \frac{1}{\sqrt{t}},$$

from which:

$$\mathcal{N}_{\mathrm{A}} = -\bar{D}_{\mathrm{A}} \left(\frac{\mathrm{d}c_{\mathrm{A}}}{\mathrm{d}\lambda}\right)_{\mathrm{K}} \int_0^t \frac{\mathrm{d}t}{\sqrt{t}} = 2\,\bar{D}_{\mathrm{A}}\sqrt{t}\left(\frac{\mathrm{d}c_{\mathrm{A}}}{\mathrm{d}\lambda}\right)_{\mathrm{K}} = 2\,\bar{D}_{\mathrm{A}}\,t\left(\frac{\mathrm{d}c_{\mathrm{A}}}{\mathrm{d}x}\right)_{\mathrm{K}}. \quad \text{(VI.21)}$$

In the same way it can be shown that the quantity of B atoms which have traversed the marker plane from right to left is equal to:

$$\mathcal{N}_{\mathrm{B}} = -2\,\bar{D}_{\mathrm{B}}\,t\left(\frac{\mathrm{d}c_{\mathrm{B}}}{\mathrm{d}x}\right)_{\mathrm{K}} = 2\,\bar{D}_{\mathrm{B}}\,t\left(\frac{\mathrm{d}c_{\mathrm{A}}}{\mathrm{d}x}\right)_{\mathrm{K}}, \quad \text{(VI.21a)}$$

and finally:

$$\boxed{\frac{\bar{D}_{\mathrm{A}}}{\bar{D}_{\mathrm{B}}} = \frac{\mathcal{N}_{\mathrm{A}}}{\mathcal{N}_{\mathrm{B}}}}. \quad \text{(VI.22)}$$

The quantities $\mathcal{N}_{\mathrm{A}}$ and $\mathcal{N}_{\mathrm{B}}$ are evaluated graphically, measuring the areas bounded by a plane fixed outside the diffusion zone, the concentration profile, and the marker plane, then taking the difference of the areas at times $t = 0$ and t. From this, one obtains (see Fig. VI.5):

$$\frac{\bar{D}_{\mathrm{A}}}{\bar{D}_{\mathrm{B}}} = \frac{\mathcal{A} - \xi_{\mathrm{K}}\,c_1}{\mathcal{B} + \xi_{\mathrm{K}}\,(1 - c_2)}, \quad \text{(VI.22a)}$$

where c_1 and c_2 are the initial concentrations in the two halves of the couple. In the case of figure VI.5, the displacement ξ_{K} of the markers is negative. The concentrations are expressed in mole fractions. If the molar volume is variable:

$$\frac{\bar{D}_{\mathrm{A}}}{\bar{D}_{\mathrm{B}}} = \frac{\mathcal{A} - \xi_{\mathrm{K}}\,c_1}{\mathcal{B} + \xi_{\mathrm{K}}\,(1 - c_2)}\,\frac{\bar{V}_{\mathrm{A}}}{\bar{V}_{\mathrm{B}}}. \quad \text{(VI.22b)}$$

3. *Study of the entire diffusion zone.* — The usual method of placing the markers only at the initial weld interface allows the determination of the intrinsic diffusion coefficients at only one concentration (that of the marker plane). Several methods have been proposed for extending these measurements over the entire range of concentrations:

– the method of *many foils*, or stacking of foils A/...A/A/A/B/B/B.../B, where all the interfaces are marked (Fig. VI.4c). The A/A and B/B interfaces begin to move only after a certain time of diffusion, and their displacement does not depend on time according to a simple law (see below);

– the *oblique interface*, created across a couple A/B with very fine markers (Fig. VI.4d), more simple to prepare than the sample consisting of many foils, has been but little used;

– the *flat disks*, a variant on Heumann's method, which requires the complete homogenization of the A/B couple (hence the geometry of a thin disk). It is useful for dilute alloys, for which the marker movement in a semi-infinite sample would be too small. It is inconvenient because very long anneals (several months) are required (Heumann, 1977);

– *incremental couples* are also useful for improving the precision of the measurements of $\tilde{D}$ and the gradient. For example, one could prepare 0-10%, 7-17%, 15-25%, etc. couples, or for lower concentrations: 0-5%, 3-8%, 6-11%,...

II.3 Marker movement. The Kirkendall interface. — Although the application of Darken's or Heumann's equations to the Kirkendall interface poses no problems, the same is not true for other planes, because their concentration must vary with time. A study of their movement and identification of the singular character of the marker plane has been carried out by Cornet and Calais (1972, 1974).

Begin with the equation for the velocity of markers bound to the lattice:

$$v = \left(\bar{D}_{\mathrm{A}} - \bar{D}_{\mathrm{B}}\right)\left(\mathrm{d}N_{\mathrm{A}}/\mathrm{d}x\right).$$

Introduce the reduced variable $\lambda = x/\sqrt{t}$; then N_{A}, $\bar{D}_{\mathrm{A}}$, and $\bar{D}_{\mathrm{B}}$ are functions of λ. For any plane x:

$$\begin{aligned}\frac{\partial x}{\partial t} &= \frac{1}{\sqrt{t}}\left(\bar{D}_{\mathrm{A}} - \bar{D}_{\mathrm{B}}\right)\left(\mathrm{d}N_{\mathrm{A}}/\mathrm{d}\lambda\right)\\ &= \frac{1}{2\sqrt{t}}F(\lambda),\end{aligned} \tag{VI.23}$$

with the reduced velocity $F(\lambda) = 2\left(\bar{D}_{\mathrm{A}} - \bar{D}_{\mathrm{B}}\right)\left(\mathrm{d}N_{\mathrm{A}}/\mathrm{d}\lambda\right)$:

$$\frac{\partial x}{\partial\sqrt{t}} = F(\lambda). \tag{VI.24}$$

Knowing $F(\lambda)$, equation (VI.24) can be used to calculate numerically the velocity of any plane located initially at x_0.

Consider a marker, the displacement of which is linear in $\sqrt{t}$:

$$x = A\sqrt{t} + B,$$

and

$$v = \frac{A}{2\sqrt{t}}.$$

For identification with (VI.23), $A = F(\lambda)$. Since A is a constant, $F(\lambda) =$ constant, $\lambda =$ constant, and $N_A =$ constant along the trajectory. The initial

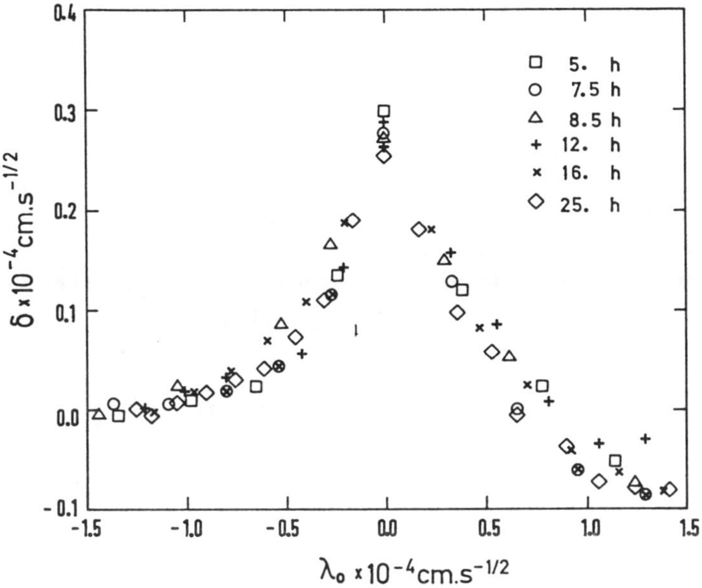

Fig. VI.6. — Interdiffusion of Au and Ag at 915°C for different annealing times. The reduced displacement of the markers, $\delta = (x - x_0)/\sqrt{t}$, is plotted as a function of their reduced initial position $\lambda_0 = x_0/\sqrt{t}$. Note that the curve of $\delta(\lambda_0)$ is independent of time. (After Monty 1972).

conditions thus require that $x_0 = 0$. A marker, the displacement of which is proportional to $\sqrt{t}$, remains at a constant concentration, its initial position is at $x = 0$ (Matano interface), and its concentration is given by the equation $F(\lambda) = \lambda$. This marker is thus in the Kirkendall interface.

Consider the converse. By logarithmic differentiation:

$$d\lambda/\lambda = dx/x - d\sqrt{t}/\sqrt{t},$$

whence, following (VI.24):

$$\frac{dx}{x} = -\frac{F\,d\lambda}{\lambda(\lambda - F)}. \tag{VI.25}$$

This equation has two singularities, the first for $\lambda = 0$, a trivial case in which the marker crosses the Matano interface, the second for $F(\lambda) = \lambda$, i.e.:

$$\frac{\partial x}{\partial \sqrt{t}} = F = \frac{x}{\sqrt{t}},$$

from which, by integration:

$$x = \lambda_K \sqrt{t} + x_0$$

with the constant of integration showing that $\lambda_K = F(\lambda_K)$. Thus if $\lambda = F$ at any given instant, λ and F remain constant along the entire path so that $N_A = N_K$. The initial position is necessarily $x_0 = 0$. *The condition $F(\lambda) = \lambda$ thus defines the Kirkendall interface.*

It is possible to calculate the path of the other markers. They satisfy the following equation, obtained from (VI.25) by integration (Cornet and Calais, 1972, Levasseur and Philibert, 1967):

$$x - x_0\,(dx/dx_0) = F\sqrt{t} = 2\,vt, \tag{VI.26}$$

or

$$\lambda - \lambda_0\,(d\lambda/d\lambda_0) = F. \tag{VI.26a}$$

This equation allows the determination of the velocity of the planes from one experiment ("many-foils" or oblique interface) without a series of experiments of different duration.

The solution of this equation has the form (Cornet, 1974):

$$(x - x_0)/\sqrt{t} = \varphi_0\left(x_0/\sqrt{t}\right), \tag{VI.27}$$

or

$$\delta = \varphi_0(\lambda_0) = \varphi(\lambda)$$

in reduced coordinates. Graphs of $\delta(\lambda)$ or $\delta(\lambda_0)$ are invariant with diffusion time (see Eq. (VI.26)), a result well confirmed by experiment (Fig. VI.6).

The plots of the displacements $x(x_0)$ or $\delta(\lambda)$ have a singularity at $x_0 = 0$ (the Kirkendall interface). But the extremum of F does not necessarily occur at $\lambda = \lambda_K$, so that the Kirkendall plane generally corresponds neither to an extremum in velocity nor to an extremum in displacement (Fig. VI.7). It is only

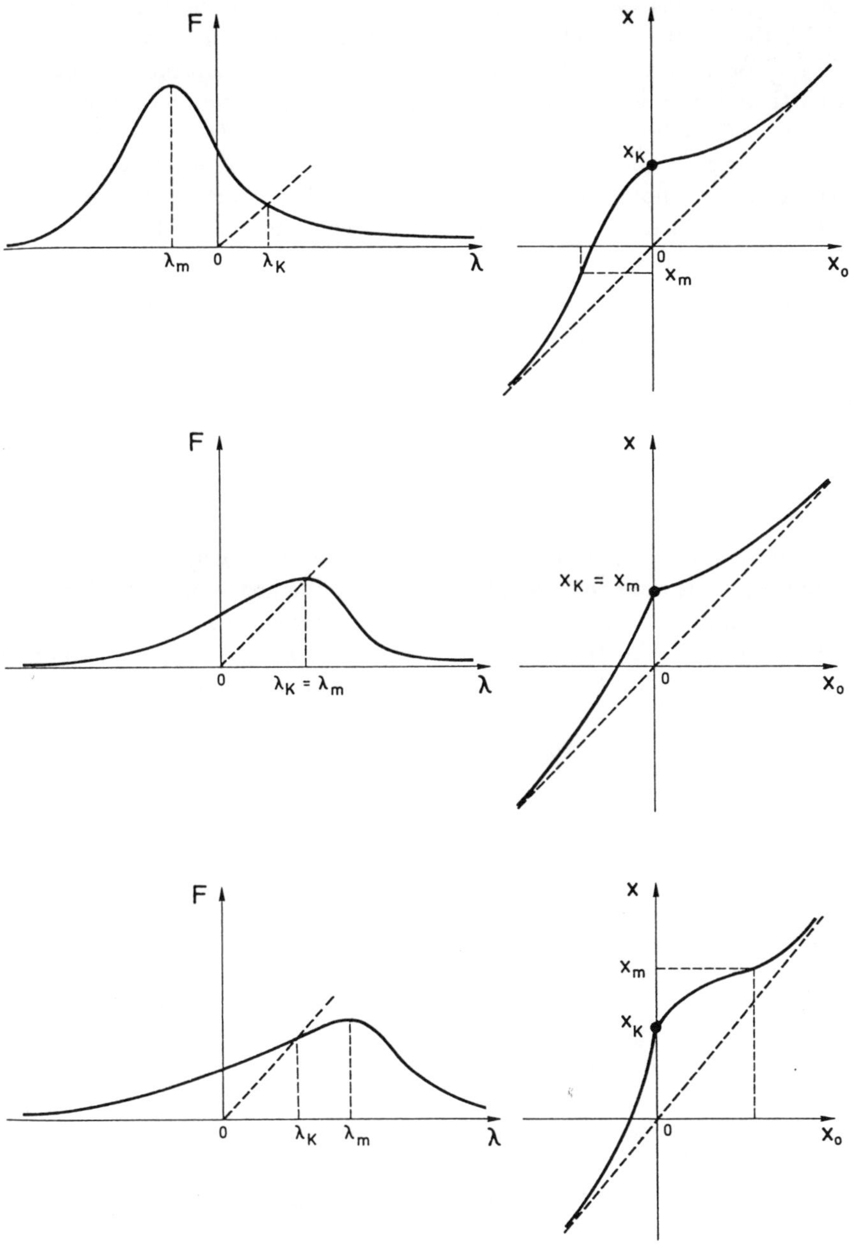

Fig. VI.7. — The Kirkendall effect. Correspondence between plots of reduced velocity $F(\lambda)$ and the plots of the position x of marker planes as a function of their initial position x_0. Three cases for different relative positions λ_m(maximum of F) and of λ_K. It was assumed that $F > 0$, $\lambda_K > 0$. After Cornet and Calais (1972).

in the particular case when $F(\lambda_K)$ is an extremum that the Kirkendall plane moves at the maximum velocity; one can deduce that then (and only then) $\bar{D}_A/\bar{D}_B$ = constant and that the fluxes J_A and J_B are simultaneously extrema. This particular case is quite often found, at least approximately, because in a number of solid solutions (FeNi, CuNi, AuCu, AuAg), $D_{A^*}/D_{B^*} \approx$ constant over a wide range of concentrations. In the general case, v_K is not an extremum, and the fluxes are not simultaneously extrema. The concentration on marker planes other than the Kirkendall plane varies with time, and the displacement of these planes does not follow a simple time law.

In analyzing the experiments, the displacements are first plotted in reduced coordinates: $\delta = (x - x_0)/\sqrt{t}$ as a function of λ_0, or more simply, λ. From this, $F(\lambda)$ is obtained either graphically or numerically. Then $\bar{D}_A$ and $\bar{D}_B$ can be calculated using two equations, one being equation (VI.17) and the other equation (VI.23) giving F, or an equation in $\bar{D}_A/\bar{D}_B$, analogous to Heumann's equation (VI.22a) for the Kirkendall plane.

II.4 Sources and sinks for vacancies. Kirkendall porosity. — The flux of vacancies, J_v, is not conserved; it varies along the abscissa. Consider a small element of volume dx bounded by two planes parallel to the diffusion front. The difference between the fluxes entering and leaving this element per unit time is $-\left(\partial J_v^0/\partial x\right)\mathrm{d}x$. The balance equation is then written:

$$\partial n_v/\partial t = -\mathrm{div}\, J_v^0 + \Re_v ,$$

which for a steady state reduces to:

$$-\left(\partial J_v^0/\partial x\right) + \Re_v = 0, \qquad \text{(VI.28)}$$

with the *source term for vacancies*:

$$\Re_v = \partial J_v^0/\partial x, \qquad \text{(VI.29)}$$

which gives the number of vacancies created ($\Re_v > 0$) or annihilated ($\Re_v < 0$) per unit volume per unit time (Fig. VI.8). These values can be such that the equilibrium concentration is never attained in one part of the diffusion couple ($\Re_v < 0$ and large in absolute value); vacancies present in supersaturated solution can precipitate as pores concurrently with their annihilation at dislocations. A zone of porosity is thus observed in many diffusion couples; this zone is located on the side of the more rapidly diffusing element (Fig. VI.8).

The Kirkendall effect is a direct proof that diffusion takes place via point defects, vacancies or interstitials, and not by an exchange mechanism. Its practical importance lies in the appearance of porosity, which can manifest itself, for example, in the course of *sintering* a mixture of A and B powders, from which arises a swelling that can eventually compensate for the shrinkage due to the sintering. Kirkendall porosity can be a source of degradation of fiber composite materials when interdiffusion between the fiber and the matrix materials takes place (e.g., boron fibers in a metallic matrix). Thus a "diffusion barrier" must be established in the form of an adequate coating for the fibers. Porosity can also appear as a consequence of the selective vaporization of a volatile component, e.g., the dezincification of brass. Diffusion in grain boundaries followed by diffusion from those boundaries into the volume of the grains can give rise to a Kirkendall effect, with the induced flux of vacancies causing grain-boundary motion. This is accompanied by a variation of composition behind the moving boundary, which can have inauspicious effects ("de-alloying") due to migration induced by diffusion (see below, Ch. VII, part 1, § VI).

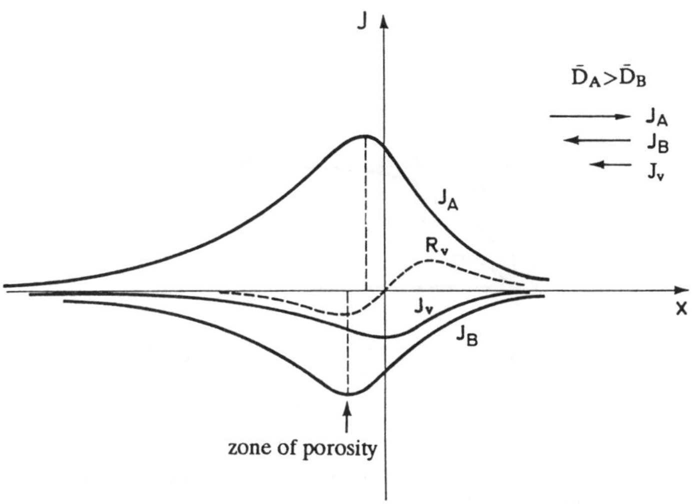

Fig. VI.8. — Interdiffusion of A and B with $\bar{D}_A > \bar{D}_B$. Determination of the vacancy source term.

III. Chemical diffusion in compounds.

When a compound deviates from the stoichiometric composition, it can be subject to diffusion under a composition gradient (or, more accurately, a gradient of chemical potential). For this reason, the process is often called *chemical diffusion*, an ambiguous term, since the interdiffusion of A and B is often referred to by this term. (3)

(3) Whence the expression "coefficient of chemical self-diffusion", sometimes used for the case of the non-stoichiometric compounds discussed in this paragraph.

III.1 CHEMICAL DIFFUSION COEFFICIENT. — Let $A_{1-\delta}$ X be the formula of the compound, with δ designating the deviation from stoichiometry. The composition at the surface is modified, e.g., by changing the partial pressure of A or of X. How does the concentration profile change so that the composition passes from δ_1 to δ_2 ? (Fig. VI.9). Since this concerns a gas-solid diffusion couple, in the frame of reference of the solid we have:

$$J_A = -\tilde{D}\,\frac{\partial c_A}{\partial x} = \tilde{D} n\,\frac{\partial \delta}{\partial x}, \qquad \text{(VI.30)}$$

since $c_A = n\,(1-\delta)$; n is the number of sites per unit volume on the A sublattice. This serves to introduce a chemical diffusion coefficient $\tilde{D}$ which allows the expression of a flux proportional to the gradient of deviation from stoichiometry.

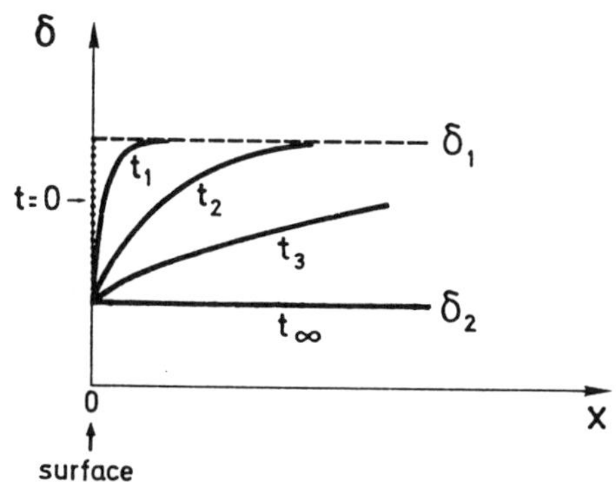

Fig. VI.9. — Chemical diffusion: evolution of the composition from the surface for a non-stoichiometric compound $A_{1-\delta}X$.

This system is in fact equivalent to the solid-solid system:

$$A_{1-\delta_1}X/A_{1-(2\delta_2-\delta_1)}X$$

with the interface composition remaining constant at $\bar{\delta} = \delta_2$.

Darken's equation (VI.17) can thus be applied here. In many cases, e.g., $D_{A^\bullet} >> D_{X^\bullet}$, the gradient is limited to the constituent A diffusing on its own sublattice, so that:

$$\begin{aligned}\tilde{D} &= N_X\ D_{A^\bullet}\left[1+\frac{\partial\ \mathrm{Log}\ \gamma_X}{\partial\ \mathrm{Log}\ N_X}\right] \\ &= N_X\ D_{A^\bullet}\,\frac{\partial\,[\mu_X/kT]}{\partial\ \mathrm{Log}\ N_X}.\end{aligned} \qquad \text{(VI.31)}$$

Care must be taken with expressions involving mole fractions. In Darken's equation, the mole fractions are equal to the number of atoms (or ions) of a species divided by the total number of atoms. The number of atoms is:

$$\begin{aligned} n_{\text{total}} &= \text{no. of sites on a sublattice} && \times [1 + (1-\delta)] \\ n_{\text{A}} &= \quad '' \qquad\qquad '' && \times (1-\delta) \\ n_{\text{X}} &= \quad '' \qquad\qquad '' && \times 1. \end{aligned}$$

From this, the atomic fractions which figure in Darken's relation are:

$$\begin{aligned} N_{\text{A}} &= \frac{1-\delta}{2-\delta} \\ N_{\text{X}} &= \frac{1}{2-\delta}. \end{aligned} \tag{VI.32}$$

The thermodynamic factor is calculated next. The gas-solid equilibrium for an X_2 gas is $1/2\ X_2 \rightleftharpoons X_x$, where X_x stands for an X atom on an X sublattice site. This leads to:

$$\mathrm{d}\mu_{\text{X}} = \frac{1}{2}\, kT \;\mathrm{d}\;\mathrm{Log}\; P_{\text{X}_2}, \tag{VI.33}$$

whence:

$$\begin{aligned} \frac{1}{kT}\,\frac{\partial \mu_{\text{X}}}{\partial\;\mathrm{Log}\;N_{\text{X}}} &= \frac{1}{kT}\,\frac{\partial \mu_{\text{X}}}{\partial\;\mathrm{Log}\;P_{\text{X}_2}} \cdot \frac{\partial\;\mathrm{Log}\;P_{\text{X}_2}}{\partial\delta} \cdot \frac{\partial\delta}{\partial\;\mathrm{Log}\;N_{\text{X}}} \\ &= \frac{1}{2}\,\frac{\partial\;\mathrm{Log}\;P_{\text{X}_2}}{\partial\delta}\,(2-\delta)\,, \end{aligned}$$

following (VI.32) and (VI.33).

Darken's equation then becomes:

$$\tilde{D} = N_{\text{X}}\; D_{\text{A}^\bullet} \cdot \frac{2-\delta}{2} \cdot \frac{\partial\;\mathrm{Log}\;P_{\text{X}_2}}{\partial\delta}, \tag{VI.34}$$

or

$$\boxed{\tilde{D} = \frac{1}{2}\, D_{\text{A}^\bullet} \cdot \frac{\partial\;\mathrm{Log}\;P_{\text{X}_2}}{\partial\delta}}. \tag{VI.35}$$

For a compound $A_a\ X_b$, the factor 1/2 is replaced by $b/2a$.

To go farther, either thermodynamic data, namely the relation $\delta = f\,(P_{\text{X}_2})$, have to be known, or a model of the defects responsible for the deviation from stoichiometry has to be used.

Example: Nickel oxide, $Ni_{1-\delta}O$, in which the deviation from stoichiometry is due to nickel vacancies with an effective charge of $-\alpha\,|e|\;(\alpha = 0, 1,\ \text{or}\ 2)$. Recall that:

$$\delta = [V_{Ni}] \propto P_{O_2}^m, \qquad \frac{\partial \text{ Log } \delta}{\partial \text{ Log } P_{O_2}} = m,$$

with:

$$m = \frac{1}{2(\alpha + 1)} \tag{VI.36}$$

(see derivation in App. III). Here it has been assumed that only one type of atomic defect is present. The expression (VI.35) becomes:

$$\tilde{D} = \frac{1}{2} D_{Ni^\bullet} \frac{1}{m\delta}.$$

But $D_{Ni^\bullet}/\delta = D_{Ni^\bullet}/[V_{Ni}] = D_{V_{Ni}}$, the diffusion coefficient of the nickel vacancies, from which:

$$\boxed{\tilde{D} = (1/2\ m)\, D_V = (1 + \alpha)\, D_V}, \tag{VI.37}$$

an expression which connects the chemical diffusion coefficient to the diffusion coefficient of the defects, here vacancies with an effective charge of $-\alpha\,|e|$. Equation (VI.37) becomes more complicated if the vacancy has several charge states in the system under consideration (α changes with T or P_{O_2}).

When D_X is not negligible with respect to D_A, a more cumbersome relation is obtained, which is met in the literature in many forms, especially that proposed by C. Wagner (1930) as a consequence of an entirely different derivation (see Ch. X, § II).

III.2 Ambipolar diffusion and the Nernst electric field. — Equation (VI.37) can be arrived at in a more direct manner by considering the flux of the point defects responsible for the deviation from stoichiometry. If these are electrically neutral, equation (VI.30) gives the flux of these defects, their concentration being $\pm n\delta$ (for either interstitials or vacancies), so that $\tilde{D} = D_{d^\times}$, where $d^\times$ indicates neutral defects. If the defects are electrically charged, an electric field, the Nernst field, is created, and flux equations have to be written for all the charged particle species.

Example 1: Consider an oxide with two types of atomic point defects: metal ion vacancies having a single negative charge V_M', oxygen ion interstitials having one negative charge, O_i', and these charges compensated by electron holes $h^\bullet$ [4] (Fig. VI.10).

---

[4] Kröger-Vink notation is used; see Appendix II.

If the subscripts V and I designate metal vacancies and oxygen interstitials, respectively, the electroneutrality condition on the concentrations (expressed in quantities per unit volume) is:

$$n_{\mathrm{h}} = n_{\mathrm{V}} + n_{\mathrm{I}}, \tag{VI.38}$$

and the condition on the fluxes of no net current flow is:

$$J_{\mathrm{h}} = J_{\mathrm{V}} + J_{\mathrm{I}}. \tag{VI.39}$$

Since the mobility is given by the Nernst-Einstein equation ([5]), the flux of each type of particle is given by:

$$J_i = -D_i \, \nabla n_i + n_i \, q_i \, \frac{E D_i}{kT} \qquad i = \mathrm{h},\ \mathrm{V},\ \mathrm{I}, \tag{VI.40}$$

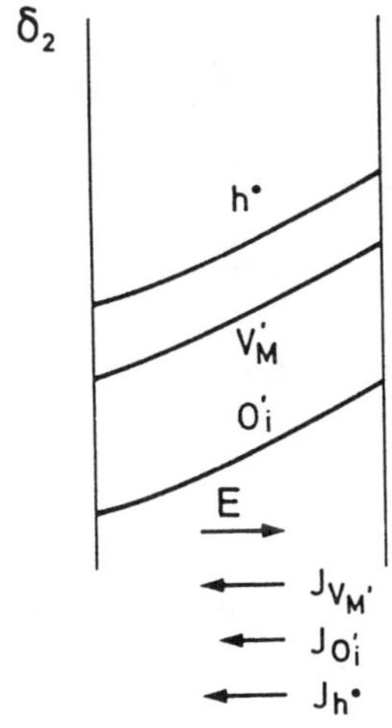

Fig. VI.10. — Ambipolar diffusion. The Nernst field increases the diffusion of the negatively charged defects and reduces that of the electron holes.

where E is the Nernst field, and q_i the electric charge. The value of E is obtained from the three equations (VI.40) and (VI.39); knowing that $q_{\mathrm{h}} = -q_{\mathrm{V}} = -q_{\mathrm{I}} = q$,

$$\frac{qE}{kT} = \frac{D_{\mathrm{h}} \, \nabla n_{\mathrm{h}} - D_{\mathrm{V}} \, \nabla n_{\mathrm{V}} - D_{\mathrm{I}} \, \nabla n_{\mathrm{I}}}{n_{\mathrm{h}} \, D_{\mathrm{h}} + n_{\mathrm{V}} \, D_{\mathrm{V}} + n_{\mathrm{I}} \, D_{\mathrm{I}}}, \tag{VI.41}$$

([5]) The symbol ∇ is used for ease of writing the concentration gradients.

whence:

$$J_{\rm h} = -D_h \frac{(n_{\rm V}\ D_{\rm V} + n_{\rm I}\ D_{\rm I})\,\nabla n_{\rm h} + n_{\rm h}\,(D_{\rm V}\ \nabla n_{\rm V} + D_{\rm I}\ \nabla n_{\rm I})}{n_{\rm h}\ D_{\rm h} + n_{\rm V}\ D_{\rm V} + n_{\rm I}\ D_{\rm I}}. \qquad \text{(VI.42)}$$

This expression can be simplified if one type of point defect is dominant.

1) *Vacancies predominant*, $n_{\rm V} >> n_{\rm I}$, $n_{\rm V} \approx n_{\rm h}$,

$$J_{\rm h} = J_{\rm V} = -D_{\rm h}\ D_{\rm V}\ \frac{\nabla n_{\rm h} + \nabla n_{\rm V}}{D_{\rm V} + D_{\rm h}}.$$

The two concentration gradients are equal to $\nabla(n\delta)$, so that:

$$J_{\rm h} = J_{\rm V} = -\tilde{D}\ \nabla(n\delta), \qquad \text{(VI.43)}$$

with:

$$\tilde{D} = \frac{2\ D_{\rm h}\ D_{\rm V}}{D_{\rm h} + D_{\rm V}}. \qquad \text{(VI.44)}$$

Since the holes are much more mobile than the vacancies V'_{M}:

$$\tilde{D} \approx 2\ D_{\mathrm{V}'_{\mathrm{M}}}. \qquad \text{(VI.44a)}$$

2) *Interstitial predominant*, $n_{\rm V} << n_{\rm I} \approx n_{\rm h}$.
As above, it is found that:

$$\tilde{D} \approx 2\ D_{\mathrm{O}'_{\mathrm{i}}}. \qquad \text{(VI.45)}$$

Example 2: Vacancies with charge $-\alpha\,|e|$ and electron holes $\mathrm{h}^\bullet$:

$$n_{\rm V} = n\delta \qquad \text{(VI.46)}$$

$$n_{\rm h} = \alpha n_{\rm V} \ \text{ and } \ J_{\rm h} = \alpha J_{\rm V} \ \text{ (electroneutrality)}. \qquad \text{(VI.47)}$$

The fluxes are written as:

$$\begin{cases} J_{\rm V} = -D_{\rm V}\ \nabla n_{\rm V} - \dfrac{n_{\rm V}\ D_{\rm V}}{kT}\ \alpha\,|e|\ E \\ J_{\rm h} = -D_{\rm h}\ \nabla n_{\rm h} + \dfrac{n_{\rm h}\ D_{\rm h}}{kT}\ |e|\ E. \end{cases} \qquad \text{(VI.48)}$$

By application of the electroneutrality conditions (VI.46) and (VI.47):

$$\frac{|e|\ E}{kT} = \frac{D_{\rm h} - D_{\rm V}}{D_{\rm h} + \alpha D_{\rm V}}\ \frac{\nabla n_{\rm V}}{n_{\rm V}}, \qquad \text{(VI.49)}$$

so that:

$$J_V = -\frac{(1+\alpha)\ D_h\ D_V}{D_h + \alpha D_V} \nabla n_V$$

$$J_V \approx -(1+\alpha)\, D_V\ \nabla n_V = -(1+\alpha)\, D_V\ \nabla(n\delta). \qquad \text{(VI.50)}$$

As a general result:

$$\boxed{\tilde{D} = (1+\alpha)\, D_d}\ , \qquad \text{(VI.51)}$$

where $\pm\alpha\,|e|$ indicates the charge of the defects d, of either sign, as the above derivations apply equally well to positively charged point defects and electrons.

Evidently, the difference between the mobilities of the electronic and atomic defects gives rise to an electric field, the Nernst field, which enhances the diffusion of the point defects and reduces the diffusion of the electronic defects, thus compensating for the difference in their mobilities; this is a case of ***ambipolar diffusion*** (two diffusing species of different signs) and the diffusion coefficient $\tilde{D}$ calculated here is identical to the ambipolar or effective diffusion coefficient (see § IV). The coupling due to the Nernst field is expressed by the factor $(1+\alpha)$ in equation (VI.51) for the chemical diffusion coefficient. The equations obtained above show that the electric field is proportional to the difference between the diffusivities of the species with opposite signs and the gradient of the deviation from stoichiometry.

III.3 Application to the oxidation of a pure metal. — The chemical diffusion coefficient is the one that enters into the process of chemical reaction with a gas, such as oxidation, sulfidation,... i.e., which governs the growth of the layer of reaction product (oxide, sulfide...) on a metal in a reactive atmosphere.

Consider a metal M that forms a metal-deficient oxide $M_{1-\delta}O$ according to the oxidation reaction:

$$\frac{1}{2}\,O_2 \rightleftharpoons O_O + V_M^{\alpha'} + \alpha h^{\bullet},$$

where the vacancy can be ionized (negative charge $-\alpha\,|e|$ with $\alpha = 1,\ 2,\ 3...$). This reaction takes place at the surface and continues only due to the migration of metal ion vacancies and electron holes across the oxide layer (Fig. VI.11) under the influence of the gradient of chemical potential. At equilibrium, the chemical potential takes the value corresponding to the metal/oxide and oxide/atmosphere interfaces at the given temperature. There is thus a composition gradient, i.e., a gradient in deviation from stoichiometry, across the oxide layer, since at every point the electroneutrality condition requires that:

$$\delta = \left[V_M^{\alpha'}\right] = [h^\bullet]/\alpha, \tag{VI.52}$$

where $[h^\bullet]$ is the mole fraction of electron holes (see App. III).

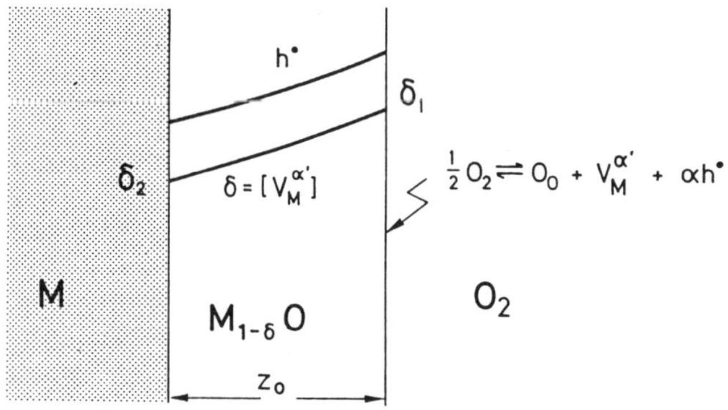

Fig. VI.11. — Growth of an oxide layer on a pure metal M. It is assumed that the oxide $M_{1-\delta}O$ is metal deficient due to the presence of metal-ion vacancies V_M'.

This is a case of *ambipolar diffusion* (negatively charged vacancies and positive holes), the difference in the mobilities of the two species giving rise to a Nernst field. Therefore:

$$J_V = -\tilde{D}\, \partial(n\delta)/\partial z, \tag{VI.53}$$

with:

$$\tilde{D} = (1+\alpha)\, D_V. \tag{VI.53a}$$

In the simplest case, that of a single majority defect, D_V, α, and $\tilde{D}$ are independent of the composition. The gradient in equation (VI.53) is thus $n(\delta_1 - \delta_2)/z_0$, where z_0 is the thickness of the oxide layer:

$$J_V\, z_0 = \tilde{D} n (\delta_1 - \delta_2),$$

and δ_1 and δ_2 are the deviations from stoichiometry at the two interfaces (Fig. VI.11). A "molecule" of oxide MO is formed for each vacancy created. For $J_V\, dt$ vacancies per unit area, a volume of oxide equal to $J_V \Omega\, dt$ (Ω is the molar volume of the oxide) is formed, with unit area and thickness dz_0, so that:

$$\frac{dz_0}{dt} = J_V \, \Omega = \tilde{D} \, \frac{\delta_1 - \delta_2}{z_0}, \tag{VI.54}$$

since $n\Omega = 1$.

This equation leads to the ***parabolic law*** for oxide layer growth:

$$\boxed{\frac{dz_0}{dt} = \frac{K_p}{z_0}} \text{ and } z_0 = (2\,K_p t)^{1/2}, \tag{VI.55}$$

with the ***parabolic rate constant for oxidation*** in $cm^2 s^{-1}$ given by ([6]):

$$K_p = \tilde{D}\,(\delta_1 - \delta_2). \tag{VI.56}$$

We return to this derivation in chapter X, § II, starting from the general flux equations.

Notes:
1) In the general case, both cationic and anionic defects contribute to mass transport and therefore to the growth of oxide layers. In most cases, one type of defect is predominant, and the chemical diffusion coefficient is related to the defect diffusion coefficient by an equation of the type (VI.51).
2) The value of δ_1 is determined by the temperature and the oxygen partial pressure of the given atmosphere and that of δ_2 by the temperature. Often δ_2 is negligible with respect to δ_1, and:

$$K_p \approx \tilde{D}\delta_1 = (1+\alpha)\,D_{V_M}\delta_1 = (1+\alpha)\,D_{M^\bullet}, \tag{VI.57}$$

where $D_{M^\bullet}$ is the self-diffusion coefficient of the cation (since δ is the mole fraction of vacancies) at the oxygen partial pressure of the ambient atmosphere. The ambipolar character of the diffusion, and the resulting existence of a Nernst field, is expressed by the ***factor*** $(1+\alpha)$ ***multiplying the self-diffusion coefficient*** of the diffusing species.

IV. The effective diffusion coefficient.

How to express the transport of material in an alloy or compound under the influence of a thermodynamic force, at ***constant composition*** over the entire sample ? Sometimes a ***molecular flux*** and a molecular diffusion coefficient are used to express the fact that the process takes place as if an entity $A_\alpha B_\beta$ of fixed composition (a "molecule" ([7])) were migrating. This type of behavior is

---

([6]) The constant K_p is often defined by $z_0^2 = K_p t$. Note the difference of a factor of 2 between these definitions.
([7]) Or better, a "formula unit".

found in viscous creep (Herring-Nabarro), sintering, coalescence of pores, the smoothing of surfaces,..., in alloys or compounds.

If the composition is to remain constant, then at every point:

$$\frac{J_A}{J_B} = \frac{n_A}{n_B} = \frac{\alpha}{\beta}. \tag{VI.58}$$

The molecular flux is then ([8]):

$$J_m = \frac{J_A}{\alpha} = \frac{J_B}{\beta}. \tag{VI.59}$$

In an ionic crystal, the same condition is imposed by the condition of electroneutrality. To fulfill this condition in the most general case, even though $D_A \neq D_B$, the fluxes must be coupled. This is possible if a supplementary thermodynamic force $F_{A,B}$ is introduced (Li, Blakely, and Feingold 1960). This adds a drift term that opposes any tendency of the system to deviate from its composition.

The fluxes expressed with the intrinsic diffusion coefficients are written, in the absence of the drift term, in the form:

$$J_A = -\bar{D}_A \frac{\partial n_A}{\partial x} = -D_{A^\bullet} \left(1 + \frac{\partial \text{ Log } \gamma_A}{\partial \text{ Log } N_A}\right) \frac{\partial n_A}{\partial x} \tag{VI.60}$$

$$= -\frac{n_A D_{A^\bullet}}{kT} \frac{\partial \mu_A}{\partial x}. \tag{VI.61}$$

Then, introducing the coupling forces, and using Nernst-Einstein equation:

$$< v >= FD/kT$$

one obtains for fluxes

$$\begin{cases} J_A = -\dfrac{n_A D_{A^\bullet}}{kT} \left[\dfrac{\partial \mu_A}{\partial x} + F_A\right] \\ J_B = -\dfrac{n_B D_{B^\bullet}}{kT} \left[\dfrac{\partial \mu_B}{\partial x} + F_B\right]. \end{cases} \tag{VI.62}$$

The relation (VI.58) between J_A and J_B is not sufficient to determine the two extra terms F_A and F_B introduced above, and another relation is required. It can be obtained from the principle of *minimum entropy production.*

The rate of entropy production is given as a function of the fluxes J_i and

([8]) For the molecular flux to be defined correctly, α and β have to be the smallest integers corresponding to the formula unit.

forces X_i by (see Ch. VIII):

$$\sigma = \sum_i J_i \, X_i, \tag{VI.63}$$

which becomes here:

$$\sigma = \frac{n_A \, D_{A^\bullet}}{kT} \left[\frac{\partial \mu_A}{\partial x} + F_A \right]^2 + \frac{n_B \, D_{B^\bullet}}{kT} \left[\frac{\partial \mu_B}{\partial x} + F_B \right]^2 . \tag{VI.64}$$

The coupling forces F_A and F_B are not independent. One relation is obtained by minimizing σ:

$$\frac{\mathrm{d}\sigma}{\mathrm{d}F_A} = 0. \tag{VI.65}$$

This leads to:

$$J_A + J_B \, \frac{\mathrm{d}F_B}{\mathrm{d}F_A} = 0. \tag{VI.66}$$

This relation, combined with the condition on the flux, yields:

$$\frac{n_A}{n_B} + \frac{\mathrm{d}F_B}{\mathrm{d}F_A} = 0,$$

whence, by integration:

$$n_A F_A + n_B F_B = 0, \tag{VI.67}$$

after the constant of integration is set equal to zero (the boundary condition is that F_A and F_B are to be simultaneously zero). Thus F_A and F_B can be eliminated from the flux equations, which yields, after a brief calculation:

$$\begin{cases} J_A = -\dfrac{1}{kT} \, n_A \, \dfrac{D_{A^\bullet} \, D_{B^\bullet}}{\alpha D_{B^\bullet} + \beta D_{A^\bullet}} \left[\alpha \, \dfrac{\mathrm{d}\mu_A}{\mathrm{d}x} + \beta \, \dfrac{\mathrm{d}\mu_B}{\mathrm{d}x} \right] \\ J_B = -\dfrac{1}{kT} \, n_B \, \dfrac{D_{A^\bullet} \, D_{B^\bullet}}{\alpha D_{B^\bullet} + \beta D_{A^\bullet}} \left[\alpha \, \dfrac{\mathrm{d}\mu_A}{\mathrm{d}x} + \beta \, \dfrac{\mathrm{d}\mu_B}{\mathrm{d}x} \right] . \end{cases} \tag{VI.68}$$

These results can be simplified to read:

$$J_m = \frac{J_A}{\alpha} = \frac{J_B}{\beta} = -\frac{n_A}{\alpha} \, \frac{D_m}{kT} \, \frac{\partial \mu_m}{\partial x}, \tag{VI.69}$$

where μ_m is the "molecular" chemical potential:

$$\mu_m = \alpha \mu_A + \beta \mu_B .$$

Since $n_m = n_A/\alpha$, we have:

$$\boxed{J_m = -\frac{n_m}{kT} D_m \frac{\partial \mu_m}{\partial x}}, \tag{VI.70}$$

an expression analogous to the Darken-Dehlinger equation written for a gradient of chemical potential (see above, Eq. (VI.61)), with:

$$D_m \,(\text{or } D_{\text{eff.}}) = \frac{D_{A^\bullet} D_{B^\bullet}}{\alpha D_{B^\bullet} + \beta D_{A^\bullet}}, \tag{VI.71}$$

or

$$\boxed{\frac{1}{D_m} = \frac{\alpha}{D_{A^\bullet}} + \frac{\beta}{D_{B^\bullet}}}, \tag{VI.71a}$$

an expression known as Herring's formula.
The thermodynamic factor does not enter explicitly into the expression for D_m because this is the coefficient of proportionality between the flux and a gradient of chemical potential, which can be caused by variation of the curvature of a surface (sintering, coalescence...), by a stress (Herring-Nabarro creep, climb of dislocations...) (see Ch. X, § III, IV, VII).

Notes:

1) If $D_{A^\bullet} >> D_{B^\bullet}$, then $D_m \approx D_{B^\bullet}/\beta$, i.e., the flux of material is governed by the diffusion of the *slower* component.

2) N.B. the difference between the intrinsic or interdiffusion coefficients, and the effective diffusion coefficient. In interdiffusion, the two constituents migrate in opposite directions. In material transport, they migrate in the same direction. For ionic crystals, the term ***ambipolar diffusion*** is used, to indicate that the cations and anions diffuse simultaneously. The other cases of ambipolar diffusion, in which two oppositely charged species (e.g. point atomic and electronic defects, such as $V_M^{'}$ and $h^\bullet$) diffuse simultaneously, have been discussed in the preceding paragraph. In all cases, the difference in the flux of the two species of opposite sign gives rise to an electric (Nernst) field which couples the two fluxes. Thus the same derivation applies.

3) It is interesting to carry out this calculation for a binary oxide of the metal M with the general formula $M_{|Z_O|} O_{|Z_M|}$, where $Z_M e$ and $Z_O e$ are the charges on the two species. The crystal is considered purely ionic, so that the Nernst field arises due to the difference in the ionic mobilities (see App. IV).

APPENDIX I

Variable molar volume. Problem of the frame of reference.

A concentration-dependent molar volume changes as the two constituents are mixed in the course of interdiffusion. The molar volume therefore varies with the distance in the sample; the number of sites per unit volume is no longer constant. The validity of the time-dependent diffusion equation (I.14), and the Boltzmann-Matano formula, as well as of the Darken equation, then comes into question.

THE FRAME OF REFERENCE. — Component fluxes must be measured with respect to some frame of reference; the diffusion coefficient thus depends on the choice of reference frame. Reference frames based on a mean velocity of all the components, such as the velocity of the center of mass, are particularly convenient for relating results obtained using different mean velocity frames (Brady, 1975, see also Ch. VIII, App. III).

Consider the case of unidirectional diffusion. Let R^0 be the frame of reference fixed in one end of the sample (also called the laboratory frame). In R^0, we measure the distance x and the flux J_i^0. Let v^R be the velocity (which generally varies with x) measured with respect to R^0 of a frame of reference R in which the fluxes are J_i^R; then:

$$J_i^0 = J_i^R + c_i\ v^R \tag{A.1}$$

(see Ch VIII, § I). The flux is expressed in number of moles per unit area per unit time, concentrations, c_i, in number of moles per unit volume.

REVIEW OF SOME THERMODYNAMIC RELATIONSHIPS. — Molar volume V_m, partial molar volumes $\bar{V}_i$:

$$V_m = N_A\ \bar{V}_A + N_B\ \bar{V}_B \tag{A.2}$$
$$dV_m = \bar{V}_A\ dN_A + \bar{V}_B\ dN_B \tag{A.3}$$
$$0 = N_A\ d\bar{V}_A + N_B\ d\bar{V}_B \tag{A.4}$$
$$1 = c_A\ \bar{V}_A + c_B\ \bar{V}_B \tag{A.5}$$
$$0 = \bar{V}_A\ dc_A + \bar{V}_B\ dc_B \tag{A.6}$$

Relation between concentrations and mole fractions:

$$c_i = N_i\,(c_{\mathrm{A}} + c_{\mathrm{B}}) = N_i/V_{\mathrm{m}} \tag{A.7}$$

$$\mathrm{d}c_i = (\bar{V}_j/V_{\mathrm{m}}^2)\,\mathrm{d}N_i \qquad i \neq j. \tag{A.7a}$$

To obtain equation (A.7a), equation (A.7) is differentiated, $\mathrm{d}c_{\mathrm{B}}$ is eliminated with the use of equation (A.6), and the terms are rearranged following equation (A.2).

MEAN VOLUME FRAME OF REFERENCE, $\mathrm{R^V}$. — The frame of reference $\mathrm{R^V}$ moves with respect to $\mathrm{R^0}$ "with the volume", so that the net flux expressed in volume of material traversing a plane normal to the diffusion direction is zero everywhere and at all times:

$$\bar{V}_{\mathrm{A}}\,J_{\mathrm{A}} + \bar{V}_{\mathrm{B}}\,J_{\mathrm{B}} = 0. \tag{A.8}$$

The velocity v^{V} of $\mathrm{R^V}$ with respect to $\mathrm{R^0}$ is obtained by combining equation (A.8) with equation (A.1) written for A and B:

$$v^{\mathrm{V}} = \bar{V}_{\mathrm{A}}\,J_{\mathrm{A}}^0 + \bar{V}_{\mathrm{B}}\,J_{\mathrm{B}}^0. \tag{A.9}$$

Writing Fick's law for each flux, equation (A.8) becomes:

$$\bar{V}_{\mathrm{A}}\,D_{\mathrm{A}}^{\mathrm{V}}\,(\partial c_{\mathrm{A}}/\partial x) + \bar{V}_{\mathrm{B}}\,D_{\mathrm{B}}^{\mathrm{V}}\,(\partial c_{\mathrm{B}}/\partial x) = 0.$$

Following equation (A.6), this relation is true only if:

$$D_{\mathrm{A}}^{\mathrm{V}} = D_{\mathrm{B}}^{\mathrm{V}} = \tilde{D}^{\mathrm{V}}. \tag{A.10}$$

The diffusion coefficient is *unique*: it is the *interdiffusion coefficient* $\tilde{D}^{\mathrm{V}}$ in the frame of reference $\mathrm{R^V}$.

Apply the conservation equation to J_{A}^0:

$$\frac{\partial c_{\mathrm{A}}}{\partial t} = \frac{\partial}{\partial x}\left[\tilde{D}^{\mathrm{V}}\,\frac{\partial c_{\mathrm{A}}}{\partial x} - c_{\mathrm{A}}\,v^{\mathrm{V}}\right]. \tag{A.11}$$

The classic time-dependent diffusion equation cannot be used in $\mathrm{R^V}$, unless the partial molar volumes are constant, which leads to $v^{\mathrm{V}} = 0$ (*vide infra* Eq. (A.30)).

MEAN MOLAR FRAME OF REFERENCE, $\mathrm{R^N}$ (or number-fixed frame of reference). — This frame of reference is defined so that the net flux in number of moles per unit area per unit time is zero across any plane:

$$J_{\mathrm{A}}^{\mathrm{N}} + J_{\mathrm{B}}^{\mathrm{N}} = 0. \tag{A.12}$$

Using equation (A.1), one finds:

$$v^{\mathrm{N}} = V_{\mathrm{m}} \left(J_{\mathrm{A}}^{0} + J_{\mathrm{B}}^{0} \right). \tag{A.13}$$

Equation (A.1) applied to R^{N} and R^{V} yields:

$$J_i^{\mathrm{V}} = J_i^{\mathrm{N}} + c_i \, v^{\mathrm{N/V}}. \tag{A.14}$$

We expand the expression for J_i^{V}; following equation (A.7):

$$J_i^{\mathrm{V}} = -\tilde{D}^{\mathrm{V}} \frac{\partial c_i}{\partial x} = -\frac{\tilde{D}^{\mathrm{V}}}{V_{\mathrm{m}}} \frac{\partial N_i}{\partial x} - c_i \left[V_{\mathrm{m}} \, \tilde{D}^{\mathrm{V}} \cdot \partial \left(1/V_{\mathrm{m}}\right) / \partial x \right].$$

The quantities J_i^{N} and $v^{\mathrm{N/V}}$ can be identified by comparison with equation (A.14):

$$\begin{cases} J_i^{\mathrm{N}} & = -\dfrac{\tilde{D}^{\mathrm{V}}}{V_{\mathrm{m}}} \dfrac{\partial N_i}{\partial x} \\ v^{\mathrm{N/V}} & = -V_{\mathrm{m}} \, \tilde{D}^{\mathrm{V}} \cdot \partial \left(1/V_{\mathrm{m}}\right) / \partial x. \end{cases} \tag{A.15}$$

The fluxes given in equation (A.15) fulfill the requirement of equation (A.12) because $N_{\mathrm{A}} + N_{\mathrm{B}} = 1$.

Following equation (A.7a):

$$J_i^{\mathrm{V}} = -\tilde{D}^{\mathrm{V}} \frac{\partial c_i}{\partial x} = -\tilde{D}^{\mathrm{V}} \frac{\bar{V}_j}{V_{\mathrm{m}}^2} \frac{\partial N_i}{\partial x},$$

which, together with equation (A.15), yields:

$$J_i^{\mathrm{V}} = \left(\bar{V}_j / V_{\mathrm{m}} \right) J_i^{\mathrm{N}},$$

Choosing an abscissa ξ in R^{N} such that (Onishi and Shimozaki, 1983):

$$\mathrm{d}\xi = \frac{1}{V_{\mathrm{m}}} \, \mathrm{d}x, \tag{A.16}$$

and defining a diffusion coefficient by:

$$J_{\mathrm{A}} = -\tilde{D}_{\xi} \left(\partial N_{\mathrm{A}} / \partial \xi \right), \tag{A.17}$$

we find that this is unique, because, from comparison with equation (A.15):

$$\tilde{D}_\xi = \tilde{D}^{\mathrm{V}} / V_{\mathrm{m}}^2. \tag{A.17a}$$

Thus it has been shown that:

$$\left(\frac{\partial N_i}{\partial t}\right) = \frac{\partial}{\partial \xi} \left(\tilde{D}_\xi \, \frac{\partial N_i}{\partial \xi}\right). \tag{A.18}$$

The classic diffusion equation is thus recovered in R^{N}, with a realistic choice of variables: N_i, ξ, and an interdiffusion coefficient $\tilde{D}_\xi$ which can be calculated by the Boltzmann-Matano method. (See also Exercise 17.) The following formula, known as the Sauer and Freise-den Broeder formula, is easily derived :

$$\tilde{D} = \frac{V_{\mathrm{m}}}{2t} \left(\frac{\partial Y}{\delta x}\right)^{-1} \left[(1-Y) \int_{-\infty}^{x} \frac{Y}{V_{\mathrm{m}}} \, \mathrm{d}x + Y \int_{x}^{\infty} \frac{1-Y}{V_{\mathrm{m}}} \, \mathrm{d}x\right], \tag{A.19}$$

where

$$Y = \left[N_{\mathrm{i}}(x) - N_i(-\infty)\right] / \left[N_i(+\infty) - N_i(-\infty)\right].$$

Mean mass (or barycentric) frame of reference. — This frame of reference is defined so that the net flux of mass across every plane is zero:

$$M_{\mathrm{A}} \, J_{\mathrm{A}}^{\mathrm{M}} + M_{\mathrm{B}} \, J_{\mathrm{B}}^{\mathrm{M}} = 0, \tag{A.20}$$

where M_{A} and M_{B} denote the molar masses. We use the mass fractions w_i so that:

$$c_i = w_i \, \rho / M_i, \tag{A.21}$$

where ρ is the density (mass per unit volume). Differentiating as above:

$$\begin{aligned} J_i^{\mathrm{V}} = -\tilde{D}^{\mathrm{V}} \frac{\partial c_i}{\partial x} &= -\frac{\rho}{M_i} \, \tilde{D}^{\mathrm{V}} \, \frac{\partial w_i}{\partial x} - c_i \left(\frac{\tilde{D}^{\mathrm{V}}}{\rho} \, \frac{\partial \rho}{\partial x}\right) \\ &= J_i^{\mathrm{M}} - c_i \, v^{\mathrm{M/V}}. \end{aligned} \tag{A.22}$$

It is easily seen that the flux

$$J_i^{\mathrm{M}} = -\frac{\rho}{M_i} \, \tilde{D}^{\mathrm{V}} \, \frac{\partial w_i}{\partial x} \tag{A.23}$$

follows equation (A.20).

Define the abscissa ξ' in R^M:

$$d\xi' = \rho \, dx, \tag{A.24}$$

then equation (A.18) becomes:

$$\left(\frac{\partial w_i}{\partial t}\right) = \frac{\partial}{\partial \xi'}\left(\tilde{D}_{\xi'} \, \frac{\partial w_i}{\partial \xi'}\right), \tag{A.25}$$

i.e., the classic diffusion equation with the variables w_i, ξ', and the interdiffusion coefficient:

$$\tilde{D}_{\xi'} = \rho^2 \, \tilde{D}^V, \tag{A.26}$$

again calculable by the Boltzmann-Matano method.

THE LABORATORY FRAME OF REFERENCE R^0. — From the expression for the fluxes in R^V, with the diffusion coefficient $\tilde{D}$ (the superscript V has been dropped for ease of writing), the flux J^0 can be calculated from equation (A.1) if the velocity v^V with respect to R^0 is known:

$$J_i^0 = -\tilde{D} \, \frac{\partial c_i}{\partial x} + c_i \, v^V. \tag{A.27}$$

Applying the conservation equation in R^0:

$$\frac{\partial c_i}{\partial t} = -\frac{\partial J_i^0}{\partial x}$$

we obtain for the constituent B:

$$\frac{\partial c_B}{\partial t} = \frac{\partial}{\partial x}\left(\tilde{D} \, \frac{\partial c_B}{\partial x}\right) - c_B \, \frac{\partial v^V}{\partial x} - \frac{\partial c_B}{\partial x} \, v^V. \tag{A.28}$$

Using equation (A.6):

$$-\frac{\bar{V}_A}{\bar{V}_B} \, \frac{\partial c_A}{\partial t} = -\frac{\partial}{\partial x}\left(\tilde{D} \, \frac{\bar{V}_A}{\bar{V}_B} \, \frac{\partial c_A}{\partial x}\right) - c_B \, \frac{\partial v^V}{\partial x} + \frac{\bar{V}_A}{\bar{V}_B} \, \frac{\partial c_A}{\partial x} \, v^V,$$

combining with:

$$\frac{\partial c_A}{\partial t} = \frac{\partial}{\partial x}\left(\tilde{D} \, \frac{\partial c_A}{\partial x}\right) - c_A \, \frac{\partial v^V}{\partial x} - \frac{\partial c_A}{\partial x} \, v^V,$$

this becomes:

$$\frac{\partial v^{\mathrm{V}}}{\partial x} = \bar{V}_{\mathrm{A}} \frac{\partial}{\partial x} \left(\tilde{D} \frac{\partial c_{\mathrm{A}}}{\partial x} \right) - \bar{V}_{\mathrm{B}} \frac{\partial}{\partial x} \left(\tilde{D} \frac{\bar{V}_{\mathrm{A}}}{\bar{V}_{\mathrm{B}}} \frac{\partial c_{\mathrm{A}}}{\partial x} \right). \tag{A.29}$$

Integrating by parts:

$$v^{\mathrm{V}} = \bar{V}_{\mathrm{A}} \tilde{D} \frac{\partial c_{\mathrm{A}}}{\partial x} - \int_{-\infty}^{x} \tilde{D} \frac{\partial c_{\mathrm{A}}}{\partial x} \, \mathrm{d}\bar{V}_{\mathrm{A}} - \bar{V}_{\mathrm{B}} \tilde{D} \frac{\bar{V}_{\mathrm{A}}}{\bar{V}_{\mathrm{B}}} \frac{\partial c_{\mathrm{A}}}{\partial x} + \int_{-\infty}^{x} \tilde{D} \frac{\bar{V}_{\mathrm{A}}}{\bar{V}_{\mathrm{B}}} \frac{\partial c_{\mathrm{A}}}{\partial x} \, \mathrm{d}\bar{V}_{\mathrm{B}}$$

and regrouping terms:

$$v^{\mathrm{V}} = - \int_{-\infty}^{x} \frac{\tilde{D}}{c_{\mathrm{B}} \bar{V}_{\mathrm{B}}} \left(\frac{\partial \bar{V}_{\mathrm{A}}}{\partial c_{\mathrm{A}}} \right) \left(\frac{\partial c_{\mathrm{A}}}{\partial x} \right)^2 \mathrm{d}x. \tag{A.30}$$

If the partial molar volumes are constant, $v^{\mathrm{V}} = 0$, and R^0 and R^{V} coincide.

Defining pseudo coefficients θ_i in R^0:

$$J_i^0 = -\theta_i \frac{\partial c_i}{\partial x} = -\tilde{D} \frac{\partial c_i}{\partial x} + c_i \int_{-\infty}^{x} \frac{\tilde{D}}{c_{\mathrm{A}} \bar{V}_{\mathrm{B}}} \left(\frac{\partial \bar{V}_{\mathrm{B}}}{\partial c_{\mathrm{A}}} \right) \left(\frac{\partial c_{\mathrm{A}}}{\partial x} \right)^2 \mathrm{d}x, \tag{A.31}$$

we see that $J_{\mathrm{A}}^0 + J_{\mathrm{B}}^0 \neq 0$. Thus *the Boltzmann-Matano method is not applicable in* R^0.

THE KIRKENDALL FRAME OF REFERENCE R^{K}. — This frame of reference is tied to the lattice planes so that the net flux of the number of lattice sites across any plane is zero:

$$J_{\mathrm{A}}^{\mathrm{K}} + J_{\mathrm{B}}^{\mathrm{K}} + J_{\mathrm{V}}^{\mathrm{K}} = 0.$$

Darken's derivation of this frame of reference was given in this chapter, § II.1. Here we compare the reference frames R^{V} (in which $\tilde{D}$ is defined) and R^{K}, with relative velocity $v^{\mathrm{K/V}} = v^{\mathrm{K}} - v^{\mathrm{V}}$

$$J_i^{\mathrm{V}} = J_i^{\mathrm{K}} + c_i \, v^{\mathrm{K/V}}, \tag{A.32}$$

from which, using equations (A.8) and (A.5):

$$\begin{aligned} v^{\mathrm{K/V}} &= -\left(\bar{V}_{\mathrm{A}} J_{\mathrm{A}}^{\mathrm{K}} + \bar{V}_{\mathrm{B}} J_{\mathrm{B}}^{\mathrm{K}} \right) \\ &= \bar{V}_{\mathrm{A}} \bar{D}_{\mathrm{A}} \frac{\partial c_{\mathrm{A}}}{\partial x} + \bar{V}_{\mathrm{B}} \bar{D}_{\mathrm{B}} \frac{\partial c_{\mathrm{B}}}{\partial x} \end{aligned}$$

$$\boxed{v^{K/V} = \left(\bar{D}_A - \bar{D}_B\right)\ \bar{V}_A\,(\mathrm{d}c_A/\mathrm{d}x)}\,, \tag{A.33}$$

introducing the intrinsic diffusion coefficients $\bar{D}_A$ and $\bar{D}_B$. Since:

$$J_i^V = -\tilde{D}\,\frac{\partial c_i}{\partial x},$$

we obtain:

$$\boxed{\tilde{D} = c_B\ \bar{V}_B\ \bar{D}_A + c_A\ \bar{V}_A\ \bar{D}_B}\,. \tag{A.34}$$

This *equation applies to the frame of reference* R^V. The corresponding equations for the reference frames R^N and R^M are easily deduced from it:

$$\tilde{D} = N_B\ \bar{D}'_A + N_A\ \bar{D}'_B \tag{A.34a}$$

$$\tilde{D} = w_B\ \bar{D}''_A + w_A\ \bar{D}''_B, \tag{A.34b}$$

with:

$$\bar{D}'_i = \bar{D}_i\,(\bar{V}_j/V_m) \qquad i \neq j \tag{A.35}$$

$$\bar{D}''_i = \bar{D}_i\,(\bar{V}_j\ \rho/M_j)\,, \tag{A.35a}$$

so that the intrinsic fluxes are given by relations analogous to those of interdiffusion (Eqs. (A.15) and (A.23)):

$$\begin{aligned} J_i^K &= -\bar{D}_i\,(\partial c_i/\partial x) \\ &= -\left(\bar{D}'_i/V_m\right)(\partial N_i/\partial x) \\ &= -\left(\rho\ \bar{D}''_i/M_i\right)(\partial w_i/\partial x)\,. \end{aligned} \tag{A.36}$$

No simple relation exists for either v^K *or* $\tilde{D}$ *in the* R^0 *reference frame.* Practical work has to be done in R^V, R^M, or R^N, where simple equations are available (N. B. the velocity to be used is *not* v^K but the velocity $v^{K/R}$ relative to the reference frame under consideration) or in R^0, but using the correct equations, analogous to equations (A.30) and (A.31), except if $\bar{V}_A$ and $\bar{V}_B$ are constant, in which case $R^V \equiv R^0$.

It is perhaps easier to work in the reference frame R^N than in R^V in order to determine the relative velocity $v^{K/N}$. Defining the flux as in equation (A.17), with the variables ξ and N_i:

$$(\mathrm{R^N})\begin{cases} v^{\mathrm{K/N}} = \left(\bar{D}'_{\mathrm{A}} - \bar{D}'_{\mathrm{B}}\right)(V_{\mathrm{m}})^{-2}\,(\mathrm{d}N_{\mathrm{A}}/\mathrm{d}\xi) & \text{(A.33a)} \\ \tilde{D} = N_{\mathrm{B}}\,\bar{D}'_{\mathrm{A}} + N_{\mathrm{A}}\,\bar{D}'_{\mathrm{B}}. & \text{(A.34a)} \end{cases}$$

EXPRESSIONS FOR v^{K}, $\tilde{D}$, $\bar{D}_{\mathrm{A}}$, AND $\bar{D}_{\mathrm{B}}$ IN THE REFERENCE FRAME R^0.

The Kirkendall velocity. — With $v^{\mathrm{K}}=v^{\mathrm{K/V}}+v^{\mathrm{V}}$, following equations (A.30) and (A.33):

$$v^{\mathrm{K}} = \bar{V}_{\mathrm{A}}\left(\bar{D}_{\mathrm{A}} - \bar{D}_{\mathrm{B}}\right)\frac{\mathrm{d}c_{\mathrm{A}}}{\mathrm{d}x} + \int_{-\infty}^{x} \frac{\tilde{D}}{c_{\mathrm{A}}\,\bar{V}_{\mathrm{B}}}\,\frac{\partial \bar{V}_{\mathrm{B}}}{\partial c_{\mathrm{A}}}\left(\frac{\partial c_{\mathrm{A}}}{\partial x}\right)^2 \mathrm{d}x. \qquad \text{(A.37)}$$

Interdiffusion coefficient. — Starting from the conservation equation applied to equation (A.27):

$$\frac{\partial c_i}{\partial x} = \frac{\partial}{\partial x}\left(\tilde{D}\,\frac{\partial c_i}{\partial x} - c_i\,v^{\mathrm{V}}\right), \qquad \text{(A.38)}$$

with v^{V} given by equation (A.30). Prager's method consists of applying the Boltzmann transformation $\lambda = x/\sqrt{t}$; equation (A.38) is integrable, and one proceeds by the method of successive approximations beginning with a $\tilde{D}$ determined by the Matano method.

One can also apply the function change defined in equation (A.31) to the intrinsic fluxes:

$$J_i^0 = -\theta_i\,\frac{\partial c_i}{\partial x} = -\bar{D}_i\,\frac{\partial c_i}{\partial x} + c_i\,v^{\mathrm{K}}. \qquad \text{(A.39)}$$

The conservation equation applied to J_i^0 leads to a determination of θ_i by the Boltzmann-Matano method. The calculations finally yield the formulas first given by Balluffi:

$$\begin{cases} \tilde{D} & = -\dfrac{1}{2t}\left(\dfrac{\partial c_{\mathrm{A}}}{\partial x}\right)^{-1}\left[1 - c_{\mathrm{A}}\left(\bar{V}_{\mathrm{A}} - \bar{V}_{\mathrm{B}}\right)\displaystyle\int_{c_{\mathrm{A}}(-\infty)}^{c_{\mathrm{A}}(x)} x \; \mathrm{d}c_{\mathrm{A}} + \right. \\ & \qquad \left. + c_{\mathrm{A}}\, \bar{V}_{\mathrm{B}} \displaystyle\int_{c_{\mathrm{A}}(-\infty)}^{c_{\mathrm{A}}(x)} \dfrac{\bar{V}_{\mathrm{A}} - \bar{V}_{\mathrm{B}}}{\bar{V}_{\mathrm{B}}}\, x \; \mathrm{d}c_{\mathrm{A}}\right]. \\ \bar{D}_i & = \dfrac{1}{2\,t}\left(\dfrac{\partial c_i}{\partial x}\right)^{-1}\left[2\,t\,v^{\mathrm{K}}\,c_i - \displaystyle\int_{c_i(-\infty)}^{c_i(x)} x \; \mathrm{d}c_i\right]. \end{cases} \qquad \text{(A.40)}$$

The formula for $\tilde{D}$ is equivalent to the Sauer-Freise-den Broeder formula (A.19) (Van Loo, 1970).

APPENDIX II

Kröger-Vink notation.

This notation is most often used as a shorthand notation for atomic point defects and electronic defects in compounds. Consider a compound AB with the species A and B forming separate sublattices; the following species are considered:

A_A	atom A on an A-site,
V_A	vacant A-site,
A_i	A atom in an interstitial site,
A_B	A atom on a B-site (antisite defect)
e	free electron,
h	electron hole.

Effective electric charges in units of |e| with respect to the normal charge of the site under consideration are indicated as follows:

$M^{\times}$	neutral,
$M^{\bullet}$, $M^{\bullet\bullet}$,	singly positive, doubly positive ... charge,
$M^{'}$, $M^{''}$,	singly negative, doubly negative ... charge.

The equations of chemical equilibrium can be written for defects the same way as for the usual chemical species, if the following rules are obeyed:

1) electrical neutrality, checked by summing the superscripts indicating electrical charge ($\mathrm{x} = 0$, $^{\bullet} = +1$, $^{'} = -1$);

2) conservation of sites: the ratio of the species on A sites and on B sites is equal to the ratio given by stoichiometry, i.e., by the compound formula such as $A_\alpha B_\beta$; this rule is verified by adding the subscripts weighted by the reaction stoichiometric coefficients;

3) conservation of mass, verified from the sum of the chemical symbols, with $V = 0$, no matter what index it carries.

The application of the last two rules explains why the interstitial is denoteds by A_i, not by I_A.

APPENDIX III

Deviation from stoichiometry in a binary oxide.

For each formation "reaction" of a type of defect, a corresponding mass-action law can be written. The condition of electrical neutrality furnishes a supplementary equation; the latter is often simplified by neglecting the minority defects in a defined domain of temperature or oxygen partial pressure (the nature of the majority defects depends on these two variables).

Example: The non-stoichiometric oxide $M_{1-\delta}O$, with defects V_M^x, $V_M^{'}$, V_O^x, $V_O^\bullet$, $e^{'}$, and $h^\bullet$. The equations for the defect equilibria are written in terms of mole fractions:

$$[V_M^x]\,[V_O^x] = K_S \tag{B.1}$$

$$[V_M^x] = K^F\,P_{O_2}^{1/2} \tag{B.2}$$

$$[V_M']\,[h^\bullet] = K_M^I\,[V_M^x] \tag{B.3}$$

$$[V_O^\bullet]\,[e'] = K_O^I\,[V_O^x] \tag{B.4}$$

$$[e']\,[h^\bullet] = K_I. \tag{B.5}$$

The first equation is that of the Schottky equilibrium:

$$V_M^x + V_O^x \rightleftharpoons 0.$$

The second is for oxidation, knowing that $[O_O] = 1$:

$$1/2\ O_2 \rightleftharpoons O_O + V_M^x.$$

The third is for ionization of the cation vacancy V_M^x:

$$V_M^x \rightleftharpoons V_M' + h^\bullet.$$

The fourth is for ionization of the anion vacancy V_O^x:

$$V_O^x \rightleftharpoons V_O^\bullet + e',$$

and the fifth for the thermal equilibrium of the electronic charge carriers.

The condition of electrical neutrality:

$$[V'_M] + [e'] = [V^{\bullet}_O] + [h^{\bullet}] \tag{B.6}$$

furnishes a sixth equation. Thus the concentration of the six types of defects can be calculated as a function of P_{O_2} and T, via the equilibrium constants for the above equations.

The deviation from stoichiometry is given by:

$$\delta = [V^{x}_M] + [V'_M] - [V^{x}_O] - [V^{\bullet}_O]\,. \tag{B.7}$$

At high P_{O_2}, the majority defects are V'_M. The equation for charge balance simplifies to:

$$[V'_M] \approx [h^{\bullet}]\,, \tag{B.6a}$$

from which, following equations (B.2), (B.3) and (B.6a):

$$[V'_M] = \left(K^{I}_M \cdot K^{F}\right)^{1/2} \left(P_{O_2}\right)^{1/4}.$$

In nickel oxide, $Ni_{1-\delta}O$, nickel-ion vacancies are normally the majority atomic defects, but they can be singly or doubly ionized: $V^{\alpha'}_M$ with $\alpha = 1$ or 2. The equation for charge neutrality is then:

$$\left[V^{\alpha'}_M\right] = \alpha\,[h^{\bullet}]\,.$$

Since formally $V^{x}_M \rightleftharpoons V^{\alpha'}_M + \alpha h^{\bullet}$, we obtain:

$$\left[V^{\alpha'}_M\right][h^{\bullet}]^{\alpha} = K^{I\alpha}_M\,[V^{x}_M]\,.$$

The calculation leads to:

$$\left[V^{\alpha'}_M\right] = \left(K^{I\alpha}_M \cdot K^{F}\,\alpha^{\alpha}\right)^{\frac{1}{\alpha+1}} \left(P_{O_2}\right)^{\frac{1}{2\,(\alpha+1)}}.$$

For the details, the reader is referred to, e.g., the paper by C. Monty, Confolant 1977, chapter 12, p. 345.

N.B.: To write the law of mass action with a dimensionless equilibrium constant, the chemical activities are used, or, for ideal solutions, the mole fractions. This poses a problem for the electronic species, the chemical potentials of which are written:

$$\mu_e = \mu^0_e + kT\ \mathrm{Log}\,(n/N_e)\,,$$
$$\mu_h = \mu^0_h + kT\ \mathrm{Log}\,(p/N_h)\,,$$

where n and p are the numbers of electrons and holes, and N_e and N_h the number of accessible states, respectively, per unit volume.

For electron band conduction, N_e equals the "density of effective states" in the conduction band:

$$N_e = 2\left(m_e^* \; kT/\left(2\;\pi\hbar^2\right)\right)^{3/2},$$

with an analogous expression for N_h, with effective masses m_e^* and m_h^*. The intrinsic equilibrium is then written:

$$np = N_e \; N_h \; \exp\left(-E_g/kT\right),$$

where E_g is the band gap.

For a thermally activated conduction mechanism (small polaron), the same equation applies, except that N_e and N_h here mean the number of sites for the carriers e' and $h^{\bullet}$ per unit volume, i.e., the number of corresponding ions. Thus the preexponential factors for the two mechanisms will differ considerably, since $N_e \approx 10^{19}$ cm^{-3} for band conduction, and $\approx 10^{22}$ for a small polaron.

Not knowing the conduction mechanism, the physical chemist may write the equations for defect equilibria with p and n, or with mole fractions $[h^{\bullet}]$ and $[e']$ obtained by conventionally dividing p or n by the concentration of cations and anions. Care must be taken with the preexponentials which this introduces in front of the equilibrium constants, and which can depend on the temperature. For more details, the reader is referred to Kröger, *op. cit.*, §7.11 and 9.6-9.8.

APPENDIX IV

Ambipolar diffusion in a binary oxide.

Consider a binary oxide with formula $M_{|Z_O|}O_{|Z_M|}$, where $Z_M\,|e|$ and $Z_O\,|e|$ indicate the charges on the two components. During mass transport, the fluxes of metal and of oxygen are in the same direction; they obey the relation:

$$J_m = \frac{J_M}{|Z_O|} = \frac{J_O}{|Z_M|}. \tag{C.1}$$

The flux of each species is written in terms of concentration gradients and electromigration:

$$\begin{aligned} J_i &= -\bar{D}_i \; \frac{\partial n_i}{\partial x} + n_i \; \frac{Z_i\,|e|}{kT} \; E_d \; D_{i\bullet} \\ &= -\;\frac{n_i \; D_{i\bullet}}{kT}\left[\frac{\partial \mu_i}{\partial x} - Z_i \; |e| \; E_d\right]. \end{aligned} \tag{C.2}$$

The relation between the fluxes J_{M} and J_{O} becomes:

$$|\mathrm{Z_M}|\,\frac{n_{\mathrm{M}}\,D_{\mathrm{M}^\bullet}}{kT}\left[\frac{\partial\mu_{\mathrm{M}}}{\partial x}-Z_{\mathrm{M}}\;|e|\;E_{\mathrm{d}}\right]=|\mathrm{Z_O}|\,\frac{n_{\mathrm{O}}\,D_{\mathrm{O}^\bullet}}{kT}\left[\frac{\partial\mu_{\mathrm{O}}}{\partial x}-Z_{\mathrm{O}}\;|e|\;E_{\mathrm{d}}\right],$$

from which the Nernst field is:

$$eE_{\mathrm{d}}=\frac{|\mathrm{Z_M}|\;n_{\mathrm{M}}\;D_{\mathrm{M}^\bullet}\,(\partial\mu_{\mathrm{M}}/\partial x)-|\mathrm{Z_O}|\;n_{\mathrm{O}}\;D_{\mathrm{O}^\bullet}\,(\partial\mu_{\mathrm{O}}/\partial x)}{|\mathrm{Z_M}|\;Z_{\mathrm{M}}\;n_{\mathrm{M}}\;D_{\mathrm{M}^\bullet}-|\mathrm{Z_O}|\;Z_{\mathrm{O}}\;n_{\mathrm{O}}\;D_{\mathrm{O}^\bullet}}.\tag{C.3}$$

The molecular flux $J_{\mathrm{m}}=J_{\mathrm{M}}/\,|\mathrm{Z_O}|$ can thus be calculated by substituting this value of E_{d}, and using the chemical potential of the oxide, given by:

$$\begin{aligned}\mathrm{d}\mu_{\text{oxide}}&=|Z_{\mathrm{M}}|\;\mathrm{d}\mu_{\mathrm{O}}+|Z_{\mathrm{O}}|\;\mathrm{d}\mu_{\mathrm{M}}\\&=Z_{\mathrm{M}}\;\mathrm{d}\mu_{\mathrm{O}}-Z_{\mathrm{O}}\;\mathrm{d}\mu_{\mathrm{M}}.\end{aligned}\tag{C.4}$$

A lengthy, but not difficult, calculation then yields:

$$J_{\mathrm{m}}=\frac{J_{\mathrm{M}}}{|\mathrm{Z_O}|}=-\frac{n_{\mathrm{O}}\;D_{\mathrm{O}^\bullet}\;n_{\mathrm{M}}\;D_{\mathrm{M}^\bullet}}{Z_{\mathrm{M}}\;|\mathrm{Z_M}|\;n_{\mathrm{M}}\;D_{\mathrm{M}^\bullet}-\;Z_{\mathrm{O}}\;|\mathrm{Z_O}|\;n_{\mathrm{O}}\;D_{\mathrm{O}^\bullet}}\,\frac{1}{kT}\,\frac{\partial\mu_{\text{oxide}}}{\partial x}.\tag{C.5}$$

This expression has the classic form, and involves an effective diffusion coefficient. To apply it, the "molecule" must first be defined. Take for example MgO: $\mathrm{Z_{Mg}}=|\mathrm{Z_O}|=2$. But the formula is not $\mathrm{Mg_2O_2}$. In the latter case, a factor of 2 is introduced twice into our calculation: in $\mathrm{d}\mu_{\text{oxide}}$ and in $J_{\mathrm{m}}=J_{\mathrm{M}}/\,|\mathrm{Z_O}|$. The flux of MgO is obtained by multiplying the result of the formula (C.5) by 4:

$$J_{\mathrm{MgO}}=-\frac{D_{\mathrm{O}^\bullet}\;D_{\mathrm{Mg}^\bullet}}{D_{\mathrm{O}^\bullet}+D_{\mathrm{Mg}^\bullet}}\,\frac{n}{kT}\,\frac{\partial\mu_{\mathrm{MgO}}}{\partial x},\tag{C.6}$$

with $n_{\mathrm{O}}=n_{\mathrm{Mg}}=n$ "molecules" of MgO per unit volume. The electrical coupling of the ambipolar diffusion indeed leads to the *molecular or ambipolar diffusion coefficient*:

$$D_{\mathrm{m}}=D_{\text{ambi}}=\frac{D_{\mathrm{O}^\bullet}\;D_{\mathrm{Mg}^\bullet}}{D_{\mathrm{O}^\bullet}+D_{\mathrm{Mg}^\bullet}}\tag{C.7}$$

which measures the transport of material, i.e., of both cations and anions, in the same direction.

The expression for the Nernst field then reduces to:

$$eE_{\mathrm{d}}=\frac{1}{2}\,\frac{D_{\mathrm{Mg}^\bullet}\,(\partial\mu_{\mathrm{Mg}}/\partial x)-D_{\mathrm{O}^\bullet}\,(\partial\mu_{\mathrm{O}}/\partial x)}{D_{\mathrm{Mg}^\bullet}+D_{\mathrm{O}^\bullet}}.\tag{C.8}$$

For an ideal solution, the two gradients of chemical potential are equal, and the field is truly proportional to the difference of the two self-diffusion coefficients.

Since $D_{\mathrm{Mg}^\bullet} >> D_{\mathrm{O}^\bullet}$:

$$eE_\mathrm{d} \approx \frac{1}{2} \cdot \frac{\partial \mu_{\mathrm{Mg}}}{\partial x}. \qquad \text{(C.8a)}$$

Bibliography

I. GENERAL REFERENCES

ADDA Y. and PHILIBERT J., *op. cit.*, Ch. X and XIX.

KOFSTAD P., *Non-stoichiometry, Diffusion and Electrical Conductivity in Binary Oxides* (Wiley-Interscience, New York) 1972.

KRÖGER F. A., *The Chemistry of Imperfect Crystals*, 2nd ed., 3 vols. (North Holland, Amsterdam) 1973.

MANNING J. R., *op. cit.* (1968), Ch. V.

WAGNER C., Equations for Transport in Solid Oxides and Sulfides of Transition Metals, *Prog. Solid State Chem.* **10** (1975) 3-16.

II. SPECIFIC REFERENCES

CORNET J. and CALAIS D., *J. Phys. Chem. Solids* **33** (1972) 1675-1684.

CORNET J., *ibid.* **35** (1974) 1247-1252.

DARKEN L. S., *Trans. AIME* **175** (1948) 184-194.

DEN BROEDER F. J. A., *Scripta Metall.* **3** (1969) 321-326.

HEUMANN Th., *Z. Naturforsch.* **32a** (1977) 54-56.

HEUMANN Th., *Dimeta-82, op. cit.* (1983) 117-144.

LEVASSEUR J. and PHILIBERT J., *Phys. Status Solidi* **21** (1967) K1-5.

LI C. Y., BLAKELY J. M., and FEINGOLD A. H., *Acta Metall.* **14** (1960) 1397-1402.

MEYER R. O., *Phys. Rev.* **181** (1969) 1086-1094.

MONTY C., Thesis, Université Paris-Sud, Orsay (1972).

SMIGELSKAS A. D. and KIRKENDALL E. O., *Trans. AIME* **171** (1947) 130-142.

III. REFERENCES TO APPENDIX I

BALLUFFI R. W., *Acta Metall.* **8** (1960) 871-873.

BRADY J. B., *Am. Jour. Sci.* **275** (1975) 954-983.

CRANK J., *op. cit.*, Ch. 10.

KIRKWOOD J., BALDWIN R., DUNLOP P., GOSTING L. and KEGELES G., *J. Chem. Phys.* **33** (1960) 1505-1513.

ONISHI N. and SHIMOZAKI T., *Dimeta-82, op. cit.*, 405-408.

PRAGER S., *J. Chem. Phys.* **21** (1953) 1344.
SAUER F. and FREISE V., *Z. Elektrochem.* **66** (1962) 353.
TRIMBLE L. E., FINN D., and COSGAREA A., *Acta Metall.* **13** (1965) 501-507.
VAN LOO F. J. J., *Acta Metall.* **18** (1970) 1107-1111.

CHAPTER VII

Diffusion in media of lower dimensionality

In the present context, a "medium of lower dimensionality" means a part of a material the spatial extent of which is limited by its very nature. These media, as opposed to the three-dimensional media treated up to now, are characterized by a reduced dimensionality, of order 2 or 1: examples are surfaces (external or internal), interfaces of all kinds (grain boundaries, subgrain boundaries, interphase interfaces,...), dislocations. At low temperature, they can constitute the *only* possible path for the transport of material (e.g., electromigration in thin films).

The defects of the material, whether point or extended defects, can exercise an attractive interaction vis-à-vis the atoms of the diffusing species. The latter will thus have a tendency to segregate to these defects, or in other words, to be "trapped" by them.

However, if the defect is extended, it can *per se* constitute a path of easy diffusion, because its structure is less dense or less ordered than the perfect lattice. For this reason, the defect can "short-circuit" the normal volume diffusion in the crystal. The interface of a small precipitate or an inclusion, the surface of a pore, or a dislocation loop usually play the role of traps, but they can also constitute small short circuits (Fig. VII.1a and b). On the other hand, the dislocations which form a two-dimensional lattice, grain boundaries, and surfaces can all trap certain atoms (or vacancies) or play the role of short-circuits for diffusion (Figs. VII.1c, d and e). From an atomistic point of view, the jump frequencies around the defects must be considered (see Fig. VII.1f for the definition of the frequencies): if $\Gamma' < \Gamma_0 < \Gamma_i$, the defect is a genuine trap, and if $\Gamma_0 << \Gamma_i$, it is a deep trap. However, if $\Gamma_0 < \Gamma_i < \Gamma'$, the defect will be a short-circuit. For self-diffusion, even though $\Gamma_0 = \Gamma_i$, the defects act as short-circuits as long as $\Gamma_0 = \Gamma_i << \Gamma'$.

These phenomena can be studied from two points of view: either in order to determine the effect of these "short-circuiting paths" on the measurement of volume diffusion, or because of the intrinsic interest in the lower-dimensional structures themselves, to investigate their characteristic diffusion parameters, and, in specific models, the defects which give rise to atomic migration. These two points of view are, to be sure, not antithetical.

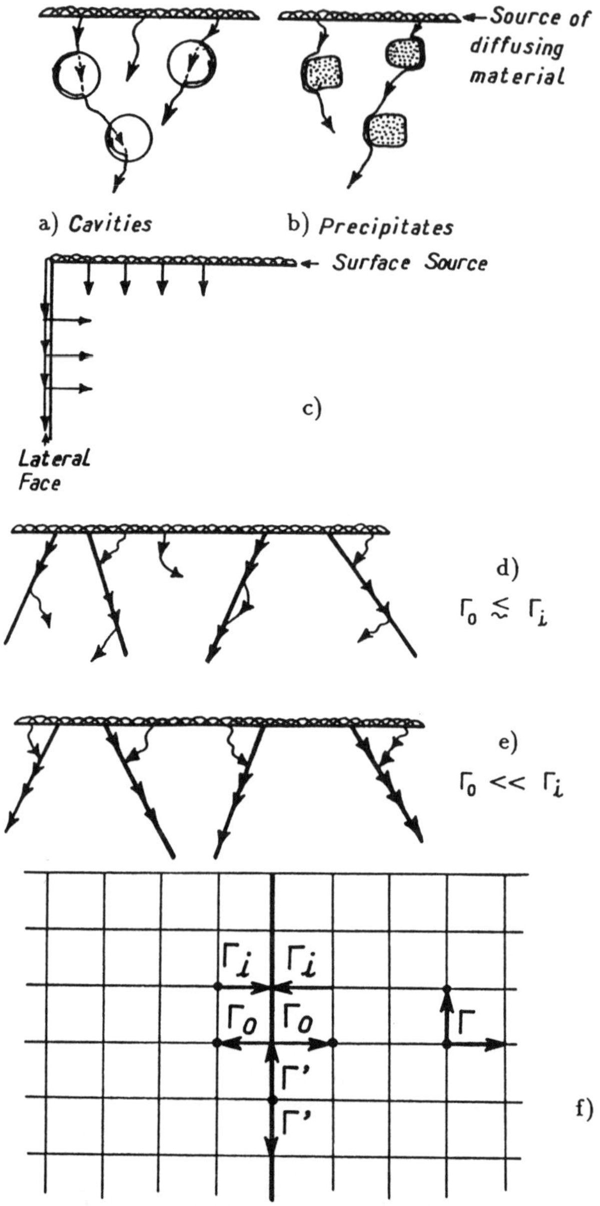

Fig. VII.1. — The behavior of extended defects as traps or as short-circuiting paths. a) and b): The interface between the matrix and a cavity (a) or a precipitate (b) can act as either a trap or a short-circuiting path. c) A lateral surface can act as a rapid diffusion path for a diffusant deposited as a surface source (a problem in measuring diffusion coefficients, Ch. IX, § II.2). d) and e): $\Gamma_i > \Gamma_0$ leads to trapping (or segregation), the extent depending on the relative sizes of Γ_i and Γ_0. f) Four-frequency model for diffusion in an extended defect; the jump frequencies are: Γ, in the bulk; Γ' in the defect; Γ_0, from the defect into the bulk, and Γ_i, from the bulk into the defect. $\Gamma' > \Gamma$ leads to short-circuiting diffusion.

Dislocation cores and interfaces make up the most frequently encountered internal short-circuits, and can be treated by similar phenomenological approaches. Our discussion will be a synthesis of these.

PART 1. — INTERNAL SHORT-CIRCUITS (dislocations, interfaces)

I. Phenomenology.

I.1 FISHER'S MODEL (Fig. VII.2). — The classic treatment is based on Fisher's model (1951): the dislocation core is represented by a cylindrical tube of radius a in which the diffusion coefficient, $D' > D$ (D is the "volume" or lattice diffusion coefficient, i.e. the diffusion coefficient in the "good" crystal). In the same way, an interface is represented by a slab of thickness δ, with a diffusion coefficient $D' > D$.

An important parameter is either the distance between the short-circuiting paths, $\Lambda = 1/\sqrt{\pi \rho_{\mathrm{d}}}$, where ρ_{d} is the density of dislocations (total length of dislocation line per unit volume), or the average grain or subgrain diameter, d.

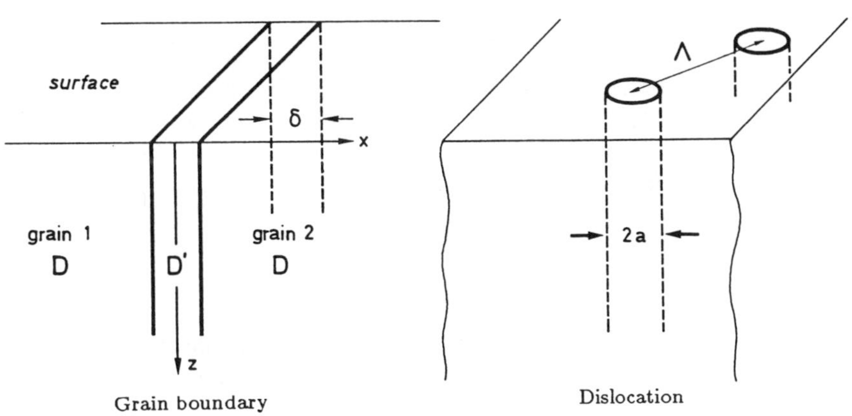

Fig. VII.2. — Fisher's model for grain boundaries or dislocations.

I.2 REGIMES OF DIFFUSION. — Consider the case of unidirectional diffusion from a surface (constant surface concentration or thin-layer source) in a polycrystalline (or polygonized) specimen or in a single crystal with a network of dislocations (Fig. VII.3). The experimental quantity determined is the concentration profile of the diffusing element $\bar{c}(z)$, where $\bar{c}$ is the *average* concentration in a section at depth z. The concentration in the sections is no longer homogeneous owing to the preferential diffusion down the grain boundaries or dislocations, but the sectioning methods yield only an average concentration. Three regimes of kinetics can be distinguished, depending on the values of the parameters of the structure; these are traditionally designated by the letters A, B, and C (Harrison, 1961).

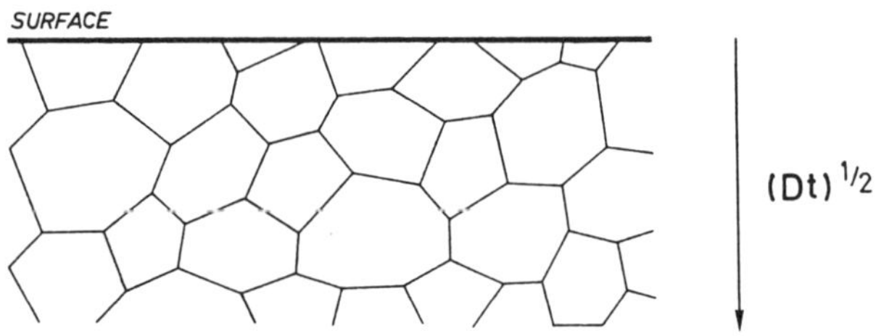

Fig. VII.3. — Diffusion in a heterogeneous medium, made up of "good" crystal and a three-dimensional network of grain boundaries or dislocations.

Type-A kinetics (Fig. VII.4).

$$(Dt)^{1/2} >> \Lambda \qquad \text{(VII.1)}$$

or

$$(Dt)^{1/2} >> d. \qquad \text{(VII.1a)}$$

The penetration in the volume is greater than the mesh size of the network of short-circuiting paths. The concentration profiles normally follow the solution of Fick's equation for a homogeneous system (Gaussian in the case of a thin layer). But the diffusion coefficient measured is in fact an apparent one, given by *Hart's equation:*

$$D_{\text{app}} = (1 - f)\, D + f D', \qquad \text{(VII.2)}$$

where f is the volume fraction of sites situated in the short-circuiting paths (defined as the tubes or slabs mentioned above). For a dislocation network:

$$f = \pi a^2 \rho_{\text{d}}, \qquad \text{(VII.3)}$$

whereas for a polycrystal, $f \approx 3\delta/d$. Hart's equation can be derived from the general equation (Ch. II, Eq. (II.31)), by taking into account the different frequencies and lengths of successive jumps.

This equation is still valid for $\sqrt{Dt} \geq \Lambda$ or d, as long as the penetration is large enough, i.e., B kinetics are not reached (these involve less restrictive conditions than Eqs. (VII.1) and (VII.1a)). According to Le Claire and Rabinovitch, the condition for the validity of Hart's equation (VII.2) for dislocations is $\sqrt{Dt} > 5\Lambda$.

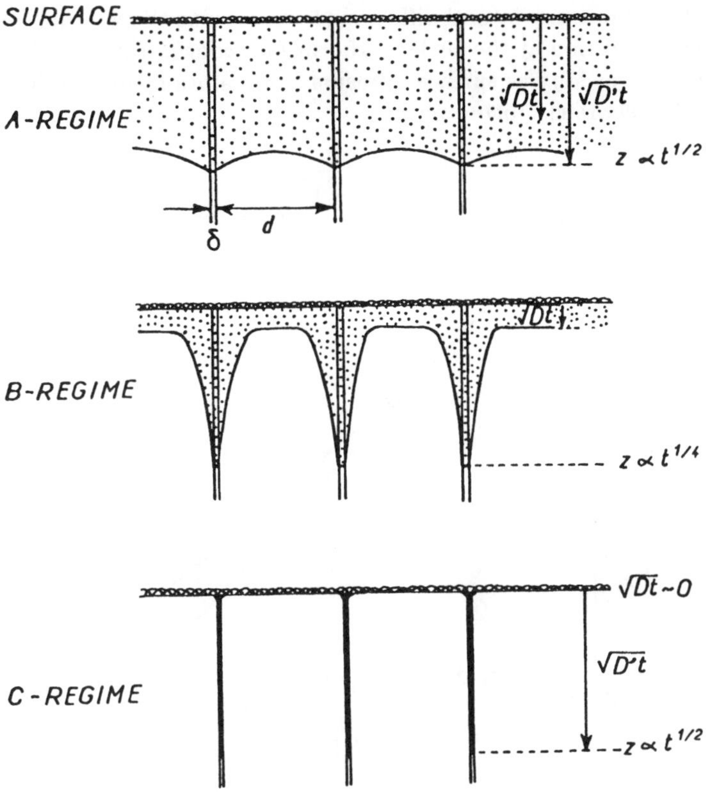

Fig. VII.4. — Harrison's (1961) three regimes of diffusion along short-circuiting paths. The lines at the lower boundary of the dotted regions are isoconcentration contours.

Since $D' >> D$, and $Q' < Q$, where Q' denotes the activation energy (see below), the effects of the short-circuiting paths manifest themselves as soon as the temperature is low enough, with $D_{\mathrm{app}} > D$. Thus a deviation from Arrhenius behavior is seen at low temperatures. For polycrystalline specimens, this deviation can be very clear. In single crystals, the effect of dislocations is more obvious in the case of solute diffusion, if the solute segregates to the dislocations (Fig. VII.5).

Type-C kinetics.
This is the opposite case of type-A kinetics:

$$(Dt)^{1/2} << \delta,\ a. \qquad \text{(VII.4)}$$

The penetration in the volume is negligible compared to the cross-sectional dimension of the short-circuiting path. The diffusional flux is practically contained in the channels constituted by the tubes (dislocations) or slabs (grain boundaries), without any lateral diffusion (leakage) into the volume. Whatever the boundary condition at the surface, the surface concentration c_s is constant, independent of t, as long as surface diffusion can maintain this condition at the entrance to the short-circuiting paths.

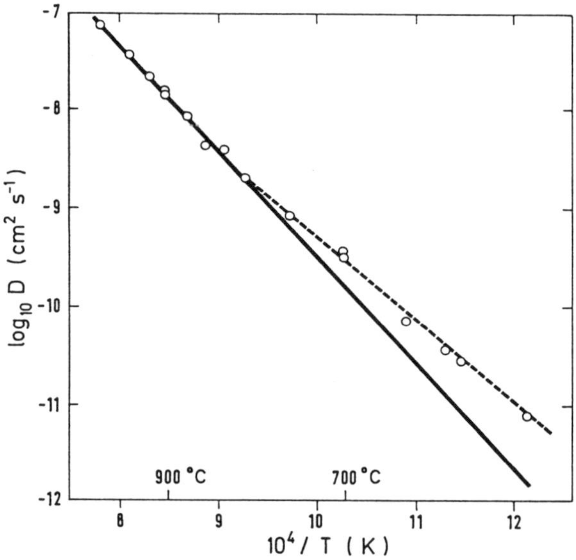

Fig. VII.5. — Arrhenius plot for the diffusion of sulfur in single-crystal copper, showing a deviation due to preferential diffusion along dislocations at low temperature. After Moya *et al.* (1969).

The solution of Fick's equation is thus:

$$\bar{c} = \text{const} \cdot \text{erfc}\left(z/2\sqrt{D't}\right),$$

where the constant is proportional to c_s and to the volume fraction of short-circuit material ($3\delta/d$ for boundaries, $\pi a^2 \rho_d$ for dislocations).

Type-C kinetics are of great interest, as they allow a *direct measurement of D'*, and not $D'\delta$ or $D'a$, which are obtained from type-B kinetics. Unfortunately, they correspond to very difficult experimental conditions (very small penetration depths and very little radioisotope in the short-circuiting path), except perhaps for the case of the diffusion of solutes with very low solubility due to their strong segregation.

Type-B kinetics.

This is the intermediate case, the classic case for studies of grain-boundary diffusion:

$$a << (Dt)^{1/2} < \Lambda/2, \tag{VII.5}$$

or

$$\delta << (Dt)^{1/2} < d/2. \tag{VII.5a}$$

In this case, there is simultaneous diffusion in the volume, starting from the surface, and diffusion along the short-circuiting paths followed by leakage, i.e., lateral diffusion into the grains from the short-circuiting paths (Fig. VII.4). The conditions on the dimensions d, Λ and the duration of the diffusion are that the lateral diffusion from two neighboring short-circuits should not overlap.

The mathematical solution of the problem is complicated, but simplified expressions can be obtained for certain experimental conditions. The quantity $\bar{c}(z)$ contains three components:

- C_{I} due to volume diffusion from the surface,
- C_{II} due to volume diffusion laterally from the short-circuit,
- C_{III} due to diffusion in the short-circuiting path itself.

The contribution of C_{III} is generally negligible. Further, beyond a certain depth, C_{I} is negligible with respect to C_{II}.

In the simplest cases, the diffusion profile $\bar{c}(z)$ can thus be decomposed into two parts:

- the first, for low penetrations, which yields D (or D_{app} in a polycrystalline sample, see below, §II.4);
- the second, for large penetrations, called the "tail" of the curve, which yields $D'\delta$ or $D'a^2$.

These two parts are separated by the use of graphical representations based on the analytical solutions of the problem.

Note that the classification into three regimes has been extended to the case of mobile boundaries (Cahn and Balluffi, 1979). If V is the velocity of the boundary:

	Type-A kinetics	$(Dt)^{1/2} > d$
or		$Vt > d$
	Type-B kinetics	with two sub-regimes
	Type-C kinetics	unchanged.

II. Analytical solutions.

In the framework of the Fisher-type model (Fig. VII.2), the diffusion equations are written for the two regions with diffusivities D and $D' > D$, respectively, with the appropriate conditions of continuity at the interface: $c' = c$ for self-diffusion, $c' = \kappa c$ for solute diffusion, where κ is the solute segregation factor (see below, § II.5).

II.1 Grain boundaries. — These are represented by a slab of thickness δ parallel to the $(y,\ z)$ plane. The problem becomes two-dimensional, the equation for diffusion in the grains is written:

$$\begin{aligned}\frac{\partial c}{\partial t} &= D\nabla^2 c, \qquad |x| > \delta/2 \\ &= D\left(\frac{\partial^2 c}{\partial x^2} + \frac{\partial^2 c}{\partial z^2}\right),\end{aligned} \tag{VII.6}$$

and for the intergranular slab at the interface slab/grain:

$$\frac{\partial c'}{\partial t} = D' \frac{\partial^2 c'}{\partial z^2} + \frac{D}{\delta/2}\left(\frac{\partial c}{\partial x}\right)_{\delta+/2}, \qquad |x| = \delta/2. \tag{VII.7}$$

The first term on the right-hand side of equation (VII.7) is the usual term for the divergence of the flux in the penetration direction z. The second term expresses the lateral diffusion from the slab into the grains. The slab is assumed to be thin enough so that the concentration in it does not depend on x; the quantity of diffusant lost from the half-slab of thickness $\delta/2$ is $-D\,(\partial c/\partial x)$ per unit area.

The solution of this system of equations requires some simplifying assumptions: type-B kinetics are considered, i.e., a single slab between two semi-infinite media, and perpendicular to the surface. We try to calculate $c(x,\ z,\ t)$, from which some quantities can be deduced that in turn can be compared to experimentally measured quantities:

- $\bar{c}(z,\ t)\,\Delta z$, the "average concentration" in a section of thickness Δz and unit width Δy, a quantity measured by the tracer-sectioning technique;
- the shape of the isoconcentration contours (which can be determined by microprobe analysis), especially the angle of the junction of the contour with the plane of the boundary;
- the depth of penetration for a certain concentration, which depends on the sensitivity of the analytical technique.

An approximate solution of equations (VII.6) and (VII.7) was calculated by Fisher (1951). Exact solutions have been given by Whipple (1954) for constant surface concentration, and by Suzuoka (1961) for the thin-layer source; these solutions are given in Appendix I. The expressions are quite cumbersome. The use of numerical tables or computers allows their convenient use, notably for the tracer-sectioning technique, and especially for the selection of the experimental conditions.

Suzuoka's solution for the thin layer source has the form:

$$\bar{c}(z,\ t) = \frac{Q_0}{(\pi D t)^{1/2}} \left| \exp\left(-\frac{\eta^2}{4}\right) + \frac{C_{\text{II}} + C_{\text{III}}}{\lambda} \right], \qquad \text{(VII.8)}$$

with $\bar{c}$ the average concentration ([1]) in a section dz of unit area at depth z, with the dimensionless variables:

$$\eta = z/(Dt)^{1/2}$$
$$\lambda = (d/2)/(Dt)^{1/2}.$$

Whipple's solution for constant surface concentration c_0 is:

$$\bar{c}(z,\ t) = c_0 \left\{ \operatorname{erfc} \frac{\eta}{2} + \frac{2\eta}{\sqrt{\pi}\lambda} (C_{\text{II}} + C_{\text{III}}) \right\}. \qquad \text{(VII.9)}$$

The factor λ is introduced for a polycrystal with average grain size d. The boundaries along which diffusion takes place were assumed to be perpendicular to the surface; the average length of the boundary contours is $\approx 2/d$ per unit area. In the original formulation, a single boundary of unit length was considered; this solution is obtained by replacing $2/d$ by unity.

The term C_{III} is generally negligible. For small penetrations, the term C_{I} is dominant, and the familiar gaussian or error function is recognized in equations (VII.8) or (VII.9). The volume diffusion coefficient, D, (or D_{app}, which can be somewhat different from the true D) can be calculated from the corresponding part of the depth profile.

For large enough penetrations, C_{I} becomes negligible, and the contribution C_{II} can be measured. At this stage, Le Claire's β parameter (1963) is introduced:

$$\beta = \frac{D' - D}{D} \frac{\delta}{2(Dt)^{1/2}} \approx \frac{D'}{D} \frac{\delta}{2(Dt)^{1/2}}. \qquad \text{(VII.10)}$$

The isoconcentration contours are more acute, the larger the value of β (Fig. VII.6). To distinguish the contributions of C_{I} and C_{II} on a concentration profile, β must be large enough; according to Le Claire, the condition is $\beta > 10$.

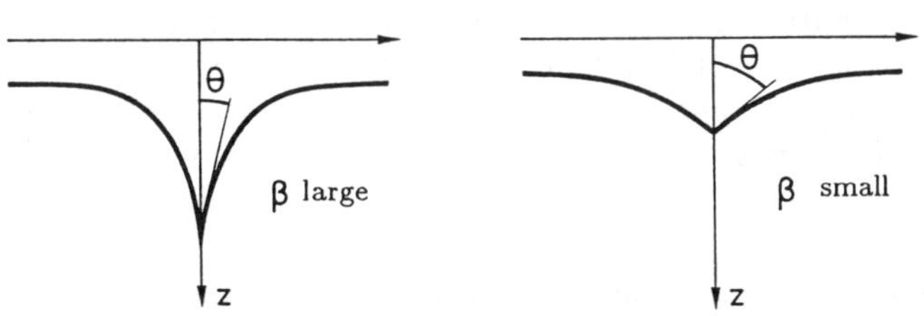

Fig. VII.6. — Effect of the parameter β on the shape of the isoconcentration contours.

([1]) For a bicrystal, $\bar{c}$ is the quantity of tracer in a slice dz.

Whipple's solution shows that under these conditions, $C_{\text{II}}\ \beta^{1/2}$ depends only on $\eta\beta^{-1/2}$; thus a universal representation of concentration profiles is possible in reduced coordinates.

However, following Le Claire's (1963) remark, it is better to consider the function:

$$\partial \text{ Log } c / \partial \left(\eta\beta^{-1/2}\right)^{6/5},$$

which is practically independent of $\eta\beta^{-1/2}$ for $\beta > 10$, and is equal to -0.78 when $\eta\beta^{-1/2} > 3$.

Differentiating $\eta\beta^{-1/2}$ with respect to z, we obtain:

$$D'\delta = \left(\frac{4\ D}{t}\right)^{1/2}\left[\frac{\partial\left(\eta\beta^{-1/2}\right)}{\partial z}\right]^{-2},$$

from which:

$$D'\delta = \left[-\frac{\partial \text{ Log } \bar{c}}{\partial z^{6/5}}\right]^{-5/3}\left(\frac{4\ D}{t}\right)^{1/2}\left[-\frac{\partial \text{ Log } \bar{c}}{\partial\left(\eta\beta^{-1/2}\right)^{6/5}}\right]^{5/3},$$

and with the value 0.78 for the third factor between brackets on the right-hand side:

$$\boxed{D'\delta = 0.66\left(\frac{4\ D}{t}\right)^{1/2}\left[-\frac{\partial \text{ Log } \bar{c}}{\partial z^{6/5}}\right]^{-5/3}}. \qquad \text{(VII.11)}$$

The last factor is determined from the experimental results if the tail of the penetration plot is such that Log $\bar{c} \propto z^{6/5}$.

Similarly, in the framework of the Suzuoka solution,

$$-\frac{\partial \text{ Log } \bar{c}}{\partial\left(\eta\beta^{-1/2}\right)^{6/5}} = 0.72\,\beta^{+0.008}, \qquad \text{(VII.12)}$$

a value very slightly different from the Whipple solution when $\beta > 10$.

The slope of the "tail" part of the profile, $-\partial$ Log $\bar{c}/\partial z^{6/5}$ varies as $t^{-0.3}$ (Eq. (VII.11)). Further, for the Suzuoka solution, the intersection of this part of the profile with the concentration axis depends only slightly on t (see exercice 53). These are *criteria of validity,* the verification of which is very useful. The sectioning-counting technique is of interest because it allows the isolation of the $z^{6/5}$ component which leads to a *linear profile for both types of boundary conditions.* Some penetration curves calculated for a sectioning experiment are shown in figure VII.7. The first part of the curve would be linear if plotted *vs.* z^2, allowing the evaluation of an effective volume diffusion coefficient, slightly larger than that measured in a single crystal, as far as the contribution of $\bar{c}_{II}$ is not negligible. The error expected on D is smaller than 10 % , if the grain size is large enough: $d > 4\,(Dt)^{1/2}$ (Kaur and Gust, 1988).

In practice, the choice of diffusion time is guided by the need to obtain a large enough value of β, and by the sensitivity of the analysis technique. In effect, the phenomenon of grain boundary diffusion in the absence of lateral diffusion (type-C kinetics) is very difficult to observe; thus the time cannot be decreased too drastically in an effort to increase β.

The experiment can also lead to very high values of β (> 100). In this case the condition $\eta\beta^{-1/2} > 3$ cannot be met, and the solution (VII.11) is no longer applicable. Recourse must be had to a numerical calculation to establish, for example, $\mathrm{Log}\,(C_{II}/c_0)$ *vs.* η parametrized by β ([2]).

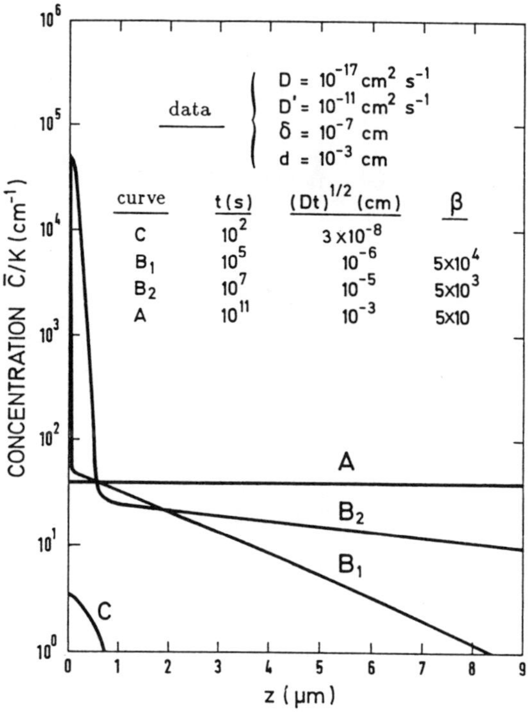

Fig. VII.7. — Calculated diffusion profiles for a polycrystalline material with grain size 10 μm for different diffusion times. Thin-layer source initial condition. For the definition of K, see Appendix I, equation (A.6). After Atkinson (1983).

([2]) For elements with small solubility in the matrix, the measurements are possible only for penetrations η that are not too high. If β is too large, the condition on $\eta\beta^{-1/2}$ is difficult to meet.

Note that the solutions for type-B kinetics yield only the grain boundary product $D'\delta$. Only type-C kinetics permit the direct determination of D', but practically only in a very small range of temperature, and only if the experiment succeeds. And, at least for self-diffusion, it is assumed that δ does not vary with temperature. The grain-boundary width, δ, or the dislocation-core radius, a, can only be obtained from a combination of experiments in the B-type ($D'\delta$) and C-type (D') regimes.

The residual activity (Gruzin) method can also be used to measure grain-boundary diffusion coefficients; it also is not sensitive to the boundary condition at the surface.

II.2 Subgrain boundaries. — The preceding solutions and techniques can be applied to subgrains (or sub) boundaries, i.e., boundaries of small misorientation. The array of dislocations that makes up the sub-boundary can be simulated by a slab of thickeness δ_{eff}, (Fig. VII.8) so that:

$$\delta_{\text{eff}} < (Dt)^{1/2} < d/2. \tag{VII.13}$$

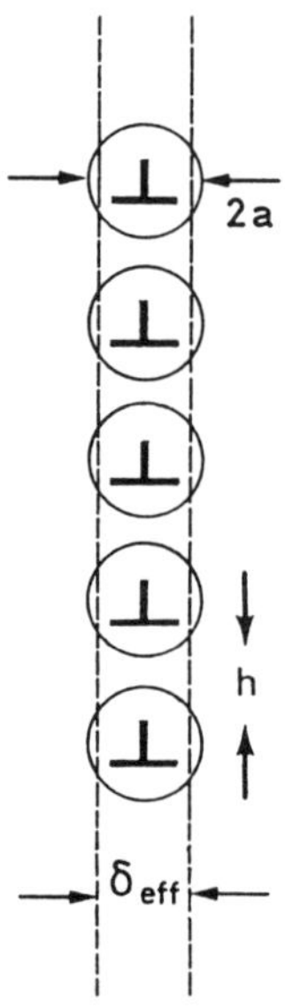

Fig. VII.8. — Model of a subgrain boundary.

The first inequality expresses that the volume diffusion is large enough to maintain a uniform concentration in the slab representing the subboundary, (inside the slab, type-B kinetics predominate) and cause significant leakage into the subgrains. The second inequality expresses the fact that the diffusion fluxes from the different boundaries of a subgrain of diameter d do not overlap.

Two equivalent approaches are possible:

1. Direct application of the Fisher-Suzuoka-Le Claire type solution, which yields a value of $D'\delta$, with:

$$\begin{aligned} \delta &= 2a \\ D' &= (1-f')\,D + f'\,D_{\mathrm{d}} \approx f'\,D_{\mathrm{d}}, \end{aligned} \qquad \text{(VII.14)}$$

where D_{d} is the diffusion coefficient in the dislocations, and f' the fraction of sites in the dislocation cores:

$$f' = \frac{\pi a^2 \cdot h^{-1}}{2\,a} = \frac{\pi}{2}\,\frac{a}{h}, \qquad \text{(VII.14a)}$$

and h is the distance between dislocations. Recall that the misorientation θ of the subgrains is given by:

$$\frac{b}{h} = 2\,\sin\,(\theta/2),$$

where $h \approx b/\theta$ if θ is small.

2. Application of the solution valid for grain boundaries, as before, but with:

$$D'\,\delta = D_{\mathrm{d}}\,\delta_{\mathrm{eff}}$$

where δ_{eff} is defined so that the area $\delta_{\mathrm{eff}} \times 1$ is equal to the total cross section of all the channels in a unit length of subgrain:

$$\delta_{\mathrm{eff}} = \pi a^2 \cdot (1/h). \qquad \text{(VII.15)}$$

The two approaches do yield identical results. Evidently these correspond to a flux parallel to the dislocations. A strong anisotropy in diffusion along the boundary can be expected, which should decrease as the misorientation increases. This technique has been used mostly to determine the diffusion coefficient D_{d} along the dislocations that make up the subboundaries.

II.3 INTERFACES BETWEEN DISSIMILAR PHASES. — The case of "interphase interfaces" requires special analysis. Let us consider the diffusion of component A in the interface of an α/β couple. Application of Fisher's model is complicated by the fact that the volume diffusion coefficients in the two phases are different (Fig. VII.9), although the equations are easily solved in the framework of the Fisher approximation. An analogous expression for $\bar{c}(z)$ is obtained with $2D^{1/2}$ replaced by $D_{\alpha}^{1/2} + D_{\beta}^{1/2}$, with penetration linear in the coordinates Log $\bar{c}$ *vs.* z.

This approach assumes that the concentration of the diffusing species is the same at both interfaces $x = \pm\delta/2$. Clearly $c_{\mathrm{A}}^{\alpha}(-\delta/2) \neq c_{\mathrm{A}}^{\beta}(+\delta/2)$. A logical assumption is that the chemical potentials of A in the three phases α, β, and the interface δ are equal. This yields the same solution as before, but with $2D^{1/2}$ replaced by:

$$\frac{c_A^\alpha}{c_A^i} D_\alpha^{1/2} + \frac{c_A^\beta}{c_A^i} D_\beta^{1/2}, \tag{VII.16}$$

with:

$$c_A^i = \frac{c_A^\alpha + c_A^\beta}{2}.$$

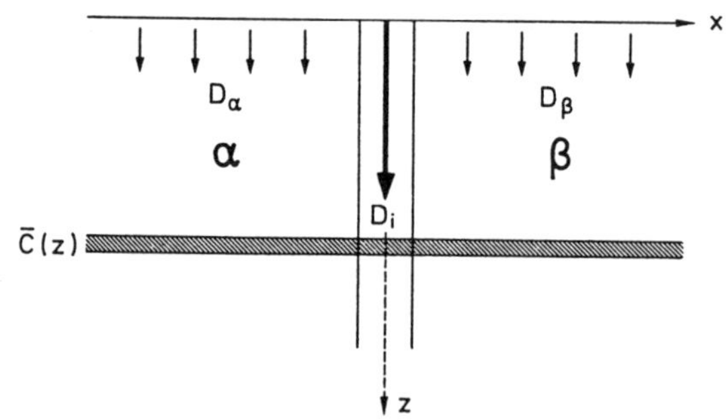

Fig. VII.9. — Diffusion in an interphase boundary.

II.4 Dislocations. — In the case of B-type kinetics, a Fisher-type model is used: dislocations normal to the surface, represented by tubes of radius a, in which the diffusion coefficient is D_d. Different approximate solutions have been proposed; those associated with a Whipple-Suzuoka calculation are rather difficult to handle.

Exact solutions of the problem and a detailed discussion of their application have been given by Le Claire and Rabinovitch (1981-1983) for the usual boundary conditions of an isolated dislocation and for dislocations perpendicular to the surface, such that the diffusion fields from individual dislocations do not overlap. The expression for c is the sum of two terms: the first, $\exp(-z^2)$ or $\mathrm{erfc}(z)$, represents volume diffusion from the surface; the second represents diffusion along the dislocations, proportional to their total cross-sectional area, $\pi a^2 \rho_d$, where ρ_d is the density of dislocations. For the thin layer source, the solution is analogous to Suzuoka's for grain boundaries. If the solution is expressed in terms of the reduced penetration $\eta = z/\sqrt{Dt}$, one finds:

$$\bar{c}(\eta) = \frac{Q_0}{\sqrt{\pi D t}} \left[\exp\left(-\frac{\eta^2}{4}\right) + \pi a^2 \rho_d Q^{II} \right], \tag{VII.17}$$

where Q^{II} is given by a double integral. There is a similar solution for the constant surface concentration:

$$\bar{c}(\eta) = c_0 \left[\operatorname{erfc} \frac{\eta}{2} + \pi a^2 \rho_{\text{d}} Q^{\text{I}}\right] . \qquad \text{(VII.17a)}$$

When the reduced penetration depth η is large enough, only the "dislocation tail" is observed on the depth profile; its slope is given by:

$$\frac{\partial \operatorname{Log} \bar{c}}{\partial \eta} = \frac{\partial \operatorname{Log} Q^{\text{I}}}{\partial \eta} = \frac{\partial \operatorname{Log} Q^{\text{II}}}{\partial \eta} = -\frac{A(\alpha)}{(\alpha\beta)^{1/2}}, \qquad \text{(VII.18)}$$

with the reduced variables: $\alpha = a/\sqrt{Dt}$ and $\beta = \dfrac{D_{\text{d}} - D}{D}\,\alpha$. $A(\alpha)$ is of order unity, varies slowly with α, and is quasi-independent of $\alpha\beta$ (Fig. VII.10). It is remarkable that the slope of the profile is identical in the two cases, which makes it insensitive to the (poorly understood) processes taking place at the surface. Equation (VII.18) predicts that $\partial \operatorname{Log} \bar{c}/\partial z$ does not depend on time

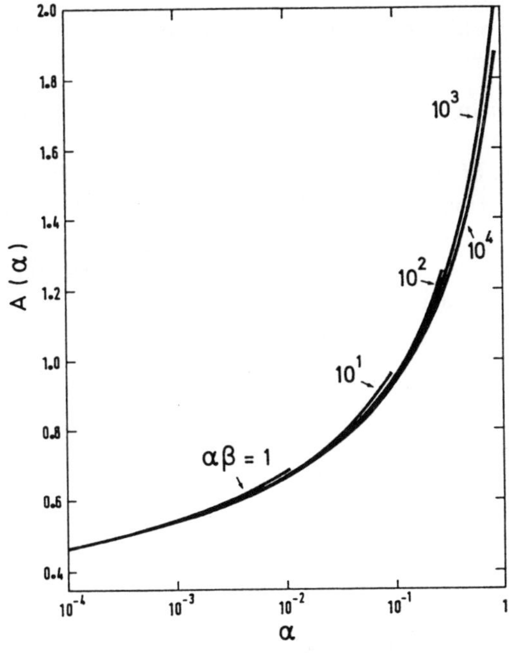

Fig. VII.10. — Diffusion along dislocations. Variation of the function $A(\alpha)$ of equation (VII.18) calculated by Le Claire and Rabinovitch (1981).

as long as $A(\alpha)$ is insensitive to variation in $\alpha\beta$:

$$\boxed{\frac{\partial \,\mathrm{Log}\, \bar{c}}{\partial z} = -\frac{A}{\left[\left(\frac{D_{\mathrm{d}}}{D} - 1\right) a^2\right]^{1/2}}} \,. \tag{VII.19}$$

The quantity D_{d}/D can be calculated from a measured slope if the value of a is known; it is doubtless of the order of a few Å. In principle, extrapolation of the tail to $z = 0$ allows the calculation of a if the dislocation density ρ_{d} is known, but the uncertainty can reach several orders of magnitude. Conversely, ρ_{d} can be calculated with an assumed value of a.

The first part of the profile follows volume diffusion kinetics. If such data are plotted conventionally, a straight line results, and an apparent diffusion coefficient:

$$D_{\mathrm{app}} = D\left(1 + \pi\, a^2\, \rho_{\mathrm{d}}\, \mathrm{M}\right)$$

can be calculated from its slope. Values of the enhancement factor M have been calculated by Le Claire and Rabinovitch.

The diffusion profile is thus quite similar to the one for grain-boundary diffusion, with three differences (Fig. VII.11):

– Log $\bar{c}$ is *linear* in z (indistinguishable in practice from linear in $z^{6/5}$),

– the slope of the straight line, $\partial \,\mathrm{Log}\, \bar{c}/\partial z$ from which D_{d} is calculated is *quasi-independent of the diffusion time* – this is a sensitive test for diffusion down isolated dislocations as opposed to dislocations grouped in subboundaries, where the time dependence is $t^{-0.3}$ (see Sect. II.1).

– the intersection of the extrapolation to $z = 0$ of the linear part with the log concentration axis varies strongly with the diffusion time.

The solutions (VII.17) and (VII.17a) apply only if $\varepsilon = \sqrt{\pi\, a^2\, \rho_{\mathrm{d}}} < \alpha$ (dislocations sufficiently distant from one another). For longer diffusion times, ε could be greater than α, with a tendency toward profiles typical of A-type kinetics with an apparent diffusion coefficient given by Hart (1957):

$$D_{\mathrm{app}} \approx D\left[1 + \varepsilon^2\,(\Delta - 1)\right],$$

with $\Delta = D_{\mathrm{d}}/D$. This approximation is valid for $\varepsilon \geq 4\alpha$ according to a calculation based on a "hexagonal" network of dislocations parallel to one another and perpendicular to the surface (Le Claire and Rabinovitch, 1983).

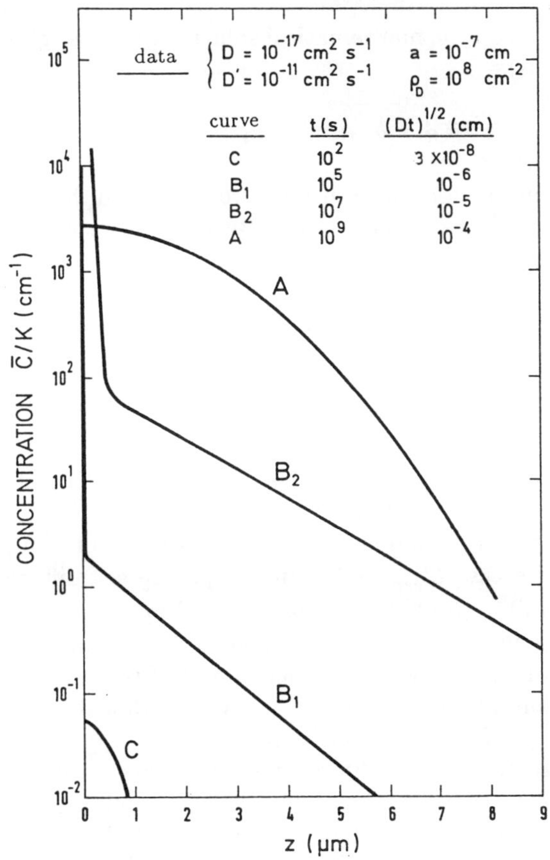

Fig. VII.11. — Calculated diffusion profiles for a single crystal containing 10^8cm/cm^3 dislocations (thin layer geometry). For the definition of K, see Appendix I, equation (A.6). After Atkinson (1983).

II.5 SOLUTE DIFFUSION. — In the case of self-diffusion, it is assumed that the parameters δ and a of the Fisher model are purely geometrical. The same assumption cannot be applied to the diffusion of solute atoms because of their interaction, elastic or electronic, with the boundary (short range) or with the dislocations (long range interaction goes as $1/r$).

The condition at the interface between the slab or tube and the volume of the grain was simply c (boundary, dislocation) $= c$ (volume) at $x = \pm\delta/2$ or $a/2$. If there is segregation of the diffusing atoms, it is assumed that the concentration c in the boundary or dislocation is constant in the volume of the short-circuiting path, and is in thermodynamic equilibrium with the volume concentration at the interface:

$$c_{\mathrm{v}}\,(x = \pm\delta/2) = c_{\mathrm{b}} \exp\left(U/kT\right). \qquad \text{(VII.20)}$$

The Fisher or Suzuoka-type mathematical solution remains valid for an effective thickness (or radius):

$$\delta_{\text{eff}} = \delta \cdot \kappa = \delta_0 \exp(-U/kT) \qquad \text{(VII.21)}$$

where δ_0 is the value of δ for self-diffusion, and κ is called the segregation factor. For a very slightly soluble element, the diffusing species can be considered to be completely trapped in the boundary, which corresponds to C-type kinetics (this chapter, § I.2).

The activation energy obtained from the variation of $D'\delta\kappa$ with temperature is thus

$$Q'_{\text{exp}} = Q' + U. \qquad \text{(VII.22)}$$

In the case of positive segregation (i.e., enrichment) of the diffusing element at the boundary (dislocations), $U < 0$ and $Q'_{\text{exp}} < Q'$. A low value of the activation energy may thus indicate segregation of the diffusing element.

In the opposite case $Q'_{\text{exp}} > Q'$. In comparing self-diffusion $Q^{\text{A}}_{\text{exp}} = Q'$, to solute diffusion $Q^{\text{B}}_{\text{exp}} = Q' + U$, U is obtained by difference, assuming that the true activation energy (i.e., characteristic of true grain-boundary diffusion, without segregation effects) is quite independent of the nature of the diffusing element. This assumption is probably an oversimplification, unless A and B are similar elements (e.g., both transition metals).

We emphasize that *solute diffusion experiments yield the quantity* $P = \kappa D' \delta$*, and not* $D' \delta$*, as in self-diffusion.* This fact often passes unperceived, and a number of publications should be corrected on this account. In the case of diffusion in alloys, κ can vary with composition as Henry's law (Eq. (VII.20)) is no longer obeyed (e.g. due to saturation effects), and segregation can also have an effect on self-diffusion.

III. Atomic models.

Fisher's model has been very successful, and is the basis of all the experimental determinations. Nevertheless, it is vulnerable to criticism from several points of view:

– the thickness of the boundary or the diameter of the dislocations are *ad hoc* parameters, in a model very unlike the atomic structure of grain boundaries as it is known today;

– the diffusion coefficient D' is purely phenomenological. It has no meaning except for macroscopic fluxes. Does the atomic theory of diffusion (Ch. II) apply to D' ? The number of types of atomic jumps in the "perfect" crystal and in a boundary are completely different: this number is small in the first case because of the regular structure, but what can be said about the second ? Is Fick's law still valid ? Does the correlation factor have meaning ?

– is the assumption of local equilibrium at the boundary/volume interface realistic during a diffusion process ?

Among the tentative atomic models, we cite that of Benoist and Martin (1975). The model is based on the presence of a certain number of jump frequencies in a bicrystal (Fig. VII.12). For a cubic lattice and a simple boundary (twin), 4 frequencies are taken into account:

Γ	in the perfect crystal,
Γ'	in the plane of the boundary,
Γ_0	from the plane of the boundary towards the crystal,
Γ_i	from the crystal into the plane of the boundary.

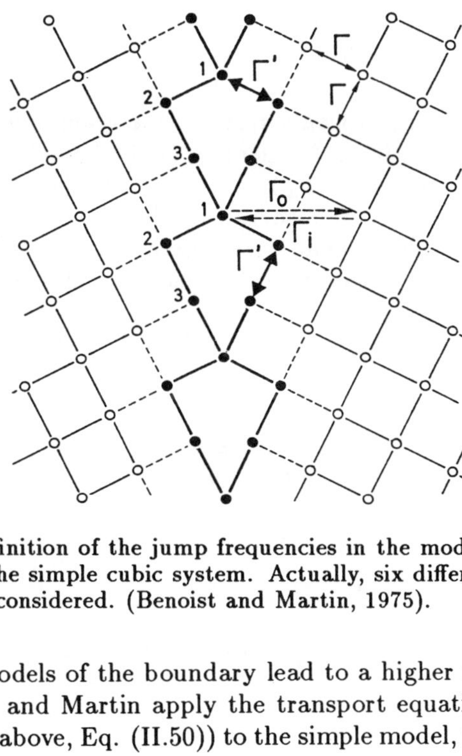

Fig. VII.12. — Definition of the jump frequencies in the model of a coherent (012) twin boundary in the simple cubic system. Actually, six different frequencies of the type Γ′ have to be considered. (Benoist and Martin, 1975).

More elaborate models of the boundary lead to a higher number of jump frequencies. Benoist and Martin apply the transport equation (balance over all atomic jumps, see above, Eq. (II.50)) to the simple model, neglecting the effects of correlation between jumps.

At the end of their analysis, they arrive at the following conclusions:

– For long diffusion times, Suzuoka's solution is valid, and the concentration profile depends only on the variable:

$$\beta = \frac{\Gamma_i}{\langle\Gamma_0\rangle}\,\frac{\langle\Gamma'\rangle}{2\Gamma\,(\Gamma t)^{1/2}}. \qquad \text{(VII.23)}$$

The average frequencies $\langle \Gamma_0 \rangle$ and $\langle \Gamma' \rangle$ are calculated on the basis of various frequencies Γ according to the structure of the boundary. An analytical calculation is possible for boundaries with a periodic structure, such as tilt boundaries. Equation (VII.23) is to be compared to Le Claire's:

$$\beta = \kappa \cdot \delta_0 \, \frac{D'}{2D\sqrt{Dt}}, \tag{VII.10a}$$

where δ has been replaced by $\kappa\,\delta_0$, and κ is the segregation factor. Thus the condition of equilibrium at the interface is justified so that the segregation factor $\kappa = \Gamma_i / \langle \Gamma_0 \rangle$.

– The parameter δ does not appear explicitly. It can be considered as a unit of length, giving the scale of the boundary with respect to the perfect crystal. It is not related to the actual atomic relaxations in the grain boundary, but only to the *number* of high-diffusivity paths which are available for the diffusing particle.

– The model gives rise to ***anisotropy*** in the grain-boundary diffusion because of the non-uniform jump frequencies in the boundary. The values of β parallel $(\beta_{\parallel})$ and perpendicular $(\beta_{\perp})$ with respect to the tilt axis are not equal; their ratio depends only on the structure of the boundary, but the individual parameters $\beta_{\parallel}$ and $\beta_{\perp}$ also depend on the exchange jumps between the lattice and the boundary. Hence the interest in experimental measurements of the anisotropy ratio.

– For short diffusion times, simple analytical solutions are obtained:
if, for self-diffusion,

$$\begin{aligned} \Gamma_i &= \Gamma_0 &&= \Gamma \quad \text{and} \\ \tau &= \Gamma t &&<< 1 \end{aligned}$$

the solution is:

$$\bar{C}_{\mathrm{II}}(z,\ t) = \exp(-2\,\tau)\ \frac{\exp\left(-\eta^2/4\,\Delta\right)}{\sqrt{\tau\Delta}}, \tag{VII.24}$$

with $\Delta = \Gamma'/\Gamma$.

For solute diffusion:

$$\bar{C}_{\mathrm{II}}(z,\ t) = \exp(-2\,\frac{\Gamma_0}{\Gamma}\,\tau)\ \frac{\exp\left(-\eta^2/4\,\Delta\right)}{\sqrt{\tau\Delta}}. \tag{VII.24a}$$

The solutions indicate a Gaussian penetration along the boundaries (type C kinetics) with an exponential attenuation factor, which in the case of solute diffusion comes from a "leakage" of the diffusing species across the grain boundary (the term $\exp(-2\,\Gamma_0\,t)$).

– This model explains in a simple way the double role played by grain boundaries, which act either as traps when $\Gamma_i > \Gamma_0$, or as short-circuits when $\Gamma' > \Gamma$.

This approach can yield an analytical expression only for boundaries having a simple atomic structure. For a general analysis, for example, including the variation of D' with misorientation across the boundary or the anisotropy of D', recourse must be had to numerical simulation (see below).

– Finally, it appears that the phenomenological Fisher model serves as an adequate basis for the analysis of experimental results, because the atomic definition of the diffusion coefficient along the grain boundary is unaffected by the exchange of atoms between the boundary and the crystal.

IV. Effect of temperature.

Consider a vacancy in a boundary. Its binding energy (the formation energy in the boundary minus the formation energy in the perfect crystal) can take a whole series of values depending on the type of site. In fact, certain binding energies are much higher than others; thus two classes of vacancy jumps can be considered: *i*) between low-energy sites and *ii*) toward sites of energy closer to that of the perfect lattice. Vacancy migration can be limited to the first set. In other words, certain sites of high binding energy can be considered as veritable traps for the vacancy. The total number of jumps on sites of the first class per unit time can be evaluated; for a certain domain of temperature, one obtains:

$$\tilde{w}_{\mathrm{b}} = \tilde{w}_{\mathrm{b}}^{0} \exp\left(-\tilde{E}_{\mathrm{b}}^{\mathrm{m}}/kT\right), \qquad \text{(VII.25)}$$

where $\tilde{E}_{\mathrm{b}}^{\mathrm{m}}$ is an effective migration energy. The concentration of vacancies on these sites can also be evaluated as

$$\tilde{X}_{\mathrm{b}}^{v} = \tilde{X}_{0}^{v} \exp\left(-\Delta\tilde{G}_{\mathrm{b}}^{\mathrm{f}}/kT\right), \qquad \text{(VII.26)}$$

where $\Delta\tilde{G}_{\mathrm{b}}^{\mathrm{f}}$ is a weighted average of the Gibbs free energy of formation on the sites under consideration. In fact, only the energies can be calculated by simulation, and hypotheses have to be constructed in order to obtain the entropies. Thus finally:

$$D' = D'_0 \exp\left(-E'/kT\right),$$

with

$$E' = \tilde{E}_{\rm b}^{\rm f} + \tilde{E}_{\rm b}^{\rm m}$$
$$D'_0 = \beta_{\rm b}\, f_{\rm b}\, \tilde{a}_{\rm b}^2\, \tilde{w}_{\rm b}^0 \exp\left(\tilde{S}_{\rm b}^{\rm f}/k\right), \tag{VII.27}$$

where $\beta_{\rm b}$ is a geometric factor, $f_{\rm b}$ the correlation factor, $\tilde{a}_{\rm b}$ an effective jump distance, and:

$$\tilde{w}_{\rm b}^0 = \tilde{Z}_{\rm b}\, \tilde{v}_{\rm b} \exp\left(\tilde{S}_{\rm b}^{\rm m}/k\right),$$

with $\tilde{Z}_{\rm b}$ the average number of neighboring sites towards which the vacancy can jump.

Most of these quantities are markedly anisotropic, because of the atomic structure of the boundary. The simulations show that in FCC and BCC metals, only the vacancies contribute to diffusion, because the interstitials are trapped. Further, there is evidence of ***paths of preferential migration*** passing over certain sites in both boundaries and dislocations (see Biscondi, 1984); these paths are connected to the description of the boundary in terms of "structural units" which repeat periodically (see Balluffi, 1982).

Simulations by molecular dynamics, often limited to model materials (Lennard-Jones potentials) have shown 1) the formation of Frenkel pairs below $0.5\, T_{\rm M}$, 2) the predominance of the vacancy mechanism, although the degree of relaxation around the vacancy appears rather variable, 3) a marked anisotropy (Balluffi, 1982; Guillopé *et al.*, 1983), 4) multiple jumps of a vacancy, 5) cyclic processes, ... Molecular dynamics calculations notably avoid the description of the diffusion in terms of jump frequencies and correlation factors, terms which, perhaps, lose their significance at high temperature (progressive change of the structure of the boundary, delocalization of the defects,...). In following the atomic trajectories, molecular dynamics allows the calculation of D' from information on the dynamics of the system, with the use of the velocity self-correlation function (see Ch. II, § VII, Eq. (II.47)). The definition of which sites belong to the boundary is somewhat arbitrary.

V. Experimental methods and results.

V.1 EXPERIMENTAL METHODS. — The most used experimental technique at this time is sectioning and counting, followed by an analysis of the diffusion profile $\bar{c}(z)$ according to the analytical expressions given in the preceding paragraphs. In view of the considerable difference in the required annealing times (see Figs. VII.7 and VII.11), it is not possible to observe the three types of kinetics on the same sample at the same temperature.

In type-B kinetics, the desired information is obtained from the "tail" of the curve, at low concentrations, i.e., in the part where the measurements are the most sensitive to the artifacts which could give rise to similar tails (pores, microcracks). Conversely, the absence of a tail linear in $z^{6/5}$ or z does not prove the absence of short-circuits, because *the tail can occur at a concentration level below the limit of detection* of the experimental technique. In some cases the tail is not linear in $z^{6/5}$, but can be analyzed in two or three linear parts, corresponding to different short circuits (subboundaries, general or special boundaries, dislocations... (see Gupta and Rosenberg, 1975)). It is recommended to verify the origin of the tail by an image of a corresponding section (autoradiography, SIMS map...). Finally, the experimental conditions should be varied over a wide range in order to verify that the analysis is carried out in the range of validity of the model (e.g., the value of β, variation of the slope of the profile with diffusion time, intersection with the ordinate, ... see the remarks in § II.1 and II.4).

Besides the sectioning-counting method and its variants, the isoconcentration contours (see Fig. VII.4) can be established, for example, with an electron microprobe, and $D'\delta\,\kappa$ can be determined from the angle θ formed by the isoconcentration contour and the boundary (App. I.3). The maximum penetration z at the boundary can also be measured, but it depends on the sensitivity of the analytical technique. This measure is used to evaluate the anisotropy of diffusion or its dependence on the misorientation (App. I.4).

In the case of polycrystalline samples, migration of the boundaries is possible, causing an under- or over estimate of $D'\delta$, depending on the method used for analyzing the data. The error varies with the temperature. It can be evaluated starting from the equation

$$\partial \,\mathrm{Log}\, \bar{c}/\partial z = -\,(V/D'\delta)^{1/2}\,,$$

(Glaeser and Evans, 1986), where V is the velocity of the grain boundary.

Regime C can also be studied by *"permeation" experiments*. The tracer is deposited on a face of a slice of the material under study, and its accumulation is followed on the opposite face by a counting technique or autoradiography for radiotracers, or by Auger spectroscopy for solute diffusion (Wuttig and Birnbaum, 1966, "surface accumulation method", see exercice #54).

The role of triple junctions as very fast short-circuits is still debated.

V.2 Experimental results. For a critical review, see Kaur *et al.* (1989). — Most of these are on metals (the majority on Ag and Ni). Here are the essentials:

1. *Metals.* — In all systems studied, $D'/D \approx 10^3$ to 10^6. Extrapolation to the melting point leads to a useful comparison:

$$\text{At } T_{\mathrm{M}}: \qquad D \approx 10^{-8}\ \mathrm{cm^2\ s^{-1}}$$

$$D'\delta \approx 10^{-12}\ \text{cm}^3\ \text{s}^{-1}\ \text{(grain boundaries)}$$
$$D'a^2 \approx 10^{-19}\ \text{cm}^4\ \text{s}^{-1}\ \text{(dislocations)}$$
from which $$D' \approx 10^{-5}\ \text{cm}^2\ \text{s}^{-1}$$

i.e., a value characteristic of diffusion in liquids if $\delta = 10^{-7}$ cm.

The lateral dimension of the short-circuiting paths is of the order of 10^{-7} cm, according to numerous indications, but no value of δ has been deduced from a diffusion experiment, and many of the results are open to criticism.

The activation energy Q' is less than Q. For grain boundaries, $Q'/Q \approx 0.5$. For subgrain boundaries, similar but somewhat higher values are obtained, $Q'/Q \approx 0.6$, or 0.7 when the dislocations are dissociated ([211] tilt boundaries or [111] twist boundaries in FCC); stacking faults are not good short-circuits.

Intergranular diffusion is anisotropic. For subgrain boundaries, the dislocation model predicts $D_{\|} > D_{\perp}$, because diffusion is easier along the dislocations, and the *anisotropy* decreases as the misorientation θ increases (the dislocation cores overlap). As $D'\delta = D_{\text{d}}\pi a^2/h$, the activation energy $Q' = Q_{\text{d}}$ is independent of θ, and the preexponential increases with θ (h is the distance between dislocations in the subboundary). These predictions have proven true and remain valid even for large misorientations. D' increases with θ at small misorientations, then either remains constant whatever the values of θ ([100] tilt boundaries in FCC), or goes through minima for certain coincidence boundaries ([110] tilt boundaries, Fig. VII. 13).

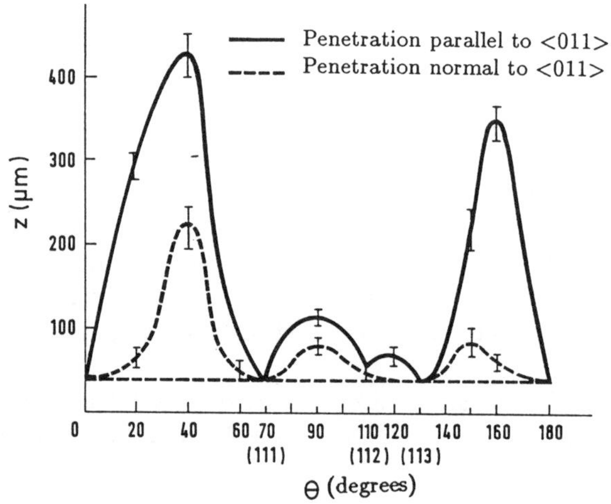

Fig. VII.13. — Variation of the penetration depth of zinc in aluminum bicrystals as a function of misorientation angle θ around the <001> axis. Anneal 720 h at 250°C. After Herbeuval *et al.* (1973).

The variation of D' with misorientation, θ, is similar to that of the grain-boundary energy γ_b; all these quantities manifest a correspondence with the structure of the boundary. Borisov's relation (1964) is apropos:

$$\gamma_b = \frac{\delta \cdot kT}{2\,\Omega}\,\text{Log}\,(D'/D),$$

where Ω is the atomic volume. In spite of the absence of a physically based derivation, this relation expresses a correlation between γ_b and D' which can be qualitatively explained: Low energy boundaries are characterized by an ordered structure with a small excess volume above that of the undistorted lattice, so the lower the grain-boundary energy, the larger the formation and motion enthalpies of point defects, and, finally, the lower the values of D'/D. Borisov's relation actually predicts the correct value of γ_b for FCC metals (Gupta, 1984).

The anisotropy in the plane of the boundary persists no matter how large the misorientation, but Q' does not vary with θ. This result suggests a structure made up of a sequence of polyhedra or structural units, in which "channels" appear.

The solute diffusion coefficient is rarely determined because the segregation coefficient κ is rarely known, and the experiment yields only the parameter $P = \kappa D'\delta$ (see Eq. (VII.21)). We quote the results for the diffusion of sulfur in silver (E. Moya, 1975):

$$P = \kappa D'\delta = 10^{-6}\,\exp(-19\;500/RT).$$

Since κ could be measured independently ([3]):

$$\kappa = 1.56\,\exp(15\;000/RT),$$

it follows that:

$$D' = 12.8\,\exp(-35\;000/RT)(\text{cm}^2{\cdot}\text{s}^{-1})$$

(see also Cabané-Brouty *et al.*, 1982).

If the solute segregates to the boundaries, the solute atoms are more or less trapped there; thus the solute diffusion coefficient can be expected to decrease as the solute content increases. For self-diffusion, such considerations about energies are not sufficient, and the variation of D' with content of solute can only be understood on the basis of models of the atomic structure of the boundary (arrangement of the solute and solvent atoms in the boundary). Eventually, the effects of cosegregation have to be taken into account (J. Cabané, 1984).

The mechanism of diffusion is probably vacancy type. Measurements under hydrostatic pressure yield an activation volume of $(1.1 \pm 0.2)\ \Omega$ for silver (cf. 0.9 for volume diffusion) (Martin *et al.*, 1967). The isotope effect in the same

---

([3]) $c_b\delta$ can be measured by Auger electron spectroscopy and $P = (c_b/c_v)\,\delta \cdot D'$.

metal is $f' \Delta K' = 0.46$ for polycrystals and [100] twist bicrystals (Robinson and Peterson, 1972). These results argue in favor of vacancies with very small relaxation, and definitely localized in a given lattice site, in accord with the models obtained by simulation.

The new simulations which have shown the periodic arrangements of structural units may inspire new determinations of diffusion coefficients and especially of their anisotropy.

2. *Other materials.* — In ceramic materials (oxides, carbides), diffusion along short-circuiting paths (boundaries, dislocations) resembles the case of metals; the effective size is no greater than a few interatomic distances, except in the presence of precipitates of second phase, or of an amorphous film. We specifically cite the case of Cr_2O_3 and NiO, where D' has been measured directly (C-type kinetics) and δ determined as $\approx 10^{-7}$ cm for boundaries and subboundaries; $D'/D \approx 10^4$ for dislocations, $\approx 10^5$ for high-angle boundaries, and $Q'/Q = 0.8$ and 0.7, respectively (Atkinson, 1984). It appears that in ionic crystals both the cations and the anions diffuse more rapidly along short-circuiting paths, but there are not enough experimental results. The contradictory or very scattered results, especially on the alkali halides, are attributed to strong segregation and/or precipitation, not revealed by the authors. In the superionic conductors, the boundaries behave as obstacles to rapidly diffusing solutes.

The anisotropy of diffusion in tilt boundaries in MgO decreases with increasing misorientation; on the other hand, the activation energy is independent of misorientation.

In molecular crystals, preferential diffusion in boundaries and subboundaries has been measured in some cases: in "plastic" crystals, for subboundaries, $D'/D \approx 10^5$ to 10^6 at temperatures near T_M. There is also a noticeable difference of this ratio between molecular crystals of high entropy of melting and those of low entropy (plastic crystals).

VI. Diffusion-induced grain-boundary migration (DIGM).

In a sample subjected to an interdiffusion process, migration of the grain boundaries in a direction normal to the diffusional flux can be observed (Fig. VII.14). The composition of the parts of the specimen swept by the boundaries simultaneously changes, the direction of the change depending on the individual case.

Example: Dezincification of an Fe-Zn alloy, or zincification of pure Fe, by heating in vacuum or a zinc atmosphere, respectively. The dezincified or zincified zones form transverse bands bounded by the initial and final positions of the boundary. These zones clearly extend only to a certain depth, depending on the diffusion time (Fig. VII.14).

This phenomenon can be interpreted as a sort of Kirkendall effect relating to grain-boundary diffusion. Assume that for the system under study, $D < D'$; interdiffusion in the volume is negligible. If in the boundaries $J'_A \neq J'_B$, the

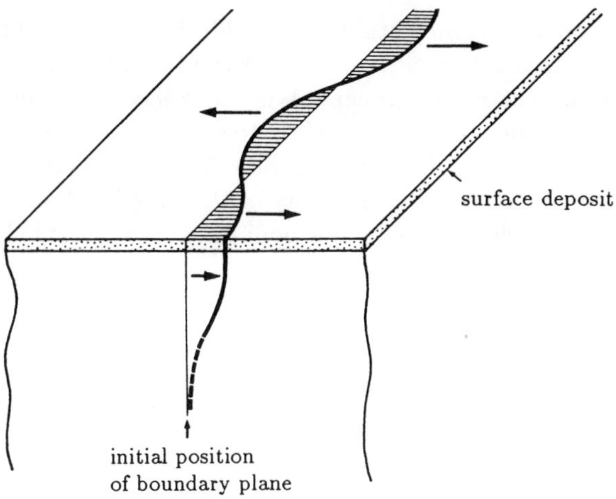

Fig. VII.14. — Diffusion-induced grain-boundary migration. The boundary will migrate to the left or right.

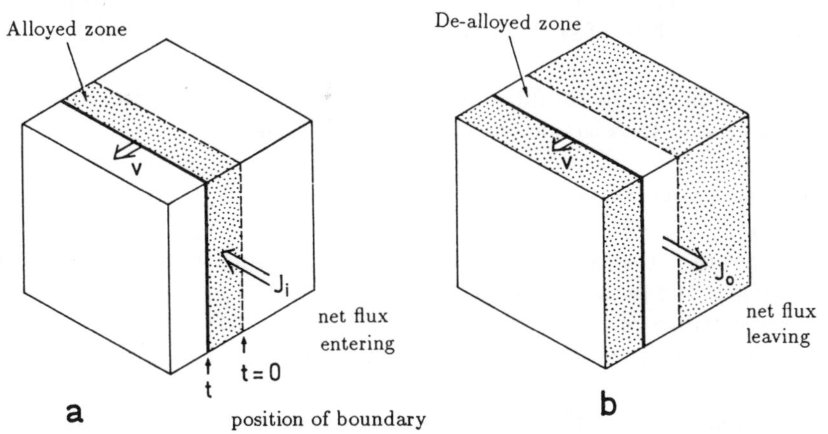

Fig. VII.15. — Diffusion-induced grain-boundary migration. It is assumed that D (volume) $<<$ D' (boundary), D' (solute) $>$ D' (solvent). a) Solute atoms are deposited by grain-boundary diffusion toward the inside of the sample; the boundary migrates leaving behind it an enriched zone. b) The solute atoms are lost as a result of grain-boundary diffusion toward the outside of the sample; the boundary migrates leaving behind an impoverished zone. After R. W. Balluffi (1982).

difference in fluxes requires point defects to come into play, for example a flux of vacancies J'_v. The boundary must then behave as a source or sink of vacancies, presuming a mechanism of diffusion by defects, even in the bad crystal structure of the grain boundary. This is in agreement with the theoretical studies (simulations) of vacancy stability in grain boundaries. The loss or gain of material is accompanied by a migration of the boundary and by an impoverishment or enrichment in component A in the zone swept by the boundary if $|J'_A| > |J'_B|$, according to whether J'_A is toward the exterior or the interior of the sample (Fig. VII.15). The driving force for the migration would be the variation of the chemical Gibbs free energy of the alloy with change of composition. Climb of the dislocations in the boundary, accompanied by glide, can explain the migration.

PART 2. — SURFACE DIFFUSION

The surface of a crystal constitutes a two-dimensional medium which appears *a priori* to lend itself to analysis and to simple studies of atomic mobility. In fact, however, our knowledge about surface diffusion is less advanced than it is for lattice diffusion, for several reasons:

- less theoretical understanding of the medium and its structural defects;
- the possibility of several (if not many) diffusion mechanisms in competition;
- the difficulty in establishing the relation between macroscopic experimental data and the elementary processes of diffusion.

I. The structure of surfaces.

Three types of surfaces can be distinguished according to the crystallographic orientation of the plane of the surface:

- *singular surfaces*: of low Millers indices: {100}, {110}, and {111} for cubic structures. At absolute zero, their structure has the minimum roughness in the hard-sphere description;
- *vicinal surfaces*: obtained from the preceding by rotation about an axis contained in the plane of the surface;
- *random surfaces*: for which the surroundings of the atoms are perturbed over a thickness of several monolayers; the surface roughness is considerable and no simple atomic description is possible.

The terrace-ledge-kink (TLK) model can be used to describe singular and vicinal, but not random, surfaces (Fig. VII.16). The surface is made up of singular planes (terraces), separated by straight ledges (or steps) or by zig-zag ledges, i.e., ledges with kinks. The poles of the surfaces are located on the sides of the standard stereographic triangle. The poles of random surfaces are on the inside of the stereographic triangle.

The plane of the surface is denoted by its Miller indices, but the notation of Lang *et al.* is preferable for specifying the surface more exactly (Fig. VII.17); e.g., $m(111) \times (111)$ means a surface constituted of (111) terraces m atoms in width and monatomic (111) ledges.

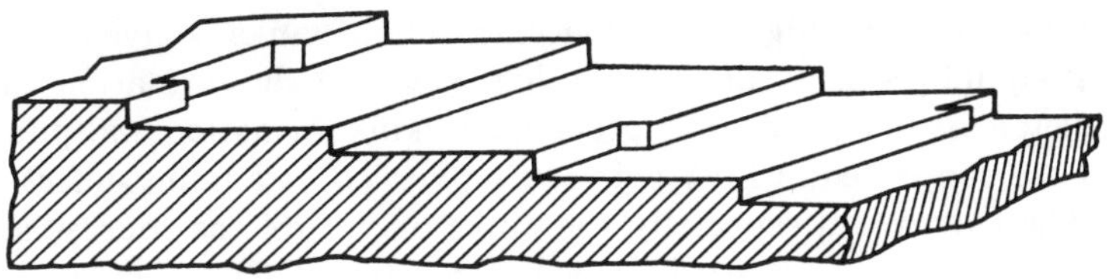

Fig. VII.16. — The surface of a crystal in the TLK model.

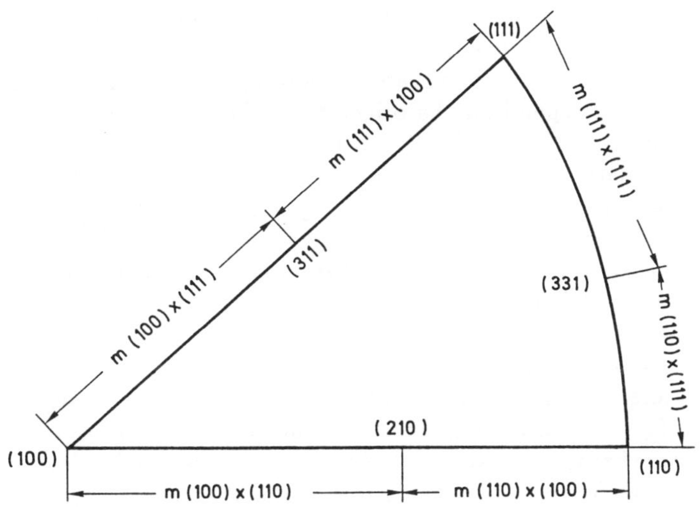

Fig. VII.17. — Notation for the surfaces of a cubic crystal, according to Lang, Joyner, and Somorjai (1972).

The surface displays the ledges and kinks which are geometrically necessary to determine its orientation according to the TLK model; in addition, there are random ledges and kinks resulting from the surface preparation:

– emergence of screw dislocations;

– error in the orientation of the surface. A misorientation of 1° corresponds to a ledge every 100 Å. For a surface oriented within 15′, there will be one ledge every 400 Å, on average;

– a microrelief resulting from the manipulations of polishing, ion milling, annealing,..., which on the atomic scale translates to a population of ledges of the same order as the above.

The TLK model describes a surface at absolute zero. As the temperature increases, kinks appear on the ledges (created in pairs of opposite sign) and point defects appear on the terraces. No new ledges appear, because their formation energy is too high. The concentration of kinks is given by the laws of thermal equilibrium; it can be very high because the formation energy, at least for metals, is quite low. The processes of kink formation and point defect formation are connected; a kink pair can be viewed as a point defect stuck at a ledge. The ledges and especially the kinks play a role as sources and sinks for point defects. The TLK model is valid only to $\approx 0.6T_M$ because of surface roughening, reconstruction, and premelting transitions.

In a pure material, the point defects are *surface vacancies* and *adatoms.* They are created and destroyed at kinks (Fig. VII.18b and c); they can equally well be produced in pairs on the terraces (Fig. VII.18a). Contrary to the case of vacancies and interstitials in metals, the enthalpies of formation of surface vacancies and adatoms are comparable, according to pair potential calculations; they do, however, depend on the orientation of the surface. For example, for the $\{111\}$ surface of copper, both enthalpies are of order 1 eV, whereas for $\{100\}$, the formation enthalpy of the surface vacancy is less than that of the adatom. When the kinks are "poisoned" by impurities, the defects are created in pairs on the terraces.

As a general rule, the enthalpies of formation are a little smaller than those in the volume of the crystal. It is difficult to predict which type of defect is predominant.

The above description applies to metals. In the case of the elemental semiconductors with the diamond structure (Ge, Si), the description of the singular surfaces is more complex: two types of $\{111\}$ surface are expected, depending on whether they are created between planes a and B (cutting of three bonds) or between B and b (cutting of 1 bond) ([4]). The dangling bonds thus produced can join ("reconstruction"); such reconstruction makes description of ledges and kinks difficult. In the AB semiconductors of type III-V, cutting along $\{111\}$ produces two chemically different surfaces, one A face, one B face. (These crystals have no center of symmetry.)

In simple A^+B^- ionic crystals, the singular surfaces are neutral (e.g., $\{100\}$ NaCl) or charged with a long range coulomb field (e.g., $\{111\}$ NaCl).

([4]) The stacking of (111) planes in the diamond structure is AaBbCcAa...

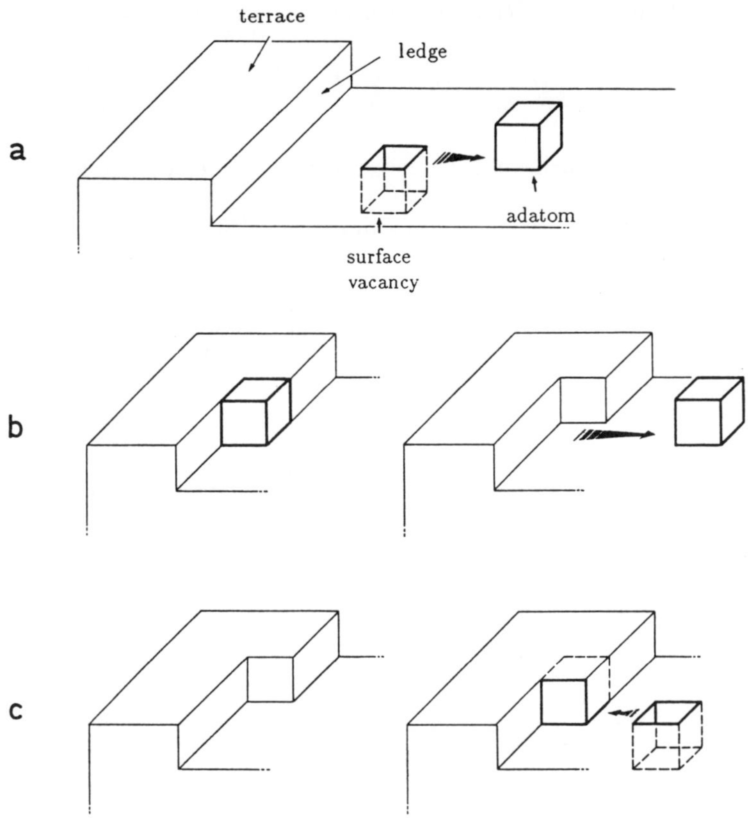

Fig. VII.18. — Creation of point defects on a surface: a) adatom-surface vacancy pair; b) adatom leaving kink; c) surface vacancy.

II. Mechanisms of diffusion.

II.1. SELF-DIFFUSION. — The theory of random walk applies to a two-dimensional medium. For point defects of type *d*, equation (II.19) of chapter II is written:

$$D_d = \frac{1}{2} \sum_{i=1}^{p} \Gamma_i \; x_i^2, \qquad \text{(VII.28)}$$

the sum is over p types of jump, of frequency Γ_i and projected length x_i. The surface self-diffusion coefficient is given by:

$$D_s = \sum_d N_d D_d. \tag{VII.29}$$

It is difficult to go much farther if several types of defects are mobile simultaneously.

At low temperature, elementary jumps, i.e., one interatomic distance, of surface vacancies and adatoms are expected. When the temperature rises, multiple jumps (several interatomic distances) of the adatoms are possible, as well as delocalized jumps of the adatoms (no measurable residence time at any site). Near T_M, even a pseudo-liquid behavior is possible. Such diverse behavior has been observed in molecular-dynamics simulations (Fig. VII.19).

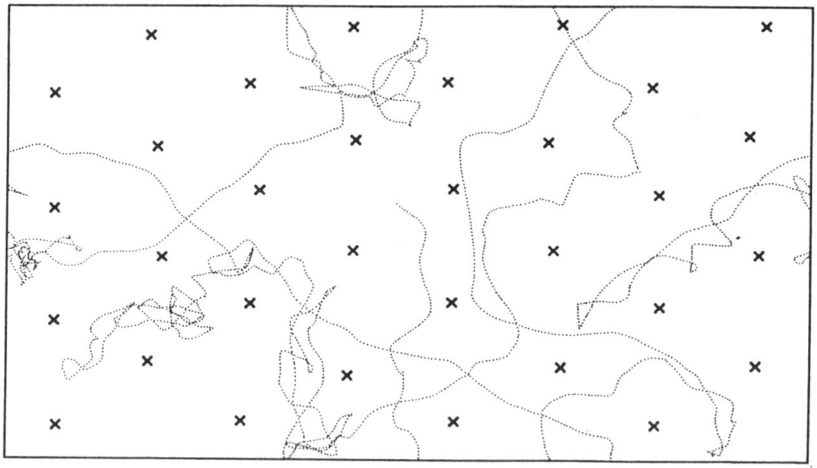

Fig. VII.19. — Path of a delocalized adatom on a (111) FCC surface. Molecular-dynamics calculation with periodic boundary conditions and 10^4 steps. After De Lorenzi *et al.* (1982).

In the case of a close-packed metal like copper, calculations predict that the migration enthalpy is higher for the surface vacancy than for the adatom, except on {100}. Over a wide range of temperature ($\lesssim 0.75\ T_M$) elementary jumps of adatoms are expected to play the dominant role, so that D_s can be written, according to (VII.28) with the total jump frequency equal to that of the adatoms:

$$D_s = N_{ad}\ D_{ad} = (1/4)\ N_{ad}\ \Gamma_{ad}\ d^2,$$

and follows an Arrhenius law:

$$D_s = \text{const } a^2 \tilde{\nu} \exp\left(\frac{\Delta S_f^s + \Delta S_m^s}{k}\right) \exp\left(-\frac{\Delta H_f^s + \Delta H_m^s}{kT}\right). \quad \text{(VII.30)}$$

For a $\{100\}$ surface, D_s could be expressed as the sum of two exponentials corresponding to surface vacancies and adatoms.

In a more rigorous treatment, all the different types of jumps would have to be considered, including exchanges with the layer directly below the surface. The atomic model of Benoist and Martin, described above for grain-boundary diffusion, can also be successfully applied to surface diffusion (Fig. VII.20). The analysis of the random walk of a point defect jumping from a surface site into a nearest neighbor site on a surface with a periodic structure in a simple cubic crystal can be carried out on the basis of an average frequency Γ of jumps in the volume, with $D = \Gamma\lambda^2$, the frequency Γ_i of the jump from a site in the volume to a site on the surface, and the average frequencies $\langle\Gamma_0\rangle$ and $\langle\Gamma_s\rangle$ of jumps from a surface site into a volume site or another surface site, respectively. Also, the average distance between surface sites, $\bar{\lambda} \neq \lambda$. These average frequencies can be expressed as functions of the diverse jump frequencies of the ad-defects. The parameter β is then found to enter into the continuum model in a manner analogous to equation (VII.10) for grain boundaries:

$$\beta = \frac{\Gamma_i}{\langle\Gamma_0\rangle} \frac{\langle\Gamma_s\rangle}{\bar{\Gamma}} \frac{\bar{\lambda}}{(Dt)^{1/2}}. \quad \text{(VII.31)}$$

The calculation can thus be applied to a vicinal surface and permits the analysis of the anisotropy of diffusion.

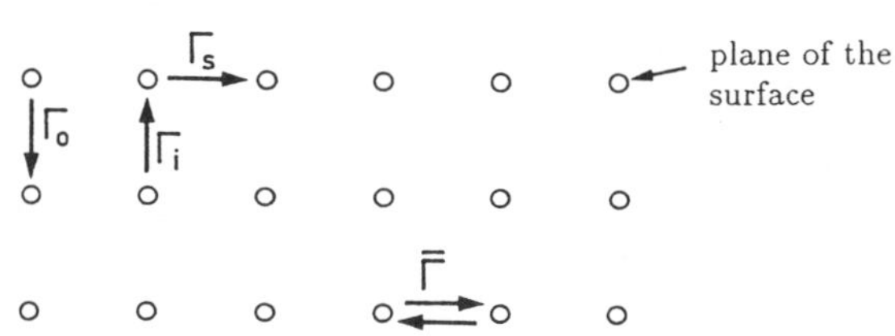

Fig. VII.20. — Benoist's and Martin's model.

On a *vicinal surface,* the *anisotropy* of diffusion can be estimated in the following manner (B. Peraillon *et al.*, 1983). Thermal equilibrium is assumed to be maintained locally, which implies that the root mean square displacement $\langle R^2\rangle^{1/2}$ is much larger than the distance between the kinks. There is a trapping of defects on the ledges; the mole fractions of adatoms on the ledges, N_L, and on the terraces, N_T, obey the relation:

$$N_L/N_T = \exp\left[(\Delta H_T - \Delta H_L)/kT\right],$$

where ΔH_T and ΔH_L are the enthalpies of binding of an adatom to a terrace and a ledge, respectively.

In a dynamic equilibrium, the divergence of the atomic flux perpendicular to the ledges should be zero in order not to change the structure of the surface. From this, it follows that:

$$N_L \Gamma_{L\perp} = N_T \Gamma_{T\perp}.$$

It is natural to assume that the adatoms move isotropically on the terraces: $\Gamma_{T\perp} = \Gamma_{T\parallel} = \Gamma_T$. Under these conditions, surface diffusion is governed by the following parameters:

- the jump frequency on the terraces Γ_T;
- the jump frequency along the ledges $\Gamma_{L\parallel}$;
- the ratio N_T/N_L, the equilibrium distribution of defects between terraces and ledges.

For a vicinal surface with parallel ledges equidistant from each other with spacing λ (total length of ledge per unit area $= \lambda^{-1}$):

$$\begin{cases} D^*_{s\perp} = \dfrac{1}{4}\, N_T\, \Gamma_T\, d^2 \\ D^*_{s\parallel} = \dfrac{1}{4}\, N_T\, \Gamma_T\left(1 - \lambda^{-1}\, a_0\right)\, d^2 + \dfrac{1}{2}\, \lambda^{-1}\, a_0\, N_L\, \Gamma_{L\parallel}\, d^2, \end{cases} \qquad \text{(VII.32)}$$

where d is the jump and a_0 the interatomic distance. The presence of the ledges thus involves only a weak anisotropy; for a misorientation of 3°, $\lambda^{-1}\, a_0 \approx 0.1$, and the anisotropy is only 10% .

The above analysis is based on the assumption that only one type of defect, the adatoms, is active; eventually, the contributions of surface vacancies and multiple defects have to be included. Finally, this analysis fails when the adatoms make jumps of variable length or are delocalized (quasi-free particle), which, according to computer simulations, appears to be the case for the {111} surface above 0.4 T_M.

The atomic structure of the surface should cause an *anisotropy* of surface diffusion in two cases:

- anisotropy of the structure of certain singular surfaces (the structure called "canals" or "rails" on the {110} face in FCC). Field-ion microscopy and molecular dynamics have shown evidence for an exchange mechanism (Fig. VII.21); the adatom situated initially in a channel forms a di-adatom in the saddle point configuration with one atom in the <110> string;
- anisotropy due to the presence of ledges on vicinal surfaces, but this seems quite small (see above).

II.2 Solute diffusion. — The subject here is the diffusion of a chemisorbed species. At average temperatures on a singular surface, the adatoms follow a random walk:

$$D = \langle X^2 \rangle /2\,t = \langle L^2 \rangle /4\,t \tag{VII.33}$$

and equation (VII.28) applies.

In the general case, the difficulty of the analysis stems from the process of trapping or detrapping the solute atoms on certain sites. There does not exist a correct atomic model for this type of diffusion. The tight binding method allows the comparison of the energies of solute adatoms in diverse sites on the surfaces of BCC metals, especially near ledges (Jardin *et al.*, 1984).

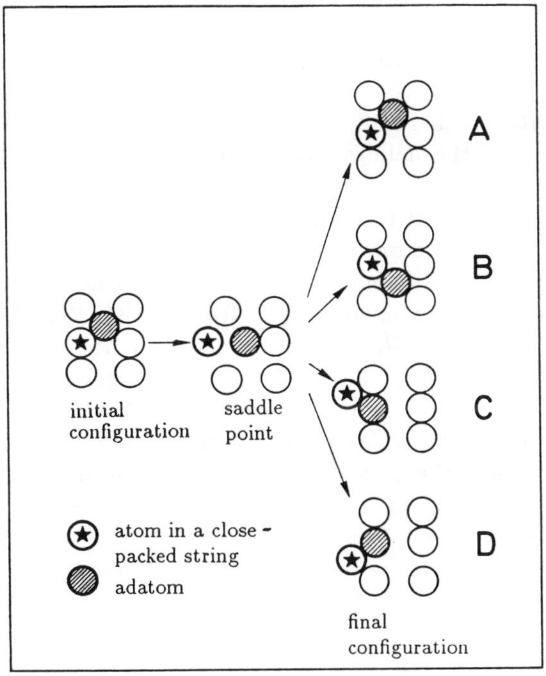

Fig. VII.21. — Adatom exchange mechanism on a (110) surface in FCC. This surface is made up of <110> channels (vertical in the figure).

When the fraction of coverage is quite high, the traps can be considered saturated, so that:

$$D = \frac{1}{4}\,\Gamma\, d^2, \tag{VII.33a}$$

but only if the atoms of the absorbate do not interact.

III. Experimental methods and results.

Several classes of methods can be used in the experimental study of surface diffusion. Unfortunately, they cannot be cross-checked, as they measure different phenomena, or are operational in different ranges of temperature.

III.1 FIELD-ION MICROSCOPY. — The image of a very fine tip of a crystal (radius of curvature some hundreds of Å) is formed by the atoms of a rare gas (the "imaging" gas, the pressure of which is very low). The rare gas atoms are ionized in the neighborhood of the surface, then ejected and accelerated towards the screen under a very strong local electric field (≈ 1 V/Å). The magnification is of the order of 10^6. The observations are carried out at very low temperature (liquid H_2 or He).

The *migration of adatoms* on a singular crystal face can be observed. First a few atoms are deposited (by evaporation from a source), their positions are marked, then, with the field off, the needle is heated for a certain time, and finally, after cooling, the field is turned on and the crystal face is observed anew.

This method is limited to a small number of systems (refractory metals: W, Mo, Ta, Re, Ir and Pt on W, Rh on Rh) and to temperatures less than $0.2\ T_M$. The mean square displacement, $\langle X^2(t) \rangle$, of adatoms on terraces can be measured directly. For example, on Rh{111} (FCC), $\Delta H_m = 0.15$ eV. However, for W (BCC), ΔH_m varies from 0.8 eV for a {211} surface with channels to 1.8 eV on {111}. Pairs of adatoms migrate with a smaller ΔH_m than do isolated adatoms. Note two possible difficulties: the small size of the crystal faces can involve deviations from a random motion; on the other hand, the adatoms and the crystal face are not in thermodynamic equilibrium because the concentration of adatoms is several orders of magnitude higher than at equilibrium.

Adatoms can also pass over a ledge. Some ledges are transparent; others, reflecting.

III.2 DIFFUSION OF RADIOTRACERS. — The redistribution of tracer atoms initially deposited on one part of the surface is studied according to one of the schemes of figure VII.22. The tracer atoms migrate on the surface, and into the volume of the crystal. By serial sectioning perpendicular to the surface (Fig. VII.23) and counting the activity, a profile $\bar{c}(y,\ t)$ is established.

The procedure is analogous to that used for grain-boundary diffusion. The Fisher model can also be used; surface diffusion is confined to a surface layer of thickness δ. A solution of the diffusion equations for the case of figure VII.22c has been given by Suzuoka; it leads to a representation of log $\bar{c}$ as a function of y. If p is the slope of the straight line obtained from the graph:

$$p = \partial \log \bar{c} / \partial y,$$

then we have:

$$\begin{aligned} \beta &= \frac{\partial \log \bar{c}}{\partial \left(\eta \beta^{-1/2}\right)} \frac{1}{p^2 \; Dt} \\ &\approx (0.9)^2 / p^2 \; Dt, \end{aligned} \tag{VII.34}$$

with:

$$\begin{aligned} \beta &= \left[\frac{D_s}{D} - 1\right] \frac{\delta}{(Dt)^{1/2}}, \\ \eta &= y / (Dt)^{1/2} \end{aligned} \tag{VII.35}$$

where D is the volume diffusion coefficient.

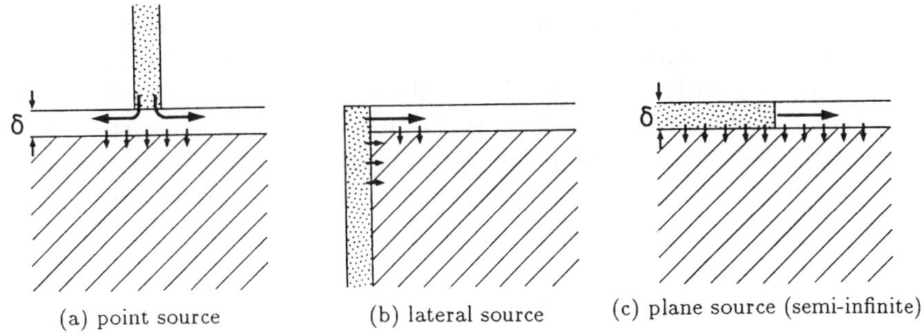

Fig. VII.22. — Surface diffusion. Experiments using radiotracers.

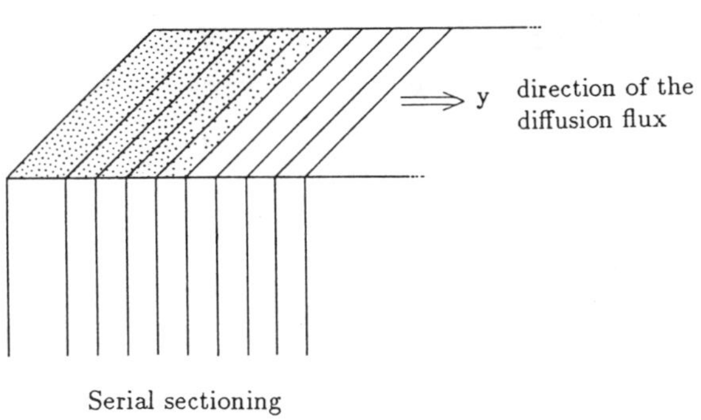

Fig. VII.23. — The sectioning method.

Advantages of the method:

– a very thin layer of tracer (≤ 1 monolayer) is sufficient, which allows the use of a nearly unperturbed surface;

– the surface can be geometrically very well defined (checked by low energy electron diffraction (LEED)); its cleanliness can be assured by ion bombardment or thermal treatment in ultra-high vacuum, and monitored by Auger electron spectroscopy;

– it is easy to study the anisotropy of diffusion.

Inconveniences:

– it is difficult to overcome the process of evaporation-condensation, which limits the measurements to $T < (0.7 \text{ or } 0.8)\ T_M$;

– the experiments require a rather sophisticated and expensive set-up;

– only the product $D_s\delta$ is measured, at least by the Fisher-type analysis;

– impurities can affect D_s by several orders of magnitude.

The experimental results obtained on some metals (Cu, Ag, Au, Ni, Fe) are badly scattered. It is definite that D_s varies with orientation of the surface, and that the anisotropy varies with this orientation. The values of D_s measured between 0.6 T_M and T_M lie in the range 10^{-7} to 10^{-4} cm^2 s^{-1} (Fig. VII.24). The anisotropy ratio $D_{\|}/D_{\perp}$ (with respect to the ledges) is small; in copper at 0.6 T_M it varies between 1.2 and 6.4, depending on the orientation of the surface. The diffusivity perpendicular to the ledges is close to that measured on a singular surface corresponding to the terraces and is insensitive to the density of the ledges. The diffusivity parallel to the ledges is accelerated for only certain orientations of these, in relation to their structure (Cousty, 1981).

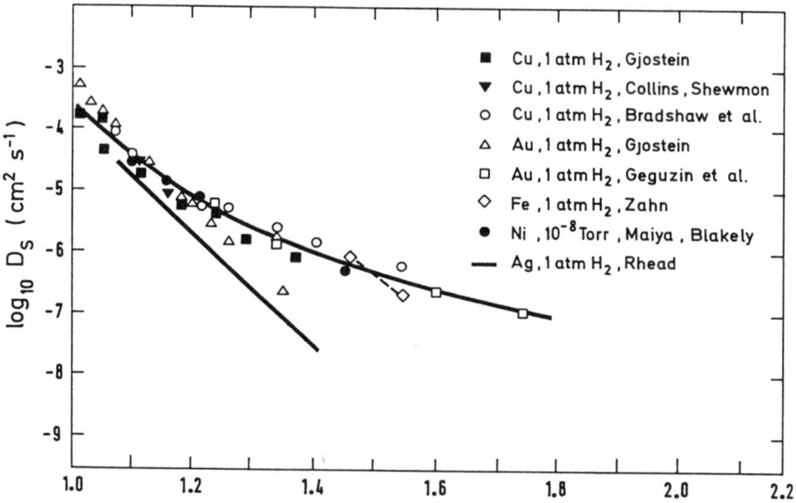

Fig. VII.24. — Arrhenius plot for surface self-diffusion of several FCC metals. Compilation by Gjostein (1963).

Gjostein has grouped the ensemble of results on a unique Arrhenius plot, with the reduced reciprocal absolute temperature T_M/T as abscissa, in order to show a general tendency: for FCC metals, a change of slope appears at about $0.75\ T_M$ (Fig. VII.24):

$$T > 0.75\ T_M \qquad D_s = 7.4 \times 10^2 \exp(-15\ T_M/T)$$
$$T < 0.75\ T_M \qquad D_s = 14 \times 10^{-2} \exp(-7\ T_M/T).$$

The existence of two regions is not well understood, and various reasons are invoked for it:

- influence of evaporation-condensation processes;
- contribution of surface vacancies;
- surface pre-melting;
- migration of delocalized defects;
- contribution of multi-defects;
- thermally activated formation of kinks.

Table I compares the experimental results and simulations for the surfaces of FCC metals.

III.3 Topographic methods. — These consist of following the change in the shape of a solid that results from material transport, which can occur by four processes (Fig. VII.25):

- surface self-diffusion;
- volume self-diffusion;
- evaporation-condensation (in vacuum);
- evaporation, diffusion in the gas phase, condensation.

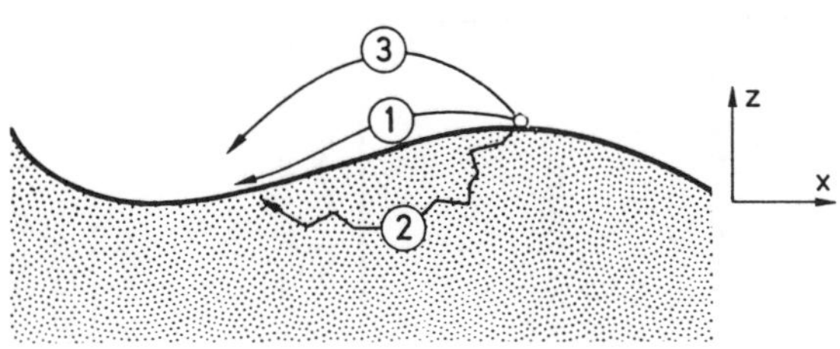

Fig. VII.25. — Mechanism of material transport near a surface: 1) surface diffusion; 2) volume diffusion; 3) evaporation-condensation.

The driving force for this transport is the gradient in the curvature of the surface... or sometimes a gradient of temperature, or of pressure (bubble migration).

In effect, the difference of chemical potentials of an atom and of a vacancy

Table I. — *Surface Self-Diffusion (FCC Structure), (after Ghaleb [1984])*

Method	Field Ion Microscopy [1,2]	Molecular dynamics [3,4]	Macroscopic Studies	
			Radiotracers [5]	Work function measurement [6]
	$T \leq 0.2\, T_M$	$0.3\, T_M \leq T \leq 0.6\, T_M$	Cu, $0.6\, T_M$	Ni $0.12\, T_M \leq T \leq 0.34\, T_M$
Isotropic surfaces	– the mobility of the adatoms depends strongly on the surface roughness : large on (111), weaker on (100) ; – no vacancies are observed.	– the surface vacancies on (111) are only somewhat mobile, whereas the adatoms behave almost as free particles ; – the mobility of the two types of defect is of the same order on (100) ; – multiple jumps over several interatomic distances are observed for both defects ; – D_s of an adatom on (111) > D_s of an adatom on (100).	$D_s\,(111) \approx D_s\,(100)$	trapping of adatoms at random ledges on (100)
Vicinal surfaces	– the adatoms are trapped by dense ledges and migrate along them.	– the adatoms, which are responsible for the mobility, are trapped at the ledges and migrate along them ; – the anisotropy depends on the atomic structure of the ledges.	– $D_{s\perp}$ = const, $\forall$ ledge density – <110> ledges $D_{s//} > D_{s\perp}$ – <001> ledges, $D_{s//} \approx D_{s\perp}$	
(110) surface	– the dimensionality of the path depends on the chemical nature of the metal – evidence for the exchange mechanism.	– the dimensionality of the path depends on the parameters used in the L.J. potential ; – evidence for the exchange mechanism (Fig. VII.21) ; – multiple jumps along the channels observed for adatoms when T increases.	– $D_{s//} > D_{s\perp}$	

[1] Kellog G.L. *et al.*, *Surf. Sci.* **70** (1978) 485.
[2] Ehrlich G. *et al.*, *Ann. Phys. Chem.* **21** (1980) 603.
[3] Cousty J. *et al.*, *Surf. Sci.* **107** (1981) 586.
[4] De Lorenzi *et al.*, *Surf. Sci.* **116** (1982) 391.
[5] Cousty J. Thesis, Orsay (1981).
[6] Schrammen P. *et al.*, *Surf. Sci.* **130** (1983) 203.

situated below a surface is given by Herring's formula:

$$\mu - \mu_v = \mu_0 + \Omega \left\{ \gamma_s \left[\frac{1}{R_1} + \frac{1}{R_2} \right] + \frac{1}{R_1} \frac{\partial^2 \gamma_s}{\partial n_1^2} + \frac{1}{R_2} \frac{\partial^2 \gamma_s}{\partial n_2^2} \right\}, \tag{VII.36}$$

where γ_s is the surface tension, R_1 and R_2 the principal radii of curvature, $\partial^2 \gamma_s / \partial n_i^2$ the second derivatives of γ_s with respect to the normals in the principal planes of curvature. If these derivatives are zero, Herring's formula reduces to that of Gibbs-Thomson:

$$\mu - \mu_v = \mu_0 + \Omega \gamma_s \left(\frac{1}{R_1} + \frac{1}{R_2} \right). \tag{VII.36a}$$

In the above formulae, the curvature is taken as positive for a convex surface.

Any gradient of curvature involves a gradient of chemical potential, and thus a flux of material. The crystal modifies its form so as to minimize its total surface energy.

Mullins (1957) has established the equations regulating the kinetics of the evolution of the surface for the case of four different profiles: an isolated groove, a series of parallel and equidistant grooves, a groove at the intersection of a grain boundary with a surface (grain-boundary grooving or "thermal etching") and the rounding of a tip. His calculation rests on the following assumptions (see App. II):

- the four mechanisms listed above act independently;
- the loss of material by evaporation is negligible;
- the surface tension is isotropic;
- the curvature is small;
- the atomic structure of the surface need not be taken into account.

Consider the case of a small sinusoidal undulation of the surface, described by:

$$z(x,\ 0) = z_m \sin \frac{2\ \pi x}{\lambda} = z_m \sin \omega x. \tag{VII.37}$$

According to Mullins, this profile decays with time following exponential kinetics (see App. II):

$$\boxed{z(x,\ t) = z(x,\ 0) \exp(-\beta t)}, \tag{VII.38a}$$

with:

$$\boxed{\beta = A\omega^2 + (C + A')\omega^3 + B\omega^4}, \tag{VII.38b}$$

each of the coefficients $A,\ B, \ldots$ corresponding to a mechanism of material transport:

A, evaporation-condensation in a vacuum;

A', the same, with diffusion in the gas phase;

B, surface diffusion;
C, volume diffusion.
The coefficients A, A', B, and C depend on temperature:

$$A = \frac{\Omega}{(kT)^{3/2}} \frac{p_0 \gamma_s}{\sqrt{2\pi M}} \quad \text{(VII.39a)}$$

$$B = \gamma_s \delta D_s \cdot \Omega / kT \quad \text{(VII.39b)}$$

$$C = \gamma_s D \cdot \Omega / kT, \quad \text{(VII.39c)}$$

where Ω is the atomic volume, M the atomic mass, γ_s the surface tension, and p_0 the pressure of the ambient atmosphere (e.g., the vapor pressure of the material).

By choosing ω, i.e., the wavelength λ of the undulations, one or the other process can be favored. With $\lambda << 2\pi B/C$, surface diffusion is dominant.

In practice, parallel grooves spaced 5 to some tens of μm apart are scribed by photolithography. The profile can be represented by a Fourier series. Since the decay is exponential in ω^n, the harmonics decay rapidly, and, after a transient, the evolution of a sinusoidal profile is followed, usually with an interference microscope.

Advantages:

- the method is easy enough to use, once the technique of ruling the grooves is mastered, and it is applicable to metals and ceramics;
- it allows the study of the influence of adsorbates of different kinds and in different, controlled quantities;
- the actual transport of material, which is at the base of certain metallurgical processes such as sintering or spheroidization, is studied;
- the method can be applied to alloys and compounds, by introducing the effective diffusion coefficient (see Ch. VI, §IV) into B and C.

Inconveniences:

- the product $\gamma_s \delta\ D_s$ is measured; however, a judicious choice of ω allows the determination of B/C, and therefore of $\delta D_s/D$;
- the orientation of the surface is not well defined because it varies along the profile of the grooves; this makes difficult the study of structural effects;
- the density of ledges can be very high (one ledge every few atomic spacings), and the transport can be governed by complex processes.

Spectacular effects of certain adsorbates have been shown by Rhead and Delamare (1969, 1971):
Bi or S on Cu, Ag, and Au can increase D_s by several orders of magnitude (to $D_s \approx 1\ \text{cm}^2\ \text{s}^{-1}$, or a jump frequency near the Debye frequency !)

III.4 LASER-INDUCED THERMAL DESORPTION (LITD). — A short (100 ns) focused laser pulse gives rise to desorption of chemical species adsorbed on a well-defined area (10-200 μm). The desorbed molecules are detected by mass spectrometry. Between laser pulses, the adsorbate diffuses along the surface and refills the area desorbed by the laser pulse. The in-diffused quantity is then measured with the following laser pulse. LITD is the method of choice for studying adsorbed gases ; it allows the determination of the surface diffusion coefficient, the isotope effect, the influence of fractional coverage, the effect of the crystal structure of the surface, and the effect of adsorbed impurities (Grunze *et al.*, 1988).

III.5 OTHER METHODS. — We mention the measurement of the electrical resistivity of thin films, which is sensitive to micro-roughness due to the ad-defects and distinguishes the migration of adatoms on terraces, their trapping at ledges, and their annihilation at kinks. We mention also the measurement of the work function, another quantity sensitive to microroughness, and the scanning tunneling microscope.

APPENDIX I

Grain-boundary diffusion.

1. WHIPPLE'S SOLUTION (CONSTANT SURFACE CONCENTRATION c_0). — In the x, z plane in the reduced coordinates $\xi = [x-(\delta/2)]/(Dt)^{1/2}$ and $\eta = z/(Dt)^{1/2}$:

$$\left\{\begin{aligned} & c(\xi,\ \eta) = c_1 + c_2 \\ & \frac{c_1}{c_0} = \operatorname{erfc}\left(\frac{\eta}{2}\right) \\ & \frac{c_2}{c_0} = \frac{\eta}{2\ \pi^{1/2}} \int_1^{\Delta} \exp\left(-\frac{\eta^2}{4\ \sigma}\right) \operatorname{erfc}\left[\frac{1}{2}\left(\frac{\Delta-1}{\Delta-\sigma}\right)^{1/2}\left(\frac{\sigma-1}{\beta}+\xi\right)\right] \frac{\mathrm{d}\sigma}{\sigma^{3/2}}, \end{aligned}\right. \tag{A.1}$$

where σ is a variable of integration, $\Delta = D'/D$, and $\beta = \frac{\delta}{2}(\Delta - 1)/\sqrt{Dt}$.

Since Δ is very large, the equation can be simplified a little:

$$\frac{c}{c_0} = \operatorname{erfc}\left(\frac{\eta}{2}\right) + \frac{\eta}{\pi^{1/2}} \int_0^1 \exp\left(-\frac{\eta^2\ \tau^2}{4}\right) \operatorname{erfc}\left[\frac{1}{2}\left(\frac{1}{\tau^2 \beta} - \frac{1}{\beta} + \xi\right)\right] \mathrm{d}\tau, \tag{A.2}$$

where $\tau = \sigma^{1/2}$ is a new variable of integration.

The average concentration, $\bar{c}$, in a slice at depth z, for unit boundary length $y = 1$, can be written as:

$$\bar{c}(z) = c_1 + \int_{-\infty}^{+\infty} c_2 \, \mathrm{d}x = c_1 + 2\int_0^{\infty} c_2 \, \mathrm{d}x. \tag{A.3}$$

Keeping in mind equation (A.1), equation (A.3) becomes for Δ very large:

$$\begin{aligned} \bar{c}(\eta) &= c_0 \operatorname{erfc}\left(\frac{\eta}{2}\right) + \frac{\eta (Dt)^{1/2} c_0}{\pi^{1/2}} \int_{\sigma=1}^{\infty} \int_{\xi=0}^{\infty} \frac{\mathrm{d}\sigma}{\sigma^{3/2}} \exp\left(-\frac{\eta^2}{4\,\sigma}\right) \times \\ &\times \operatorname{erfc}\left[\frac{1}{2}\left(\frac{\sigma-1}{\beta} + \xi\right)\right] \mathrm{d}\xi. \end{aligned} \tag{A.4}$$

If the volume-diffusion term is neglected, equation (A.4) becomes:

$$\bar{c} = \frac{2\,\eta (Dt)^{1/2} c_0}{\pi^{1/2}} \int_1^{\infty} \exp\left(-\frac{\eta^2}{4\,\sigma}\right) \left[\frac{1}{\pi^{1/2}} \exp\left(-v^2\right) - v \operatorname{erfc} v\right] \frac{\mathrm{d}\sigma}{\sigma^{3/2}}, \tag{A.5}$$

with

$$v = \frac{\sigma - 1}{2\,\beta}.$$

2. Suzuoka's solution (thin layer).

$$\left\{\begin{aligned} & c = c_1 + c_2 \\ & c_1 = \frac{K}{(\pi D t)^{1/2}} \exp\left(-\frac{\eta^2}{4}\right) \\ & c_2 = \frac{K}{(\pi D t)^{1/2}} \frac{\partial^2}{\partial \eta^2} \int_1^{\Delta} \exp\left(-\eta^2/4\,\sigma\right) \operatorname{erfc}\left[\frac{1}{2}\left(\frac{\Delta-1}{\Delta-\sigma}\right)^{1/2} \times \right. \\ & \qquad\qquad \left. \times \left(\frac{\sigma-1}{\beta} + \xi\right)\right] \frac{\mathrm{d}\sigma}{\sigma^{1/2}} \end{aligned}\right. \tag{A.6}$$

The average concentration $\bar{c}$ in a thin layer parallel to the surface is:

$$\bar{c}(z) = c_1 + \int_{-\infty}^{+\infty} c_2 \, \mathrm{d}x = c_1 + \bar{c}_2, \tag{A.7}$$

i.e.,

$$\begin{aligned} \bar{c} &= \frac{K}{(\pi D t)^{1/2}} \exp\left(-\frac{\eta^2}{4}\right) + \frac{K}{\pi^{1/2}} \frac{\partial^2}{\partial \eta^2} \int_{-\infty}^{+\infty} \int_1^{\Delta} \exp\left(-\frac{\eta^2}{4\,\sigma}\right) \times \\ &\times \operatorname{erfc}\left[\frac{1}{2}\left(\frac{\Delta-1}{\Delta-\sigma}\right)^{1/2} \left(\xi + \frac{\sigma-1}{\beta}\right)\right] \frac{\mathrm{d}\sigma}{\sigma^{1/2}} \, \mathrm{d}\xi. \end{aligned} \tag{A.8}$$

Neglecting the volume diffusion term, equation (A.8) becomes:

$$\bar{c}_2 = \frac{K}{\pi^{1/2}} \int_1^{\Delta} \left(\frac{\eta^2}{\sigma} - 2\right) \exp\left(-\frac{\eta^2}{4\,\sigma}\right) \left(\frac{\Delta - \sigma}{\Delta - 1}\right)^{1/2} \times$$

$$\times \left[\frac{1}{\pi^{1/2}} \exp\left(-v^2\right) - v \operatorname{erfc} v\right] \frac{d\sigma}{\sigma^{3/2}}, \tag{A.9}$$

with

$$v = \left(\frac{\Delta - 1}{\Delta - \sigma}\right)^{1/2} \left(\frac{\sigma - 1}{\beta}\right).$$

3. Angle of the equiconcentration contour with the plane of the boundary. — Whipple's solution gives:

$$\operatorname{cotg} \theta = \left(\frac{\eta\beta}{4}\right)^{1/3} - \frac{1}{3}\left(\frac{1}{2\,\eta\beta}\right)^{1/3}. \tag{A.10}$$

Suzuoka's solution does not lend itself well to the evaluation of θ because the curvature of the contours changes sign as the depth increases (the source is exhausted, and the region around the boundary, at the surface and a little below it, becomes impoverished in diffusant).

4. Penetration depth. — From the definitions given above:

$$\eta\beta^{-1/2} = \frac{z}{(Dt)^{1/4}} \left(\frac{2\,D}{D'\delta}\right)^{1/2}.$$

According to Suzuoka's solution, for $\beta > 10$, c/c_0 depends only on $\eta\beta^{-1/2}$. Then for a given concentration c_1, $\eta\beta^{-1/2}$ always has the same value and the penetration depth z_1 corresponding to c_1 is such that $D'\delta \propto z_1^2/t^{1/2}$ and $z_1 \propto t^{1/4}$.

APPENDIX II

Evolution of the profile of a surface by material transport.

1. Mullins' model (1957). — This is a phenomenological model, proposed for a pure metal. The chemical potential in the neighborhood of a surface varies with its curvature, leading to the gradients of chemical potential that give rise to the flux of material.

Herring's relation (Eq. (VII.36)) gives the difference of the chemical potential of an atom and of a vacancy in the neighborhood of a surface characterized by its principal radii of curvature R_1 and R_2. Mullins' model is founded on the following approximations:

– the effect of the orientation of the surface is negligible; γ_s the surface tension is taken as a constant, as is D_s the surface diffusion coefficient;

– the curvatures are small; for axial symmetry, the profile of the surface is $z = z(x,\ t)$. Its curvature:

$$\frac{1}{R} = -\frac{\partial^2 z/\partial x^2}{\left[1+(\partial z/\partial x)^2\right]^{3/2}},$$

if it is not too large, is approximated by:

$$1/R \approx -\partial^2 z/\partial x^2. \qquad \text{(A.11)}$$

Herring's equation reduces, with $\gamma_s = \gamma$, to:

$$\mu - \mu_v = \mu_0 + \frac{\Omega\gamma}{R} = \mu_0 - \Omega\gamma\left(\partial^2 z/\partial x^2\right). \qquad \text{(A.12)}$$

Material transport by surface diffusion. — Following the phenomenological equations (Ch. VIII, Eq. (VIII.18)), the flux of A in pure A is given by:

$$J_A = -\frac{L_{AA}}{T}\nabla(\mu-\mu_v) = -\frac{D_s}{\Omega kT}\nabla(\mu-\mu_v) \qquad \text{(A.13)}$$

$$\approx \frac{D_s\,\gamma}{kT}\frac{\partial^3 z}{\partial x^3}, \qquad \text{(A.13a)}$$

since the gradient in our problem follows the curvelinear abscissa s, and $\mathrm{d}s \approx \mathrm{d}x$ (small curvature).

The material-balance equation is written, modeling the surface as a layer of thickness δ, in which the diffusion coefficient is D_s. Consider an element of area $\mathrm{d}x \cdot 1$. The number of atoms entering is $J_A\delta\mathrm{d}t$, the number leaving is $(J_A + \frac{\partial J}{\partial x}\,\mathrm{d}x)\,\delta\mathrm{d}t$, thus the net number is $-\delta(\partial J/\partial x)\mathrm{d}x\ \mathrm{d}t$, which produces a variation in the volume $\mathrm{d}V = -\Omega\delta(\partial J/\partial x)\mathrm{d}x\ \mathrm{d}t$. This variation of the volume over an area of $\mathrm{d}x \cdot 1$ produces an advance (or retreat) of the profile $\mathrm{d}z$ such that $\mathrm{d}V = \mathrm{d}z\ \mathrm{d}x \cdot 1$. Thus the rate of advance of the profile is:

$$\frac{\partial z}{\partial t} = -\Omega\delta\frac{\partial J}{\partial x} = -\frac{\delta D_s}{kT}\Omega\gamma\frac{\partial^4 z}{\partial x^4}.$$

The motion of the profile obeys a differential equation:

$$\boxed{\frac{\partial z}{\partial t} = -B\frac{\partial^4 z}{\partial x^4}} \qquad \text{(A.14)}$$

with:

$$B = \delta D_s \frac{\Omega\gamma}{kT}. \tag{A.14a}$$

Material transport by volume diffusion. — As above, the flux equation is given by:

$$J = -\frac{D}{\Omega kT} \nabla (\mu - \mu_v), \tag{A.15}$$

where D is the volume self-diffusion coefficient. To calculate J, the quantity $(\mu - \mu_v)$ must be known at every point of the medium. The assumption is made that *the surfaces are the only sources and sinks for vacancies.* Under these conditions, the conservation equation for vacancies in the volume is div $J_v = 0$, from which div $J = 0$, whence:

$$\nabla^2 (\mu - \mu_v) = 0. \tag{A.16}$$

This condition imposes a steady state in the entire volume. The solution of equation (A.16) with the boundary conditions given by equation (A.12) gives $(\mu - \mu_v)$as a function of x and z, and its gradient $\partial (\mu - \mu_v) / \partial z$ as well.

The material-balance equation is then written:

$$\frac{\partial z}{\partial t} = \Omega J_{\perp} (z = 0) = -\frac{D}{kT} \left[\frac{\partial}{\partial z} (\mu - \mu_v) \right]_{z=0}.$$

Equation (A.16) is solved by a Fourier transformation, and after some calculations an integro-differential equation in R is obtained (Mullins, 1960):

$$\frac{\partial(1/R)}{\partial t} = -\frac{2\,C}{\pi} \int_0^{\infty} \omega^3 \cos \omega x \, \mathrm{d}\omega \int_0^{\infty} R^{-1}(\xi,\, t) \cos \omega\xi \, \mathrm{d}\xi, \tag{A.17}$$

with:

$$C = D \frac{\Omega\gamma}{kT}. \tag{A.17a}$$

Material transport by evaporation-condensation. — There is a preferential evaporation from the convex parts of the surface, and a preferential condensation on the concave parts. Two cases have to be distinguished:

(1) *an atmosphere of gas*: the evaporated metal atoms have to diffuse in the gas phase. The kinetics found are those for volume diffusion; equation (A.17) applies, replacing the coefficient C by:

$$A' = \rho_0 \, D_G \Omega^2 \, \gamma / kT, \tag{A.18}$$

where D_G is the diffusion coefficient of the metal atoms in the gas, and ρ_0 is the number of metal vapor atoms per unit volume of the gas above a plane surface.

(2) *vacuum*: the Gibbs-Thomson equation is applied; this gives the variation of the vapor pressure with the curvature for a cylindrical surface:

$$\frac{p-\bar{p}}{\bar{p}} = \frac{\Omega\gamma}{kT}\,\frac{1}{R}, \tag{A.19}$$

where $\bar{p}$ is the pressure over a plane surface.

The kinetic theory of gases gives the number of atoms evaporated per unit time and unit area from a surface in equilibrium with its vapor:

$$\dot{n} = p/\,(2\,\pi M kT)^{1/2}\,,$$

where M is the atomic mass.

The difference in the evaporation rates from a curved surface and from a plane surface is given by:

$$\Delta\dot{n} = \Delta p/(2\,\pi M kT)^{1/2} = \frac{1}{R}\,\frac{\bar{p}\gamma\Omega}{(2\,\pi M)^{1/2}(kT)^{3/2}},$$

from which the rate of advance of the surface is $-\Omega\,\Delta\dot{n}$ and

$$\frac{\partial z}{\partial t} = -\frac{\bar{p}\gamma\Omega^2}{(2\,\pi M)^{1/2}(kT)^{3/2}}\,\frac{1}{R},$$

i.e.:

$$\boxed{\frac{\partial z}{\partial t} = +A\,\frac{\partial^2 z}{\partial x^2}}, \tag{A.20}$$

with:

$$A = \frac{\Omega^2}{(kT)^{3/2}}\,\frac{\bar{p}\gamma}{(2\,\pi M)^{1/2}}. \tag{A.20a}$$

2. APPLICATION OF MULLINS' EQUATIONS. — The solution of the three equations (A.14), (A.17), and (A.20) depends on the boundary conditions of the problem.

a) Parallel and equidistant grooves. — This is the easiest case to treat mathematically. Any periodic profile $z(x,\ t)$ can be described by a Fourier series. Thus it suffices to do the calculation for a sinusoidal profile with period $\lambda = 2\pi/\omega$:

$$z(x,\ t=0) = a\,\sin\,\omega x,$$

and to find a solution by separation of variables (see Ch. I, App. I.):

$$z(x,\ t) = a \sin \omega x \cdot \zeta(t). \tag{A.21}$$

For the three cases considered above:

$$\begin{cases} \mathrm{d}\zeta/\mathrm{d}t = -\ B\omega^4\ \zeta(t) \\ \mathrm{d}\zeta/\mathrm{d}t = -\ C\omega^3\ \zeta(t) \\ \mathrm{d}\zeta/\mathrm{d}t = -\ A\omega^2\ \zeta(t). \end{cases} \tag{A.22}$$

If these three mechanisms (or four including diffusion in the vapor phase) operate independently, their effects are additive:

$$\mathrm{d}\zeta/\mathrm{d}t = -\left[A\omega^2 + (C + A')\,\omega^3 + B\omega^4\right]\zeta(t).$$

The solution of this equation is obtained by inspection:

$$\zeta(t) = \exp\ \left\{-\left[A\omega^2 + (C + A')\,\omega^3 + B\omega^4\right] t\right\}. \tag{A.23}$$

The profile decays exponentially with time. Surface diffusion can be the dominant mechanism for small values of λ (the term $B\omega^4$ being dominant over the terms in ω^2 and ω^3). If the initial profile is not sinusoidal, the higher-order terms of the Fourier series decay much more rapidly than the first, and the simple case of a sinusoidal profile is quickly reached. If the evaporation-condensation mechanisms can be neglected (temperature not too high), the results can be treated in the following way. The decay term β is measured experimentally from $\beta t = -\mathrm{Log}[z(x,\ t)/z(x,\ 0)]$ for experiments with different times t. The experiments are repeated with samples scribed with lines spaced at different wavelengths. For a given temperature, β/ω^3 is plotted as $f(\omega)$; this graph should be linear with slope $= B$ and intercept on the ordinate axis $= C$ (or A').

b) Grain boundary grooving. — Where a grain boundary intersects a surface, a groove forms at equilibrium, following the line of intersection at which the surfaces of the two contiguous grains form a dihedral angle θ, given by the equilibrium of the surface tensions $\gamma_{\mathrm{j}} = 2\ \gamma_{\mathrm{s}} \cos\theta$ (the phenomenon is often called "thermal attack").

Mullins' equations have to be solved with the initial condition:

$$z(x,\ 0) = 0$$

and the boundary condition:

$$\partial z(0,\ t)/\partial x = \ \mathrm{cotg}\ \theta. \tag{A.24}$$

– *Evaporation-Condensation.* — Equation (A.20) is exactly Fourier's equation; analytical solutions are listed for a large number of cases. Here:

$$z(x,\ t) = -2\ \mathrm{cotg}\ \theta (At)^{1/2}\ \mathrm{ierfc}\left[x/2(At)^{1/2}\right].$$

The form of the profile depends only on θ. The depth of the groove varies as the square root of the time:

$$z(0,\ t) = 1.13 \text{ cotg } \theta (At)^{1/2}.$$

– *Surface diffusion.* — The solution of equation (A.14) with the same boundary conditions as above, plus the additional condition of small curvature ($\partial^3 z(0,\ t)/\partial x^3 = 0$), requires the Laplace transformation (see Ch. I, App. I). Mullins (1957) found:

$$z(x,\ t) = \text{cotg } \theta (Bt)^{1/4}\ F\left[x/(Bt)^{1/4}\right]$$

where $F(u)$ can be expanded in a power series in $u = x/(Bt)^{1/4}$. The profile has a hump on both sides of the boundary, as can be expected from the conservation of matter (matter transported from the boundary to its neighborhood; the same is not the case for transport by evaporation-condensation, which leads to flat profiles on both sides of the groove). A universal profile is obtained, dependent only on θ, expressed in reduced coordinates $z(Bt)^{-1/4}$ *vs.* $x(Bt)^{-1/4}$. The depth p of the groove can be obtained from it:

$$p = 0.973 \cdot \text{cotg } \theta (Bt)^{1/4},$$

as well as its width (distance between the humps):

$$w = 4.6(Bt)^{1/4}.$$

– *Volume diffusion.* — This case is more difficult to treat. According to Mullins (1960):

$$p = 1.01 \cdot \text{cotg } \theta (Ct)^{1/3}$$
$$w = 5(Ct)^{1/3}.$$

The characteristic exponents for each mechanism, 1/2, 1/3, and 1/4, are again found, a form of *Herring's scaling law.*

Remarks:

1) Mullins' equations have been applied to other problems, such as the blunting and shrinkage of a point (the case of pointed "filaments" for field emission) and more generally the evolution of the form of surfaces of revolution: spheroidization of cylinders, sintering of spheres (evolution of the profile of the neck, see Ch. X, § III.1), coalescence of pores or precipitates, decay of an isolated scratch,... In most of these cases, the solution of Mullins' equations requires resorting to numerical methods (finite differences).

2) The limits of the validity of Mullins' model and equations should not be forgotten: γ_s and D_s independent of the surface orientation, small curvature,

neglect of the atomic structure of the surface (and thus the problem of the surface sources and sinks for vacancies).

3) Among the experimental difficulties encountered in measurements of D and D_s by this method, the following must be pointed out: the uncontrolled presence of evaporation, the frequently encountered faceting of the surface (a manifestation of the anisotropy of the surface tension), and the difficulty of assuring the cleanliness of the surface.

4) The method can be extended to alloys and compounds; an effective diffusion coefficient is determined as defined in chapter VI, § IV; it often reduces to the diffusion coefficient of the slower species.

Bibliography

I. GENERAL REFERENCES

ADDA Y. and PHILIBERT J., *op. cit.* **2** Ch. XII and XIII.

BLAKELY J. M., *Introduction to the Properties of Crystal Surfaces* (Pergamon Press) 1973.

KAUR I. and GUST W., *Fundamentals of Grain and Interface Boundary Diffusion* (Ziegler Press Stuttgart) 1988.

KAUR I., KOZMA L. and GUST W., *Handbook of Grain and Interphase Boundary Diffusion Data,* 2 vols. (Ziegler Press Stuttgart) 1989.

NEUMANN G. and NEUMANN G. M., *Surface Self-Diffusion of Metals,* Diff. and Defect Monograph series, # 1 (Trans. Tech. Publ., Aedermannsdorf) 1972.

MARTIN G. and PERAILLON B., Les Joints intergranulaires dans les Métaux, *J. Phys. Colloq.* **36** (1975) C4-165-C4-190.

Metal Surfaces: Structure, Energetics, and Kinetics, A.S.M. Seminar (1962).

Interfaces et Surfaces en Métallurgie, Ecole d'Eté de Gassin 1973 (Trans. Tech. Publ., Aedermannsdorf) 1975, esp. Ch. 10, by V. Lévy, p. 243-280.

Grain Boundary Structure and Kinetics, A.S.M. Materials Science Series Seminar (1979), esp. the articles by G. Martin and B. Peraillon, p. 239-295, and by N. L. Peterson, p. 209-238.

II. SPECIFIC REFERENCES.

ATKINSON A., *Solid State Ionics* **12** (1984) 309-320.

BALLUFFI R. W., *Metall. Trans.* **13A** (1982) 2069-2095.

BENOIST P. and MARTIN G., *Thin Solid Films* **25** (1975) 181-197; *J. Phys. Colloq. France* **36** (1975) C4-213-C4-223.

BERNARDINI J., GAS P., HONDROS E. D., and SEAH M. P., *Proc. Roy. Soc. London* **A379** (1982) 159-178.

BISCONDI M., in *Physical Chemistry of the Solid State: Applications to Metals and their compounds,* P. Lacombe, ed. (Elsevier, Amsterdam) (1984) p. 225-241.
BORISOV V. T., GOLIKOV V. M., and SHCHERBEDINSKIY, *Fiz. Met. i Metalloved.* **17** (1964) 881-885, English Transl. #6, pp. 80-86.
CABANÉ-BROUTY F. and BERNARDINI J., *J. Physique Colloq.* **43** (1982) C6-163-173.
CABANÉ J., Ecole d'été de Carry-le Rouet (1984) *op. cit.* (Les Editions de Physique) p. 161-197.
CAHN J. and BALLUFFI R. W., *Scr. Metall.* **13** (1979) 499-502.
COUSTY J., Thesis, University of Paris XI, Orsay (1981).
DELAMARE F. and RHEAD G. E., *Surface Science* **28** (1971) 267-283.
FISHER J. C., *J. Appl. Phys.* **22** (1951) 74.
GHALEB D., Thesis, University of Paris VI (1983).
GJOSTEIN N. A., *Metal Surfaces*, A.S.M. (1963) p. 99-154.
GLAESER A. M. and EVANS J. W., *Acta. Metall.* **34** (1986) 1545-1552.
GRUNZE M., KREUZER H. J. and WEIMER J. J., *Springer Ser. Surf. Sci.*, Springer : Berlin, **12** 1988.
GUILLOPÉ M., PONTIKIS V., and CICCOTTI G., *Dimeta-82, op. cit.*, p. 415-417.
GUPTA D. and ROSENBERG R. *Thin Solid Films* **25** (1975) 171-180.
GUPTA D. and OBERSCHMIDT J., *Diffusion in Solids, Recent Developments* (1984) *op. cit.* p. 121-140.
HARRISON L. G., *Trans. Faraday Soc.* **57** (1961) 1191-1199.
HART E. W., *Acta Metall.* **5** (1957) 597-598.
HERBEUVAL I., BISCONDI M., and GOUX C., *Mém. Sci. Rev. Metall.* **70** (1973) 39-46.
JARDIN J. P., DESJONQUERES M. C., and SPANJAARD D., *J. Physique. Colloq.* **45** (1984) C9-9-C9-12.
LANG B., JOYNER R. W., and SOMORJAI G. A., *Surf. Sci.* **30** (1972) 440-453.
LE CLAIRE A. D., *Brit. J. Appl. Phys.* **14** (1963) 351-366.
LE CLAIRE A. D. and RABINOVITCH A., *J. Phys. C, Solid State Physics* **14** (1981) 3863, **15** (1982) 3345, *erratum* 5727; **16** (1983) 2087-2104.
DE LORENZI G., JACUCCI G., and PONTIKIS V., *Surf. Sci.* **116** (1982) 391-413.
MARTIN G., BLACKBURN D. and ADDA Y., *Phys. Stat. Solidi* **23** (1967) 223-228.
MOYA E., Thesis, University of Marseille (1975), # AO 11535.
MOYA F., MOYA-GONTIER G. E. and CABANÉ-BROUTY F., *Phys. Stat. Solidi* **35** (1969) 893-901.
MULLINS W. W., *J. Appl. Phys.* **28** (1957) 333-339, and *Trans. AIME* **218** (1960) 354-361.
NICHOLS F. A. and MULLINS W. W., *J. Appl. Phys.* **36** (1965) 1826-1835.
PERRAILLON B., PONTIKIS V. and ADDA Y., *Dimeta-82, op. cit.*, 163-180.

RHEAD G. E., *Surf. Sci.* **15** (1969) 353-357.
ROBINSON J. T. and PETERSON N. L., *Surf. Sci.* **31** (1972) 586-616.
SUZUOKA T., *Trans. Japan. Inst. Metals* **2** (1961) 25.
WHIPPLE R. T. P., *Philos Mag.* **45** (1954) 1225.
WUTTIG M. and BIRNBAUM H. K., *Phys. Rev.* **147** (1966) 495.

CHAPTER VIII

Phenomenological theory of diffusion

In the first chapter, the fundamental law of diffusion, Fick's law, was introduced as a postulate on the basis of its formal analogy with other physical laws. The drift terms were added without justifying the nature of the driving force. Even though the atomic models based on the random walk of the diffusing species have confirmed these laws (Ch. II), it is interesting to examine their basis in a rigorous manner, in order to be able to derive the exact relations corresponding to experimental cases that differ in the number and nature of the components, the types of forces, etc... Since diffusion is by its very nature an irreversible process, it is the *Thermodynamics of Irreversible Processes* (T.I.P.) that allows a completely general approach to the problem.

I. Review of the Thermodynamics of Irreversible Processes (T.I.P.).

The physical quantity that characterizes irreversible processes is the *entropy production*, since in a system that is the site of an irreversible transformation, be the system open or closed, entropy is produced:

$$\mathrm{d}S = \frac{\mathrm{d}Q}{T} + \mathrm{d}S_{\mathrm{irr}},$$

($\mathrm{d}Q$ is the exchange of heat with the surroundings (bath)), or, if t is the time:

$$\frac{\mathrm{d}S}{\mathrm{d}t} = \frac{\mathrm{d}S_{\mathrm{e}}}{\mathrm{d}t} + \sigma; \qquad \text{(VIII.1)}$$

$\sigma = \mathrm{d}S_{\mathrm{irr}}/\mathrm{d}t$ is the rate of entropy creation and $\mathrm{d}S_{\mathrm{e}}/\mathrm{d}t$ is the rate of inflow of entropy. Under the assumption of local equilibrium, σ is calculated by application of Gibbs' relation to a small volume of material. Then the conservation law for the diverse components and for the energy is written $\partial c_i/\partial t = -\mathrm{div}\ J_i$. As the entropy is not conserved, such a relation does not apply to $\mathrm{d}S/\mathrm{d}t$: a non-conserved part, the internal creation of entropy has to be added:

$$\frac{\mathrm{d}S}{\mathrm{d}t} = -\mathrm{div}\ J_{\mathrm{s}} + \sigma. \qquad \text{(VIII.2)}$$

The fundamental result of the calculation is the following: the rate of entropy creation σ can be written in the form:

$$\boxed{\sigma = \mathbf{J}_q \cdot \mathbf{X}_q + \sum_k \mathbf{J}_k \cdot \mathbf{X}_k} \tag{VIII.3}$$

i.e, the sum of the products of the fluxes $\mathbf{J}$ and the corresponding forces $\mathbf{X}$, with:

$\mathbf{J}_q$: the flux of heat,
$\mathbf{J}_k$: the flux of component k,
$\mathbf{X}_q = \nabla(1/T)$ with $\nabla \equiv$ gradient operator (vector),
$\mathbf{X}_k = \mathbf{F}_k/T - \nabla(\mu_k/T)$,

where $\mathbf{F}_k$ is the external force applied to component k (e.g., an electric field). One can also write:

$$\mathbf{X}_k = -\nabla(\tilde{\mu}_k/T) = -\left[\nabla(\mu_k/T) - \mathbf{F}_k/T\right],$$

with the generalized chemical potential $\tilde{\mu}_k$. If $\mathbf{F}_k$ derives from an electrical potential Φ, $\mathbf{F}_k = -\nabla\Phi$, then $\tilde{\mu}_k = \mu_k + q_k\Phi$. This is the electrochemical potential (q_k is the electrical charge on component k).

If all the forces $\mathbf{X}_k$ are zero, $\sigma = 0$; there is then no irreversible process, and the fluxes are thus zero. The fluxes are expressed as a function of the forces so that this condition is fulfilled ([1]). With the further assumption of linearity, we can write:

$$\begin{aligned} \mathbf{J}_i &= \sum_{k=1}^{p} L_{ik}\,\mathbf{X}_k + L_{iq}\,\mathbf{X}_q \qquad i,\ k = 1,\ 2,\ \dots,\ p. \\ \mathbf{J}_q &= \sum_{k=1}^{p} L_{qk}\,\mathbf{X}_k + L_{qq}\,\mathbf{X}_q. \end{aligned} \tag{VIII.4}$$

The coefficients L_{ij}, called the *phenomenological coefficients*, are independent of the forces. The diagonal coefficients have a simple physical meaning (atomic mobility, thermal conductivity,...). The off-diagonal coefficients are related to coupling or interference of the various processes and describe such phenomena as thermomigration or Soret effect, Thomson effect, Dufour effect, etc. On the basis of the principle of microscopic reversibility, and under the condition that the fluxes and forces are calculated from the entropy production, one can prove the *Onsager relations*:

$$L_{ik} = L_{ki}. \tag{VIII.5}$$

([1]) More precisely, each flux is expressed as a function of the forces of the same rank tensor: rank 2 (vector) in the case of equations (VIII.3-VIII.4). If chemical reactions also take place in the system, scalar terms (rank 1) appear: reaction rate and the corresponding force of supersaturation.

In a system in *mechanical equilibrium*, the forces must fulfill the Gibbs-Duhem equation:

$$\sum_{k=1}^{p} \rho_k \left(\nabla \tilde{\mu}_k\right)_T = 0, \tag{VIII.6}$$

where ρ_k is the concentration of species k in mass/unit volume. Because of this relation between the forces, the entropy production can be written:

$$\sigma = \mathbf{J}_q \cdot \mathbf{X}_q + \sum_{k=1}^{p-1} \mathbf{J}_k \cdot \left(\mathbf{X}_k - \mathbf{X}_p\right), \tag{VIII.7}$$

from which each flux can be expressed as a function of $(p-1)$ independent forces for an isothermal system:

$$\mathbf{J}_i = \sum_{k=1}^{p-1} L_{ik} \left(\mathbf{X}_k - \mathbf{X}_p\right). \tag{VIII.8}$$

All the material fluxes are defined relative to the *velocity of the center of mass*:

$$\omega = \sum_k \frac{\rho_k \mathbf{v}_k}{\rho}, \tag{VIII.9}$$

which represents the average velocity ($\mathbf{v}_k$ is the velocity of the component k in the laboratory frame of reference $\mathbf{R}^0$ (see Ch. VI, App. I, Ch. VIII, App. III)). It follows that the fluxes are defined in the frame of reference $\mathbf{R}^\omega$ by:

$$\mathbf{J}_k = \rho_k \left(\mathbf{v}_k - \omega\right)$$

from which

$$\sum_k \mathbf{J}_k = 0. \tag{VIII.10}$$

In general, the measurements are referred to the frame of reference $\mathbf{R}^0$, in which the fluxes are defined by:

$$\mathbf{J}_k^0 = \rho_k \; \mathbf{v}_k = \mathbf{J}_k + \rho_k \; \omega. \tag{VIII.11}$$

II. The application of T.I.P. to diffusion in solids.

Some precautions are necessary before this formalism can be applied. The point defects have to be taken into account; this means a component with zero mass for the vacancies. On the other hand, vacancies and interstitials differ fundamentally from chemical species because they can be created or annihilated locally.

These remarks lead to choosing for the frame of reference in which the fluxes **J** are expressed one that is connected to the *average atomic velocity*, a quantity that is measurable experimentally (e.g., by the use of inert markers), rather than a reference frame connected to the mean velocity of the center of mass. For a vacancy mechanism, to which the following discussion is limited:

$$\mathbf{v} = \frac{n_v\ \mathbf{v}_v + \sum_{k=1}^{p} n_k\ \mathbf{v}_k}{n}, \tag{VIII.12}$$

where $\mathbf{v}_v$ and $\mathbf{v}_k$ are the average velocities of the vacancies and k atoms respectively, measured in the laboratory frame of reference $\mathbf{R}^0$.

The entropy balance, based on Gibbs' equation, leads to:

$$\boxed{\sigma = \mathbf{J}_q \cdot \nabla(1/T) - \sum_{k=1}^{p} \mathbf{J}_k \cdot \nabla\left(\frac{\tilde{\mu}_k - \tilde{\mu}_v}{T}\right) - \frac{\mathcal{R}_v\ \mu_v}{T}}. \tag{VIII.13}$$

From this, the phenomenological relations can be deduced for the vector and scalar quantities in a system with p material components; the vacancies are the $(p+1)$st component, with the fluxes expressed in number of particles per unit area per unit time:

$$\mathbf{J}_k = L_{kq}\ \nabla(1/T) - \sum_{i=1}^{p} L_{ki}\ \nabla\left(\frac{\tilde{\mu}_i - \tilde{\mu}_v}{T}\right) \tag{VIII.14.1}$$

$$\mathbf{J}_q = L_{qq}\ \nabla(1/T) - \sum_{i=1}^{p} L_{qi}\ \nabla\left(\frac{\tilde{\mu}_i - \tilde{\mu}_v}{T}\right) \tag{VIII.14.2}$$

$$L_{ik} = L_{ki},\ L_{qk} = L_{kq} \tag{VIII.14.3}$$

$$\mathcal{R}_v = -\alpha\mu_v / T \tag{VIII.14.4}$$

$$\mathbf{J}_v = -\sum_{k=1}^{p} \mathbf{J}_k. \tag{VIII.14.5}$$

The last equation follows from the definition of the mean atomic velocity (VIII.12). In effect, in the mobile frame of reference **R**:

$$\begin{aligned}\mathbf{J}_k &= n_k(\mathbf{v}_k - \mathbf{v}) \quad k = 1, 2, \ldots, p \\ \mathbf{J}_v &= n_v(\mathbf{v}_v - \mathbf{v}),\end{aligned}$$

from which

$$\mathbf{J}_v + \sum_{k=1}^{p} \mathbf{J}_k = 0.$$

The creation or annihilation of the vacancies appears in the form of a term (Eq. (VIII.14.4)) analogous to a chemical reaction rate. The conservation equation for a conserved material species in the frame of reference $\mathbf{R}^0$ is written (see Ch. I, Eq. (I.3)):

$$\frac{\partial n_k}{\partial t} = -\mathrm{div}\ (n_k\ \mathbf{v}_k) = -\mathrm{div}\ \mathbf{J}_k^0,$$

whence

$$\sum_k \frac{\partial n_k}{\partial t} = -\mathrm{div} \sum_k n_k\ \mathbf{v}_k.$$

Since n (the total number of sites per unit volume) $= n_v + \sum_{k=1}^{p} n_k =$ const ([2]), the equation above becomes, with the definition of $\mathbf{v}$ (Eq. (VIII.12)):

$$\begin{aligned}\frac{\partial n_v}{\partial t} &= -\mathrm{div}\ (n_v\ \mathbf{v}_v) + \mathrm{div}\,(n\mathbf{v}) \\ &= -\mathrm{div}\ \mathbf{J}_v^0 + \mathcal{R}_v,\end{aligned} \qquad \text{(VIII.15)}$$

an equation valid for any non-conserved species, in which the *source term for vacancies* (number of vacancies created or destroyed per unit volume in unit time):

$$\boxed{\mathcal{R}_v \equiv \mathrm{div}\ (n\mathbf{v})} \qquad \text{(VIII.16)}$$

is stated explicitly. The coefficient α in equation (VIII.14.4) depends on the density and efficiency of the vacancy sources and sinks. If the concentration of

([2]) This hypothesis is not always correct, especially in the case of interdiffusion when the molar volume varies with the composition, and therefore with the distance (see Ch. VI, App. I, and this chapter, App. III).

vacancies, n_v does not differ greatly from its equilibrium value $\bar{n}_v$ (see Appendix I):

$$\mu_v = kT \text{ Log } \frac{n_v}{\bar{n}_v} \simeq kT \, \frac{n_v - \bar{n}_v}{\bar{n}_v},$$

and

$$\mathcal{R}_v = \alpha k \, \frac{\bar{n}_v - n_v}{\bar{n}_v}. \qquad \text{(VIII.17)}$$

This equation has the form of a chemical rate equation for first-order kinetics (see Ch. I, Eq. (I.25)); it expresses the rate of annihilation (or creation) as a function of the vacancy super- (or under-) saturation, with:

$$\alpha = \bar{n}_v / k\tau_v; \qquad \text{(VIII.17a)}$$

here τ_v is the mean life time of the vacancies.

Remarks:

1) In a system subjected to an electrical potential, the electronic components, free electrons or electron holes, which form the $(p+2)$ nd component, will eventually have to be considered:

$$\begin{cases} \mathbf{J}_k = L_{kq} \, \nabla(1/T) - \sum\limits_{i=1}^{p} L_{ki} \, \nabla\left(\dfrac{\tilde{\mu}_i - \tilde{\mu}_v}{T}\right) - L_{ke} \, \nabla\left(\dfrac{\tilde{\mu}_e}{T}\right) \\ \mathbf{J}_e = L_{eq} \, \nabla(1/T) - \sum\limits_{i=1}^{p} L_{ei} \, \nabla\left(\dfrac{\tilde{\mu}_i - \tilde{\mu}_v}{T}\right) - L_{ee} \, \nabla\left(\dfrac{\tilde{\mu}_e}{T}\right) \\ \mathbf{J}_q = L_{qq} \, \nabla(1/T) - \sum\limits_{i=1}^{p} L_{qi} \, \nabla\left(\dfrac{\tilde{\mu}_i - \tilde{\mu}_v}{T}\right) - L_{qe} \, \nabla\left(\dfrac{\tilde{\mu}_e}{T}\right) \\ \mathbf{J}_v = -\sum\limits_{k=1}^{p} \mathbf{J}_k . \end{cases} \qquad \text{(VIII.18)}$$

2) The Onsager relations between the phenomenological coefficients do not hold if the velocity of the reference frame is chosen incorrectly.

3) The above equations are based on the assumption that all species, including the vacancies, have the same atomic volume. It is possible to lift this restriction by introducing the partial molar volumes (see Ch. VI, App. I).

III. Applications of the phenomenological equations.

The phenomenological equations are very powerful, but they lead quickly to very cumbersome expressions, with numerous phenomenological coefficients for which the physical meaning is not always clear at first glance. Only a few simple cases are treated in the following pages.

Detailed expressions for the coefficients L_{ij} in terms of the elementary characteristics of the system (concentrations of defects, jump frequencies, interactions with solute atoms,...) are furnished by statistical theories and models (random walk, Monte Carlo method, the kinetic method, see Ch. III, § V).

III.1 DIFFUSION OF A RADIOACTIVE TRACER IN A PURE MATERIAL. — Three species must be taken into account: the A atoms of the pure material, the A* atoms of the tracer, and the vacancies. Let n_A, n_{A^*}, and n_v be their numbers per unit volume, and μ_A, μ_{A^*}, and μ_v be their chemical potentials. The fluxes are then written:

$$\left\{\begin{aligned} \mathbf{J}_A &= -\frac{L_{AA}}{T}\,\nabla(\mu_A-\mu_v) - \frac{L_{AA^*}}{T}\,\nabla(\mu_{A^*}-\mu_v) \\ \mathbf{J}_{A^*} &= -\frac{L_{A^*A}}{T}\,\nabla(\mu_A-\mu_v) - \frac{L_{A^*A^*}}{T}\,\nabla(\mu_{A^*}-\mu_v) \\ \mathbf{J}_A &+ \mathbf{J}_{A^*} + \mathbf{J}_v = 0. \end{aligned}\right. \tag{VIII.19}$$

If the concentration of vacancies is everywhere at thermal equilibrium, $n_v = \bar{n}_v$, then $\nabla\mu_v = 0$ and according to equation (VIII.17) $\mathcal{R}_v = 0$. Then, by equation (VIII.16), $v = \text{const.} = 0$ ([3]) and by equation (VIII.15) $J_v^0 = 0$, and $J_v = 0$. Thus $\mathbf{R} = \mathbf{R}^0$ and the flux equations apply directly to the laboratory frame of reference:

$$\left\{\begin{aligned} \mathbf{J}_A^0 &= -\frac{L_{AA}}{T}\,\nabla\mu_A - \frac{L_{AA^*}}{T}\,\nabla\mu_{A^*} \\ \mathbf{J}_{A^*}^0 &= -\frac{L_{A^*A}}{T}\,\nabla\mu_A - \frac{L_{A^*A^*}}{T}\,\nabla\mu_{A^*}. \end{aligned}\right. \tag{VIII.20}$$

([3]) The constant is zero because $v = 0$ outside the diffusion zone.

Set $n_{\mathrm{A}} + n_{\mathrm{A}^*} = n' = \text{const}$. For the ideal AA* solution:

$$\mu_{\mathrm{A}^*} = \mu^0_{\mathrm{A}^*} + kT\,\mathrm{Log}\,(n_{\mathrm{A}^*}/n')$$

$$\mu_{\mathrm{A}} = \mu^0_{\mathrm{A}} + kT\,\mathrm{Log}\left(\frac{n' - n_{\mathrm{A}^*}}{n'}\right),$$

whence for unidirectional diffusion:

$$\left\{\begin{aligned} \mathbf{J}^0_{\mathrm{A}} &= k\left(\frac{L_{\mathrm{AA}}}{n'} - \frac{L_{\mathrm{AA}^*}}{n_{\mathrm{A}^*}}\right)\frac{\partial n_{\mathrm{A}^*}}{\partial x} \\ \mathbf{J}^0_{\mathrm{A}^*} &= k\left(\frac{L_{\mathrm{A}^*\mathrm{A}}}{n'} - \frac{L_{\mathrm{A}^*\mathrm{A}^*}}{n_{\mathrm{A}^*}}\right)\frac{\partial n_{\mathrm{A}^*}}{\partial x}. \end{aligned}\right. \tag{VIII.21}$$

Since $\mathbf{J}^0_{\mathrm{A}} + \mathbf{J}^0_{\mathrm{A}^*} = 0$,

$$\frac{L_{\mathrm{A}^*\mathrm{A}^*}}{n_{\mathrm{A}^*}} - \frac{L_{\mathrm{A}^*\mathrm{A}}}{n'} = \frac{L_{\mathrm{AA}}}{n'} - \frac{L_{\mathrm{AA}^*}}{n_{\mathrm{A}^*}}. \tag{VIII.22}$$

The diffusion coefficient of the tracer is defined by Fick's law:

$$\mathbf{J}^0_{\mathrm{A}^*} = -\,D^{\mathrm{A}}_{\mathrm{A}^*}\,(\partial n_{\mathrm{A}^*}/\partial x);$$

thus

$$\begin{aligned} D^{\mathrm{A}}_{\mathrm{A}^*} &= k\left[\frac{L_{\mathrm{A}^*\mathrm{A}^*}}{n_{\mathrm{A}^*}} - \frac{L_{\mathrm{A}^*\mathrm{A}}}{n'}\right] \\ &\approx k\,\frac{L_{\mathrm{A}^*\mathrm{A}^*}}{n_{\mathrm{A}^*}} \quad \text{since } n_{\mathrm{A}^*} << n'. \end{aligned} \tag{VIII.23}$$

It then follows from equation (VIII.22):

$$D^{\mathrm{A}}_{\mathrm{A}^*} = k\left[\frac{L_{\mathrm{AA}}}{n'} - \frac{L_{\mathrm{AA}^*}}{n_{\mathrm{A}^*}}\right]. \tag{VIII.24}$$

The first term on the RHS defines the true self-diffusion coefficient, since this is the only term in the absence of the tracer. As is well known, it is not experimentally measurable:

$$D^{\mathrm{A}}_{\mathrm{A}} = kL_{\mathrm{AA}}/n',$$

from which

$$D^{\mathrm{A}}_{\mathrm{A}^*} = f_0\,D^{\mathrm{A}}_{\mathrm{A}}, \tag{VIII.25}$$

with

$$f_0 = 1 - \frac{L_{A^\bullet A}}{L_{AA}} \frac{n'}{n_{A^\bullet}}, \qquad \text{(VIII.26)}$$

Regrouping terms and using equations (VIII.22) and (VIII.23), we easily obtain that:

$$\boxed{\frac{1-f_0}{f_0} = \frac{L_{A^\bullet A}}{L_{A^\bullet A^\bullet}}}. \qquad \text{(VIII.27)}$$

The coupling between fluxes is thus at the origin of the correlation factor: in addition to the defects, two material species must be present in order to have coupling, and thus correlation. This is again the rule stated in chapter III that three types of particles are required in order to have correlation between successive jumps.

III.2 INTERDIFFUSION OF A AND B. — The system consists of three species: A atoms, B atoms, and point defects, for example, vacancies. The flux equations are then written:

$$\begin{cases} \mathbf{J}_A = -\dfrac{L_{AA}}{T} \nabla(\mu_A - \mu_v) - \dfrac{L_{AB}}{T} \nabla(\mu_B - \mu_v) \\[2ex] \mathbf{J}_B = -\dfrac{L_{BA}}{T} \nabla(\mu_A - \mu_v) - \dfrac{L_{BB}}{T} \nabla(\mu_B - \mu_v) \\[2ex] \mathbf{J}_v = -\mathbf{J}_A - \mathbf{J}_B \\[2ex] \mathcal{R}_v = -\alpha\mu_v/T. \end{cases} \qquad \text{(VIII.28)}$$

If the vacancy sources and sinks are sufficiently numerous and efficient to maintain the vacancy concentration very near equilibrium, $n_v \approx \bar{n}_v$ at every point, although it may vary with composition, and $\mu_v \approx 0$, negligible with respect to μ_A and μ_B. Further, the chemical potentials μ_i can be expressed as a function of the mole fraction N_i and the thermodynamic activity coefficient γ_i:

$$\mu_i = \mu_i^0 + kT \text{ Log } N_i \, \gamma_i.$$

The expressions for the fluxes then become:

$$\begin{cases} \mathbf{J}_A = -k \left[\dfrac{L_{AA}}{n_A} - \dfrac{L_{AB}}{n_B}\right] \left[1 + \dfrac{\partial \text{ Log } \gamma_A}{\partial \text{ Log } N_A}\right] \nabla n_A \\[3ex] \mathbf{J}_B = -k \left[\dfrac{L_{BB}}{n_B} - \dfrac{L_{BA}}{n_A}\right] \left[1 + \dfrac{\partial \text{ Log } \gamma_B}{\partial \text{ Log } N_B}\right] \nabla n_B. \end{cases} \qquad \text{(VIII.29)}$$

These equations have the form of Fick's law, with the *intrinsic diffusion coefficients* given by:

$$\begin{cases} \bar{D}_{\rm A} = k\left[\dfrac{L_{\rm AA}}{n_{\rm A}} - \dfrac{L_{\rm AB}}{n_{\rm B}}\right]\Phi \\ \\ \bar{D}_{\rm B} = k\left[\dfrac{L_{\rm BB}}{n_{\rm B}} - \dfrac{L_{\rm BA}}{n_{\rm A}}\right]\Phi \end{cases} \qquad \text{(VIII.30)}$$

where Φ is the thermodynamic factor $[1 + \partial \text{ Log } \gamma/\partial \text{ Log } N]$, identical for the two species A and B according to the Gibbs-Duhem relation.

The terms in brackets are not equal to the self-diffusion coefficients. One has in effect:

$$\bar{D}_{\rm A} = k\frac{L_{\rm AA}}{n_{\rm A}}\left[1 - \frac{L_{\rm AB}}{L_{\rm AA}}\frac{n_{\rm A}}{n_{\rm B}}\right]\Phi, \qquad \text{(VIII.31)}$$

an expression which can be written in the form $D_{\rm A}\cdot R_{\rm A}\Phi$ (see below Eq. (VIII.34)), in which an extra factor, $R_{\rm A}$, which is due to the coupling of the fluxes of A and of B, appears. But this simple approach does not give an useful expression for $R_{\rm A}$: an atomistic model is necessary, as shown by equations (VI.6) for Manning's random alloy model. Just as the cross-terms for the self-diffusion of a tracer correspond to the effect of geometric correlation, here they represent the effects of "physical" correlation, known, following Manning, as the "vacancy wind". The physical significance of this will be treated below (§ III.3).
Alternatively, from equation (VIII.28), the following relation is easily derived:

$$J_{\rm A} = -\left[L_{\rm AA} - L_{\rm AB}\frac{L_{\rm AA} + L_{\rm AB}}{L_{\rm BB} + L_{\rm AB}}\right]\frac{\nabla \mu_{\rm A}}{T} - \frac{L_{\rm AB}}{L_{\rm BB} + L_{\rm AB}} J_v .$$

The flux of A-atoms is made up of a direct term coming from the driving force $\nabla \mu_{\rm A}/T$, and an indirect vacancy wind term induced by the flux of vacancies J_v (Lidiard, 1986).

If the vacancy concentration does not differ much from that at equilibrium, as has been assumed in neglecting μ_v with respect to $\mu_{\rm A}$ and $\mu_{\rm B}$, $\partial n_v/\partial t$ is everywhere negligible ([4]). The result of equations (VIII.15) and (VIII.16) is then:

$$\text{div } \mathbf{J}_v^0 = \mathcal{R}_v = \text{div } n\mathbf{v}$$

([4]) At a given position, the composition varies with time in the course of the interdiffusion. Then n_v also varies even if the vacancies remain at their equilibrium concentration. This is, however, a second-order effect.

and by integration, $\mathbf{J}_v^0 = n\ \mathbf{v}$. Since $\mathbf{J}_v = \mathbf{J}_v^0 - n_v\mathbf{v}$, $n_v << n$ implies $\mathbf{J}_v \approx \mathbf{J}_v^0 = n\ \mathbf{v}$ and following equation (VIII.14.5):

$$\mathbf{J}_\mathrm{A} + \mathbf{J}_\mathrm{B} = -\mathbf{J}_v = -n\mathbf{v},$$

whence:

$$\boxed{\mathbf{v} = \left(\bar{D}_\mathrm{A} - \bar{D}_\mathrm{B}\right)\ \nabla N_\mathrm{A}}. \tag{VIII.32}$$

Passing from the mobile frame of reference which is fixed in the lattice to the laboratory frame of reference $\mathbf{R}^0$ which is fixed in one of the ends of the diffusion couple, one finds:

$$\mathbf{J}_\mathrm{A}^0 = \mathbf{J}_\mathrm{A} + n_\mathrm{A}\mathbf{v}$$

$$= -\tilde{D}\ \nabla n_\mathrm{A},$$

with:

$$\boxed{\tilde{D} = N_\mathrm{A}\ \bar{D}_\mathrm{B} + N_\mathrm{B}\ \bar{D}_\mathrm{A}}. \tag{VIII.33}$$

The equations of Dehlinger, Darken, and Manning, introduced and discussed in chapter VI, are thus found again. In Manning's random-alloy model with two jump frequencies, it can be shown that (a simple derivation is given by Bocquet, 1987):

$$\left\{\begin{aligned}
&L_\mathrm{AA} = \frac{n_\mathrm{A}\ D_{\mathrm{A}^\bullet}}{k}\ [1 + 2\ N_\mathrm{A}\ D_{\mathrm{A}^\bullet}\ \mathcal{V}]\\
&L_\mathrm{BB} = \frac{n_\mathrm{B}\ D_{\mathrm{B}^\bullet}}{k}\ [1 + 2\ N_\mathrm{B}\ D_{\mathrm{B}^\bullet}\ \mathcal{V}]\\
&L_\mathrm{AB} = \mathrm{L}_\mathrm{BA} = \frac{2\ n_\mathrm{A}\ n_\mathrm{B}\ D_{\mathrm{A}^\bullet}\ D_{\mathrm{B}^\bullet}\ \mathcal{V}}{(n_\mathrm{A} + n_\mathrm{B})\, k}\\
&\mathcal{V}^{-1} = M_0\,(N_\mathrm{A}\ D_{\mathrm{A}^\bullet} + N_\mathrm{B}\ D_{\mathrm{B}^\bullet})\ \text{with}\ M_0 = \frac{2\ f_0}{1 - f_0},
\end{aligned}\right. \tag{VIII.34}$$

so the relations (VI.6) given in chapter VI are again found. Following these relations, the vacancy-wind term $\mathcal{V}$ makes up the total cross terms and further contributes to the diagonal terms L_AA and L_BB. This is neglected in the usual form of Darken's equations, and comes from the fact that a force X_A contributes to the flux both of A (the direct mass flow term) and of B (cross term).

The factors R_i of equation (VIII.31) are easily worked out:

$$\begin{cases} R_A = 1 + 2\,\nu N_A\,(D_{A^*} - D_{B^*}) \\ R_B = 1 + 2\,\nu N_B\,(D_{B^*} - D_{A^*})\,, \end{cases}$$

from which the complete expression for the interdiffusion coefficient is obtained as:

$$\tilde{D} = (N_A\ D_{B^*} + N_B\ D_{A^*})\ \Phi \left[1 + 2\,\nu\,\frac{N_A\ N_B\,(D_{A^*} - D_{B^*})^2}{N_A\ D_{B^*} + N_B\ D_{A^*}}\right].$$

The second term in the square brackets expresses the existence of the "vacancy wind" which results from the coupling of the fluxes of A and B, *via* the flux of vacancies.

N. B.: equation (VIII.32) and the following expressions are based on the assumption of a constant molar volume.

Particular cases :

a) *Self-diffusion.* — B is an isotope of A. The expressions for the L_{ij} simplify, and the correlation factor $f_0 = M_0/(M_0 + 2)$ appears. With $n_{A^*} << n_A$ and $A^* \equiv B$, equations (VIII.34) become:

$$\begin{cases} L_{AA} \approx \dfrac{n_A}{k}\,\dfrac{D_{A^*}}{f_0} = \dfrac{n_A}{k}\,D_A^A \\ L_{A^*A^*} \approx \dfrac{n_{A^*}}{k}\,D_{A^*} \\ L_{AA^*} = L_{A^*A} \approx \dfrac{n_{A^*}}{k}\,D_{A^*}\,\dfrac{1 - f_0}{f_0}\,, \end{cases}$$

which are clearly compatible with the definitions of the preceding paragraph (Eqs. (VIII.23) and (VIII.27)), with D_{A^*} written for ease instead of $D_{A^*}^A$.

b) *Dilute AB alloys.* — For dilute alloys, $n_B \to 0$ and $\Phi \to 1$. It can be shown that L_{BB}/n_B approaches a finite value, whereas L_{AB}/n_A, which is of order n_B, goes to zero, a result easily derived from equations (VIII.34) but of more general significance. Then, from the second equation in (VIII.30):

$$\lim_{n_B \to 0} \bar{D}_B = k\frac{L_{BB}}{n_B} = D_B,$$

which is the solute diffusion coefficient at infinite dilution. From the first equation in (VIII.30), we have:

$$\lim_{n_B \to 0} \bar{D}_A = \lim k\left[\frac{L_{AA}}{n_A} - \frac{L_{AB}}{n_B}\right] \neq k\frac{L_{AA}}{n_A}.$$

The cross coefficient L_{AB}/n_B cannot be neglected. The limit of $\bar{D}_A$ is not equal to the self-diffusion coefficient, not even to within the correlation factor.

The equations (VIII.34) cannot be used to describe the phenomenological coefficients more explicitly because of possible solute atom B-vacancy interactions. One can start from the five-frequency model for the FCC structure, calculate the flux by the kinetic method (Ch. III, § V, 2) and determine the L_{ij} by identification. According to Howard and Manning:

$$\left\{\begin{aligned} kL_{AA} &= -\frac{a^2}{12}\, n_p \left[\frac{40\, w_1\, w_3 + 40\, w_3^2 + 14\, w_2\, w_3 + 4\, w_1\, w_2}{w_1 + w_2 + 7/2\, w_3} + \right. \\ &\qquad \left. + \frac{7\, w_3\, w_0}{w_4}\right] + a^2\,(n_v - n_p)\; w_0 \\ kL_{BB} &= \frac{a^2}{12}\, n_p \cdot \frac{w_2\,(w_1 + 7/2\, w_3)}{w_1 + w_2 + 7/2\, w_3} \\ kL_{AB} &= kL_{BA} = \frac{a^2}{12}\, n_p \cdot \frac{w_2\,(-2\, w_1 + 7\, w_3)}{w_1 + w_2 + 7/2\, w_3}, \end{aligned}\right. \qquad \text{(VIII.35)}$$

where n_v is the total number of vacancies per unit volume and n_p is the number of vacancies associated with solute B atoms:

$$n_p = p\; n_B, \text{ with } p = 12\; N_v\; (w_4/w_3)\,,$$

and N_v is the mole fraction of vacancies in pure solvent A. With n_B $(<< n_A) \to 0$,

$$\bar{D}_B \to k\, \frac{L_{BB}}{n_B} = \frac{a^2}{12}\, p\; w_2\; f_2,$$

we again find the solute diffusion coefficient at infinite dilution as the limit of the intrinsic diffusion coefficient. On the other hand, $\bar{D}_A$ does not approach the self-diffusion coefficient of the solvent under these conditions (see above).

In effect, for an infinitely dilute AB alloy ($N_B \to 0,\; N_A \approx 1$), equation (VIII.30) become, for $\Phi = 1$:

$$\bar{D}_A\,(0) \approx \frac{D_{A^\bullet}\,(0)}{f_0} - k\, \frac{L_{AB}\,(0)}{n_B},$$

since

$$D_A^A \approx kL_{AA}/n_A,$$

$$\bar{D}_B\,(0) \approx k\, \frac{L_{BB}\,(0)}{n_B} = D_{B^\bullet}\,(0)\,.$$

Heumann's relation:

$$\frac{L_{AB}(0)}{L_{BB}(0)} = \frac{D_{A^\bullet}(0)}{D_{B^\bullet}(0)} \left[\frac{1}{f_0} - \frac{\bar{D}_A(0)}{D_{A^\bullet}(0)}\right]$$

is deduced from the above equations and is an expression which allows the vacancy-wind factor for infinite dilution, L_{AB}/L_{BB}, to be calculated from experimental results. It can be shown that if the molar volume is not constant, the above expression should be corrected by multiplying $\bar{D}_A(0)/D_{A^\bullet}(0)$ by the factor $\bar{V}_B/\bar{V}_A$ ($\bar{V}_i$ is the partial molar volume).

III.3 FLUX OF MATERIAL ARISING FROM A FLUX OF POINT DEFECTS: SEGREGATION INDUCED BY QUENCHING OR IRRADIATION. — A net flux of vacancies, often called metaphorically "the vacancy wind", can have several different origins:

– *a gradient in the vacancy concentration*: in quenching or irradiation experiments, where a supersaturation of vacancies is produced, with gradients established around the sinks (surfaces, grain boundaries, dislocations); also sintering, Herring-Nabarro creep,...

– *a driving force*, chemical (interdiffusion, gradient of chemical potential), electrical (electromigration), or thermal (thermomigration) in origin.

A flux of self-interstitials is produced during irradiation because of the migration of these defects to sinks.

The conservation laws require that the flux of vacancies be balanced by an equal flux of atoms (conservation of the number of atomic sites outside the sources and sinks). However, in a binary solution, the corresponding atomic fluxes J_A and J_B are not in the same ratio as the concentrations n_A and n_B because the vacancies do not exchange with A and B atoms at the same frequency. Assume that there is a positive interaction between vacancies and solute B atoms (attraction). Depending on the relative values of the exchange frequencies w_A and w_B of the vacancy with A or B atoms, two cases can be envisaged: 1) the fluxes of vacancies and of B atoms are in the same direction ($w_A > w_B$), which causes *an enrichment of solute* near the vacancy sinks (Fig. VIII.1a); 2) the fluxes of vacancies and of solute atoms are in opposite directions ($w_A < w_B$) which causes *a depletion* of solute near the sinks (Fig. VIII.1b). In the first case, the solute is dragged along due to the migration of vacancy-solute atom B complexes. Analogous processes take place with interstitials, thanks to mixed dumbbells. At higher temperatures, another phenomenon is possible: *the inverse Kirkendall effect*. Here the flux of vacancies towards the sinks gives rise to opposing fluxes of A and B atoms, and there is a depletion or enrichment of B near the sinks depending whether $D_B > D_A$ or $D_B < D_A$. There is an analogous process with interstitials (J_A and J_B are in the same direction as J_i).

In a homogeneous alloy (substitutional solid solution), the migration of vacancies at supersaturation toward sinks gives rise to composition gradients.

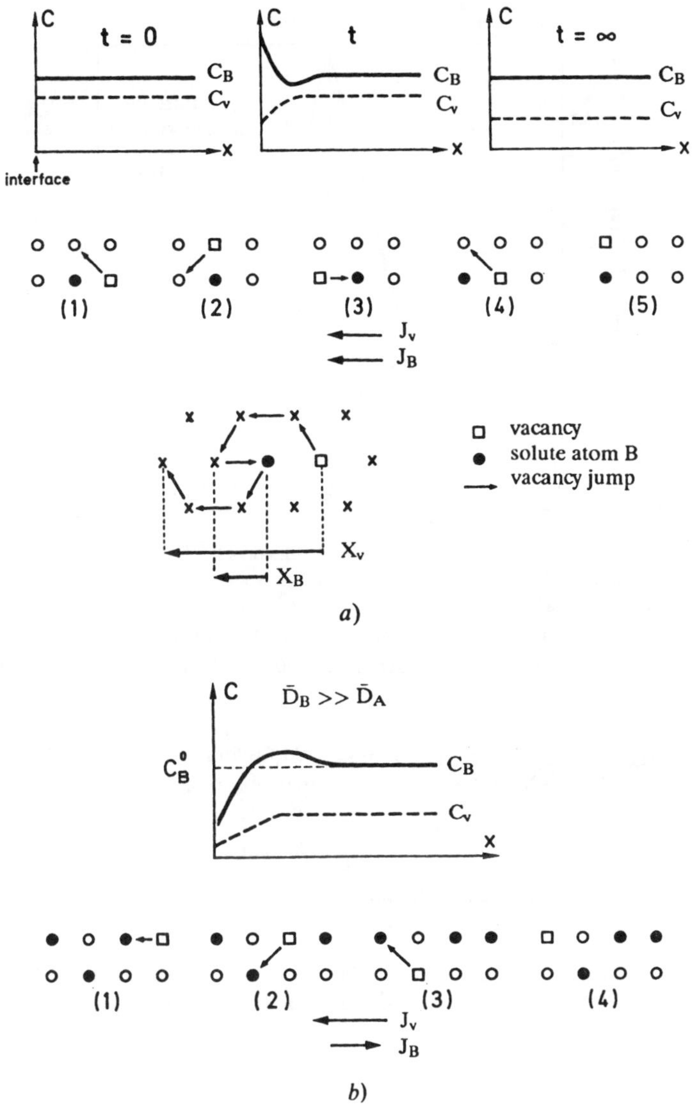

Fig. VIII.1. — Segregation of a solution $\underline{A}B$ by a vacancy flux: a) strong vacancy-solute B binding, direct segregation: b) weak vacancy-B atom binding, inverse segregation.

After air cooling, alloys of either metals or oxides show a hardening (and embrittlement) of the grain boundaries measurable by microhardness over a width of several tens of micrometers (Fig. VIII.2). In quenched aluminum alloys, the vacancies precipitate in the form of pores at the metal-oxide interface (Fig. VIII.3); the enrichment or depletion of the alloying elements was measured

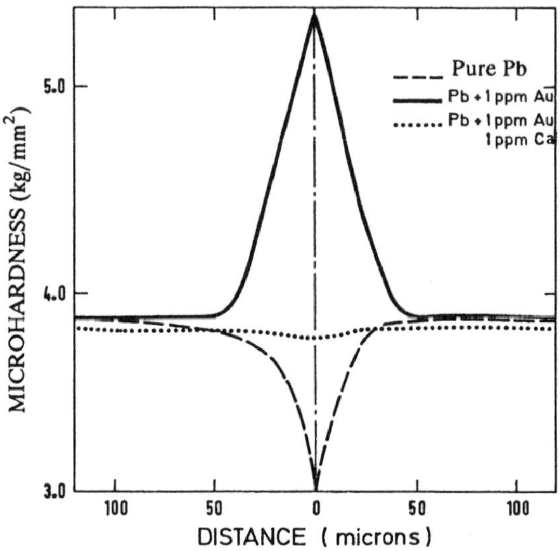

Fig. VIII.2. — Microhardness profile across a grain boundary in pure lead containing 1 ppm of impurity after quenching in air from 300°C. After Aust *et al.* (1968).

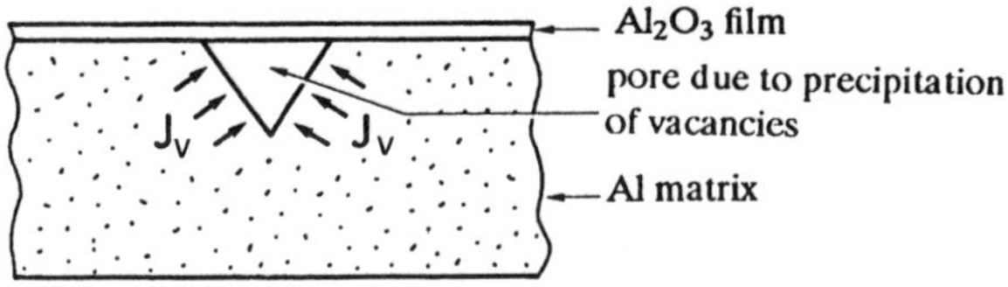

Fig. VIII.3. — Anthony's experiment: formation of pores at the Al/oxide layer interface by precipitation of vacancies during cooling following a high temperature anneal.

near the pores with an electron microprobe (Anthony, AlZn and AlGe alloys). In irradiated alloys, segregation to the neighborhood of the sinks is sometimes observed (e.g., enrichment in silicon at dislocations in irradiated NiSi alloys, enlargement or shrinking of the stacking fault ribbon of dissociated dislocations by segregation of solute during annealing after a quench ([5])). Under irradiation, defects are continuously being generated, and steady-state profiles are observed. Large effects are expected for solutes of small atomic size which can form "dumbbell" complexes with interstitials and for the inverse Kirkendall

([5]) Not to be confused with the Suzuki effect (equilibrium segregation).

effect which leads to segregation to sinks (e.g., Si in nickel alloys). Solutes of large atomic size are more likely to migrate away from the sinks under the effect of a vacancy flux (e.g., Al and Ti in nickel alloys).

These effects are called ***non-equilibrium or kinetic segregation***, because they disappear after a sufficiently long holding time following the quench (Fig. VIII.1a), as opposed to the equilibrium segregation which is predicted by the Gibbs adsorption isotherms for surfaces, boundaries, or dislocations, and is limited to much shorter distances (tens or hundreds of Ångströms). The effects can be analyzed by two approaches: phenomenological or kinetic (balance equations).

The phenomenological equations can be used to treat the problem quantitatively in the case of a dilute alloy, for which the L_{ij} coefficients can be expressed as a function of the jump frequencies. The calculation is, however, somewhat complicated because isolated vacancies, isolated solute atoms, and vacancy- solute pairs all have to be taken into account, and the corresponding gradients of chemical potential have to be written down in order for the forces on the A and B atoms to be determined.

The flux of solute B for an initially homogenous ($\nabla N_{\rm B} = 0$) dilute alloy with FCC structure can be written, after simplification, in the framework of the five-frequency model, using equations (VIII.35) for the L_{ij} (Anthony):

$$J_{\rm B} = \left[1 + \frac{L_{\rm BA}}{L_{\rm BB}}\right] \frac{N_p}{N_v} D_p \nabla n_v \qquad \text{(VIII.36)}$$

where N_v and N_p are the total mole fraction of vacancies and the mole fraction of solute-vacancy pairs, respectively, D_p is the diffusion coefficient of the pairs, and $N_{\rm B} >> N_v, N_p$.

The flux of solute is proportional to the diffusion coefficient of the pairs and to the fraction of vacancies paired to solute atoms; solute atoms, in effect, cannot move unless a nearest neighbor site is vacant. The factor $(1 + L_{\rm AB}/L_{\rm BB})$ can be either positive or negative. Since $L_{\rm BA}/L_{\rm BB} = (-2w_1 + 7w_3)/(w_1 + 7/2w_3)$, this factor can be negative for tight binding, where $w_3 << w_1$, so that the solute atoms and the vacancies migrate in the same direction as pairs. For weak binding or high temperatures, the opposite effect is expected, i.e. the vacancy flux produces opposed vacancy and solute fluxes. The ratio of the fluxes is a useful expression:

$$\frac{J_{\rm B}}{J_v} = \left(1 + \frac{L_{\rm AB}}{L_{\rm BB}}\right) \frac{N_{\rm B}\, D^{\rm A}_{\rm B^\bullet}}{(D^{\rm A}_{\rm A^\bullet}/f_0) + N_{\rm B}\, D^{\rm A}_{\rm B^\bullet}}. \qquad \text{(VIII.37)}$$

This ratio can be determined experimentally in a homogenous alloy, and $L_{\rm AB}/L_{\rm BB}$ can be determined from it if the solvent self-diffusion coefficient $D^{\rm A}_{\rm A^\bullet}$ and the solute diffusion coefficient $D^{\rm A}_{\rm B^\bullet}$ are known.

Neither J_{B} nor J_v are directly accessible experimentally, but the resulting segregation is measured by means of the concentration profiles near the sinks, or by the quantities $\int JS\,\mathrm{d}t$ of vacancies and atoms accumulating during the course of the experiment (here S is the sink area) (Fig. VIII.3). The first type of measurement is more usual, but the second is more interesting, because it alone allows the direct application of equation (VIII.37). Because of the displacement of the lattice planes, the ratio J_{B}^0/J_v^0 is the relevant quantity. Since $v = J_v/n$, we have:

$$\frac{1}{N_{\mathrm{B}}}\,\frac{J_{\mathrm{B}}^0}{J_v^0} = \frac{1}{N_{\mathrm{B}}}\,\frac{J_{\mathrm{B}}}{J_v} + 1.$$

It is this last expression that indicates whether enrichment or depletion takes place.

The fact that the "vacancy wind" is expressed by the presence of the cross term L_{AB} is easily understood. In effect, the choice of the phenomenological coefficients is arbitrary. The coefficients L_{AA}, $L_{\mathrm{AB}} = L_{\mathrm{BA}}$, and L_{BB} were chosen by writing the forces as $(\mu_{\mathrm{A}} - \mu_v)$ and $(\mu_{\mathrm{B}} - \mu_v)$. However, one could just as well write:

$$J_v = -\frac{L_{v\mathrm{A}}}{T}\,\nabla\mu_{\mathrm{A}} - \frac{L_{v\mathrm{B}}}{T}\,\nabla\mu_{\mathrm{B}} - \frac{L_{vv}}{T}\,\nabla\mu_v.$$

Identifying terms, we obtain:

$$L_{\mathrm{A}v} = L_{v\mathrm{A}} = -(L_{\mathrm{AA}} + L_{\mathrm{AB}})$$

$$L_{\mathrm{B}v} = L_{v\mathrm{B}} = -(L_{\mathrm{BB}} + L_{\mathrm{BA}})$$

$$L_{vv} = L_{\mathrm{AA}} + L_{\mathrm{BB}} + 2\,L_{\mathrm{AB}}.$$

These coefficients describe the coupling between atomic and vacancy fluxes.

In non-stoichiometric oxides, a vacancy flux can be generated by imposing an oxygen potential gradient on the sample. Such a vacancy flux will give rise to a flux of solute atoms, which will drift toward the low or high oxygen potential side depending on the value of the solute-vacancy binding energy. Such an impurity drift has been observed for Fe solute in $Co_{1-\delta}O$ (Martin and Dorris, 1987).

III.4 Electromigration in a substitutional binary alloy. — Let $\mathbf{E} = -\nabla\Phi$ be the applied electric field. At a uniform temperature, the thermodynamic force is:

$$\mathbf{X}_i = -\left[\frac{1}{T}\,\nabla\mu_i - \frac{q_i}{T}\,\mathbf{E}\right] \qquad \text{(VIII.38)}$$

$$= -\frac{1}{T}\,\nabla\tilde{\mu}_i, \qquad \text{(VIII.38a)}$$

with the electrochemical potential

$$\tilde{\mu}_i = \mu_i + q_i\ \Phi,$$

where q_i is the charge on atoms of species i. For electrons, $\tilde{\mu}_e = E_F - |e|\ \Phi$, with E_F being the Fermi energy.

Consider the case of a vacancy mechanism and purely electronic conduction in a metal. Since the vacancies are not charged, the fluxes are written:

$$\left\{\begin{array}{l} \mathbf{J}_A = -\dfrac{L_{AA}}{T}\ \nabla(\tilde{\mu}_A - \mu_v) - \dfrac{L_{AB}}{T}\ \nabla(\tilde{\mu}_B - \mu_v) + \dfrac{L_{Ae}}{T}\ q_e\ \mathbf{E} \\ \\ \mathbf{J}_B = -\dfrac{L_{BA}}{T}\ \nabla(\tilde{\mu}_A - \mu_v) - \dfrac{L_{BB}}{T}\ \nabla(\tilde{\mu}_B - \mu_v) + \dfrac{L_{Be}}{T}\ q_e\ \mathbf{E} \\ \\ \mathbf{J}_v = -(\mathbf{J}_A + \mathbf{J}_B) \\ \\ \mathbf{J}_e = \dfrac{L_{ee}}{T}\ q_e\ \mathbf{E}, \end{array}\right. \tag{VIII.39}$$

and $i = q_e\ J_e$ is the current density, with $q_e = -|e|$.

Writing the different gradients out explicitly, and assuming that the vacancies are at thermal equilibrium ($\nabla\mu_v$ negligible), one finds (e.g., for J_B):

$$\begin{aligned} \mathbf{J}_B = & -k\left[\frac{L_{BB}}{n_B} - \frac{L_{BA}}{n_A}\right] \cdot \left[1 + \frac{\partial\ \text{Log}\ \gamma_B}{\partial\ \text{Log}\ N_B}\right]\ \nabla n_B \\ & + \frac{1}{T}\ [L_{BB}\ q_B + L_{BA}\ q_A + L_{Be}\ q_e]\ \mathbf{E}. \end{aligned} \tag{VIII.40}$$

The first term is just $-\bar{D}_B\ \nabla n_B$ defined above ($\bar{D}_B$ is the intrinsic diffusion coefficient).

Assume a dilute $\underline{A}B$ alloy. The flux of solute can be written in simplified form, using the fact that $n_B << n_A$:

$$\begin{aligned} \mathbf{J}_B &= -k\left[\frac{L_{BB}}{n_B} - \frac{L_{BA}}{n_A}\right]\ \nabla n_B + \frac{1}{T}\ L_{BB}\left[q_B + \frac{L_{BA}}{L_{BB}}\ q_A + \frac{L_{Be}}{L_{BB}}\ q_e\right]\ \mathbf{E} \\ &= -\ D_B^A\ \nabla n_B + \frac{n_B\ D_B^A}{kT}\ [\underbrace{q_B - q_B^*}_{Z_B^*\ |e|} + \frac{L_{BA}}{L_{BB}}\ \underbrace{(q_A - q_A^*)}_{Z_A^*\ |e|}]\ \mathbf{E}, \end{aligned} \tag{VIII.41}$$

having set:

$$\left\{\begin{array}{l} -L_{\mathrm{Ae}}\, q_{\mathrm{e}} = L_{\mathrm{AA}}\, q_{\mathrm{A}}^{*} + L_{\mathrm{AB}}\, q_{\mathrm{B}}^{*} \\ -L_{\mathrm{Be}}\, q_{\mathrm{e}} = L_{\mathrm{BA}}\, q_{\mathrm{A}}^{*} + L_{\mathrm{BB}}\, q_{\mathrm{B}}^{*} \end{array}\right. \tag{VIII.42}$$

$$\text{or} \left\{\begin{array}{l} -L_{\mathrm{Ae}} = L_{\mathrm{AA}}\, Z_{\mathrm{A}}^{*} + L_{\mathrm{AB}}\, Z_{\mathrm{B}}^{*} \\ -L_{\mathrm{Be}} = L_{\mathrm{BA}}\, Z_{\mathrm{A}}^{*} + L_{\mathrm{BB}}\, Z_{\mathrm{B}}^{*}. \end{array}\right.$$

The electromigration of B is characterized by an effective valence $Z_{\mathrm{B}}^{**} = Z_{\mathrm{B}}^{*} + (L_{\mathrm{BA}}/L_{\mathrm{BB}})\, Z_{\mathrm{A}}^{*}$; the fluxes of A and B are coupled via the vacancy wind, which manifests itself in the term $L_{\mathrm{BA}}/L_{\mathrm{BB}}$. This can vary between +2 and −2 and can result in a large correction to Z_{B}^{*}.

A particular case:

Self-electromigration: B = A*. — The equation for the flux of tracer A* is written, substituting A* for B in equation (VIII.41):

$$\mathbf{J}_{\mathrm{A}^*} = -D_{\mathrm{A}^*}^{\mathrm{A}}\, \nabla n_{\mathrm{A}^*} + \frac{n_{\mathrm{A}^*} D_{\mathrm{A}^*}^{\mathrm{A}}}{kT}\, Z_{\mathrm{A}}^{*} \left[1 + \frac{L_{\mathrm{AA}^*}}{L_{\mathrm{A}^*\mathrm{A}^*}}\right] |e|\ \mathbf{E} \tag{VIII.43}$$

$$\mathbf{J}_{\mathrm{A}^*} = -D_{\mathrm{A}^*}^{\mathrm{A}}\, \nabla n_{\mathrm{A}^*} + \frac{n_{\mathrm{A}^*}}{kT}\, \frac{D_{\mathrm{A}^*}^{\mathrm{A}}}{f_0}\, Z_{\mathrm{A}}^{*}\ |e|\ \mathbf{E} \tag{VIII.44}$$

where f_0 is the correlation factor for self-diffusion (see Eq. (VIII.27)). The effective charge of the solvent Z_{A}^{*} (see Tab. I) is determined from a self-electromigration experiment, and the quantity $Z_{\mathrm{B}}^{*} + (L_{\mathrm{BA}}/L_{\mathrm{BB}})\, Z_{\mathrm{A}}^{*}$ is determined from the electromigration of the solute. Thus Z_{B}^{*} can be determined only if $L_{\mathrm{BA}}/L_{\mathrm{BB}}$ is known.

It has been noted that the electric force acting on the atoms is the resultant of two effects: $Z_i^{*}\,|e| = q_i - q_i^{*}$, i.e.,

- an electrostatic effect directly on the ion: $q_{\mathrm{A}}\mathbf{E}$, $q_{\mathrm{B}}\mathbf{E}$,
- an effect of the charge carriers (electrons or holes) which can transmit to an ion at the saddle point a fraction of their momentum; one speaks of the "electron wind" or the effect of friction, $-q_{\mathrm{A}}^{*}\,\mathbf{E}$, $-q_{\mathrm{B}}^{*}\,\mathbf{E}$, connected to the coupling terms L_{Ae} and L_{Be}. The effective charge is a function of the excess resistivity due to an atom at the saddle point:

$$q_i^{*} = \frac{q^i}{2}\left[100\frac{\Delta\rho_i^{\mathrm{sp}} + \Delta\rho_i^{\mathrm{gs}}}{\rho^0} - \mathbf{F}_0\right],$$

where $\Delta\rho_i^{\mathrm{sp}}$ and $\Delta\rho_i^{\mathrm{gs}}$ are the residual resisitivites (in $\mu\Omega$-cm per at. %) of the i atoms at the saddle point and in the ground state, i.e., on a normal site, respectively, ρ^0 is the resistivity of the matrix, and $\mathbf{F}_0$ is a correction for the presence of the neighboring vacancy.

Electromigration is thus the method of choice for the study of the saddle point. Depending on the particular case, the resultant force can be directed towards the anode (carried by the electron wind) or towards the cathode (the "hole wind"). In normal metals, the direct term q_A is negligible. A selection of values of Z_A^* is given in table I; the values for a given element can be quite scattered.

Table I. — *Effective Valences for Self-Electromigration in Metals.*

(From Pratt and Sellors, 1973)

Element	Ag	$\beta - \mathrm{Zr}$	γ−U	Na	Zn ($\parallel c$)	Zn ($\perp c$)	Al	Sn
Z_A^*/f	−20.5	+0.3	−1.6	−2.4	−2.0	−4.4	−12 to − 30	−80

In self electromigration, the vacancy flux is not zero, $\mathbf{J}_v = -\mathbf{J}_A$, leading to a displacement of the lattice planes with a velocity $\mathbf{v}$:

$$\mathbf{v} = -\frac{1}{kT}\frac{D_{A^*}^{A}}{f_0} Z_A^* \; |e| \; \mathbf{E}. \tag{VIII.44a}$$

In the laboratory frame of reference, the flux of tracer becomes:

$$\mathbf{J}_{A^*}^{0} = \mathbf{J}_{A^*} + n_{A^*} \; \mathbf{v} = -D_{A^*}^{A} \; \nabla n_{A^*}.$$

For a thin layer in a sandwich, the tracer concentration profile in the laboratory frame of reference is a Gaussian centered on the plane located at the zero of the abscissa. The distance between the center of the Gaussian and the marker plane thus gives a measure of v, and thus of Z_A^*.

In solute electromigration at infinite dilution, the velocity of the lattice is still given by equation (VIII.44a), the vacancy flux not having been modified appreciably if $n_B << n_A$. The flux of B* in the laboratory frame of reference is:

$$\mathbf{J}_{B^*}^{0} = \mathbf{J}_{B^*} + n_{B^*} \; \mathbf{v},$$

from which, in the absence of a concentration gradient (following Eqs. (VIII.41) and (VIII.44a)):

$$\mathbf{J}^0_{\mathrm{B}^\bullet} = n_{\mathrm{B}}\, D_{\mathrm{B}^\bullet}\, \frac{e\mathbf{E}}{kT} \left[\left(Z^*_{\mathrm{B}} + \frac{L_{\mathrm{BA}}}{L_{\mathrm{BB}}}\, Z^*_{\mathrm{A}} \right) - \frac{1}{f_0}\, \frac{D_{\mathrm{A}^\bullet}}{D_{\mathrm{B}^\bullet}}\, Z^*_{\mathrm{A}} \right],$$

where the first term is just equation (VIII.41), without the term in ∇n_{B}, and the second is the velocity of the lattice. If one begins with a homogenous alloy, the flux due to electromigration gives rise to a concentration gradient which opposes the action of the electric field. After a certain time, the two fluxes balance each other exactly and $J^0_{\mathrm{B}^\bullet} = 0$. This is the steady state, characterized by a time-independent concentration profile $n^0_{\mathrm{B}^\bullet}(x)$. One can extract from this a complex quantity containing the effective charges of the solute and solvent and the diffusion coefficients. The sandwich method is simpler, because the displacement of the Gaussian with respect to the lattice can be measured with inert markers. This allows the calculation of the quantity $Z^*_{\mathrm{B}} + (L_{\mathrm{BA}}/L_{\mathrm{BB}})\, Z^*_{\mathrm{A}}$ (Fig. VIII.4).

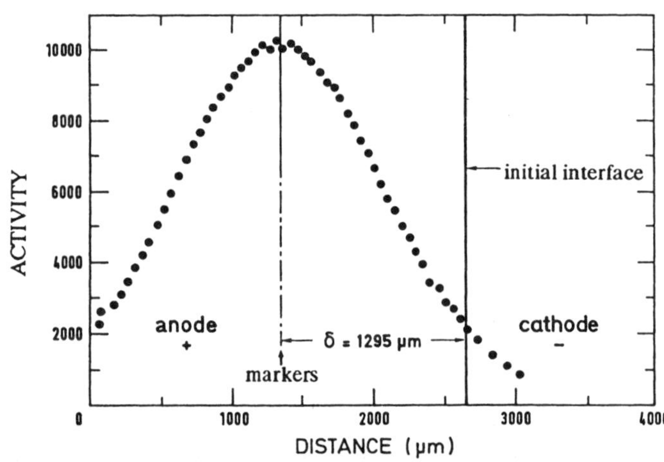

Fig. VIII.4. — Electromigration of ^{124}Sb in Ag. $T = 892°$C, $t = 2.517 \times 10^5$ s, $i = 6520$ A cm^{-2}, $Z_{\mathrm{Sb}} = +5$. After Doan and Brébec (1970).

The Nernst-Einstein relation. — Consider a homogenous alloy A̲B. The flux of solute B due to the electric field **E** is given by equation (VIII.41), and reduces to:

$$\mathbf{J}_{\mathrm{B}} = \frac{n_{\mathrm{B}}\, D^{\mathrm{A}}_{\mathrm{B}}}{kT} \left[Z^*_{\mathrm{B}} + \frac{L_{\mathrm{BA}}}{L_{\mathrm{BB}}}\, Z^*_{\mathrm{A}} \right] |e|\ \mathbf{E}. \qquad \text{(VIII.45)}$$

Since the mobility is defined by (Ch. I, Eq. (I.21)),

$$\mathbf{J}_i = n_i \; u_i \; \mathbf{E}, \qquad \text{(VIII.46)}$$

we have:

$$u_{\rm B} = \frac{D_{\rm B}^{\rm A}}{kT} \left[Z_{\rm B}^* + \frac{L_{\rm BA}}{L_{\rm BB}} \; Z_{\rm A}^* \right] \; |e| \, . \qquad \text{(VIII.47)}$$

The generalized Nernst-Einstein relation is then written:

$$\boxed{\frac{u_{\rm B}}{D_{\rm B}^{\rm A}} = \frac{Z_{\rm B}^* \; |e|}{kT} \left[1 + \frac{L_{\rm BA}}{L_{\rm BB}} \frac{Z_{\rm A}^*}{Z_{\rm B}^*} \right]} \qquad \text{(VIII.48)}$$

instead of the usual form, in which the term in brackets reduces to 1. The coupling between the fluxes of A and B introduces an extra term, called the "vacancy wind" already encountered above, and explainable in terms of the different jump frequencies.

For the case of self-diffusion:

$$\begin{aligned} \frac{u_{\rm A^\bullet}}{D_{\rm A^\bullet}^{\rm A}} &= \frac{Z_{\rm A}^* \; |e|}{kT} \left[1 + \frac{L_{\rm A^\bullet A}}{L_{\rm A^\bullet A^\bullet}} \right] \\ &= \frac{1}{f_0} \; \frac{Z_{\rm A}^* \; |e|}{kT} . \end{aligned} \qquad \text{(VIII.49)}$$

The Nernst-Einstein relation is valid under the condition that the proper self-diffusion coefficient $D_{\rm A^\bullet}^{\rm A}/f_0$ is used and not the tracer self-diffusion coefficient $D_{\rm A^\bullet}^{\rm A}$. Note that there is no coupling term $L_{\rm AB}$ for interstitial impurities. The simple Nernst-Einstein relation applies, with the effective valence Z^* of the interstitials.

Ionic crystals. — The above analysis applies to each sublattice, but the effective charges are equal to the true ionic charges $q_{\rm A}$ and $q_{\rm B}$. For self-electromigration in particular, the Nernst-Einstein equation (VIII.49) applies, with f_0 being the correlation factor for the mechanism operative in the sublattice under consideration.

Application. — Electromigration can be used to purify metals of all impurities with non-zero effective valence $Z_{\rm B}^*$, because under the influence of the electric field, the impurities segregate to one end of the specimen, leaving a purified section at the center. This method is especially applicable for interstitial impurities, or for the liquid state, because the treatment time is not prohibitively long; $t \propto 4 \, l / \, Z_{\rm B}^* D_{\rm B}$ for a bar of length l. Electromigration is also an important source of failure in the metallic interconnects on semiconductor devices (see Rosenberg and Berenbaum, 1971).

III.5 Thermomigration.

III.5.1 *Phenomenological equations.* — The reasoning is analogous to that used above for electromigration. However, there is an added complication in that the system is, *per se*, not isothermal. The chemical forces are given by:

$$-X_i = \nabla\left(\frac{\mu_i}{T}\right) = \frac{\partial\left(\mu_i/T\right)}{\partial T}\,\nabla T + \frac{1}{T}\,\frac{\partial \mu_i}{\partial n_i}\,\nabla n_i. \qquad \text{(VIII.50)}$$

The second term on the RHS is the term usually met in isothermal systems. The first term is easily transformed:

$$\frac{\partial\left(\mu_i/T\right)}{\partial T} = \frac{\left(\partial \mu_i/\partial T\right)\,T - \mu_i}{T^2} = -\frac{h_i}{T^2}$$

by applying the Helmholtz relation $G = H + T\left(\partial G/\partial T\right)$, to the partial molar free energy, μ_i. In X_i (Eq. (VIII.50), the first term is thus $h_i\;\nabla\left(1/T\right) = h_i\;X_q$. It is convenient to use a different set of forces, obtained by the linear transformation:

$$-X_i = -X_i' + h_i\;X_q$$

with

$$\begin{cases} X_i' &= -\dfrac{1}{T}\,\nabla_T\,\mu_i \\ X_q &= \nabla\left(\dfrac{1}{T}\right). \end{cases} \qquad \text{(VIII.51)}$$

so that the gradient of chemical potential is calculated at *constant temperature*, giving it a simpler physical significance. A "reduced" flux of heat is calculated from these chemical forces by substitution of the expressions for the transformed forces into equation (VIII.3) for the entropy production:

$$J_q' = J_q - \sum_{k=1}^{n-1}\left(h_k - h_n\right)\,J_k. \qquad \text{(VIII.52)}$$

III.5.2 *Application to the binary system* AB. — The fluxes are:

$$\begin{cases} J_{\rm A} = -\dfrac{L_{\rm AA}}{T}\,\nabla_T\,(\mu_{\rm A}-\mu_v) - \dfrac{L_{\rm AB}}{T}\,\nabla_T\,(\mu_{\rm B}-\mu_v) + L_{{\rm A}q}\,\nabla\left(\dfrac{1}{T}\right) \\ J_{\rm B} = -\dfrac{L_{\rm BA}}{T}\,\nabla_T\,(\mu_{\rm A}-\mu_v) - \dfrac{L_{\rm BB}}{T}\,\nabla_T\,(\mu_{\rm B}-\mu_v) + L_{{\rm B}q}\,\nabla\left(\dfrac{1}{T}\right) \\ J_v = -\,(J_{\rm A}+J_{\rm B}) \\ J'_q = -\dfrac{L_{q{\rm A}}}{T}\,\nabla_T\,(\mu_{\rm A}-\mu_v) - \dfrac{L_{q{\rm B}}}{T}\,\nabla_T\,(\mu_{\rm B}-\mu_v) + L_{qq}\,\nabla\left(\dfrac{1}{T}\right). \end{cases} \tag{VIII.53}$$

The expressions for $J_{\rm A}$ and $J_{\rm B}$ are transformed by introducing the ***heats of transport***, Q_i^*:

$$\begin{cases} L_{{\rm A}q} = L_{q{\rm A}} = L_{\rm AA}\,Q_{\rm A}^* + L_{\rm AB}\,Q_{\rm B}^* \\ \qquad\qquad\qquad\qquad\qquad L_{\rm AB} = L_{\rm BA}. \\ L_{{\rm B}q} = L_{q{\rm B}} = L_{\rm BA}\,Q_{\rm A}^* + L_{\rm BB}\,Q_{\rm B}^* \end{cases} \tag{VIII.54}$$

It is easily shown that the flux of heat is written:

$$\mathbf{J}'_q = Q_{\rm A}^*\,\mathbf{J}_{\rm A} + Q_{\rm B}^*\,\mathbf{J}_{\rm B} + (L_{qq} - L_{{\rm A}q}\,Q_{\rm A}^* - L_{{\rm B}q}\,Q_{\rm B}^*)\,\nabla\left(\frac{1}{T}\right), \tag{VIII.55}$$

an expression that justifies the name "heat of transport", since Q_i^* is the contribution of the flux of species i to the flux of heat:

$$Q_{\rm A}^* = \left(J'_q/J_{\rm A}\right) \text{ with } J_{\rm B} = 0 \text{ and } \nabla T = 0.$$

$\nabla_T\mu_{\rm A}$ and $\nabla_T\mu_{\rm B}$ can be calculated as in the case of interdiffusion, and the intrinsic diffusion coefficients can again be found. Until this point, the derivation has been identical to the one followed for electromigration, but a difference appears here. In effect (see Appendix I, Eq. (A.2a)):

$$\nabla_T\,\mu_v = \nabla_T\,\mu_v^0 + kT\,\nabla_T \mathrm{Log}\,N_v.$$

where N_v is here an *independent* thermodynamic parameter ([6]). However, if the density of vacancy sources and sinks is high enough, then $N_v = \bar{N}_v = \exp\left(-h_v^{\rm f}/kT\right)$ everywhere. Because of this supplementary relation and because

([6]) If local equilibrium obtains, $\nabla\mu_v = 0$, but as the system is not isothermal, $\nabla_T\mu_v \neq 0$.

at equilibrium $\mu_v = 0$, from which $\nabla_T \mu_v^0 = \nabla_T h_v^{\mathrm{f}}$, we find:

$$\frac{1}{T}\,\nabla_T\,\mu_v = \frac{1}{T}\,\nabla_T\,h_v^{\mathrm{f}} + k\left[\frac{h_v^{\mathrm{f}}}{k\,T^2}\,\nabla T - \frac{1}{kT}\,\nabla h_v^{\mathrm{f}}\right] = h_v^{\mathrm{f}}\,\nabla T/T^2.$$

From equations (VIII.29, 30, and 54) and with the above result, equations (VIII.53) become:

$$\begin{cases} J_{\mathrm{A}} &= -\bar{D}_{\mathrm{A}}\,\nabla n_{\mathrm{A}} - \left[L_{\mathrm{AA}}\left(Q_{\mathrm{A}}^* - h_v^{\mathrm{f}}\right) + L_{\mathrm{AB}}\left(Q_{\mathrm{B}}^* - h_v^{\mathrm{f}}\right)\right]\dfrac{\nabla T}{T^2} \\ J_{\mathrm{B}} &= -\bar{D}_{\mathrm{B}}\,\nabla n_{\mathrm{B}} - \left[L_{\mathrm{BA}}\left(Q_{\mathrm{A}}^* - h_v^{\mathrm{f}}\right) + L_{\mathrm{BB}}\left(Q_{\mathrm{B}}^* - h_v^{\mathrm{f}}\right)\right]\dfrac{\nabla T}{T^2}. \end{cases} \tag{VIII.56}$$

For the *dilute solution* AB*:

$$\bar{D}_{\mathrm{B}} \to D_{\mathrm{B}^*}^{\mathrm{A}} = k\frac{L_{\mathrm{BB}}}{n_{\mathrm{B}}}$$

$$J_{\mathrm{B}} = -D_{\mathrm{B}^*}^{\mathrm{A}}\,\nabla n_{\mathrm{B}} - \frac{n_{\mathrm{B}}\,D_{\mathrm{B}^*}^{\mathrm{A}}}{kT^2}\left[Q_{\mathrm{B}}^{**} + \frac{L_{\mathrm{BA}}}{L_{\mathrm{BB}}}\,Q_{\mathrm{A}}^{**}\right]\nabla T, \tag{VIII.57}$$

with

$$Q_{\mathrm{A}}^{**} = Q_{\mathrm{A}}^* - h_v^{\mathrm{f}} \text{ and } Q_{\mathrm{B}}^{**} = Q_{\mathrm{B}}^* - h_v^{\mathrm{f}}.$$

According to (VIII.56), the average velocity of the lattice is (see the calculation for interdiffusion, Eq. (VIII.32)):

$$\begin{aligned} \mathbf{v} = &-\frac{J_{\mathrm{A}} + J_{\mathrm{B}}}{n} = \left(\bar{D}_{\mathrm{A}} - \bar{D}_{\mathrm{B}}\right)\,\nabla N_{\mathrm{A}} \\ &+ \left[\left(L_{\mathrm{AA}} + L_{\mathrm{BA}}\right)Q_{\mathrm{A}}^{**} + \left(L_{\mathrm{AB}} + L_{\mathrm{BB}}\right)\,Q_{\mathrm{B}}^{**}\right]\frac{\nabla T}{nT^2}. \end{aligned}$$

In the case of solute diffusion at infinite dilution, the first term on the RHS is zero. It can also be shown that the term in ∇T reduces approximately to:

$$\mathbf{v} \simeq \frac{D_{\mathrm{A}^*}^{\mathrm{A}}}{f_0}\,\frac{Q_{\mathrm{A}}^{**}}{kT^2}\,\nabla T. \tag{VIII.58}$$

The thin-layer technique is not very easy to use on a sample which is under a temperature gradient since D_{B^*} varies with position in the sample. However, two other experimental methods are possible:

– measurement of the displacement of inert markers, from which $\mathbf{v}$ and Q_{A}^{**} can be obtained in a pure metal,

– solute diffusion at steady state in an initially homogenous alloy. For this case, J_{B}^0 must be known in the laboratory frame of reference. According to equations (VIII.57) and (VIII.58):

$$\begin{aligned}\mathbf{J}_{\mathrm{B}}^{0} &= \mathbf{J}_{\mathrm{B}} + n_{\mathrm{B}}\,\mathbf{v} \\ &= -D_{\mathrm{B}^*}^{\mathrm{A}}\ \nabla n_{\mathrm{B}} - \frac{n_{\mathrm{B}}}{kT^2}\left[D_{\mathrm{B}^*}^{\mathrm{A}}\ Q_{\mathrm{B}}^{**} - D_{\mathrm{A}}\ Q_{\mathrm{A}}^{**}\right]\ \nabla T,\end{aligned} \tag{VIII.59}$$

with

$$D_{\mathrm{A}} = \frac{D_{\mathrm{A}^*}^{\mathrm{A}}}{f_0} - \frac{L_{\mathrm{BA}}}{L_{\mathrm{BB}}}\ D_{\mathrm{B}^*}^{\mathrm{A}}.$$

After a certain time, the concentration gradient produced by diffusion under the action of the thermal force is exactly the opposite of the latter and $J_{\mathrm{B}}^{0} = 0$ everywhere:

$$\frac{\nabla n_{\mathrm{B}}/n_{\mathrm{B}}}{\nabla\,(1/T)} = \frac{1}{k}\left[Q_{\mathrm{B}}^{**} - \frac{D_{\mathrm{A}}}{D_{\mathrm{B}^*}^{\mathrm{A}}}\ Q_{\mathrm{A}}^{**}\right] = \frac{\bar{Q}}{k}. \tag{VIII.60}$$

A plot of Log n_{B} *vs.* $1/T$ is a straight line with slope $\bar{Q}/k$ (Fig. VIII.5).

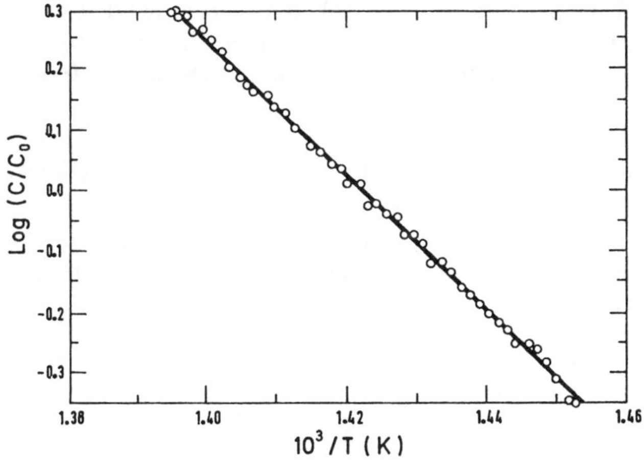

Fig. VIII.5. — Thermomigration of $^{85}Sr^{++}$ in AgCl under a temperature gradient of 330°C/cm. Steady-state profile. After Crolet (1971).

III.5.3 *Self-thermomigration.* — With $\mathrm{B} \equiv \mathrm{A}^*$, equation (VIII.57) becomes:

$$\mathbf{J}_{\mathrm{A}^*} = -D_{\mathrm{A}^*}^{\mathrm{A}}\ \nabla n_{\mathrm{A}^*} - \frac{n_{\mathrm{A}^*}\ D_{\mathrm{A}^*}^{\mathrm{A}}}{f_0\ kT^2}\ Q_{\mathrm{A}}^{**}\ \nabla T, \tag{VIII.61}$$

with the correlation factor appearing in the coupling term (Eq. (VIII.27)). According to (VIII.58) and (VIII.61):

$$\mathbf{J}_{\mathrm{A}^\bullet} = -D^{\mathrm{A}}_{\mathrm{A}^\bullet}\,\nabla n_{\mathrm{A}^\bullet} - n_{\mathrm{A}^\bullet}\,\mathbf{v}$$
$$\mathbf{J}^0_{\mathrm{A}^\bullet} = -D^{\mathrm{A}}_{\mathrm{A}^\bullet}\,\nabla n_{\mathrm{A}^\bullet}. \tag{VIII.62}$$

In a *pure material or in a dilute solution under a temperature gradient*, the velocity v has the same sign as ∇T if the heat of transport is *positive*; *the lattice planes and the inert markers then move toward the hot end*, i.e., the gradient gives rise to a flux of atoms from the hot end to the cold end. The phenomena are reversed for a negative heat of transport.

III.5.4 *Nomenclature.* — Unfortunately, a certain confusion reigns in the literature over the nature of the heat of transport, the symbol used to represent it, and the sign convention. The reduced heat of transport, here denoted by Q^{**} (Eq. (VIII.57)), is often found under the label Q^*. For the sign convention, and in accord with equation (VIII.61), it is recommended to take the + sign for Q^{**} when the atomic species under consideration migrates from the hot end to the cold end, the mass flow being defined in the lattice frame of reference.

III.5.5 *Simplified atomic theory of the heat of transport and the physical significance of the heat of transport.* — Consider the simple scheme of chapter II, § I: two neighboring planes (1) and (2) separated by a distance λ in a gradient of concentration and, further, of temperature. The net flux from left to right is (Eq. (II.3)):

$$J = -\lambda^2\,\Gamma\,\frac{\mathrm{d}n}{\mathrm{d}x} + \lambda n\,(\Gamma_{12} - \Gamma_{21})\,. \tag{VIII.63}$$

As the diffusing species, consider an interstitial impurity with jump frequency $\Gamma = \tilde{\nu}\,\exp\left(-\Delta G^{\mathrm{m}}(T)/kT\right)$. Assume that the temperature to consider in this expression is the temperature on the side from which the jump begins, say, T_1:

$$\begin{aligned}\left(\frac{\Delta G^{\mathrm{m}}}{kT}\right)_{T_2} &= \left(\frac{\Delta G^{\mathrm{m}}}{kT}\right)_{T_1} + (T_2 - T_1)\,\frac{\partial\,(\Delta G/kT)}{\partial T}\\ &= \left(\frac{\Delta G^{\mathrm{m}}}{kT}\right)_{T_1} - \frac{\Delta T}{kT^2}\,\Delta H^{\mathrm{m}}.\end{aligned} \tag{VIII.64}$$

From which

$$\Gamma_{12} - \Gamma_{21} = \tilde{\nu} \exp\left(-\frac{\Delta G^{\mathrm{m}}}{kT}\right) \left[-\frac{\Delta T}{kT^2} \, \Delta H^{\mathrm{m}}\right].$$

Then, with $\nabla T = \Delta T/\lambda$,

$$\begin{aligned} J &= -\lambda^2 \, \Gamma \, \frac{\mathrm{d}n}{\mathrm{d}x} - n \frac{\lambda^2 \, \Gamma}{kT^2} \, \Delta H^{\mathrm{m}} \, \nabla T \\ &= -D \left[\frac{\mathrm{d}n}{\mathrm{d}x} + n \, \Delta H^{\mathrm{m}} \, \frac{\nabla T}{kT^2}\right]. \end{aligned} \tag{VIII.65}$$

This expression has the same form, except for the correlation factor, as the expression of the phenomenological law with the heat of transport equal to ΔH^{m} (see Eq. (VIII.61)).

The same reasoning applies to vacancies, except that, since the energy is supplied to the jumping atom, the temperature to be considered is that of the site on which the vacancy arrives. Hence there is a change of sign, so that the heat of transport is equal to $-\Delta H_v^{\mathrm{m}}$.

This model predicts that interstitials move toward the cold end, and vacancies toward the hot end. These predictions are qualitatively verified for self-thermomigration in pure metals in that the material flux is toward the cold part, i.e., the vacancies move toward the hot end, although most often the value of the heat of transport is different from ΔH_v^{m}. In other cases, notably some interstitials, even the sign is incorrectly predicted. These disagreements can be explained by:

– the simplistic nature of the model: what is the significance of the local temperature on a scale smaller than the mean free path of "heat carriers" ? On the other hand, the jump involves several atoms located in the neighborhood of the initial and final sites, and not just one atom;

– the contribution of heat-carrying particles, electrons and phonons: q_{e}^* and q_{ph}^*. This is the analog of the friction force in electromigration. It can be shown that $q_{\mathrm{e}}^* < 0$ (the electrons coming from the cold side have a larger cross section for scattering than do the hot electrons). q_{e}^* is a few tenths of an eV and is proportional to the charge Z_{e}^* calculated for electromigration. The value of q_{ph}^* is positive and is probably negligible in metals;

– the contribution of the thermoelectric field

$$q_{\mathrm{th}}^* = (Ze) \; \eta T,$$

where Ze is the charge of the ion, and η the Thomson coefficient.

The experimental results are often uncertain and contradictory.

III.5.6 *Ionic crystals.* — In ionic crystals, thermomigration is at the origin of the *thermoelectric power* of these materials. The temperature gradient makes the probabilities of the defect jumps anisotropic. For each species of ion or for each species of mobile defect (cation and anion vacancies in alkali halides, cation vacancies and interstitials in silver halides,...), the flux for self-thermo-electromigration is, according to equations (VIII.44) and (VIII.61):

$$J_i = -n_i \; D_{i\bullet} \; \left[\frac{\nabla n_i}{n_i} - \frac{1}{f_0 \; kT} \; Z_i \; |e| \; E + \frac{Q_i^{**}}{f_0 \; kT^2} \; \nabla T\right], \; i = 1, \; 2. \quad \text{(VIII.66)}$$

On setting the total electric current equal to zero, i.e.,

$$Z_1 \; J_1 + Z_2 \; J_2 = 0,$$

the electric field E and the *homogenous thermoelectric power* $\theta = E/\nabla T$ can be calculated. Introducing the transport numbers, t_i, we obtain:

$$\theta \; |e|/k = \frac{t_1}{Z_1} \frac{\mathrm{d \; Log} \; n_1}{\mathrm{d \; Log} \; T} + \frac{t_2}{Z_2} \frac{\mathrm{d \; Log} \; n_2}{\mathrm{d \; Log} \; T} + \left(\frac{t_1}{Z_1} \; Q_1^{**} + \frac{t_2}{Z_2} \; Q_2^{**}\right) \; \frac{1}{kT} \quad \text{(VIII.67)}$$

with $t_i = n_i D_i / \sum \; n_i D_i$ following equation (VIII.66).

Measurement of the thermoelectric power allows the term in parentheses to be determined. The experimental difficulty is subtracting the contact potentials between the crystal and the electrodes from the thermoelectric power. If the fluxes and other quantities appearing in the above equations relate to defects, the thermomigration of vacancies and interstitials toward the hot part, which is observed experimentally for NaCl, KCl, TlCl, and AgCl, can be explained by negative heats of transport for the defects (see Gillan, 1983).

III.5.7 *Non-stoichiometric compounds.* — Thermomigration in non-stoichiometric compounds gives rise to a composition gradient due to the migration of majority defects. This is the case, for example, in (U, Pu) O_2 fuel pins in fast breeder reactors where the fissionable material is subjected to a very strong temperature gradient (≤ 6000 K cm^{-1}). Beginning with a hypostoichiometric oxide (oxygen deficient), a hyperstoichiometric oxide is obtained in the region closest to the cladding (the coldest zone). A steady state is established such that Log $\delta = Q^*/RT +$ const., where Q^* is a heat of transport, and δ is the deviation from stoichiometry. Note that the process of evaporation-condensation in the gas phase inside the pores (sintered material) can be at least as or even more important than thermomigration in the solid phase.

III.6 Problems connected with non-conserved species. — The discussion will be limited to the case of *vacancies*. If the number of sites per unit volume remains constant during the course of the diffusion, the local balance is (see Eq. (VIII.15)):

$$\frac{\partial n_v}{\partial t} = -\text{div } J_v^0 + \mathcal{R}_v, \qquad \text{(VIII.15)}$$

where the vacancy source term $\mathcal{R}_v$ gives the number of vacancies produced ($\mathcal{R}_v > 0$) or annihilated ($\mathcal{R}_v < 0$) per unit time per unit volume.

The vacancies are destroyed at certain "sinks" such as dislocations, grain boundaries, and surfaces, where they can equally well be created. The kinetics of creation or annihilation depend on the deviation from the equilibrium concentration, and a completely general reaction can be written for non-saturable sources or sinks:

$$\text{Vacancy} + \begin{bmatrix}\text{sink}\\ \text{source}\end{bmatrix} = \begin{bmatrix}\text{sink}\\ \text{source}\end{bmatrix},$$

with a kinetic law:

$$\mathcal{R}_v = \frac{\bar{n}_v - n_v}{\tau_v}$$

For supersaturation, $\mathcal{R}_v < 0$. If the mechanism is unique and the kinetics are indeed first order, τ_v is well defined as the lifetime of the corresponding vacancies; it is related to the mean free path by $\langle R^2 \rangle = 6\tau_v\ D_v$, where $\langle R^2 \rangle^{1/2}$ is typically the distance between sources and sinks.

The chemical potential of the vacancies is:

$$\mu_v = kT\ \text{Log}\left(n_v/\bar{n}_v\right) \approx kT\left(n_v - \bar{n}_v\right)/\bar{n}_v,$$

whence $\mathcal{R}_v = -\alpha\mu_v/T$, which is indeed the phenomenological equation deduced earlier from the thermodynamics of irreversible processes (Eq. (VIII.14)).

Equation (VIII.15) can be simplified in two particular cases:

1) div $J_v^0 = 0$. Example: creation and annihilation of vacancies in the course of irradiation. The creation is by nature homogeneous, as is the annihilation on a certain scale, if the sinks are distributed finely enough:

$$\frac{\partial n_v}{\partial t} = G_v + \left(\bar{n}_v - n_v\right)/\tau_v,$$

where G_v is the vacancy production rate. After a quenching experiment, $G_v = 0$, and the solution $n_v\,(t)$ follows immediately (see Ch. I, § VI). However, in most cases it is necessary to consider the balance in detail because of the wide variety of annihilation processes (see Ch. X, § VI).

2) $\partial n_v/\partial t = 0$. Steady state for vacancies. It is often assumed that such a state exists when the sources and sinks are sufficiently abundant and efficient:

$$\text{div } J_v^0 = \mathcal{R}_v,$$

with $J_v \approx J_v^0$. The *divergence of the vacancy flux* can have different origins. It can involve large effects on the shape and the porosity of the sample:
Porosity: If in the region of their maximum supersaturation, the vacancies are not eliminated fast enough by dislocation climb, they can form *pores*. The nucleation of pores is generally heterogeneous because most often a small number of pores is observed, which grow during the course of the process. Occluded gases favor the formation of pores because the pressure term can help to overcome the surface energy. For spherical pores of radius R, the Gibbs-Thomson equation (Ch. VII, Eq. (VII.36a)) leads to:

$$\frac{1}{\Omega} kT \operatorname{Log} \frac{n_v}{\bar{n}_v} = p + \frac{2\gamma_s}{R}.$$

For a given supersaturation, the critical radius R is the smaller, the higher the pressure p (p is taken as < 0 for internal pressure).

Change of shape: Near the surfaces that constitute sources and sinks, a swelling or shrinkage is observed, depending on whether there is sub- or supersaturation; this is particularly well seen on thin ribbons. If these are polycrystalline, the shrinkage is observed as grooves that mark the grain contours, the grain boundaries playing the role of short circuits for elimination of vacancies toward the surfaces.

In massive samples undergoing unidirectional diffusion, changes of shape are also observed if the elimination and/or production of vacancies does not take place in the ideal manner by the creation or elimination of planes perpendicular to the diffusional flux (see, e.g., Fig. VI.2). If the creation or elimination is effected isotropically, a swelling of the sample results if $\mathcal{R}_v > 0$, and the reverse occurs if $\mathcal{R}_v < 0$. The S-form observed when $\mathcal{R}_v$ changes sign along the sample (Fig. VIII.6d) originates here. This is the case of the Kirkendall effect (see Ch. VI, Fig. VI.8) or of electromigration when the temperature is not constant along the rod (Fig. VIII.6a), and the variation of the diffusion coefficients with the temperature along the rod (Fig. VIII.6b) is added to the effect of the electric field.

If the vacancies are eliminated in a perfectly isotropic manner, the displacement of the lattice planes would be only one third of that predicted by the equation $\mathbf{J}_v^0 = n\mathbf{v}$ (Fig. VIII.7). In reality, an intermediate situation is observed: $\mathbf{J}_v^0 = \alpha n\mathbf{v}$, with $1/3 \leq \alpha \leq 1$, due to the constraints generated by the specimen because the diffusion zone is not "free" to undergo an isotropic deformation, whether in Kirkendall effect or electromigration. One must calculate the velocity that the lattice planes would take on if the elimination or production of vacancies would give rise to a purely axial deformation ($\alpha = 1$). This is the corrected velocity that has to be used in equations (VIII.32), (VIII.44a) or

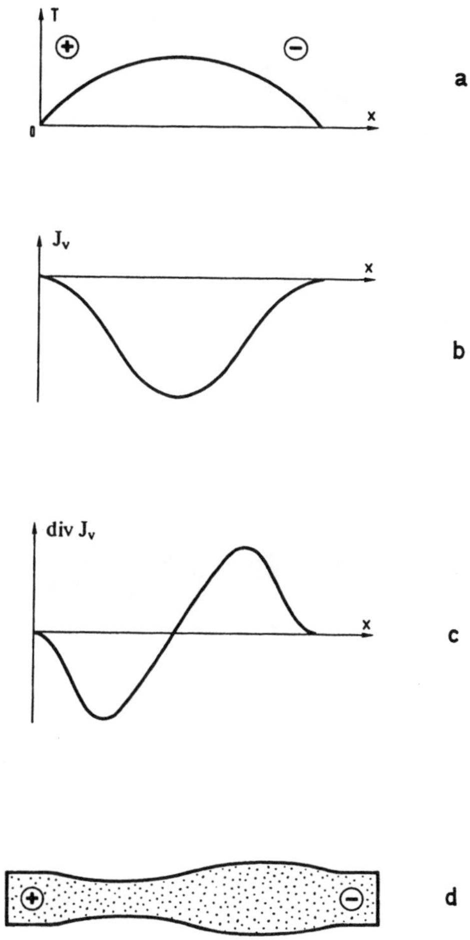

Fig. VIII.6. — Change of shape of a rod under the effect of a divergent vacancy flux produced by superposition of an electric current and a temperature gradient: a) temperature profile along the rod; b) vacancy flux (electro-, and eventually, thermomigration); c) vacancy source $\mathcal{R}_v = \text{div } J_v$; d) shape due to the creation and elimination of vacancies at the surface. After Oldham and Blackburn (1971).

(VIII.58) in order to calculate the intrinsic diffusion coefficients, the effective valence, or the heat of transport.

Applications. — In a material under a temperature gradient (*thermomigration*),

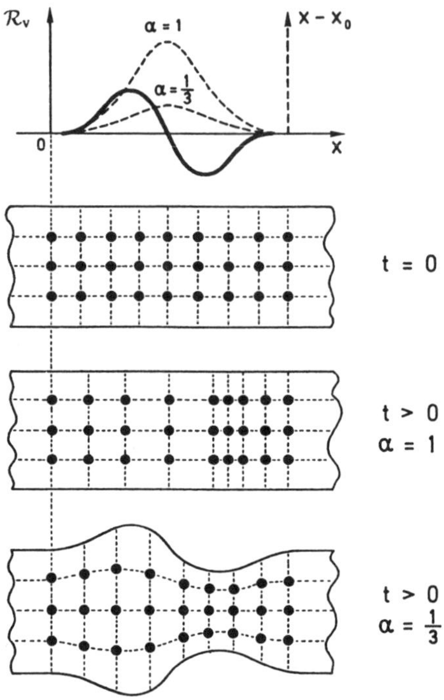

Fig. VIII.7. — Divergent flux of vacancies. Creation and annihilation of vacancies giving rise to purely axial deformation ($\alpha = 1$) or isotropic deformation ($\alpha = 1/3$). The points represent inert markers. After Monty (1972).

the divergence of the vacancy flux is not zero because the diffusion coefficient varies along the sample. In effect, porosity is observed on the hot side if the heat of transport is positive, as well as a migration of the initial porosity in the case of a sintered material. This is also why the sintered $(U, Pu)\,O_2$ oxides used as fuel pins in fast reactors, where they are under very high temperature gradients $\left(\leq 6000 \text{ K cm}^{-1}\right)$, densify at the cold part and large pores appear near the center.

In *electromigration*, the vacancy flux can have a non-zero divergence for two reasons:

– superposition of a parabolic temperature profile because the ends of the sample are cooled by the current leads (Fig. VIII.6a);

– structural heterogeneity, such as irregularities in the lattice, e.g., grain boundaries, in the case when electromigration takes place preferentially along the boundaries (low temperatures).

The latter case has been extensively studied for very thin metallic strips (thickness on the order of μm). The interconnects (stripes) in integrated circuits are just such strips, and they can fail in service due to fracture nucleated at pores or surface grooves (see, e.g., Rosenberg and Berenbaum (1971), and the IBM colloquium, edited by Gangulee *et al.*, 1975). In Al at 400 K under 10^6 A cm^{-2}, the supersaturation $(n_v - \bar{n}_v)/\bar{n}_v$ can reach unity for certain sites, especially at triple junctions where two boundaries with high diffusivities join with a boundary of low diffusivity, or on the contours of large grains embedded in a matrix of fine grains. Several remedies are possible: 1) stripes with a strong preferred orientation and elongated grains with tilt boundaries of low diffusivity (tilt axis perpendicular to the plane of the stripe); 2) addition of a solute that slows down the diffusion along grain boundaries. The micro-electronics industry has adopted the latter solution (addition of 4 % Cu) for aluminum conductors.

In *interdiffusion*, it is the variation of D with composition, different for the two components, that gives rise to the divergence of the vacancy flux. However, the supersaturation in general remains small; recall that these experiments are usually carried out at high temperature. The supersaturation $\mathcal{S} = (n_v - \bar{n}_v)/\bar{n}_v = -\mathcal{R}_v\, \tau_v/\bar{n}_v$ can be calculated. Returning to figure VI.8: $\mathcal{R}_v/n = \mathrm{d}v/\mathrm{d}x = (1/2t)(\mathrm{d}\xi/\mathrm{d}x) = (1/2t)(\mathrm{d}\delta/\mathrm{d}\lambda_0)$. The maximum of the supersaturation, i.e., the minimum in $\mathcal{R}_v$, is thus given by the minimum in $\mathrm{d}\delta/\mathrm{d}\lambda_0$, i.e., by the slope calculated at the inflection point on the decreasing side of the $\delta(\lambda_0)$ curve. The curve for the Au-Ag system in figure VI.6 thus gives $\mathcal{R}_v/n \approx 0.5 \times 10^{-5}$ s^{-1}. Then $\mathcal{S} = 0.5 \times 10^{-5}\ \tau_v/\bar{N}_v \approx \tau_v$. The supersaturation does not exceed 1 % if $\tau_v \leq 10^{-2}$ s. A calculation with $D_v \approx \tilde{D}/\bar{N}_v = 10^{-8}/0.5\times10^{-5}$ yields $< R^2 > \leq 10^{-4}$ cm^2, which requires a dislocation density of $\rho_{\mathrm{d}} \geq 10^4$ cm^{-2}, a condition generally realized in metals.

IV. Ternary systems.

Diffusion phenomena in systems with three or more chemical components (ternary or n-ary systems, $n \geq 3$) are frequently encountered in practice. The phenomenological equations facilitate the definition of the diffusion coefficients characteristic of such a system, which can be obtained experimentally.

DEFINITION OF THE DIFFUSION COEFFICIENTS. — The general equations applied to three chemical elements, denoted by the subscripts 1, 2, and 3 and vacancies yield:

$$\left\{\begin{aligned} \mathbf{J}_1 &= -\frac{L_{11}}{T}\,\nabla(\mu_1-\mu_v)-\frac{L_{12}}{T}\,\nabla(\mu_2-\mu_v)-\frac{L_{13}}{T}\,\nabla(\mu_3-\mu_v)\\ \mathbf{J}_2 &= -\frac{L_{21}}{T}\,\nabla(\mu_1-\mu_v)-\frac{L_{22}}{T}\,\nabla(\mu_2-\mu_v)-\frac{L_{23}}{T}\,\nabla(\mu_3-\mu_v)\\ \mathbf{J}_3 &= -\frac{L_{31}}{T}\,\nabla(\mu_1-\mu_v)-\frac{L_{32}}{T}\,\nabla(\mu_2-\mu_v)-\frac{L_{33}}{T}\,\nabla(\mu_3-\mu_v)\\ \mathbf{J}_v &= -(\mathbf{J}_1+\mathbf{J}_2+\mathbf{J}_3)\,. \end{aligned}\right. \tag{VIII.68}$$

If the vacancies are everywhere at equilibrium, $\mu_v \approx 0$, and $\mathbf{J}_v = n\ \mathbf{v}$ (see § III.2). The Gibbs-Duhem relation:

$$n_1\ \nabla\mu_1 + n_2\ \nabla\mu_2 + n_3\ \nabla\mu_3 = 0$$

allows the fluxes to be rewritten in the form:

$$\mathbf{J}_1 = -\frac{1}{T}\left[L_{11}-L_{13}\frac{n_1}{n_3}\right]\ \nabla\mu_1 - \frac{1}{T}\left[L_{12}-L_{13}\frac{n_2}{n_3}\right]\ \nabla\mu_2 \tag{VIII.69}$$

and analogous equations for J_2 and J_3.

If the vacancies are at equilibrium, there are only two independent concentrations, say the mole fractions N_1 and N_2. Since:

$$N_3 \approx 1-(N_1+N_2) \ \text{ and } \ N_v \equiv N_v\,(T,\ N_1,\ N_2)\,,$$

$$\nabla\mu_1 = \frac{\partial\mu_1}{\partial N_1}\ \frac{\nabla n_1}{n} + \frac{\partial\mu_1}{\partial N_2}\ \frac{\nabla n_2}{n}. \tag{VIII.70}$$

Then:

$$\begin{aligned} \mathbf{J}_1 \ &= -\frac{1}{T}\left\{\left[L_{11}-L_{13}\ \frac{n_1}{n_3}\right]\ \frac{1}{n}\ \frac{\partial\mu_1}{\partial N_1} + \left[L_{12}-L_{13}\ \frac{n_2}{n_3}\right]\ \frac{1}{n}\ \frac{\partial\mu_2}{\partial N_1}\right\}\ \nabla n_1\\ &\quad -\frac{1}{T}\left\{\left[L_{11}-L_{13}\ \frac{n_1}{n_3}\right]\ \frac{1}{n}\ \frac{\partial\mu_1}{\partial N_2} + \left[L_{12}-L_{13}\ \frac{n_2}{n_3}\right]\ \frac{1}{n}\ \frac{\partial\mu_2}{\partial N_2}\right\}\ \nabla n_2, \end{aligned} \tag{VIII.71}$$

with analogous equations for components 2 and 3. If we denote the concentrations by c_i, the system of equations is:

$$\left\{\begin{aligned} \mathbf{J}_1 &= -D^3_{11}\ \nabla c_1 - D^3_{12}\ \nabla c_2\\ \mathbf{J}_2 &= -D^3_{21}\ \nabla c_1 - D^3_{22}\ \nabla c_2\\ \mathbf{J}_3 &= -D^3_{31}\ \nabla c_1 - D^3_{32}\ \nabla c_2. \end{aligned}\right. \tag{VIII.72}$$

Thus 6 ***intrinsic diffusion coefficients*** are defined, a function of 9 phenomenological coefficients L_{ij} (of which only 6 are independent, according to the Onsager relations) and four partial derivatives $\partial\mu_i/\partial N_j$ $(i,\ j=1,\ 2)$.

These coefficients are defined in the mobile frame of reference. In practice, *4 interdiffusion coefficients* are measured: these are defined by:

$$\begin{cases} J_1^0 = -\tilde{D}_{11}^3\ \nabla c_1 - \tilde{D}_{12}^3\ \nabla c_2 \\ J_2^0 = -\tilde{D}_{21}^3\ \nabla c_1 - \tilde{D}_{22}^3\ \nabla c_2 \\ J_3^0 = -J_1^0 - J_2^0 \text{ with } \begin{cases} \tilde{D}_{ik}^3 = \tilde{D}_{ik} - \tilde{D}_{i3}\ (i,\ k=1,\ 2) \\ \tilde{D}_{12}^3 \neq \tilde{D}_{21}^3. \end{cases} \end{cases} \tag{VIII.73}$$

with the condition $\tilde{D}_{11}^3 + \tilde{D}_{22}^3 > 0$, but $\tilde{D}_{ii}^3$ can be negative, depending on the choice of component 3.

In a ternary solution, only three $\tilde{D}_{ik}^3$ are independent (Kirkaldy). In the particular case of an ideal solution:

$$N_2\,(1-N_2)\ \tilde{D}_{12}^3 - N_1\,(1-N_1)\ \tilde{D}_{21}^3 = N_1\ N_2\left(\tilde{D}_{11}^3 - \tilde{D}_{22}^3\right).$$

The *interdiffusion coefficients* can be determined in four ways:

– By fitting the experimental points to analytical solutions, which is possible if the $\tilde{D}$ do not depend on the concentration, i.e., if the gradients are not too steep ("incremental couples" with a small initial difference in concentration $\Delta c_i\ (0)$).

– By generalization of the Matano method (see Ch. I, § II.4). With $c \equiv c\,(\lambda)$, and $\lambda = x/\sqrt{t}$, application of the conservation equation for c_1 to the flux J_1^0 yields, as in equation (I.15):

$$\int \lambda\ \mathrm{d}c_1 = -2\left[\tilde{D}_{11}^3\ \frac{\mathrm{d}c_1}{\mathrm{d}\lambda} + \tilde{D}_{12}^3\ \frac{\mathrm{d}c_2}{\mathrm{d}\lambda}\right],$$

and an analogous equation for c_2.

The origin of the λ's is chosen so that:

$$\int \lambda\ \mathrm{d}c_i = 0, \qquad i = 1,\ 2.$$

The above equations can be rewritten in the form:

$$\psi_i = -\frac{1}{2}\,\frac{\mathrm{d}\lambda}{\mathrm{d}c_i}\int \lambda\ \mathrm{d}c_i,$$

with:

$$\begin{cases} \tilde{D}_{11}^3 + \tilde{D}_{12}^3\ \dfrac{\mathrm{d}c_2}{\mathrm{d}c_1} & = \psi_1 \\ \tilde{D}_{21}^3\ \dfrac{\mathrm{d}c_1}{\mathrm{d}c_2} + \tilde{D}_{22}^3 & = \psi_2. \end{cases} \tag{VIII.74}$$

Usually ψ_1, ψ_2, and dc_1/dc_2 are determined graphically or numerically.

The determination of the 4 $\tilde{D}$ coefficients thus requires *two independent diffusion couples* having a common domain of composition; the values of $\tilde{D}$ are calculated for the composition corresponding to the intersection of the two diffusion paths in the ternary diagram (Fig. VIII.8). It is thus not easy to measure the diffusion coefficients in the entire field of a ternary system.

– By the method of the square root diffusivity (Thomson and Morral, 1986), see Appendix IV.

– From the intrinsic diffusion coefficients, by a generalization of Darken's equations. This method requires Kirkendall-effect experiments.

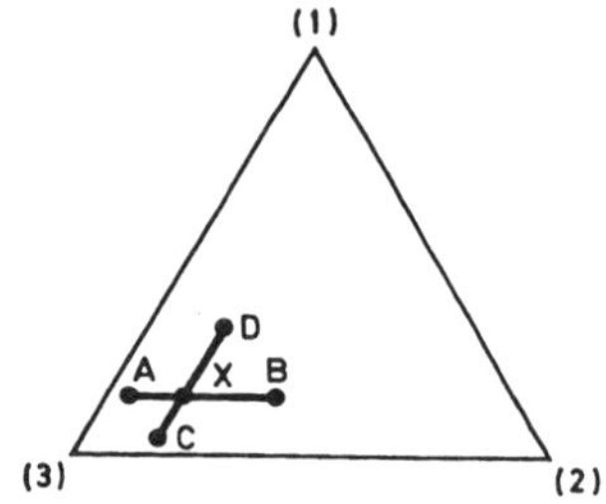

Fig. VIII.8. — Ternary diffusion. The diffusion coefficients for the composition at point X are determined from the couples A/B and C/D.

The *intrinsic diffusion coefficients* can be calculated by a generalization of Heumann's method, described for binary diffusion couples in chapter VI, § II.2, 2 (see Fig. VI.5 and Fig. VIII.9). The quantity of component i that has crossed the Kirkendall plane during the elapsed time t can be determined by measuring an area for component 1 (see Ch. VI, Eqs. (VI.21) and following equations):

$$A_1 = -2\,t\left[D^3_{11}\frac{dc_1}{dx} + D^3_{12}\,\frac{dc_2}{dx}\right], \qquad \text{(VIII.75)}$$

the gradients being measured at the Kirkendall plane. With experiments carried out on two diffusion couples (Fig. VIII.8), two values of A_1 are obtained, allowing the calculation of D^3_{11} and D^3_{12}.

The migration of components 2 and 3 is given by analogous equations, allowing the determination of the 6 intrinsic diffusion coefficients.

The Onsager relations can in principle be verified on ternary systems; unfortunately, the determination of the L_{ij} from the values of D is quite imprecise. It is particularly important to know the thermodynamics of the system with sufficient accuracy.

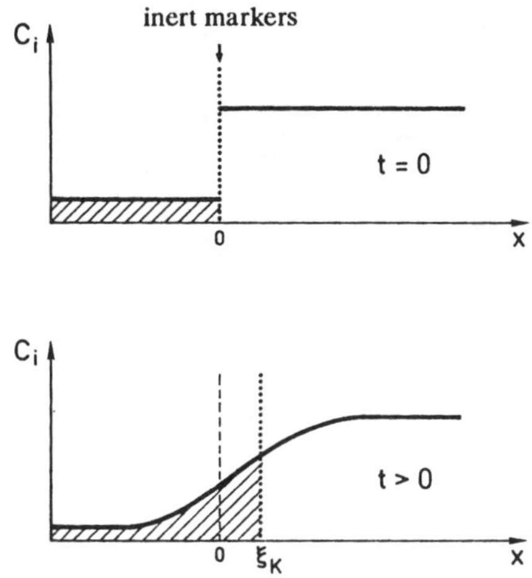

Fig. VIII.9. — The method of Guy and Philibert: the difference between the two cross-hatched areas gives the quantity of the component i that has crossed the Kirkendall plane during the diffusion time t.

The most spectacular characteristic of ternary diffusion is the evidence of "cross effects". The most useful experimental conditions for showing this are given in figure VIII.10. In case b, the concentration of component 1 is initially constant. The cross terms in the expression for the flux lead to a redistribution of this element due to the diffusion of the other two (Fig. VIII.11). A flux that "reascends" the concentration gradient ("uphill diffusion") can thus be observed; this proves that the driving force is not the concentration gradient, but the gradient of chemical potential. The departure of the concentration c_1 from its initial constant value is a measure of the off-diagonal coefficients. One measures D_{12}/D_{11}; the effect is larger, the more D_{11} is greater than D_{22}; component 1 is redistributed according to the diffusion of component 2. It is equally important that the initial gradient of component 2 be high. Finally, the diagonal coefficient for component 2 can be calculated with good precision from the concentration profile $c_2(x)$ by the Matano-Boltzmann method in the same way as for a binary system.

Such solute redistribution has been observed for substitutional components, and, more often, for interstitial ones (the latter certainly fulfill the condition $D_{11}/D_{22} > 1$).

Dilute solutions. – For dilute solutions with the solvent taken as component 3 $(n_1,\ n_2 < n_3)$, the equations (VIII.71) can be simplified. Neglecting the cross terms since they are multiplied by the small factors $n_{1,\ 2}/n_3$, we obtain from the expression for J_1:

$$\frac{D_{12}^3}{D_{11}^3} = \frac{L_{11}\,(\partial\mu_1/\partial N_2) + L_{12}\,(\partial\mu_2/\partial N_2)}{L_{11}\,(\partial\mu_1/\partial N_1) + L_{12}\,(\partial\mu_2/\partial N_1)}.$$

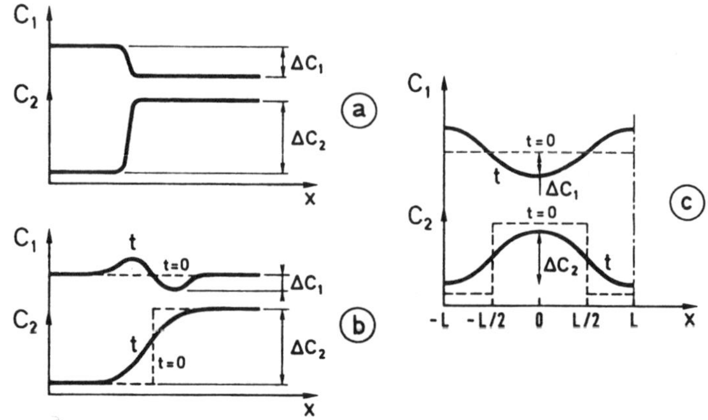

Fig. VIII.10. — Three types of ternary diffusion couples that can display the redistribution Δc_1 due to the diffusion of solute 2 (Kirkaldy *et al.*, 1965):
a) quasi-steady state for component 1 ($J_1 \approx 0$, very rapidly diffusing component) $\Delta c_1 = -(D_{12}/D_{11})\,\Delta c_2$
b) Darken's experiment (solute 1 initially homogenous)
$\Delta c_1 = -(1/2)\ K\ (D_{12}/D_{11})\,\Delta c_2$, with $K = f\,(D_{22}/D_{11}) \leq 1$
c) sandwich (couple of finite thickness, or a multilayer specimen):
$\Delta c_1 = -(1/2)\ K'\,(D_{12}/D_{11})\,\Delta\, c_2$, with $K' \leq 1$.

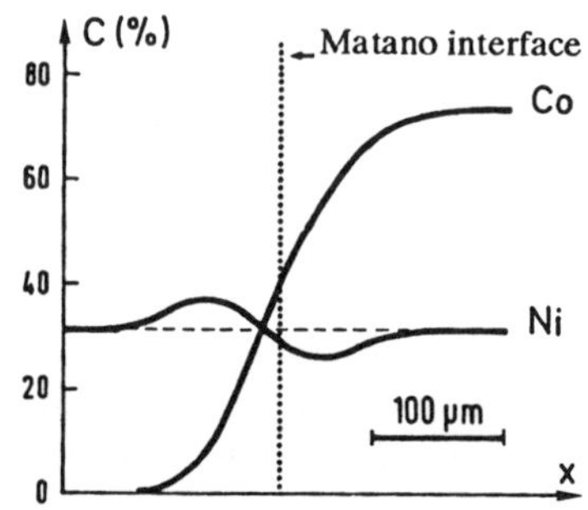

Fig. VIII.11. — Redistribution of nickel by uphill diffusion in an Fe-Ni-Co diffusion couple, in which the nickel concentration was initially constant. Annealed 72 h at 1135°C. After Sabatier and Vignes (1967).

According to C. Wagner, the chemical potentials in dilute solutions can be written as $\mu_1 - \mu_1^0 = kT\ [\text{Log}\ N_1 + e_{11}\ N_1 + e_{12}\ N_2]$, with an analogous expression for component 2, with $e_{12} = e_{21}$. Neglecting the product of the cross coefficients L_{12} and e_{12}, we have:

$$\frac{D^3_{12}}{D^3_{11}} = \frac{L_{11}\ N_1\ e_{12} + L_{12}\,(N_1/N_2)}{L_{11}\,(1 + N_1\ e_{11})}.$$

If component 1 is interstitial, the simplification can be continued further since generally $N_1 < N_2$ and the coupling term L_{12} is probably very small. Then:

$$\frac{D^3_{12}}{D^3_{11}} = \frac{N_1\ e_{12}}{1 + N_1\ e_{11}},$$

an expression which lends itself well to experimental verification if e_{11} and e_{12} are known (e.g., Fe-C-X alloys, with X = Ni, Co, Cr,...).

The same simplification is not possible for the substitutional solute 2:

$$\frac{D^3_{21}}{D^3_{22}} = \frac{N_2\left(e_{12} + \dfrac{1}{N_1}\,\dfrac{L_{21}}{L_{22}}\right)}{1 + e_{22}\ N_2}.$$

For substitutional solutions, the Kirkendall effect must be considered, and the above relationships for the D_{ij} are valid only if the Kirkendall effect can be neglected.

Note that ternary diffusion does not involve any new principles, but it does display clearly the effects of thermodynamic interaction among the components (Figs. VIII.10 and VIII.11).

The diffusion path. — The sequence of compositions that appears in a diffusion couple can be represented by a curve on an isothermal section of the ternary phase diagram. This curve is the locus of the points c_1, c_2, c_3 in the (1, 2, 3) triangle. The curve is stationary, i.e., invariant with time if the Boltzmann transformation applies, i.e., if $c_i \equiv f\,(\lambda)$, with $\lambda = x/\sqrt{t}$, which is observed experimentally.

This curve is called *the diffusion path.* It can take a form very different from the chord (the straight line joining the two points representing the alloys A and B that made up the original diffusion couple). The conservation of matter requires that the curve cut the chord at least once, hence the S-shaped diffusion path that is usually observed (Fig. VIII.12).

There is no general rule for predicting the diffusion path because its form springs from the kinetics of the process. Recall that the c_i are solutions of 2 equations which contain 4 interdiffusion coefficients that are functions of composition ! This explains why one occasionally has recourse to empirical methods for representing the curve, e.g., power functions characterized by the exponent

n and the intersection point of the curve with the chord AB (Dayananda 1985). Often, the flux of a species *changes sign* at a plane (or even several planes), becoming zero at that plane. Under these conditions, a constituent can diffuse "uphill", not only against its concentration gradient, but also up the gradient of its thermodynamic activity. The ratio of the cross to the main ternary diffusion coefficients for the zero flux component can be calculated at the zero flux plane, and with the help of thermodynamic data, some conclusions can be drawn about diffusion paths in its neighborhood. (Dayananda, 1985).

The above rules continue to apply in the case when the chord AB crosses a phase boundary. Discussion of this case is taken up in chapter X, § I, in the framework of diffusion in polyphase systems.

Among the ternary systems studied, we list (Kirkaldy, 1970): Fe-C-X, with X = Si, Ni, Mn, Cr, Co; FeCrNi, FeCoNi (all compositions), AlCuMn, CoNiCr (all compositions), CuSnZn, CuZnMn, CuAgAu (all compositions).

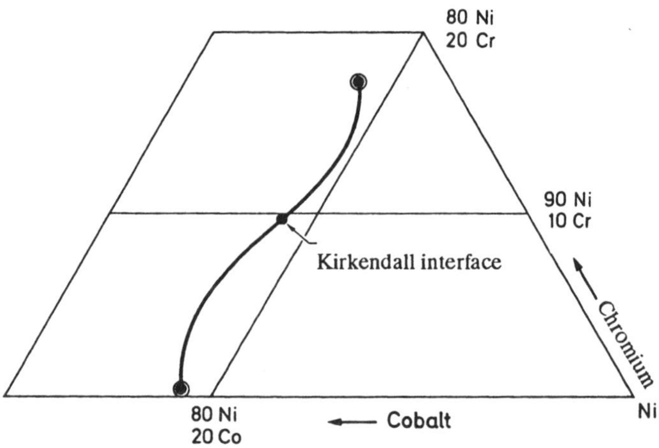

Fig. VIII.12. — Invariance of the diffusion path with time. Ni-Co-Cr system, couples annealed 72 h and 450 h at 1300°C. The diffusion path is a stationary curve; the Boltzmann transformation $\lambda = x/\sqrt{t}$ applies. ⊙ denote the compositions of the original halves of the couple. After Leroy (1966).

V. Heterogeneous solid solutions: effect of composition gradients.

Departures from Fick's laws because of strong gradients in composition were briefly discussed on a very simple basis in chapter II, § VIII. The problem should be posed in the framework of a more general theory, since the driving force for interdiffusion is the gradient of the chemical potential, not of composition. Cahn and Hilliard (1958) have given an expression for the free-energy density of a non-homogeneous solid solution. After reviewing this expression, we shall deduce from it Cahn's diffusion equation and discuss some of its applications, limiting the discussion to the case of not too high concentration gradients (Cahn's linear equation).

V.1 EXPRESSION FOR THE GIBBS FREE ENERGY. — Let $g_{\mathrm{h}}(c)$ be the Gibbs free energy of a homogeneous binary solution of concentration c, which varies along the x axis, and let $c' = \mathrm{d}c/\mathrm{d}x$, $c'' = \mathrm{d}^2c/\mathrm{d}x^2$ be the spatial derivatives of c. Expanding in Taylor's series:

$$g(c,\ c',\ c'',...) = g_{\mathrm{h}}(c) + c'\ \mathrm{d}g/\mathrm{d}c' + 1/2\ c'^2\mathrm{d}^2g/\mathrm{d}c'^2 + ...$$
$$+c''\ \mathrm{d}g/\mathrm{d}c'' + \frac{1}{2}\ c''^2\mathrm{d}^2g/dc''^2 + ...$$

an expression which can be written, neglecting higher order terms:

$$g = g_{\mathrm{h}}(c) + \kappa_{11}\partial c/\partial x + \kappa_{12}(\partial c/\partial x)^2 + \kappa_{21}\ \partial^2c/\partial x^2.$$

In a centro-symmetric medium, $\kappa_{11} = 0$, owing to symmetry. The Gibbs free energy per unit area normal to the x axis is then:

$$G = \int \left[g_{\mathrm{h}}(c) + \kappa_{12}(\partial c/\partial x)^2 + \kappa_{21}\partial^2c/\partial x^2\right]\ \mathrm{d}x.$$

Integrating the third term in the brackets by parts, we obtain the equation of Cahn and Hilliard:

$$\boxed{G = \int \left[g_{\mathrm{h}}(c) + \kappa(\partial c/\partial x)^2\right]\ \mathrm{d}x} \qquad \text{(VIII.76)}$$

with $\kappa = \kappa_{12} - \partial\kappa_{21}/\partial c$.

The quantity κ can be calculated from the regular solution model as a function of the pair interaction energies $V_{\mathrm{AA}n}$, $V_{\mathrm{BB}n}$, $V_{\mathrm{AB}n}$ between each atom and its neighbors in the nth coordination shell ($n = 1,\ 2, ...$). If the calculation is limited to nearest neighbors, κ has the same sign as the enthalpy of mixing. This purely "chemical" description is, however, not complete. All variations in composition imply a variation of the lattice parameter, $(a - a_0)/a_0 = \eta\ (c - c_0)$, with $\eta = (1/a)\ \mathrm{d}a/\mathrm{d}c$. Maintaining the coherence of the lattice then gives rise to plane strain in the planes normal to the x axis ([7]). This in turn gives rise to

([7]) This elastic deformation is the difference between the total strain resulting from coherence constraints in the specimen and the part of the strain due only to the change of the lattice parameter (i.e. stress free).

a density of elastic energy $\mathrm{Y}\eta^2 (c - c_0)^2$, where Y is an elastic modulus, a function of the compliances and thus of the crystallographic direction (Cahn, 1962). For an isotropic medium, $\mathrm{Y} = \mathrm{E}/(1-\nu)$, with E and ν denoting Young's modulus and Poisson's ratio, respectively. The term for the coherence strain must be added to equation (VIII.76):

$$\boxed{G = \int \left[g_{\mathrm{h}}(c) + \kappa (\partial c/\partial x)^2 + \mathrm{Y}\eta^2 (c - c_0)^2\right] \mathrm{d}x.} \tag{VIII.77}$$

V.2 Interdiffusion. — From chapter VI, equation (VI.19), the interdiffusion flux in a binary system $J_{\mathrm{A}}^0 = -J_{\mathrm{B}}^0$ can be written:

$$J_{\mathrm{A}}^0 = -\left(N_{\mathrm{A}}\bar{D}_{\mathrm{B}} + N_{\mathrm{B}}\bar{D}_{\mathrm{A}}\right) \frac{N_{\mathrm{A}}\,(1 - N_{\mathrm{A}})\, g_{\mathrm{h}}''}{kT}\, \partial c_{\mathrm{A}}/\partial x,$$

or:

$$J_{\mathrm{A}}^0 = -M g_{\mathrm{h}}''\, \partial c_{\mathrm{A}}/\partial x = -M\nabla\, g_{\mathrm{h}}', \tag{VIII.78}$$

with the interdiffusion coefficient $\tilde{D} = M g_{\mathrm{h}}''$. The subscript "h" means a homogeneous solution.

In a heterogeneous system, where the Gibbs free energy is expressed in the form of equation (VIII.76) or (VIII.77), let us impose an arbitrary change of concentration δc and calculate the corresponding value of δG:
Following equation (VIII.77):

$$\delta G = \int \left[g_h'(c)\ \delta c + (\partial\kappa/\partial c)(\partial c/\partial x)^2 \delta c + 2\kappa(\partial c/\partial x)\ \delta(\partial c/\partial x) + \right.$$
$$\left. +2\mathrm{Y}\ \eta^2 (c - c_0)\ \delta c\right]\ \mathrm{d}x \tag{VIII.79}$$

After inverting the order of the operators in the third term, an integration by parts ([8]) is possible. It yields:

$$\delta G = \int \left[g_{\mathrm{h}}'(c) - (\partial\kappa/\partial c)(\partial c/\partial x)^2 - 2\kappa\ (\partial^2 c/\partial x^2) + 2\mathrm{Y}\eta^2 (c - c_0)\right]\ \delta c\ \mathrm{d}x \tag{VIII.80}$$

([8]) $\int 2\,\kappa\,(\partial c/\partial x)\ \delta\,(\partial c/\partial x)\ \mathrm{d}x = \int 2\,\kappa\,(\partial c/\partial x)\ [\partial(\delta c)/\partial x]\ \mathrm{d}x =$
$= [2\kappa\ (\partial c/\partial x)\ \delta c] - \int\ [2(\partial\kappa/\partial c)(\partial c/\partial x)^2 + 2\kappa\ (\partial^2 c/\partial x^2)]\ \delta c\ \mathrm{d}x.$

The *local* variation of the Gibbs free energy, which is the driving force for diffusion, is given by the term in brackets. Taking the gradient of this term ([9]), and writing by analogy to equation (VIII.78):

$$J^0_{\mathrm{A}} = -M \left\{ g''_{\mathrm{h}}(c)\,(\partial c/\partial x) - (\partial^2\kappa/\partial c^2)\,(\partial c/\partial x)^3 - \right.$$
$$\left. -4\,(\partial\kappa/\partial c)\,(\partial c/\partial x)\,(\partial^2 c/\partial x^2) - 2\kappa\,(\partial^3 c/\partial x^3) + 2\eta^2\,\mathrm{Y}\,(\partial c/\partial x)\right\} \qquad \text{(VIII.81)}$$

The diffusion equation is obtained by applying the equation of continuity $\partial c/\partial t = -\mathrm{div}\ J^0$ to equation (VIII.81). If the inhomogeneities are not too great, the parameters M, g''_{h}, $\tilde{D}$, and κ can be treated as independent of concentration. Then only the first, fourth and fifth terms remain in equation (VIII.81), and a linear equation in the partial derivatives, Cahn's equation, is obtained:

$$\boxed{\partial c/\partial t = \tilde{D}\,(1 + 2\eta^2\,\mathrm{Y}/g''_{\mathrm{h}})\,(\partial^2 c/\partial x^2) - 2\,(\kappa/g''_{\mathrm{h}})\,\tilde{D}\,(\partial^4 c/\partial x^4)}\,, \qquad \text{(VIII.82)}$$

with $\tilde{D} = M g''_{\mathrm{h}}$.

V.3 Evolution of a modulation of composition. — How does a modulation of composition evolve by interdiffusion ? It is sufficient to consider one harmonic, since equation (VIII.82) is linear:

$$c - c_0 = A\,\sin\,\beta x = A\,\sin\,(2\pi x/\Lambda)\,. \qquad \text{(VIII.83)}$$

By application of the method of separation of variables, equation (VIII.82) yields:

$$\mathrm{d}\ \mathrm{Log}\ A/\mathrm{d}t = -\beta^2\,\tilde{D}_\Lambda, \qquad \text{(VIII.84)}$$

with

$$\tilde{D}_\Lambda = \tilde{D}\left[(1 + 2\eta^2\,\mathrm{Y}/g''_{\mathrm{h}}) + 2\,(\kappa/g''_{\mathrm{h}})\,\beta^2\right]. \qquad \text{(VIII.84a)}$$

An effective interdiffusion coefficient $\tilde{D}_\Lambda$ is thus substituted for the classical interdiffusion coefficient $\tilde{D}$. Like the latter, it can be positive or negative (see Ch. VI, §II.1), so that the modulation can attenuate or grow. Further, it depends on the wavelength of the modulation, Λ, and can change sign for

([9]) Based on the Euler-Lagrange relations (the calculus of variations), it can be shown that the term in brackets constitutes a "generalized chemical potential" which should be constant at equilibrium. In this case, the concentration profile $C\,(x)$ takes a form that minimizes G.

a certain value of Λ. In general, each Fourier component of a composition modulation evolves with time with its own relaxation time $\tau = 1/\beta^2\ \tilde{D}_\Lambda$.

In general, g''_h is positive for solid solutions with a tendency to unmixing, except in the $(c,\ T)$ region limited by the spinodal (locus of points where $g''_h = 0$). The effect of the coherence energy is to displace the chemical spinodal ($g''_h = 0$) toward lower temperatures (the coherent spinodal: $g''_h + 2\eta^2 Y = 0$). As for κ, its value and thus its sign can be predicted from the regular solution model. Three cases are possible:

1) $g''_h + 2\eta^2 Y > 0$ and $\kappa > 0$: tendency for unmixing, compositions outside the spinodal. The diffusion coefficients $\tilde{D}$ and $\tilde{D}_\Lambda$ are always positive. The modulation of the composition is attenuated with time whatever the wavelength.

2) $g''_h + 2\eta^2 Y < 0$ and $\kappa > 0$: tendency for unmixing, compositions inside the spinodal. $\tilde{D}$ is negative, but $\tilde{D}_\Lambda$, negative for long wavelengths, becomes positive for $\Lambda < \Lambda_c$, with:

$$\Lambda_c = \left[-8\pi^2\kappa/\left(g''_h + 2\eta^2\ Y\right)\right]^{1/2}. \qquad \text{(VIII.85)}$$

Thus the modulation is amplified for $\Lambda > \Lambda_c$ and attenuated for $\Lambda < \Lambda_c$. The amplification factor - $\beta^2\ \tilde{D}_\Lambda$ goes through a maximum for $\Lambda_m = \sqrt{2}\Lambda_c$; the amplitude of the modulation with wavelength Λ_m grows most rapidly (see Fig. VIII.13).

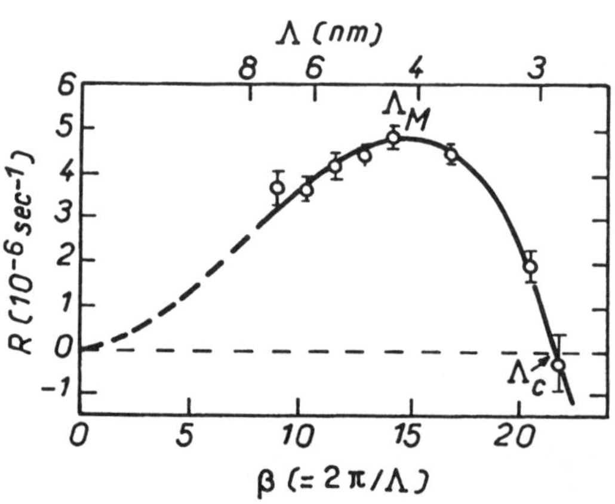

Fig. VIII.13. — Amplification factor $R = -\tilde{D}_\Lambda b^2$ for composition modulations in films with average composition $Au_{56}Ni_{44}$ annealed at 250°C. After Yang (1971), cited by Greer and Spaepen (1985).

3) $g''_h > 0$ and $\kappa < 0$: tendency for ordering. $\tilde{D}$ is always positive, but $\tilde{D}_\Lambda$ changes sign at Λ_c. The modulations are attenuated except those with short wavelength, $\Lambda < \Lambda_c$. In this case there is, in fact, one-dimensional ordering, but the continuum model no longer applies very well under these conditions ($\tilde{D}_\Lambda$ diverges for $\Lambda \to 0$). The treatment for a discrete medium (Ch. II, § VIII) must be used. This type of model leads to exponential attenuation in the simple cases: diffusion along $\langle 100\rangle$ or $\langle 110\rangle$ in the BCC structure, $\langle 100\rangle$ or $\langle 111\rangle$ in the FCC structure. According to Cook *et al.*, it is enough to replace the wave number β in equations (VIII.84) and (VIII.84a) by $\beta'^{\,2}$, where:

$$\beta'^{\,2} = \left(2/\lambda^2\right)\,\left[1 - \cos\left(2\pi\lambda/\Lambda\right)\right], \qquad \text{(VIII.86)}$$

where λ is the interplanar distance in the diffusion direction under consideration.

These models have been verified for various systems by the multilayer technique. The sample is prepared as a series of alternate layers A/B/A/B... with wavelength 1-5 nm and a total thickness 200-500 nm ([10]). The evolution of the modulation is generally followed by the satellites of the Bragg reflections which arise from the modulation (see Ch. IX, pt. 1, § III.5). The model described above predicts a linear variation of $\tilde{D}_\Lambda$ with β^2 or, for small Λ, β'^2; this prediction has been fulfilled for the Au-Ag, Cu-Pd, and Au-Ni systems. There is a strong size effect in the last of these systems: $\Delta d/d = 15.7$ %. The amplification coefficient $-\tilde{D}_\Lambda\beta^2$ in these systems passes through a maximum and then goes to zero with increasing $\beta = 2\pi/\Lambda$ (i.e. the amplitude of the modulation increases for $\Lambda > 3$ nm, Fig. VIII.13). In the Cu-Ni and Ag-Pd systems, on the other hand, higher order terms in g cannot be neglected (see the review article by Greer and Spaepen, 1985).

This model is a useful guide for predicting the stability of modulated composition materials. A low $\tilde{D}$ is a necessary condition. If $R = -\tilde{D}\beta^2$ is positive, the system will be more stable for modulation wavelengths $\Lambda > \Lambda_c$, with the risk of a coalescence process, which does not occur for $R < 0$.

([10]) The average composition of the system is adjusted via the ratio of the thickness of the A and B layers.

APPENDIX I

Chemical potential of vacancies.

Following the definition of the chemical potential:

$$\mu_v = (\partial G/\partial n_v)_{n_A,\ T,\ p}\,, \tag{A.1}$$

where G is the Gibbs free energy of crystal A, made up of n_A A atoms and n_v vacancies. The part of G that belongs to the vacancies is, if there is no interaction among the vacancies (dilute solution), $n_v\ \Delta G_v^f - T\Delta S_{conf.}$ Here ΔG_v^f is the Gibbs free energy of formation of a vacancy and $\Delta S_{conf.}$ is the configurational entropy of the crystal containing n_v vacancies.. Under the same assumption of a dilute (ideal) solution:

$$\Delta S_{conf.} = k\ \mathrm{Log}\ [(n_A + n_v)!/n_A!\ n_v!]\,.$$

Using Stirling's formula, we have:

$$\mu_v = \Delta G_v^f + kT\ \mathrm{Log}\ [n_v/(n_v + n_A)]\,, \tag{A.2}$$

which is indeed the usual form:

$$\mu_v = \mu_v^0 + kT\ \mathrm{Log}\ N_v\,. \tag{A.2a}$$

At thermodynamic equilibrium, G is a minimum, i.e., $\mu_v = 0$, from which:

$$\bar{N}_v = \frac{\bar{n}_v}{n_A + \bar{n}_v} = \exp\left(-\Delta G_v^f/kT\right). \tag{A.3}$$

With $n_v << n_A$, the expression for the chemical potential becomes:

$$\boxed{\mu_v = kT\ \mathrm{Log}\ (n_v/\bar{n}_v)}\,. \tag{A.4}$$

APPENDIX II

Diffusion in anisotropic media.

Equation (VIII.3) for an isothermal system

$$\sigma = \sum_k \mathbf{J}_k \cdot \mathbf{X}_k \tag{A.5}$$

can be written in its cartesian components:

$$\sigma = \sum_k \sum_{i=1}^{3} J_{ki} X_{ki}. \tag{A.5a}$$

It follows that each component of the flux is a linear function of the three components of the forces $\mathbf{X}_k$. For species A moving along the x_1 axis:

$$J_{A,\,1} = \sum_k \left[(L_{Ak})_{11}\; X_{k1} + (L_{Ak})_{12}\; X_{k2} + (L_{Ak})_{13}\; X_{k3} \right], \tag{A.6}$$

where X_{k1}, X_{k2}, and X_{k3} are the components of the force $\mathbf{X}_k$. In other words, the flux $\mathbf{J}_A$ is related to the forces $\mathbf{X}_k$ by k second-rank tensors:

$$\mathbf{J}_A = \sum_k \mathcal{L}_{Ak}\; \mathbf{X}_k \qquad k = \text{A, B...} \tag{A.7}$$

According to the Onsager relations, the $\mathcal{L}_{Ak}$ tensors are symmetric; the corresponding matrix thus contains six independent coefficients:

$$(\mathrm{L}_{Ak})_{12} = (\mathrm{L}_{Ak})_{21}\,, \quad \text{etc.}$$

Now consider a binary system. The equations (A.7) can be written:

$$\mathbf{J}_A = -\mathbf{D}_A\; \nabla c_A \tag{A.8}$$

where $\mathbf{D}$ is the diffusion tensor (intrinsic, self- or solute diffusion).

The $\mathbf{D}$ tensor is a second-rank tensor. The $(\mathrm{D}_A)_{i,\,j}$ matrix is symmetric and thus has six independent coefficients. It can be diagonalized by referring it to the principal axes. There are thus only three coefficients of diffusion (there are always $p \leq 6$ independent parameters, the p-3 others define the orientation of the principal axes). The number p varies according to the symmetry of the crystal system:

System	Triclinic	Monoclinic	Orthorhombic	Hexagonal Tetragonal Rhombohedral	Cubic
p	6	4	3	2	1

A diffusion experiment at a given temperature thus requires the preparation of p different samples, cut from a single crystal in the desired orientations. The p values of D_{ij} are then determined; they yield **D** which can then be referred to its principal axes (e.g., by constructing a Mohr's circle).

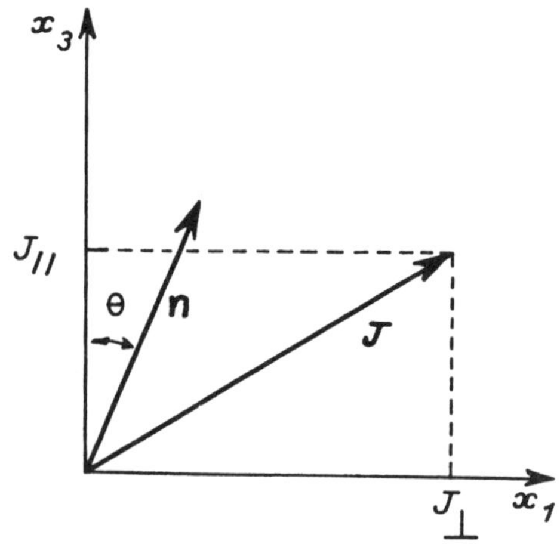

Fig. VIII.14. — Diffusion in a hexagonal lattice.

1) *Expression for* $D(\theta)$. — Consider a disc defined by the normal to its surface $\mathbf{n}(\alpha, \beta, \gamma)$. If a gradient ∇c is established along $\mathbf{n}$, the resulting flux is not directed along $\mathbf{n}$ according to equation (A.8); it is $\mathbf{J} \cdot \mathbf{n}$, the component of the flux along $\mathbf{n}$, and the diffusion coefficient:

$$D(\theta) = -\mathbf{J} \cdot \mathbf{n} / |\nabla c| \qquad (A.9)$$

that are measured. Relative to the principal axes we have:
components of $\mathbf{n}$: α, β, γ;
components of ∇c: $|\nabla c|\,\alpha$, $|\nabla c|\,\beta$, $|\nabla c|\,\gamma$;
components of $\mathbf{J}$:$-D_{11}\ |\nabla c|\,\alpha$, $-D_{22}\ |\nabla c|\,\beta$, $-D_{33}\ |\nabla c|\,\gamma$;
from which:

$$D = D_{11}\alpha^2 + D_{22}\beta^2 + D_{33}\gamma^2. \tag{A.10}$$

2) *Application to a hexagonal system.* — Take the x_3 axis along the c direction: $D_{11} = D_{22} \neq D_{33}$.
Let the direction $\mathbf{n}$ in the $x_1\ x_3$ plane make an angle θ with the x_3 axis.

Then $\alpha = \sin\theta$, $\beta = 0$, $\gamma = \cos\theta$, and:

$$D(\theta) = D_\perp \sin^2\theta + D_\parallel \cos^2\theta, \tag{A.11}$$

with $D_\perp = D_{11}$ and $D_\parallel = D_{33}$. For a polycrystal, with grains randomly oriented, the measured coefficient of diffusion is equal to:

$$D = (2/3)\,D_\perp + (1/3)\,D_\parallel.$$

In a general way, $D(\theta)$ referred to the principal axes is represented by a characteristic *quadric* (i.e., a second-degree surface). For further discussion of the tensor properties of D, the reader should consult J. F. Nye's book *Physical Properties of Crystals,* Oxford, Clarendon Press, 1957, especially chapters II and XI.

APPENDIX III

The frame of reference (Brady, 1975).

The expressions for the fluxes and forces, and therefore the values of the diffusion coefficients, depend on the frame of reference in space with respect to which these quantities are measured. These frames of reference have been discussed for binary systems in chapter VI, Appendix I, in connection with the Kirkendall effect and variable molar volumes in interdiffusion. That discussion is here generalized to n-component systems.

It is convenient to choose a frame of reference $\mathbf{R}^0$ fixed in one end of the sample. This frame of reference is commonly called the *laboratory frame of reference.* In practice, however, it is not always possible to use a unique frame of reference for all the measurements. This is especially the case when a change of volume takes place during the course of interdiffusion. Since different frames of reference lead to different diffusion coefficients, the relationships that allow changing from one frame of reference to another must be established so that the diffusion coefficients will have a precise meaning and can be compared among themselves.

The following discussion will be restricted to *unidirectional* diffusion.

To define a frame of reference *independent* of the frame of reference chosen by the observer as *his* laboratory frame of reference $\mathbf{R}^0$, a frame of reference $\mathbf{R}$ has to be determined that is connected to an *average velocity* of the n constituents participating in the diffusion:

$$v^{\mathrm{R}} = \sum_{i=1}^{n} \Pi_i^{\mathrm{R}} \, v_i^{\mathrm{R}^0} \tag{A.12}$$

where $v_i^{\mathrm{R}^0}$ is the velocity of constituent i in $\mathbf{R}^0$ and the Π_i^{R} are weighting factors such that:

$$\sum_i \Pi_i^{\mathrm{R}} = 1. \tag{A.13}$$

The fluxes are then given by:

$$J_i^{\mathrm{R}} = c_i v_i^{\mathrm{R}} = c_i \left(v_i^{\mathrm{R}0} - v^{\mathrm{R}} \right). \tag{A.14}$$

It follows that, no matter what frame of reference is defined on the basis of an average velocity, the fluxes are *not independent*:

$$\sum_i \Pi_{\mathrm{i}}^{\mathrm{R}} \, J_i^{\mathrm{R}}/c_i = 0. \tag{A.15}$$

Finally, the following relationship allows changing from one frame of reference to another:

$$J_i^{\mathrm{R}} = J_i^{\mathrm{Q}} + c_i v^{\mathrm{QR}}, \tag{A.16}$$

with

$$v^{\mathrm{QR}} = v^{\mathrm{QR}^\circ} - v^{\mathrm{RR}^\circ},$$

or more briefly:

$$v^{\mathrm{QR}} = v^{\mathrm{Q}} - v^{\mathrm{R}}.$$

Common frames of reference

1) *Mean-volume frame of reference* $\mathbf{R}^{\mathrm{V}}$. — By definition:

$$\Pi_i^{\mathrm{V}} = c_i \, \bar{V}_i, \tag{A.17}$$

with $\bar{V}_i$ the partial molar volume, and n_i the number of moles of component i per unit volume. Recall that $\sum N_i \, \bar{V}_i = V_m$, the molar volume and $\sum c_i \, \bar{V}_i = 1$, with $N_i = c_i / \sum c_i$. Then:

$$\sum_i \bar{V}_i \, J_i^{\mathrm{V}} = 0, \tag{A.18}$$

and

$$v^{\mathrm{V}} = \sum_i \bar{V}_i \, J_i^0, \tag{A.19}$$

where J_i^0 is the flux in the frame of reference $\mathbf{R}^0$.

2) *Mean-molar frame of reference or number-fixed frame of reference.*

$$\Pi_i^N = N_i, \tag{A.20}$$

from which

$$\sum_i J_i^N = 0, \tag{A.21}$$

and

$$v^N = V_m \sum_i J_i^0. \tag{A.22}$$

3) *Mean-mass or barycentric frame of reference* $\mathbf{R}^{\mathrm{M}}$

$$\Pi_i^{\mathrm{M}} = w_i, \tag{A.23}$$

with $w_i = M_i c_i / \rho$, the mass fraction. Here M_i is the molar weight (g/mole) and ρ the density (g cm^{-3}). From this:

$$\sum_i M_i \, J_i^{\mathrm{M}} = 0, \tag{A.24}$$

and

$$v^{\mathrm{M}} = (1/\rho) \sum_i M_i \, J_i^0. \tag{A.25}$$

4) S^{th} *component frame of reference* $\mathbf{R}^{\mathrm{S}}$

$$\Pi_i^{\mathrm{S}} = \delta_{is} \tag{A.26}$$

where δ_{is} is the Kronecker delta. Then:

$$J_{\mathrm{S}}^{\mathrm{S}} = 0 \tag{A.27}$$

and

$$v^{\mathrm{S}} = v_{\mathrm{s}}^0 = J_{\mathrm{S}}^0 / c_{\mathrm{S}}. \tag{A.28}$$

S usually denotes the major constituent.

Remark. — Following relation (A.15), there are only $(n-1)$ independent fluxes. It can be shown that in any frame of reference based on an average velocity, the concentration gradients are also not independent; e.g., since $\sum_i N_i = 1$, also $\sum \bar{V}_i \, \mathrm{d}c_i = 0$. This leads to a system of $(n-1)$ flux equations as a function of $(n-1)$ gradient terms. The interdiffusion coefficient is then a matrix of rank $(n-1)^2$ and not n^2. If the phenomenological coefficients L_{ij} are used, with gradients of chemical potential for the forces, the L matrix will be symmetric (Onsager's relations) and consist of only $n\,(n-1)/2$ independent coefficients.

5) *Lattice, inert-marker, or Kirkendall frame of reference.* — This frame of reference does not belong to the same category as the preceding ones. It must be introduced when the diffusion process gives rise to a *net flux* of material. To isolate this phenomenon, a frame of reference $\mathbf{R}^{\mathrm{K}}$ fixed to *inert markers* is chosen, because such a frame of reference (fixed to the lattice in the case of a crystal, but the concept is also applicable to non-crystalline materials) moves with the local flux of material. It is not connected to an average velocity of the constituents, but to the velocity of the markers v^{K}.

With respect to the $\mathbf{R}^{\mathrm{N}}$ frame of reference:

$$J_i^{\mathrm{N}} = J_i^{\mathrm{K}} + c_i v^{\mathrm{KN}}. \tag{A.29}$$

From equation (A.21):

$$v^{\mathrm{KN}} = -V_{\mathrm{m}} \sum_i J_i^{\mathrm{K}} \tag{A.30}$$

Thus there is no relation between the fluxes, contrary to the cases of the preceding frames of reference (Eqs. (A.18, 21, 24, and 27)). This is why the $\tilde{D}$ matrix in this frame of reference is made up of n^2 terms and not $(n-1)^2$. For a binary system, for example, $(n-1)^2 = 1$; in all frames of reference based on an average velocity, there is a unique diffusion coefficient $\tilde{D}$ which characterizes *the* flux $\left(J_1^{\mathrm{N}} = -J_2^{\mathrm{N}}\right)$. On the other hand, for the $\mathbf{R}^{\mathrm{K}}$ frame of reference, $n^2 = 4$, so that 4 independent diffusion coefficients are expected (in fact, there are only two if cross terms are neglected).

However, in a crystalline system it is possible to take an approach based on the idea of an average velocity, by considering point defects, generally vacancies, in addition to the material constituents. The binary system then becomes a ternary, in which v^{K} is indeed an average velocity, taken over all the sites of the lattice; one then comes back to a mean molar frame of reference, $\mathbf{R}^{\mathrm{N}}$. From equation (A.21):

$$J_1^{\mathrm{K}} + J_2^{\mathrm{K}} + J_v^{\mathrm{K}} = 0, \tag{A.31}$$

where the subscript v designates the third constituent (vacancies).

The frame of reference $\mathbf{R}^{\mathrm{N}}$ is chosen because of the simplification that results when changes of volume take place. From equation (A.16):

$$J_i^N = J_i^{\mathrm{K}} + c_i v^{\mathrm{KN}},$$

or, from equation (A.30):

$$J_i^N = J_i^{\mathrm{K}} - N_i \sum_j J_j^{\mathrm{K}}, \tag{A.32}$$

or, for a binary system:

$$J_1^N = N_2 J_1^{\mathrm{K}} - N_1 J_2^{\mathrm{K}}. \tag{A.33}$$

This relation leads to the generalized Darken equations including the partial molar volumes. However, diffusion experiments are usually not carried out in steady state; the conservation equation $\partial c/\partial t = -\partial J/\partial x$ applies in $\mathbf{R}^0$, and not in the other frames of reference $\mathbf{R}^{\mathrm{N}}$, $\mathbf{R}^{\mathrm{V}}$... unless the partial molar volumes are constant. The reader is referred to chapter VI, appendix I, for the transformations from $\mathbf{R}^0$ to $\mathbf{R}^{\mathrm{N}}$ or $\mathbf{R}^{\mathrm{V}}$, and for the relations among the diffusion coefficients.

APPENDIX IV

The square root diffusivity

The square root diffusivity is a matrix which can be useful for handling ternary (or higher order) systems (Thomson and Morral, 1986). It assumes that the diffusion coefficients D^3_{ij} (or D^n_{ij}) are constant across the diffusion zone, a condition which limits its application to not too large concentration differences. In such a case it is shown that:

– the concentration profiles $C_i(x,t)$ $i = 1, 2, ...n-1$, are given by linear combinations of error functions:

$$C_i = \sum A_{ij} \cdot \text{erfc}\left\{x/2\,(E_j t)^{1/2}\right\} \qquad j = 1,\ 2, \dots\ n-1; \qquad \text{(A.34)}$$

– the concentration gradients at the Matano interface (i.e. the initial interface) are linear functions of the initial concentration differences:

$$\text{grad } C_i = \left(1/2\sqrt{\pi t}\right) \cdot \sum p_{ij} \Delta C_j, \qquad j = 1,\ 2, \dots\ n-1; \qquad \text{(A.35)}$$

– the amount of solute i which has crossed the initial interface after time t :

$$S_i = -\left(t/\sqrt{\pi}\right) \sum r_{ij} \cdot \Delta C_j \qquad j = 1, 2 ... n-1, \qquad \text{(A.36)}$$

where the matrix $[r_{ij}]$ is called the square root diffusivity as:

$$[\mathrm{D}^n_{ij}] = [r][r] \qquad \text{(A.37)}$$

and:

$$[p] = [r]^{-1}.$$

The E_i are the $n-1$ eigenvalues of the matrix $[D]$.

Equations (A.36) offer a convenient way of calculating the r_{ij} from experimental data ($n-1$ diffusion couples, whence $(n-1) \cdot (n-1)$ equations for the $(n-1)^2$ r_{ij} unknown) and the D_{ij}. Conversely, knowing the r_{ij}, one can predict the diffusion behavior : e.g. the shape of the concentration profiles (monotonic or with extrema, see Fig. VIII.11), the existence of zero flux planes, etc. For a practical application to the NiAlCr system, see Thomson *et al.* (1990).

Bibliography

I. General references

Adda Y., and Philibert J., *op. cit.* Ch. V, X, XV, and XVI.
Brady J. B., *Am. J. Sci.* **275** (1975) 954-983.
Denbigh K. G., *The Thermodynamics of the Steady State* (1951) Methuen's Monographs (London).
Howard R. E., and Lidiard A. B., *Rep. Prog. Phys.* **27** (1964) 161.
Kirkaldy J. S., Young D. J., *Diffusion in the Condensed State* (London, 1987).
Nye J. F., *The Physical Properties of Crystals* (1957), Clarendon Press (Oxford).
Pratt J. N., and Sellors R. G., *Electrotransport in Metals and Alloys,* Diff. and Defect Monograph Series, #2 (1973), Trans. Tech. Pub., Aedermansdorf, Switzerland.
Wever H., *Elektro- und Thermotransport in Metallen* (1973) Barth (Leipzig).
Eléments de Métallurgie Physique, op. cit. Vol. **1**, Ch. I. § 10, 11, and 12 for the thermodynamics of irreversible processes.
Atomic Transport in Solids and Liquids, A. Lodding and T. Lagerwall, eds., *Z. Naturforsch.* (Tübingen) (1971) parts 1-5 and App. I.
Low Temperature Diffusion and Application to Thin Films, A. Gangulee, P. S. Ho, and K. N. Tu, eds. *Thin Solid Films* **25** (1975) (Elsevier, Sequoia. Lausanne, Switzerland).
Electro- and Thermotransport in Alloys, R. E. Hummel and H. B. Huntington, eds., AIME, New York (1977).

II. Specific references

Anthony T. R., *Diffusion in Solids, op. cit.* (1975) Ch. 7, p. 353-379.
Aust K. T., Hanemann R. E., Niessen P., Westbrook J. H., *Acta Metall.* **16** (1968) 291-302.
Bocquet J. - L., *Res Mechanica* **22** (1987) 1-44.
Cahn J. W. and Hilliard J. E., *J. Chem. Phys.* **28** (1958) 258; **31** (1959) 688.
Cahn J. W., *Acta Metall.* **9** (1961) 795-801; **10** (1962) 179-183.
Cook H. E., deFontaine D., and Hilliard J. E., *Acta Metall.* **17** (1969) 765-773.
Crolet J.-L., *Atomic Transport, op. cit.*, 44-51, or *Z. Naturforsch.* **26a** (1971) 907-914.

DAYANANDA M. A., in *Diffusion in Solids, Recent Developments* (1985) *op. cit.* 195-230.

DOAN N. V., and BREBEC G., *J. Phys. Chem. Solids* **31** (1970) 475-484.

DOAN N.V., *J. Phys. Chem. Solids* **33** (1972) 2161-2166.

GILLAN M.J, *Radiat. Effects* **74** (1983) 109-118.

GREER A. L. and SPAEPEN F., in *Synthetic Modulated Structures,* Academic Press (New York, 1985). Ch. 11, p. 419-486.

KIRKALDY J. S., *Adv. Mater Res.* **1** (1970) 56-100, H. Herman, ed. (Wiley, New York).

KIRKALDY J. S., BRIGHAM R. J., and WEICHERT D. H., *Acta Metall.* **13** (1965) 907-915.

LEROY V., Thesis, University of Liège (1966), Publ. C. R. M., Liège (1966).

LIDIARD A. B., *Acta Metall.* **34** (1986) 1487-1490.

MARTIN M. and DORRIS S., *Ber. Bunsenges. Phys. Chem.* **91** (1987) 779-785.

MONTY C., Thesis, Université Paris-Sud, Orsay (1972).

OLDHAM D. J. and BLACKBURN D. A., *Atomic Transport* (1971) *op. cit.* 151-154.

ROSENBERG R., and BERENBAUM L., *Atomic Transport* (1971) *op. cit.* 113-121.

SABATIER J. P. and VIGNES A., *Mém. Sci. Rev. Metall.* **64** (1967) 225-240.

Scripta Met. **20** (1986) # 14 contains several articles on multilayers.

THOMSON M. S., MORRAL J. E., *Acta Metall.* **34** (1986) 339-346 and 2201-2203.

THOMSON M. S., MORRAL J. E., ROMIG A. D., *Met. Trans.* **21** A (1990) 2679-2685.

CHAPTER IX

Techniques for the study of diffusion

The experimental study of diffusional phenomena is sometimes only qualitative or semi-quantitative. Such is often the case in metallurgy, where micrographic observation, with the help of the proper etching reagents, shows the existence of diffusion zones, intermediate phases, and even concentration gradients on the transverse section of an interdiffusion couple. Such observations are usefully complemented by measurements of the Vickers (or, for unidirectional diffusion, Knoop) microhardness.

However, this chapter is essentially devoted to the quantitative measurement of the diffusion coefficient, D, and its variation with temperature. With this information, one can determine the preexponential factor D_0 and the activation energy, Q. The experimental methods are divided into two categories according to whether:

1) One studies diffusion over a long distance (long here means with respect to the interatomic spacing λ) i.e., over a displacement $\langle X^2 \rangle^{1/2} >> \lambda$. One measures a flux, the integral of a flux, or a concentration profile $c(x, y, z, t)$, and deduces D from it.

2) One studies a phenomenon governed by atomic jumps and then measures the relaxation times or frequencies directly related to the jump frequency of one (or more) species of particles. The diffusion coefficient, D, is deduced from these data by a calculation based on a precise model of the phenomenon under study. Examples are the elastic after-effect, internal friction, magnetic or dielectric relaxation, NMR, Mössbauer effect, etc...

The ranges of values of the diffusion coefficient accessible to the different techniques are compared in table I.

PART 1. — DIFFUSION OVER A LONG DISTANCE

I. Methodology of the measurements. Sample preparation.

Several techniques can be used to introduce the diffusing element:

– *Surface deposit* (thin layer or sandwich). The material is deposited by various methods. The most frequently used ones are: electroplating or precipitation from a liquid solution (much used with radioactive tracers), deposition by cathodic sputtering or by evaporation from a target containing the element or tracer under study. The two latter methods allow the sample surface to be cleaned *in situ* by ion bombardment before depositing the diffusant under a proper vacuum. Deposition is also possible by the sublimation of materials deposited from solution on a foil of refractory metal, or by chemical vapor deposition (CVD).

– *Bonding* of two samples A/B. This is easy to accomplish with metals or ionic crystals by the use of a small press that allows breaking the surface oxide and carrying out the bonding at a temperature very much lower than the diffusion temperature. The process is harder to carry out with oxides or covalent materials; these materials must therefore be kept pressed together during the diffusion anneal, which requires an enclosure that allows the application of a load during the anneal, such as a controlled atmosphere creep machine.

Table I. — *Values of the diffusion coefficient accessible to different techniques.*

Method	Δx	$D\,(\mathrm{cm^2 s^{-1}})$
Lathe sectioning, grinding	0.1 - 250 μm	10^{-15} - 10^{-6}
Microtome	1 - 10 μm	10^{-13} - 10^{-8}
Chemical	10 μm	10^{-11} - 10^{-8}
Electrochemical	50 nm	10^{-16} - 10^{-13}
Sputtering	5 - 100 nm	10^{-18} - 10^{-13}
Modulated structures	0.5 - 5 nm	$\geq 10^{-22}$
Ion microprobe (SIMS)	1 - 100 nm	10^{-19} - 10^{-13}
Electron microprobe	≥ 2 μm	10^{-12} - 10^{-8}
Rutherford backscattering	50 nm	10^{-16} - 10^{-13}
Nuclear reaction analysis	20 - 100 nm	$5 \times 10^{-17} - 10^{-12}$
Nuclear Magnetic Resonance (NMR)		10^{-16} - 10^{-5}
Neutron inelastic scattering		10^{-7} - 10^{-5}
Mössbauer effect		10^{-11} - 10^{-7}
Conductivity (ionic crystals)		10^{-13} - 10^{-6}
Resistivity (semiconductors)		10^{-16} - 10^{-8}
Elastic after-effect		10^{-21} - 10^{-17}
Internal friction		10^{-16} - 10^{-11}
Magnetic anisotropy		10^{-21} - 10^{-17}

Note: For methods based on the resolution Δx in the depth measurement of a profile, the values of D are given following F. Bénière (1983); the diffusion zone of depth $\approx 4\,(Dt)^{1/2}$ is divided into at least 13 sections of thickness Δx so that $c = 10^{-2}\ c_s$ at the 13th section. The diffusion times are between 10^3 and 10^6 s.

– *Ion implantation*, with the use of an implanter that accelerates and mass separates the ions. Energies of a few keV to 100 keV allow implantations to depths of a hundred to several thousand Ångströms. The advantages of this method are absence of surface contamination, no diffusion barriers due to surface oxides, no special problems with chemical species only slightly soluble in the base material. The disadvantages are the simultaneous creation of irradiation defects and somewhat complicated boundary and initial conditions. In principle, if the implantation is not carried out along a channeling direction, the implanted profile is a gaussian centered at depth R_p (the projected range) with width $\Delta R_p \approx R_p$.

– *Introduction from a gas phase.* The diffusing species is a gas, and is deposited on the surface by the decomposition of the gas molecules, e.g., ammonia for N, methane for C, arsine and phosphine for As and P... It is assumed, and should be verified, that the surface concentration of the diffusing species in equilibrium with the gas phase equals its solubility in the sample material at the temperature of the experiment ([1]). In a better arrangement, a buffer atmosphere (e.g., $CH_4 + H_2$, $CO + CO_2, ...$) can be used to establish the chemical potential of the diffusant on the surface. Simple gases can be introduced into the sample by simply heating the sample in an atmosphere of the gas; the surface concentration introduced is given by Sieverts' law (proportional to $p^{1/2}$ for a diatomic molecule, see Appendix I). The rare gases can be introduced in a glow discharge.

– *Isotopic Exchange.* This is also a gas-phase process, very much used for introducing the stable isotope ^{18}O. The sample, an oxide, is heated to the diffusion temperature in an atmosphere enriched in the isotope ^{18}O. The exchange reaction takes place at the surface:

$$\mathrm{M}\ {}^{16}\mathrm{O} + \frac{1}{2}\ {}^{18}\mathrm{O}_2 \rightleftharpoons \mathrm{M}\ {}^{18}\mathrm{O} + \frac{1}{2}\ {}^{16}\mathrm{O}_2.$$

On the surface, ^{16}O atoms are replaced by ^{18}O until the isotopic ratio is the same in the gas and the solid. These ideal conditions are not always realized and the kinetics of oxygen penetration can be governed by the reactions taking place at the surface. This would introduce a boundary condition which involves a flux entering the solid at the surface, and leads to a diffusion profile that is different from the classical error function.

Hydrogen (and its isotopes) can be introduced by making the sample the cathode in an electrolytic cell containing an aqueous electrolyte or a molten salt (cathodic charging), by simple heating in an H_2 atmosphere, or by holding the sample under an elevated pressure of H_2. Cathodic charging allows very high fugacities (several thousand atmospheres) to be attained at the surface, and eliminates the necessity of dissociating the H_2 molecules. With electrolytic

---

([1]) Otherwise, the kinetics of the surface reaction (see Ch. I, Eq. (I.11)) have to be taken into account, and it must be shown that they do not govern the process under study.

cells on both sides of a metallic plate, the hydrogen being oxidized in the anode compartment, electrochemical permeation experiments can be carried out (see App. I for a definition of permeation). This is an excellent method, if one takes heed of surface effects, and if it is complemented by other techniques: kinetics of degassing as a function of temperature, radiotracers (tritium) and autoradiography, and local probes (proton NMR).

II. Determination of the diffusion profile $c(x, y, z, t)$.

The best experimental method for determining D is to establish the concentration-distance curve experimentally. The advantages of the method are verification that the diffusion kinetics follow the expected law, and avoidance of artifacts due to diffusion barriers or to local inhomogeneities.

It is nevertheless wise to verify the homogeneity of the diffusion, by autoradiography, for example, in the case of radiotracers, or by X-ray or ion images. The diffusion profiles should be determined by an analytical method, which may be destructive or non-destructive.

II.1 NON-DESTRUCTIVE METHODS.

a) Certain methods require a *deconvolution* procedure, e.g.:

- photoelectron spectroscopy,
- Auger electron spectrometry,
- Rutherford backscattering (generally α particles with energies of the order of one MeV) (Fig. IX.1),
- nuclear reaction analysis.

The first two are not often used because they are limited to rather small thicknesses because of the very small depth from which the detected electrons can escape (10 or some tens of Å).

The spectra obtained in the second two, the energy distribution of backscattered α particles or the particles produced by the nuclear reaction, are the result of three processes: 1) penetration of the incident particle to a depth x, with energy loss; 2) interaction with an atom at this depth (elastic backscattering or nuclear reaction); 3) ejection of a particle in the direction of the detector; the particle then traverses a path in the material and again loses energy.

If the profile $c(x)$ of the atoms that give rise to the reaction under consideration is known, the energy spectrum of the emitted particles can be calculated (convolution). The inverse operation (deconvolution) is not practical. The diffusivity is thus determined by comparison of the calculated spectra with the experimental spectrum, with D as a fitting parameter.

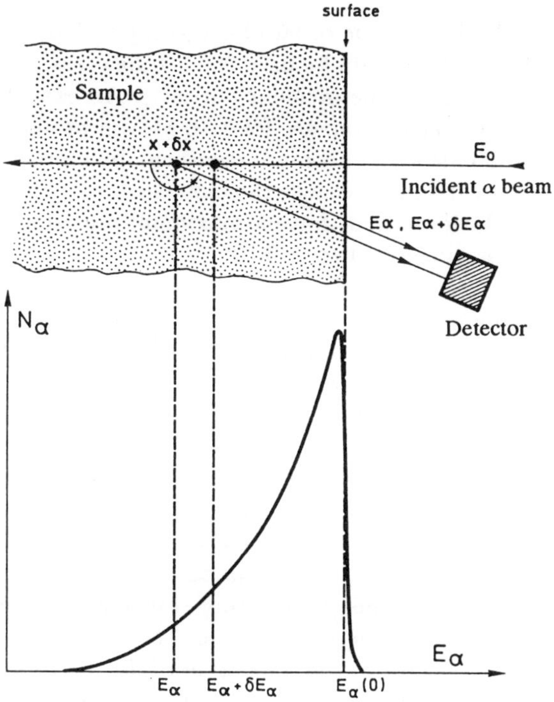

Fig. IX.1. — Rutherford backscattering of α particles:
- above, the sample, the incident beam, the detector;
- below, for the same depth scale, the energy spectrum of the backscattered particles. δE_α is the energy resolution of the detector.

In Rutherford backscattering with α particles, the analyzed layer has a depth of about 1 μm; this can be as large as 5 μm for nuclear reactions. For greater penetrations, one must proceed in steps: abrasion, analysis, abrasion,... The depth resolution is not as good as the energy resolution in the spectrum (the width of one channel) because the energy of a particle in the course of penetration is dispersed about the mean value given by the equation for energy loss ("straggling"), so the depth resolution deteriorates with depth.

Rutherford backscattering works best for the analysis of heavy atoms in a light matrix, the sensitivity being the better, the larger the difference between the masses. A heavy element can be used as an interface marker (e.g., Xe in silicide formation).

Nuclear-reaction analysis is used for elements with small atomic mass because the cross section of the reactions with light bombarding particles of energy ≈ 2 MeV is too small when $M > 40$ because of Coulomb repulsion. The method is limited in practice to the analysis for H, Li, C, N, O, F, Al, Si, and

P. The depth resolution is of the order of 0.1 μm, but can be improved to 200-500 Å if a resonance reaction can be used (i.e., if the cross section has a narrow peak for a certain energy of the incident particle). An example is the reaction $^{18}O\,(p,\alpha)\ ^{15}N$, which has a resonance for 1.76 MeV protons, with the emission of α particles of 3.896 MeV. The computations can be more difficult than those for Rutherford backscattering because the cross-sections can be anisotropic and very complicated. Both methods have a maximum precision of 10 % , because the stopping power is uncertain to ± 5 % .

b) *Microanalysis on a transverse section.* — The diffusion profile $c(x)$ or element maps $c(y,z)$ can be obtained from a transverse section containing the diffusion direction if a spot-analysis method is available:

– *Electron probe microanalysis* (Castaing's electron microprobe). The electron optics (2 or 3 lenses) focus a beam of electrons, accelerated to ≈ 5 - 40 keV, on the surface. The size of the spot is determined by the aberrations of the last lens, but the lateral resolution is limited by the scattering of electrons in the sample, and the depth resolution by their penetration. The resolution is typically ≈ 1 μm, but it can be as small as 100 Å with a very thin sample (analytical electron microscope, or scanning transmission electron microscope (STEM) ([2])). The characteristic x-rays of the chemical elements present (Moseley's law) are emitted owing to the ionization of deep atomic levels by the incident electrons. The x-rays are analyzed either by wavelength with a crystal or wave-length dispersive spectrometer (WDS), or by energy with a solid state proportional detector (EDX, energy-dispersive x-ray analysis). The intensity of the x-rays gives an absolute mass concentration when compared with standards of pure materials after certain corrections have been made, which have been verified experimentally. Computer programs for this are commercially available.

– ***Ion probe and microprobe*** (SIMS). Secondary ions are emitted owing to bombardment of the sample by a beam of primary ions (A^+, O^+, or Cs^+ of several keV energy). The secondary ions are analyzed by a mass spectrometer (sector or quadrupole). The depth resolution is very high, some tens of Ångströms. The lateral resolution may reach 1 μm, but it depends on the depth resolution and on the sensitivity. The latter depends directly on the quantity of material sputtered. Quantitative analysis poses a number of problems. The minimum detectable concentration may be as low as 1 ppm or even lower.

c) *Diffusion of gas* across a membrane at steady state: permeation, see appendix I.

II.2 DESTRUCTIVE METHODS. — These methods are based on the analysis of successive sections cut from the sample perpendicularly to the diffusion direc-

---

([2]) With these instruments, analysis by electron energy loss spectroscopy (EELS) is also available with a resolution of a few tens of Å.

tion; they are clearly limited to unidirectional diffusion. Two general categories can be distinguished:

– sectioning followed by analysis of the removed material after cutting;

– the two operations, sectioning and analysis, are carried out simultaneously in the same equipment.

a) *Methods of sectioning.*

1. *Mechanical methods*: lathe sectioning and grinding. Lathe sectioning can be carried out on a precision lathe fitted with a chuck that permits orientation of the specimen, so that the cuts are taken exactly parallel to the diffusion front. The turnings or chips are collected for analysis (counting of radioactivity, chemical analysis, etc....). A depth resolution of ten or fewer μm is attainable. A microtome lends itself well to cutting soft materials (alkali or silver halides, molecular crystals, soft metals such as Al or Pb); sections 2 μm or even 1 μm thick can be obtained in favorable cases.

Mechanical abrasion can be carried out on (usually home-made) precision grinders, and sections as thin as a few tenths of a μm can be removed by this method. The abrasive material and its backing are collected for analysis after a section has been ground off.

In the case of unidirectional diffusion, with the diffusant deposited on one face, a small amount of material should be removed from the lateral faces. This eliminates traces of diffusant that may have spread to the lateral sides by surface diffusion, and then have diffused into the volume, where it would be counted in the sections.

2. *Chemical or electrochemical peeling* can yield very regular sections. Certain metals lend themselves well to anodic oxidation. Sections as thin as a few tens of Ångströms can be removed. Continuous methods have been proposed; for example, passing the sample over a strip of blotting paper impregnated with electrolyte, followed by counting the blotting paper. Unfortunately, these techniques are not universal.

3. *Cathodic sputtering* is a simple method applicable to a wide range of materials. The sample is bombarded by argon ions from an ion gun or from a gaseous discharge in a low-pressure argon plasma under an accelerating voltage of several hundred volts to tens of kV. The rare-gas ions bombard the sample and knock out atoms from the surface (Fig. IX.2). The atoms deposit on a catcher foil or on the anode in the case of a gas discharge; a series of catcher foils has to be used, one to catch the material eroded in each section. The method can be applied to non-conducting materials if a radio-frequency field is applied (RF sputtering), or by use of an ion gun and electron flooding. The rate of erosion can be regulated by changing the voltage or current on an ion gun, or the pressure in a glow discharge ; rates range from a few Å/s to 1000 Å/min. The sputtering process has to be uniform. Any perturbation leads to a broadening of the profile; this can be measured on a standard sample that has not been diffused, which has a step-function profile. The minimum

broadening is not less than 50 Å in the best case; this leads to a pseudo-profile (Fig. IX.3) that corresponds to $2Dt = (50\ \text{Å})^2$ or $Dt \approx 10^{-13}$ cm^2. For an anneal of 10^7 s, about 3 months, this corresponds to $D \approx 10^{-20}$ cm^2 s^{-1}, which is thus the smallest value of D measurable by this technique. A more common limit is 10^{-18} cm^2 s^{-1}.

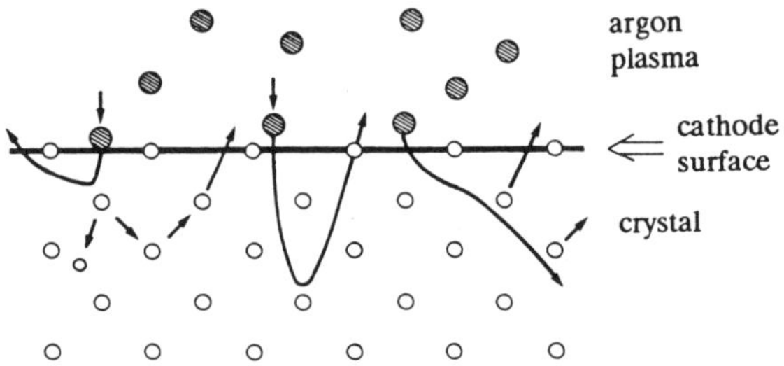

Fig. IX.2. — The mechanism of cathodic sputtering.

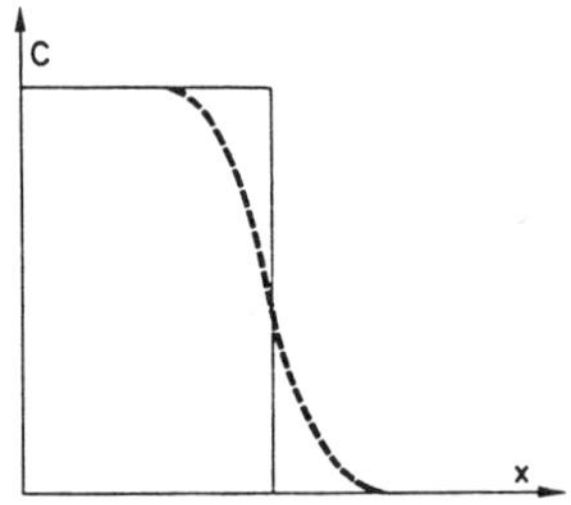

Fig. IX.3. — Broadening of a step function.

We list some reasons for the broadening of the profiles:

– instrumental effects (redeposition of the material deposited on the anode, sputtering from the "walls" of the "crater" excavated by the sputtering,...),

– the initial surface roughness,

– surface irregularities induced by the sputtering process itself, the elementary process being random and proceeding atom by atom (Fig. IX.4),

– preferential sputtering, in a given crystal direction or due to structural heterogeneities,

– direct effects of radiation (effect of the primary knock-on atoms, effects of displacement cascades, i.e., ion-beam mixing,...),

– indirect effects of radiation: diffusion accelerated by the point defects introduced by the bombardment (radiation-enhanced diffusion), or by the heating or by the electric field induced by the bombardment,

– the information depth of the analysis method when this is carried out *in situ*. The analysis can, in effect, be carried out on the removed material after cutting, or in the course of the depth profiling of the continuously newly uncovered surface of the sample.

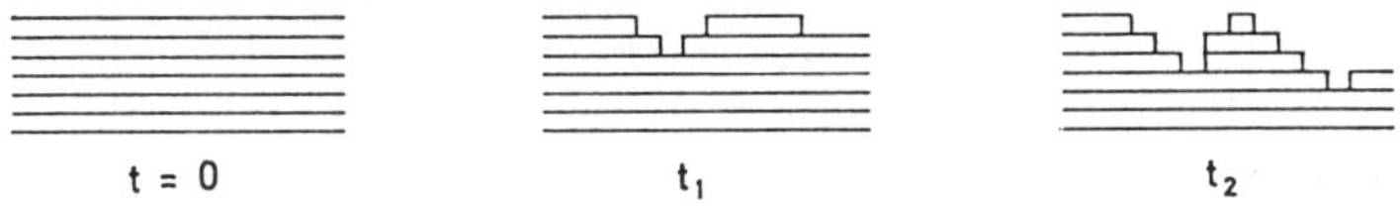

Fig. IX.4. – Progressive erosion by sputtering at 0 K. (At higher temperatures, a stepped surface would no longer be stable, but would be smoothed out by surface diffusion).

Some other difficulties are:

– change of the surface composition of alloys or compounds owing to the different sputtering yields of elements with different atomic numbers or surface binding energies,

– change of the sputtering rate with composition (especially bad in the case of a polyphase diffusion zone),

– limits to the depth attainable by sputtering (a few μm).

The method is made attractive by its universal character and the very clean conditions during sputtering. The planarity of the surface must be checked after sputtering, either with an interference microscope, or with a surface profilometer.

b) *Techniques of analysis.* — The sectioning (typically 15-20 sections) is followed or accompanied by a quantitative analysis:

– *Chemical analysis*, (classical wet analysis, measurement of the lattice parameter, etc...).

– *Assay of radioactivity* i.e., of the activity of the radiotracer (diffusing element) in the sections, with the appropriate counting techniques (detectors and spectrometers). Carrier-free tracers of high chemical purity are now commercially available for many elements.

Instead of diffusing a radioactive tracer, the diffusing impurity can sometimes be activated (e.g., solute diffusion in silicon).

The radiotracers are β or γ emitters; α emitters are used only in special cases. The β spectra are continuous below the energy corresponding to the nuclear transition; there is no reason that their absorption should follow an exponential law. The γ-ray spectra have lines of definite energies. The γ-rays are more penetrating than the β' s.

The radioactivity is measured by counting pulses with different types of detectors: gas proportional counters, solid or liquid scintillators, solid-state detectors. Since the amplitude of the signal depends on the energy of the incident particle, *energy discrimination* improves the signal to noise ratio, and eliminates some spurious counts. Anti-coincidence counters allow the reduction of the background to a few counts/min. Automatic sample changers are useful for counting the large number of samples needed.

In the case of short-lived radioisotopes, corrections have to be made for the decrease of the activity during counting.

This method also allows the study of the *isotope effect*, if the radiations from two isotopes of the same chemical element can be separated conveniently (by their energies or their half-lives).

– *In situ analysis*, during sputtering. This analysis can be carried out by Auger electron spectroscopy or optical spectroscopy in the case of chemical diffusion, or by mass spectrometry with non-radioactive isotopes ([3]). A glow discharge allows very rapid analyses ($\approx$ 1 ms per element for a 0.1 - 0.5 Å section) because of the very strong luminous intensity. It can also be combined with mass spectrometry, but that is most often used following bombardment by a primary ion beam, as in the SIMS (see above). Certain models of this apparatus can also be used to image the sputtered surface and thus check the homogeneity of the diffusion.

If the diffusional penetration is too deep for sputtering, one can:

– alternate mechanical removal with ion bombardement,

– use the ion probe with a well focussed beam along a taper section (see below, Fig. IX.6).

– *Residual activity.* Instead of analyzing the material in the removed sections, the activity remaining in the sample after each section is cut can be measured. This method, associated with the names of Gruzin and Seibel, is applicable if the radiation being detected is absorbed exponentially, so that a linear absorption coefficient, μ, can be defined. If A_n is the residual activity after the nth section (depth x_n), it can be shown that:

$$\mu A_n - \frac{\mathrm{d}A_n}{\mathrm{d}x_n} = kc\,(x_n)\,, \tag{IX.1}$$

whatever the functional form of $c\,(x_n)$. The equation can be simplified in two cases:

– radiation only slightly absorbed:

$$\frac{\mathrm{d}A_n}{\mathrm{d}x_n} = -kc\,(x_n)\,.$$

Thus if $c\,(x)$ is a Gaussian, A_n follows an error function.

([3]) This case is frequently encountered in ceramic materials, where convenient radioisotopes are unavailable for most of the base elements, either because they do not exist (O), the specific activity is too low (Al), or the half-life is too short (Si).

– radiation strongly absorbed:

$$A_n = \frac{k}{\mu}\, c\,(x_n)\,.$$

The residual activity follows the same functional form as c.

These approximations have not always been justified, nor has the exponential absorption been verified. This method is probably not as reliable as sectioning and counting the sections.

– *Evaporation of the tracer.* If there is loss of tracer due to evaporation into a vacuum (without loss of the base material; in that case the solution for a moving surface (Eq. (I.12)) is used), the classic solution for the thin layer geometry (Eq. (I.7))) is no longer valid. The outward flux of the tracer will be proportional to the concentration of the tracer at the surface:

$$D\mathrm{d}c/\mathrm{d}x \;\; (x = 0) = -K\;c\,(0)\,,$$

where K is the rate constant for the evaporation. The solution of the diffusion equation for this boundary condition is (Carslaw and Jaeger, 1959, pp. 70 and 358, Crank, 1975, p. 36):

$$\begin{aligned} c\,(x,t) = Q_0 \Big\{ &(\pi Dt)^{-1/2} \;\exp\;\left(-x^2/4Dt\right) \\ &-h\;\exp\left(h^{\,2}Dt + hx\right)\;\mathrm{erfc}\;\left[x/2\,(Dt)^{1/2} + h\,(Dt)^{1/2}\right]\Big\}\,, \end{aligned}$$

where $h = K/D$.

III. Indirect methods.

It is possible to measure the diffusion coefficient without establishing a concentration profile, but it is then more difficult to insure that the process is indeed Fickian diffusion. Many instances of non-Fickian behavior (e.g., diffusion along short-circuits such as grain boundaries and dislocations) were discovered only from the anomalies on the $c\,(x)$ curves. Therefore, it is necessary to verify in some way (e.g., the $t^{1/2}$ law) that diffusional kinetics are indeed being followed.

III.1 Radiotracers: decrease of surface activity. — The activity of the sample is measured as a function of the diffusion time. For a thin layer and exponential absorption of the radiation, the activity varies as:

$$A/A_0 = \exp\left(\mu^2\;Dt\right)\;\mathrm{erfc}\left(\mu\,\sqrt{Dt}\right), \qquad \text{(IX.2)}$$

where μ is the linear absorption coefficient. This method is not very reliable.

III.2 GAS-SOLID DIFFUSION COUPLES.

– *Thermogravimetry.* An integral measurement is carried out, e.g., the variation of the mass of the sample during the course of diffusion is measured with a thermogravimetric balance. This yields the quantity of diffusing substance $Q(t)$, from which D can be calculated. This type of experiment can go in either direction, weight gain or weight loss.

– *Isotopic exchange*: aliquots of the annealing atmosphere are taken for analysis of the isotopic ratio. The gas volume has to be very large with respect to the amount lost by diffusion into the sample. Isotope exchange with materials other than the sample (the support, the furnace tube) presents a danger of serious inaccuracy.

– Measure of the *total quantity of gas* which diffuses through a wall due to a pressure differential (Fig. IX.5); such a measurement yields a value of the *permeability*, i.e., the product of the solubility of the gas and its diffusivity.

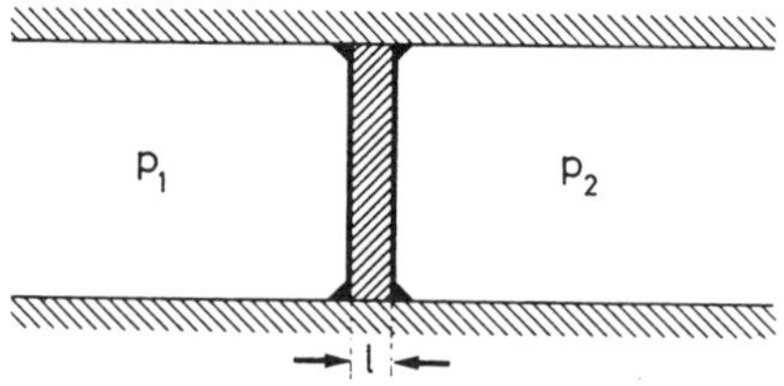

Fig. IX.5. — Diffusion of a gas across a membrane of thickness l.

After a delay time $\tau = l^2/6D$ (l is the thickness of the membrane), steady-state is established, characterized by a constant flux:

$$J_\infty = Dc_s/l, \qquad \text{(IX.3)}$$

where c_s is the solubility of the gas under pressure p.

A precise determination of D is obtained from following the permeation curve $J(t)/J_\infty$ over the entire time of the experiment (see App. I).

Permeation isotherms yield only apparent diffusion coefficients and solubilities because of *trapping* of the gas at certain sites in the material. Then the quantity of trapped gas and an estimate of the trapping energy (or energies) is obtained, but for this, it is better to study degassing kinetics as a function of temperature (see App. I).

III.3 MICROGRAPHIC METHODS. — This method involves making visible a zone of known concentration and following its motion with time. This can sometimes be done with chemical etching, which is especially important in the case of the p-n junction in semiconductors. The concentration can equally well be a phase boundary in a polyphase diffusion couple, or the boundary of a two-phase zone (matrix + precipitate) which moves owing to precipitation or dissolution of precipitates resulting from diffusion of an element.

The displacement thus measured should vary as $t^{1/2}$. The ratio x^2/t is proportional to a diffusion coefficient, and experiments at several temperatures yield an activation energy.

III.4 AUTORADIOGRAPHY uses β or X-ray radiation to record the distribution of the radioisotope on a photographic emulsion. The best resolution is obtained with low energy radiation. Gamma radiation results more in a general fogging of the emulsion than a distinctive pattern.

The best resolution (usually $\approx 5\ \mu$m) is obtained with a fine grained stripping film, or, better, with liquid emulsion. The sample is dipped into the latter and the emulsion is allowed to run off, leaving a layer with one grain thickness (the silver halide grains are $\approx$ 1200 Å). To obtain a ***microautoradiograph***, the emulsion is viewed in an electron microscope in transmission, with a carbon film deposited between the sample and the emulsion to serve both as a surface replica and a support for the emulsion. Resolution of 1 μm can be attained with ^{14}C (155 keV β), 0.1 μm with tritium (18.5 keV β). In the case of very soft radiation, the lateral resolution can be improved by the use of a taper section (Fig. IX.6); the gain is a factor of cosec α.

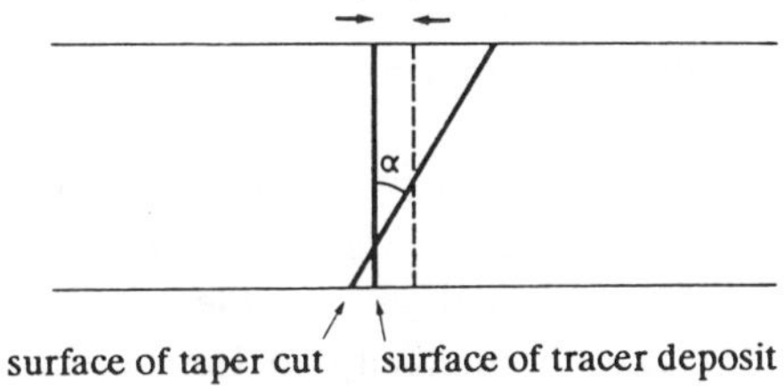

Fig. IX.6. — Principle of the taper section.

III.5 SYNTHETIC MODULATED STRUCTURES (INTERDIFFUSION). — The sample of alternate thin layers A/B/A/B... is prepared by evaporation or sputtering. Typical values are 5-50 Å for the wavelength of the modulation and 100-1000 for the number of layers (Cook and Hilliard, 1969). The amplitude of the concentration modulation after the diffusion anneals is followed via the satellites around the Bragg peaks; their intensity I varies with time according to the law:

$$d\ \mathrm{Log}\,(I/I_0)\,/dt = -2\tilde{D}\cdot B\left(\lambda^2\right)$$

(see Ch. VIII, § V). The method is also applicable to amorphous materials, the satellites around the forward diffraction spot (000) are followed. In effect, the satellites result from the reflections on the pseudo-diffraction planes created by the modulation. The diffusion in multilayer structures can also be followed by other properties, e.g., the electrical resistivity.

III.6 TRANSMISSION ELECTRON MICROSCOPY. — The climb of a dislocation can be governed by diffusion if the density and efficiency of the jogs are sufficient. The climb can be measured by following the *shrinking of prismatic dislocation loops* during the anneal. The volume self-diffusion coefficient is measured if the dislocation loop, located in the center of the thin sample, emits or absorbs point defects coming from or going to the surface. In other configurations, the diffusion along the dislocations may be measured.

Let r be the radius of the loop, b the Burgers vector of the dislocation, L the thickness of the foil, and μ the shear modulus. If $r > L$:

$$\frac{r\,(dr/dt)}{\mathrm{Log}\ (1.7\ r/b)} = -\frac{\mu D b^3\ \mathrm{Log}\,(L/b)}{1.4\ \pi kT}.$$

If $r < L$, the equation is the same but without the term Log (L/b).

III.7 ELECTRICAL RESISTIVITY. — Measurements of the electrical resistivity are used in several types of diffusion experiments:

– The resistivity is a characteristic property of a sample, and in this role can be used to monitor any change in its structure, in particular a diffusion process, e.g., interdiffusion in a multilayer sample.

– In *semiconductors*, the conductivity in the extrinsic domain measures the concentration of electrically active impurities. The concentration profile for such an impurity can be established by measuring the surface resistance of the sample by the four-point method with four contacts in line (current injected at the end points, potential measured between the two interior ones) or in a square, or between two points (point or "spreading" resistance) in which the resistivity is measured in the immediate vicinity of the contact. In the last case, the contact is a very fine point that gives rise to a strong local potential gradient. The measurements are carried out either on a taper section, or after serial sectioning (by anodic oxidation or chemical etching); the difference in conductivity between two measurements equals the conductivity of the removed layer.

– In *non-stoichiometric oxides*, the conductivity is via electronic charge carriers and is proportional to their concentration, which in turn is proportional to the concentration of charged point defects connected with deviation from stoichiometry or with doping. The measurement of the electrical conductivity is an indispensable complement to that of self-diffusion. Its variation after a change of P_{O_2} is used to determine the chemical diffusion coefficient (see Ch. VI, § III).

– *In ionic crystals*, the conductivity results from the migration of point defects, and it measures a diffusivity D_σ via the Einstein relation. The measurements are carried out with an AC bridge ($\nu \approx 10^3$ Hz) to avoid polarization effects. The electrical contact with the electrodes is aided by a layer of Pt, and by light pressure applied with a spring in the cold zone of the furnace. Figure IX.7 shows an apparatus for comparing the conductivities of two samples, e.g., a doped and an undoped one.

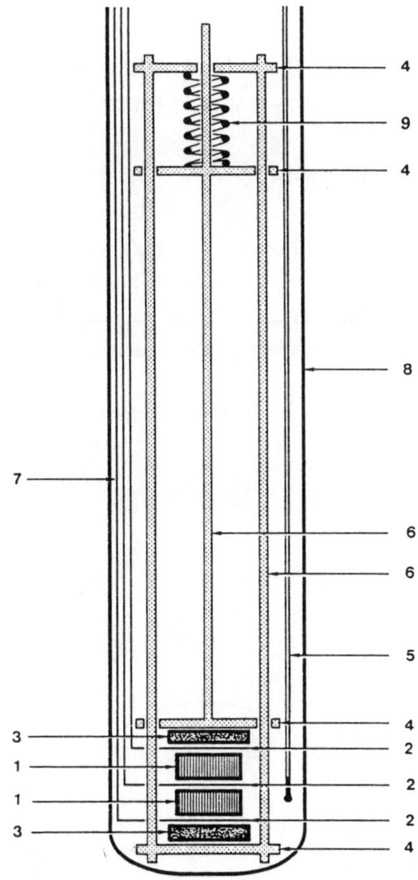

Fig. IX.7. — Apparatus for the measurement of the electrical conductivity of ionic crystals (M. Bénière, 1976). 1, specimens; 2 and 7, Pt electrodes; 4-6, Pt holder for making contact; 5, thermocouple; 8, quartz tube; 9, spring.

With oxides or metals, a 4 point measurement is more common (2 current leads and 2 potential leads). Platinum leads can be used and the contact improved with platinum paint. For semiconductor samples, a preliminary study is necessary to obtain ohmic, non-rectifying contacts.

The use of *complex impedances* $Z = Z' + iZ''$ (representation of the impedance as a function of frequency in the complex Z', Z'' plane) allows the separation of the processes: intragrain conductivity, intergrain conductivity, and electrode processes.

Recall that the complex impedance of a circuit formed of an ohmic resistance R and a capacitance C in parallel is given by:

$$Z = R(1 + iCR\omega)\,,$$

so that the real and imaginary parts of the impedance are written:

$$\text{Re } Z = R/\left(1 + R^2C^2\omega^2\right)$$
$$-\text{Im } Z = \text{R}^2C\omega/\left(1 + R^2C^2\omega^2\right).$$

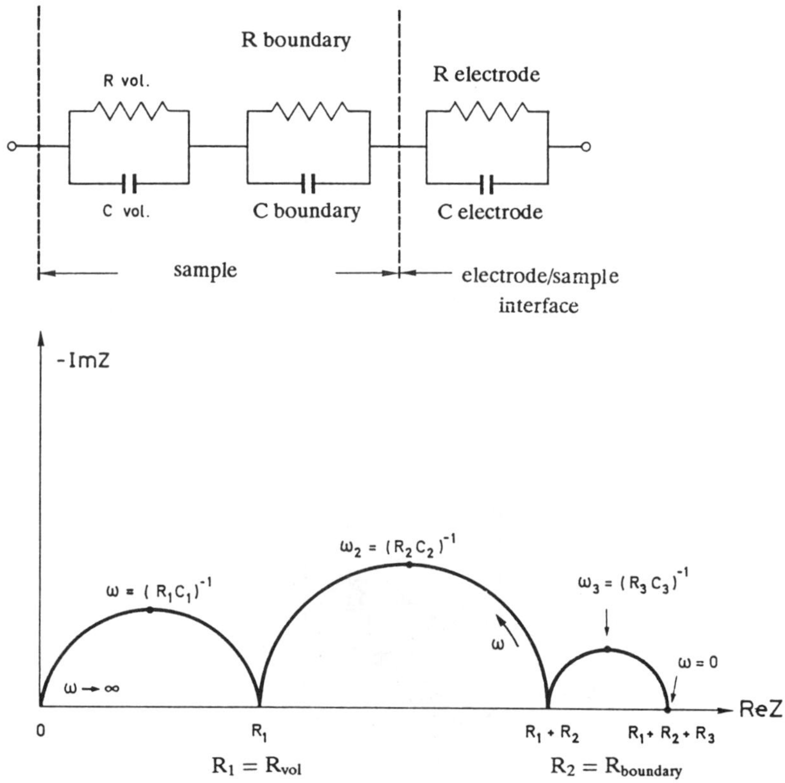

Fig. IX.8. — Measurement of complex impedance and the equivalent circuit (Cole and Cole diagram).

The values of $Z(\omega)$ plotted in the Re Z, $-$Im Z plane thus fall on a semicircle of diameter R, passing through the origin for $\omega \to \infty$, through $(R, 0)$ for $\omega = 0$, through $(R/2, R/2)$ for $RC\omega = 1$. When several circuits of this type are connected in series, the graphic representation is a series of semicircles (Fig. IX.8). An ensemble of three RC circuits in series can represent a measurement cell (sample + electrodes), each circuit representing a conductivity process (volume (1), at the boundaries (2), at the electrodes (3)). Note that the equivalent circuit is oversimplified and that other equivalent circuits could better represent the actual processes. Also, the centers of the arcs are frequently depressed below the Re Z axis.

The experiment requires a variable frequency generator ($\approx$1 Hz to 100 kHz or 1 MHz) and a synchronous detector with a standard resistance box that allows the measurement of the in-phase and quadrature components with applied voltage.

IV. Data processing.

IV.1 CONCENTRATION PROFILES. — When the solution of Fick's equation is a simple analytical expression, it is sometimes useful to change the variable to linearize the solution. In the case of a Gaussian, for example, Log c (or of any quantity proportional to c) is considered to be a function of z^2, or for the error function, $\text{erf}^{-1}(c/c_0)$ or $\text{erfc}^{-1}(c/c_0)$ to be a function of z. The slope of this straight line can then be easily determined by *linear regression*, and D can be calculated from the slope. The estimation of the slope is carried out by the method of least squares. Strictly speaking, it is more rigorous to make a weighted non-linear least-squares fit to the unlinearized functional form, but this is generally more difficult and the results for the two methods often do not differ.

Review of the method of linear least squares. — Consider that a random variable Y is related to the independent variable X by the equation:

$$Y = \alpha + \beta X + \varepsilon(X), \tag{IX.4}$$

where the errors $\varepsilon(X)$ are normally distributed ([4]). We seek the *estimators* A and B (for the parameters α and β, respectively) that define the regression line:

$$y'_i = A + Bx_i, \tag{IX.5}$$

([4]) X is chosen as the independent variable so that all error is assumed to reside in Y, this assumes that $S_y^2 >> B^2\, S_x^2$, where the S are standard deviations.

which is the best estimate for the pairs of data points x_i, y_i. These estimators are obtained by minimizing the sum of the squares of the differences:

$$\$ = \sum_{i=1}^{n} (y_i - y_i')^2 , \qquad n \text{ measurements.}$$

The conditions of minimization are:

$$\frac{\partial \$}{\partial A} = 0, \quad \frac{\partial \$}{\partial B} = 0.$$

This leads to:

$$B = \frac{\sum_i y_i (x_i - \bar{x})}{\sum_i (x_i - \bar{x})^2} \quad \text{or} \quad B = \frac{\sum x_i \, y_i - n \, \bar{x} \cdot \bar{y}}{\sum x_i^2 - n\bar{x}^2} \tag{IX.6}$$

$$A = \bar{y} - B\bar{x},$$

with

$$\begin{aligned} \bar{x} &= (1/n) \sum x_i \\ \bar{y} &= (1/n) \sum y_i . \end{aligned} \tag{IX.7}$$

The confidence interval of the estimators A and B is given by their variances:

$$\left. \begin{aligned} S_B^2 &= \frac{S_{yx}^2}{\sum_i (x_i - \bar{x})^2} \\ S_A^2 &= S_{yx}^2 \frac{\sum x_i^2}{n \sum_i (x_i - \bar{x})^2} \end{aligned} \right\} \tag{IX.8}$$

with

$$S_{yx}^2 = \frac{\sum_i (y_i - A - Bx_i)^2}{n - 2}.$$

Correlation coefficient:

$$r = \frac{\sum x_i\, y_i - n\, \bar{x} \cdot \bar{y}}{\left(\sum x_i^2 - n\bar{x}^2\right)^{1/2} \left(\sum y_i^2 - n\bar{y}^2\right)^{1/2}}.$$

If $r = 0$, the variables are independent.
If $r = 1$, there exists a linear functional relation between x and y.

Notes:

1) Since Y is a random variable due to the effect of $\varepsilon(X)$, any change of a variable bearing on Y changes the variance. The change of the variable would in principle change the "weight" attached to each measurement, and would require calculation of the results with formulas of the type (IX.6-IX.8) but containing the weights w_i. (For details, the reader is referred to a textbook on statistics).

2) In the case that the solution of the diffusion equation cannot be linearized, i.e., when a complicated function or a function that depends on several parameters (e.g., D and an evaporation rate or a rate constant and/or a surface concentration,...) must be fitted, the method of least squares is still applicable if the function can be differentiated with respect to the parameters. Non-linear least-squares programs are available for microcomputers. A fitting by eye is often useful for obtaining initial estimates of the parameters; such estimates are required as input by most such programs.

3) The experimental measurements should be carried out over a wide range of concentration and penetration. Typically, the penetration should be several times $(Dt)^{1/2}$ (see the remark at the bottom of Tab. I). The variation in concentration should extend over at least two decades for the gaussian solution.

4) For short diffusion annealing times, a correction to the diffusion time is necessary because diffusion during the heating up and cooling down of the sample is not negligible. An effective time at the diffusion annealing temperature T_0 is introduced, defined by the equation:

$$D(T_0)\; t_{\text{eff}} = \int_{t=0}^{\text{end of anneal}} D\;[T(t)]\; dt,$$

and

$$t_{\text{eff}} = \Delta t\,(\text{heating}) + t\,(\text{holding at } T_0) + \Delta t\;\;(\text{cooling}),$$

with

$$\Delta t\,(\text{heating}) = \int_{T_a}^{T_0} \frac{D(T)}{D(T_0)} \frac{\mathrm{d}T}{\mathrm{d}T/\mathrm{d}t} = \exp\left(\frac{Q}{RT_0}\right) \int_{T_a}^{T_0} \exp\left(-\frac{Q}{RT}\right) \frac{\mathrm{d}T}{\mathrm{d}T/\mathrm{d}t}, \tag{IX.9}$$

and a similar equation for Δt (cooling). Here T_a is the ambient temperature, and $\mathrm{d}T/\mathrm{d}t$ the heating or cooling rate.

To use this equation, a value of the activation energy must be known; Q obtained from values of D, calculated without a time correction, is used as a first approximation. The calculation is easily done graphically on a graph of $\exp\left[-\frac{Q}{R}\left(\frac{1}{T_0}-\frac{1}{T}\right)\right]$ *vs.* t. The heating and cooling curves can be determined with a thermocouple embedded in a dummy sample.

In the best cases, the precision of D is as good as 1 % , and the reproducibility of D between different laboratories can be as good as a few percent (see Fig. IV.14). Such reproducibility is obtained only for well-specified materials.

IV.2 Variation of D with temperature. — It is useful to recall that the measurement of the temperature of a diffusion sample is not always easy, especially in the presence of temperature gradients.

Starting with a series of determinations of D_i at temperatures T_i, D_0 and Q are calculated by the method of least squares applied to the linearized equation:

$$\mathrm{Log}_{10}\ D = \mathrm{Log}_{10}\ D_0 - \frac{Q}{2.303\ R}\ \frac{1}{T},$$

with

$\log_{10} D \to y$ or $\log_{10}(D \cdot 10^p) \to y$
$1/T \to x$ or $10^4/T \to x$
$A \to \log D_0$
$B \to Q/(2.303\ R)$
$\Delta B \to \Delta Q = \pm 2.3\ R\,\Delta B$

and ΔB is evaluated from the standard deviation S_B.

When D_0 and Q are functions of temperature, the result is not unique, as it depends on the functional form of $D(T)$ chosen for the fitting of the experimental points. In such cases, non-linear regression methods must be used.

Any error in Q involves an error in D_0. In effect, for a given D, $\log D_0 - Q/RT$ = constant, from which $\mathrm{d}\log D_0/\mathrm{d}Q = 1/RT$. The values of $\log D_0$ and Q are thus correlated.

IV.3 The interdiffusion coefficient. — The calculation is carried out by application of the Boltzmann-Matano formula (Ch. I, eq. (I.16)). A graphical method can be used to locate the Matano interface (Eq. (I.17)) and to obtain a tangent and a slope for the application of equation (I.16). The scatter can be smoothed by plotting the data on probability paper (see Exercise 1), and D can be obtained very simply if the plot is a straight line. One can equally well fit a polynomial of high enough degree to the experimental points or carry out a *spline interpolation* (a least square spline fitting allows the filtering out of the noise of the data, see Kapoor and Eagar, 1990); the numerator and denominator of equation (I.16) can then be calculated analytically. All these computations can be carried out easily on a microcomputer, assisted by a digitizing table.

With the use of modern analytical equipment, especially the electron microprobe, a very large number of points can be obtained on a concentration profile in an interdiffusion couple. This causes no difficulty in the application of the Boltzmann-Matano formula for calculating the interdiffusion coefficient, $\tilde{D}$, as far as the integrals are concerned, but the same is not true for the derivatives. The following procedure is therefore recommended:

First step: smooth the experimental points in order to decrease the erratic variation of the derivatives (Heijwegen and Visser, 1973).

Second step: fit a polynomial to the smoothed points. The choice of the degree of the polynomial is delicate: a value equal to 1/4 of the number of experimental points can be taken.

Third step: analytical calculation of the integrals and derivatives in the Boltzmann-Matano formula.

Another procedure is possible:

First step: as above.

Second step: calculate the integrals by the trapezoidal rule from the smoothed points. Calculate the derivatives from the slope of the straight line fitted by least squares to p smoothed points ($p = 4$, for example) surrounding the concentration for which the diffusion coefficient is being calculated; the abscissa is determined by linear interpolation between two smoothed points (Aubin *et al.*, 1985).

When the molar volume is variable, the formula of Sauer and Freise and den Broeder is useful (see Ch. VI, App. I). For this calculation, it is necessary:

1) to express the concentrations in mole fractions;

2) to calculate the molar volume as a function of composition from a knowledge of the lattice parameter as a function of composition.

PART 2. — METHODS BASED ON THE MEASUREMENT OF JUMP FREQUENCIES

In this part, we touch on methods which do not bring into play a flux of particles and thus are not based on Fick's laws. They are, on the contrary, sensitive to small displacements of the atoms (of the order of an interatomic spacing), i.e., sensitive to elementary jumps, and thus allow measurement of the jump frequency. These movements can be induced by an external stimulus, in the case of relaxation phenomena, or be completely random in nature, such as in NMR, Mössbauer,...

I. Relaxation induced by an external stimulus.

I.1 MECHANICAL RELAXATION. — Anelasticity is a rheological phenomenon: under the influence of an applied stress or strain, an instantaneous elastic effect ε_{el} or σ_{el} is observed, followed by an anelastic strain ε_{an} or stress σ_{an}, which varies with the time (Fig. IX.9).

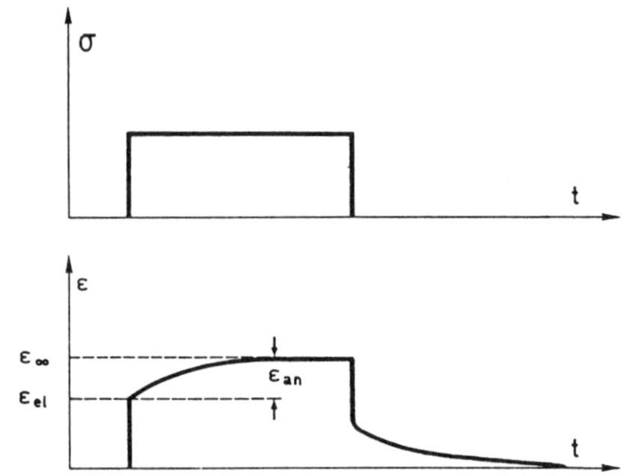

Fig. IX.9. — Anelasticity: anelastic deformation at loading and unloading.

If this variation is a simple exponential function, it is characterized by a unique relaxation time. This phenomenon is called *elastic after-effect*. In the case of a cyclic deformation, the anelasticity translates to a phase difference between the stress and strain (Fig. IX.10). The phase angle ϕ is a Lorentzian function of the frequency ω:

$$\tan\phi = \Delta \cdot \frac{\omega\tau_R}{1+\omega^2\tau_R^2},$$

where $\Delta = \varepsilon_{an}/\varepsilon_{el}$ is the strength of the relaxation and τ_R is the relaxation time characteristic of the process; tan ϕ passes through a maximum for $\omega\tau_R = 1$. This maximum is called the "*internal friction peak*". It is observed by varying the frequency of the torsion pendulum; more often, advantage is taken of the fact that τ_R is thermally activated, and the temperature is varied at fixed frequency.

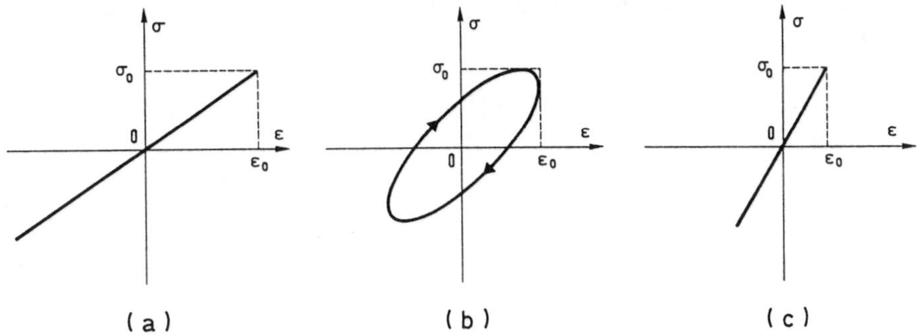

Fig. IX.10. — Principles of internal friction: a) low-frequency, relaxed modulus M; b) intermediate frequency, complex modulus; c) high-frequency, unrelaxed modulus $M' = M + \Delta M$.

If the process that gives rise to the internal friction peak is thermally activated:

$$\tau_R^{-1} = \tau_{R\infty}^{-1} \exp(-\Delta H/kT),$$

from which the internal friction, Q^{-1}, is:

$$Q^{-1} = \Delta/[2\cosh(\mathrm{Log}\,\omega\tau_{R\infty} + \Delta H/kT)].$$

An internal friction peak is observed at $T = T_{max}$ for $\omega\tau = 1$, *i.e.* :

$$\mathrm{Log}(\omega\tau_{R\infty}) = -\Delta H/kT_{max},$$

and

$$Q^{-1}(T) = \Delta/\{2\cosh[(\Delta H/k)(1/T - 1/T_{max})]\},$$

if Δ does not depend on temperature, a condition which is seldom exactly true. This underlines the desirability of *isothermal conditions during measurements involving varying the frequency,* which are necessary to yield a Debye peak.

Experimentally, one measures either the phase angle ϕ or the logarithmic decrement δ of decaying oscillations, or the loss of energy per cycle ΔW. These quantities are related by:

$$\boxed{Q^{-1} = \frac{\delta}{\pi} \approx \frac{1}{2}\frac{\Delta W}{W} \cong \tan\phi} \qquad \text{(IX.10a)}$$

where $W = (1/2)\ \varepsilon_0\sigma_0$ is the elastic energy per cycle. In computer-controlled experiments, the precision is improved by including the anharmonicity of the oscillations:

$$Q^{-1} = \delta/\pi\,(1 - \delta/2\pi + ...)\,.$$

These phenomena are of interest in the framework of this chapter if the anelastic process is caused by atomic jumps. A model is necessary to connect τ_R to the atomic jump frequency and to evaluate the strength Δ of the relaxation. The process of establishing short-range order under the action of a stress gives rise to anelasticity and internal friction.

The intensity of the relaxation depends on the structure of the entity responsible for the process: elastic dipole, local (short-range) order,... The relaxation time, on the other hand, is related to the mobility of this entity. For mechanical relaxation to be possible, this entity must be able to assume several distinct, but equivalent orientations; the application of a stress leads to "removing the degeneracy" and makes the entity pass from one orientation to another by means of atomic jumps. The jump can be of the same type as the one that comes into play in diffusion over a long distance (e.g., the Snoek peak, see below), or it can correspond to a purely local process (e.g., a solute in an off-center interstitial position, able to jump between several of these positions around an interstitial site (in the same "cage" formed by the neighboring solvent atoms)). In the second case the jump frequencies are much higher than those in the first.

These remarks apply, *mutatis mutandis*, to the other relaxation processes described below.

1. *Snoek effect.* — The Snoek effect is due to interstitial atoms in a BCC solution. The octahedral sites of the BCC structure are known not to have cubic symmetry; they are octahedra flattened in the <100> directions. Thus there exist three types of sites : [100], [010], and [001]. The interstitial atom gives rise to an anisotropic distortion of the site, thus creating a small elastic dipole elongated along one of the three <100> directions. Under the action of a mechanical stress, the interstitial atoms will jump into the nearest position such that the direction of the dipole is as closely as possible parallel to an applied tensile stress, or normal to an applied compressive stress. This results in an excess deformation in this direction, regulated by the kinetics of the redistribution of the interstitial atoms. It can be shown (App. II) that the number of atoms on one given type of site varies as $\exp(-6\Gamma_s t) = \exp(-t/\tau_R)$, where Γ_s is the frequency of jumps toward one neighboring site. In a BCC interstitial solid solution, each octahedral site has 4 nearest neighbors (Ch. II, Fig. II.7), so $\Gamma = 4\Gamma_s$, and following equation (II.24):

$$\boxed{D = a^2/36\ \tau_R}\,. \qquad \text{(IX.11)}$$

At resonance for a frequency of 1 Hz, $\omega\tau_R = 1$, and $\tau_R = 1/2\pi$, from which

$D = 4 \times 10^{-16}$ cm^{2}s^{-1} for $a = 5$ Å.

With the elastic after-effect, much longer relaxation times can be attained. With $\tau_R = 1$ h, for example, $D \approx 10^{-20}$ cm^{2}s^{-1}. This is thus an excellent method for measurements at low temperatures ([5]). Combining these results with other methods used at higher temperatures, one can confirm or reject the Arrhenius relationship for a given system (e.g., N in α-Fe, Fig. V.8).

The Snoek effect has been applied to the study of the diffusion of C, N, and O in BCC metals. It is also applicable to HCP metals, since the non-ideality of the c/a ratio gives rise to an asymmetry of the octahedral sites. Very pure and very dilute alloys must be used in these experiments in order to avoid solute-solute interactions (ordering or clustering) which would broaden the peak.

The Snoek effect also furnishes a means for studying interstitials created by irradiation and yields information especially on the configuration of dissociated interstitials (the jump gives rise to, or does not give rise to, a relaxation, according to the configuration), their mobility (local or long distance), their reaction with lattice defects (annihilation, capture, trapping, formation of defect aggregates, precipitation,...).

2. *Zener effect* (or directional ordering). — Analogous effects are produced any time that a dipole can reorient itself by atomic jumps under the effect of an imposed mechanical stress. Such dipoles are pairs of atoms (interstitial or substitutional, associated point defects, etc....) that possess lower symmetry than the lattice. Zener had attributed an internal friction peak in binary substitutional solid solutions to the reorientation of solute atom pairs, but due to the lack of a precise model, such as is available for the Snoek effect, only the activation energy could be determined. In their model, Le Claire and Lomer interpreted the relaxation on the basis of a process of changing short-range order under the influence of a stress. In reality, the phenomenon under study depends on at least two jump frequencies of the solute atoms, and it is difficult to relate the effect quantitatively to the diffusion coefficient of the solute. In a pair model valid for low solute concentrations, the activation energy is more characteristic of the rotation of the dipole than of long-range migration.

The Zener effect is most often used to study the evolution of the vacancy concentrations in solid solutions, or of self-interstitials after irradiation. In ionic crystals, the pairs of point defects constitute elastic dipoles, which give rise to a Zener effect, as well as electric dipoles (see below, § I.3).

3. *The Gorski effect.* — This effect involves the transport of solute B atoms, which have a strong size effect in the A matrix, under the effect of a gradient of deformation produced, for example, by bending. The transport produces a relaxation of the elastic stresses, by the migration of centers of dilatation from the regions in compression to those in tension. A relaxation time τ_R is

([5]) The peak height is proportional to the concentration of interstitials.

measured as in the elastic after-effect, and D is obtained from it according to:

$$D_{\mathrm{B}} = \frac{1}{\Phi \tau_{\mathrm{R}}} \left(\frac{d}{\pi}\right)^2 ,$$

where d is the dimension of the zone over which diffusion takes place (e.g., the thickness of a bar in bending) and Φ is the thermodynamic factor. This method is useful for chemical species with very high diffusivities. Wire or thin-foil samples are used in a pendulum so as to produce bending. It is chiefly used to study the diffusion of hydrogen in metals; its principal advantage is the absence of surface effects.

I.2 MAGNETIC RELAXATION. — Owing to the energy of magnetic anisotropy, a coupling between certain atoms on particular sites or certain pairs of atoms and a magnetic field exists in a ferromagnetic material (Néel). The magnetic field plays the role of the stress in anelastic processes.

1. *Magnetic after-effect.* — This is the analog of the elastic after-effect. Because one <100> direction is that of easy magnetization in α-iron, the octahedral interstitial sites are not equivalent. The three types of <100> sites are differentiated, and a redistribution of interstitial atoms among these sites takes place when the magnetization direction is changed ([6]).

Experimentally, the permeability is measured in a weak alternating magnetic field. Beginning with a structure in which the interstitials are uniformly distributed (just after demagnetization), the redistribution of the interstitial atoms in the favored sites stabilizes the magnetic structure and reduces the amplitude of displacement of the Bloch walls, which leads to a diminution of the induction and the permeability. On the one hand, in the magnetic domains, all the magnetic moments are parallel to one direction of easy magnetization; in the Bloch walls, on the other hand, the direction of the moments undergoes a transition between the directions corresponding to the two contiguous domains. Under the influence of a magnetic field, the walls are displaced, the direction of easy magnetization changes locally, and the interstitials must "follow", redistributing themselves according to the types of sites. This leads to a variation of the permeability:

$$\frac{1}{\mu} = \frac{1}{\mu_0} + \frac{1}{\mu_1} \left[1 - \exp\left(-t/\tau\right)\right] , \qquad \text{(IX.12)}$$

where t is the time elapsed since the demagnetization. The method is evidently limited to ferromagnetic materials, but, as in the case of the Snoek peak, it extends to all cases in which an elastic dipole can couple with the magnetization.

([6]) In α-iron, the C atoms are located preferentially on sites with axes of symmetry normal to the direction of easy magnetization. The magnetic anisotropy energy is not due to magnetostriction, but is of electronic origin, due to spin-orbit coupling of the 3d electrons.

A ring-shaped specimen or a wire must be used.

2. *Induced magnetic anisotropy.* — If certain atoms (or defects) are coupled to the magnetic field, their rearrangement in the magnetic field ("directional order") produces a magnetic anisotropy. After changing the direction of the applied field, the change of this anisotropy is followed with the aid of a disk shaped specimen hanging from a torsion wire. The magnetic field is applied in the plane of the disk.

The method is applicable to the directional order of solute interstitials in an AB alloy (the interstitial site is not symmetric because of the presence of B atoms) and to the directional order of AB pairs in concentrated binary alloys.

The kinetics of the establishment of order after an appropriate thermomagnetic treatment can yield the diffusion coefficient or at least the activation energy for diffusion. For binary alloys, this effect is the magnetic analog of the Zener effect.

This method is clearly limited to ferromagnetic alloys and to temperatures below the Curie point.

Remarks:

1) The elastic or magnetic relaxation may be due to the defect itself (as it is in the Snoek model) or it may be induced by defect migration (as in the Le Claire-Lomer model of the Zener effect, see § II.1.2 above). In the latter case, it is difficult to connect the measured relaxation time (equation of the type (IX.11)) to a well-defined diffusion coefficient; in an AB alloy, both atomic species can move. The relaxation time can be expected to be inversely proportional to an effective diffusion coefficient:

$$\tau_R = \beta a^2 / D_{\text{eff}}.$$

An atomic model is necessary for determining β and to express D_{eff} in terms of the self-diffusion coefficients of the constituents. The only exact model is the one for the Snoek effect for interstitial solutes in sites of non-cubic symmetry. Different formulas that weight D_{A^*} and D_{B^*} have been proposed for the Zener effect or induced anisotropy in substitutional AB alloys (see Caplain and Chambron, 1977); one must, however, be careful as several types of jumps, i.e., several jump frequencies, can be involved in the reorientation process (see the Le Claire-Lomer model above).

2) This method works very well for paint defects in alloys (see e.g. Philibert, 1986). If proper thermal or thermomagnetic treatments are followed, the temperature dependence of τ_R can yield:

– the sum of $\Delta H_v^f + \Delta H_v^m$ for the condition of thermal equilibrium,

– ΔH_v^m for isochronal annealing at different temperatures following the same annealing and quenching treatment,

– ΔH_v^f for isothermal anneals after quenches from different annealing temperatures.

Let τ_R be the relaxation time measured immediately after quenching and let $N_v\,(0)$ be the mole fraction of vacancies introduced by the quench. Then at the temperature T_r, the relaxation time just after quenching is given by:

$$\tau_R\,(0)^{-1} \propto D_{\text{eff}} \propto N_v\,(0)\,\omega_v \propto N_v\,(0)\ \exp\left(-\frac{\Delta H_v^m}{kT_r}\right).$$

If T_r is held fixed, then after quenches from different temperatures T_q:

$$\tau_R\,(0)^{-1} \propto N_v\,(0)\ \exp\left(-\frac{\Delta H_v^f}{kT_q}\right).$$

I.3 DIELECTRIC RELAXATION. — Pairs of point defects (first or second nearest neighbors) in ionic crystals can constitute small ***electric dipoles*** (e.g., cation vacancy + aliovalent cation), and can give rise, under the effect of an electric field, to phenomena analogous to those described in the preceding section à propos of elastic dipoles. The study of these defects is interesting for the information it furnishes on the low-temperature behavior of the vacancies and impurities; the results are thus complementary to the diffusion measurements carried out at higher temperatures where the pairs are more often dissociated.

Dielectric loss. – The reorientation of an electric dipole by a diffusional jump is not in phase with the electric field that excites it. There is thus a phase difference ϕ between the electric field and the polarization (or the electric displacement $\mathbf{D} = \varepsilon\ \mathbf{E}$). The formalism which describes this phenomenon – complex permittivity $\varepsilon = \varepsilon' + i\varepsilon''$, tan $\phi = \varepsilon'/\varepsilon''$ – is analogous to the formalism from rheology that was used to describe anelasticity. Thus a peak in the dielectric loss is observed as a maximum in the frequency dependence of tan ϕ for a given temperature. This peak is the analog of the Zener peak in internal friction.

Induced thermocurrent (7). – The electric dipoles are first oriented by the application of an electric field at a sufficiently high temperature, and then fixed in that orientation by rapid cooling. In the course of reheating, the dipoles progressively randomize their orientations, which leads to a variation of the polarization, **P**, and of the displacement, **D**. Following one of Maxwell's equations, this gives rise to a depolarization current:

$$\mathbf{j} = -\frac{\varepsilon\varepsilon_0}{4\ \pi}\,\frac{\mathrm{d}\mathbf{D}}{\mathrm{d}t} = -\varepsilon\varepsilon_0\,\frac{\mathrm{d}\mathbf{P}}{\mathrm{d}t}. \qquad \text{(IX.13)}$$

With an increase in temperature, this current passes through one or more maxima, corresponding to a temperature (or temperatures) where the process of dipole reorientation is active. Assume that in effect one component of the

(7) This method is most often denoted by the acronyms I.T.C. (Induced Thermo-Current) or T.S.D.C. (Thermally Stimulated Depolarization Current).

polarization follows first order kinetics:

$$\frac{dP}{dt} = -\frac{P}{\tau}, \tag{IX.14}$$

from which:

$$P = P_0 \exp(-t/\tau),$$

with a thermally activated relaxation time:

$$\tau = \tau_0 \exp(Q/RT). \tag{IX.15}$$

In the course of reheating, the rate of depolarization reaches a significant value as soon as τ is small enough. Since both τ and P decrease with rising temperature, a maximum in dP/dt (Eq. (IX.14)) is expected for a certain temperature T_m, which is determined by the condition:

$$(dj/dT)_{T_m} = 0.$$

Consider a linear rise in temperature: $T = T_0 + bt$. Then:

$$\frac{dP}{dT} = -\frac{1}{b}\frac{P}{\tau},$$

and by integration:

$$P = P_0 \exp\left[-\frac{1}{b}\int_{T_0}^{T}\frac{dT'}{\tau}\right],$$

from which:

$$j = \varepsilon\varepsilon_0\,\frac{P}{\tau} = \varepsilon\varepsilon_0\,\frac{P_0}{\tau}\,\exp\left[-\frac{1}{b}\int_{T_0}^{T}\frac{dT'}{\tau}\right]. \tag{IX.16}$$

The condition of an extremum yields, with the use of equation (IX.16):

$$\frac{Q}{RT_m^2} = \frac{1}{b\tau_0}\exp\left(-\frac{Q}{RT_m}\right). \tag{IX.17}$$

As a first approximation, one can take $\tau_0 = v_D^{-1}$ (the Debye frequency), and evaluate Q. The experiment can also be carried out at several different rates of temperature rise. Note that at low temperature, where $T << T_m$, the exponential term in equation (IX.16) ≈ 1, and:

$$\log j = \text{Const.} - Q/RT, \tag{IX.18}$$

allowing a direct determination of Q. Further, the area under the current peak equals P_0, which allows the calculation of the average dipole moment, and a determination whether the charges making up the dipoles are at first or second nearest neighbor distances. A precise atomic model is needed to connect the relaxation time and the activation energies to the frequencies of well-defined atomic jumps.

The reorientation of a dipole can take place by several different mechanisms, eventually giving rise to peaks in the depolarization current at different temperatures; a Schottky pair can reorient itself by the jump of either anion or cation, a Frenkel pair by the migration of the interstitial or of an ion on a lattice site,...

II. Nuclear methods.

II.1 Incoherent neutron scattering. — The interaction of a neutron with a scattering center (nucleus) is measured by a scattering amplitude b (in barns), which depends on the chemical species, the isotope, and its nuclear spin I ([8]). Only a monoisotopic assemblage of atoms with spin $I = 0$ would scatter neutrons in a totally coherent manner. The incoherent scattering of neutrons is connected to isotopic disorder and to nuclear spin disorder.

Recall a classical result of X-ray diffraction. The scattered intensity at the point $\mathbf{k}$ in reciprocal space is the Fourier transform of the Patterson function:

$$I(\mathbf{k}) \propto \int G(\mathbf{r}) \exp i\mathbf{k} \cdot \mathbf{r} \, d\mathbf{r}, \tag{IX.19}$$

with:

$$G(\mathbf{r}) = \int \rho(\mathbf{u}) \cdot \rho(\mathbf{u} + \mathbf{r}) \, d\mathbf{u}, \tag{IX.20}$$

where $\rho(\mathbf{r})$ is the electron density. If the medium consists of N identical atoms or basis, $I(\mathbf{k})$ is proportional to NF^2, where F is the atomic scattering factor, or the structure factor of the unit cell, and to $G(\mathbf{r})$, derived from the atomic density $\rho(\mathbf{r})$ according to equation (IX.20). In this case, $G(\mathbf{r})$ is the correlation function defined in chapter II, § VII; it is the probability of finding two atoms separated by the vector $\mathbf{r}$.

In the same way, the elastic coherent scattering for neutrons is given by:

$$I(\mathbf{k}) \propto \langle b \rangle^2 \; S(\mathbf{k}),$$

where $S(\mathbf{k}) \neq 0$ in a crystal only for discrete values of $\mathbf{k}$. For neutrons, the

([8]) For X-rays, the scattering amplitude from each atom is measured by the scattering factor $f(\mathbf{k})$.

incoherent scattering is added to this: $N\left[\langle b^2\rangle - \langle b\rangle^2\right]$, since, in general, $\langle b\rangle^2 \neq \langle b^2\rangle$. The averages are taken over all the nuclei.

When the scattering atoms are in motion, the correlation function $G(\mathbf{r},t)$ (see Ch. II, § VII.1, and Fig. II.10) has to be used. An inelastic scattering experiment yields a double Fourier transform of $G(\mathbf{r},t)$.

– *Coherent scattering.*

$$I_{\text{coh}}(\mathbf{k},\ \omega) \ \propto\ \langle b\rangle^2\ S(\mathbf{k},\omega) \tag{IX.21}$$

$$S(\mathbf{k},\omega) = \int\int G(\mathbf{r},t)\ \mathrm{e}^{i\mathbf{k}\cdot\mathbf{r}}\ \mathrm{e}^{-i\omega t}\ \mathrm{d}\mathbf{r}\ \mathrm{d}t, \tag{IX.22}$$

with the equations of conservation of momentum and energy (Fig. IX.11a):

$$\begin{cases} \mathbf{k} = \mathbf{k}_1 - \mathbf{k}_0 \\ \hbar\omega = \dfrac{\hbar^2\ k_1^2}{2\ m} - \dfrac{\hbar^2\ k_0^2}{2\ m}. \end{cases} \tag{IX.23}$$

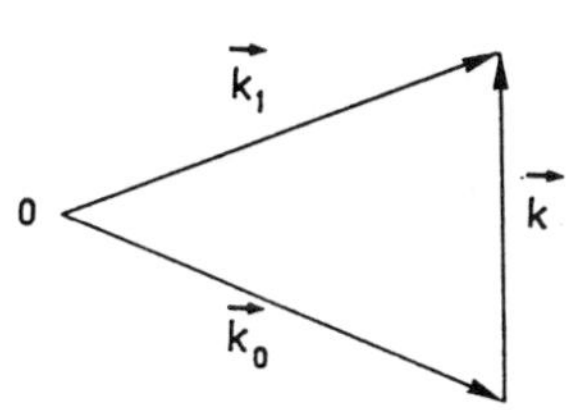

a)

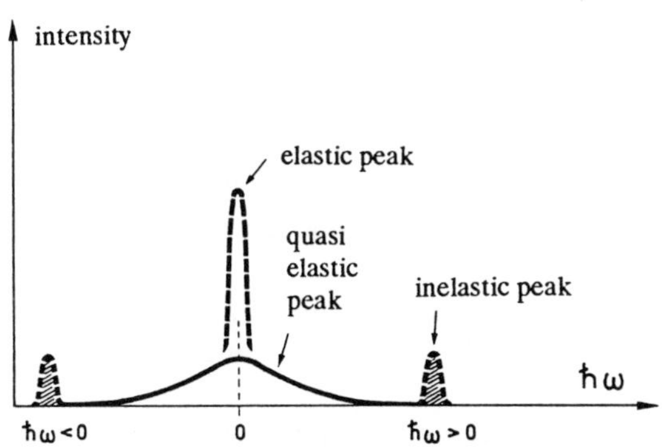

b)

Fig. IX.11. — Neutron scattering: a) scattering of a neutron from wave vector $\mathbf{k}_0$ to wave vector $\mathbf{k}_1$; b) energy spectrum for scattering of neutrons.

– *Incoherent scattering.* — In an assembly of incoherent scatterers, the waves scattered by the same nucleus at ($\mathbf{u}$, $t = 0$) and at ($\mathbf{u} + \mathbf{r}$, t) can interfere. This is incoherent inelastic scattering, which is proportional to the Fourier transform of the "self" part of the correlation function only (the self-correlation function G_s):

$$I_{\mathrm{inc}}(\mathbf{k},\omega) \propto \left[\langle b^2\rangle - \langle b\rangle^2\right] \cdot S_s(\mathbf{k},\ \omega) \tag{IX.24}$$

$$S_s(\mathbf{k},\ \omega) = \int\!\!\int G_s(\mathbf{r},t)\ \mathrm{e}^{i\mathbf{k}.\mathbf{r}}\ \mathrm{e}^{-i\omega t}\ \mathrm{d}\mathbf{r}\ \mathrm{d}t. \tag{IX.25}$$

It is thus possible in principle to determine the diffusion coefficient D from a measurement of $S_s(\mathbf{k},\omega)$ (see Eq. (II.44), Ch. II), which is accessible if the incoherent scattering is strong and the coherent scattering is negligible. Only two elements correspond to these conditions, hydrogen ([9]) and vanadium. Otherwise, different mixtures of isotopes must be used to separate the coherent and incoherent contributions to scattering. However, for monoisotopic nuclei with non-zero spin, incoherent scattering takes place with an inversion of the spin; the separation of S_{coh} and S_{inc} is then possible with a polarized neutron spectrometer. This is the case for H, D, ^{7}Li, As, Na and V.

Assume that due to atomic diffusion, $G_s(\mathbf{r},t)$ takes the form of a Gaussian (see Eq. (II.40), Ch. II):

$$G_s(\mathbf{r},t) = (4\pi Dt)^{-3/2}\ \exp\left[-r^2/4Dt\right]; \tag{IX.26}$$

its space Fourier transform is also a Gaussian:

$$S_s(\mathbf{k},t) = \exp\left(-k^2Dt\right). \tag{IX.27}$$

This is an exponential function of time, the time Fourier transform of which is a Lorentzian:

$$I_{\mathrm{inc}}(\mathbf{k},\omega)\ \propto\ \frac{Dk^2}{\left(Dk^2\right)^2 + \omega^2}. \tag{IX.28}$$

The width of the line $I(\omega)$ measures Dk^2. Since the line is centered at $\omega = 0$, small values of ω are involved, whence the name ***quasi-elastic*** scattering.

In addition to this peak, inelastic peaks are observed for coherent as well as incoherent scattering, due to the absorption or emission of phonons (Fig. IX.11b).

([9]) For hydrogen, $4\pi\langle b\rangle^2 \approx 1.79$ barn, $4\pi\left\langle b^2\right\rangle \approx 81.5$ barns.

In any case, the scattered intensity is attenuated by the Debye-Waller factor $\exp\left(-\frac{1}{3}\langle u^2\rangle \mathbf{k}^2\right)$, which results from thermal vibrations with a mean square amplitude $\langle u^2\rangle$.

The width of the scattering peak is limited by the instrument resolution. The measurement requires a monochromatic beam of neutrons, with the wavelength adjustable in various ways (e.g., by the Doppler effect due to the movement of the monochromator: 2.08 meV $\pm$ 10 μeV, or by the thermal expansion of the monochromating crystal). With a precision of measurement ≈ 1 μeV, atomic motion on the scale 10^{-9} to 10^{-7} s can be measured, i.e., $D \approx 10^{-6} - 10^{-8}$ cm^{2}s^{-1}. The material under study should not absorb neutrons.

Application of the method. — The time scale accessible to these measurements restricts the method to fast diffusers: hydrogen in metals, superionic conductors (α – AgI, β – Al_2O_3, see Ch. V, § VIII), intercalation compounds, liquid crystals,... The measurements are carried out in the absence of a concentration gradient, on homogeneous systems at equilibrium.

The scale of distances of this method ($10^{-10} - 10^{-5}$ cm) permits the observation of microscopic details of the diffusion process, even in the absence of mass transport: the jump length, and, in single crystals, the jump direction, observation of rotational motion or the change of shape of the molecules, spatial - temporal analysis of the fractions of interatomic distance involved in the geometry of the jump (e.g., for a proton, the type of site, the shape of the "thermal cloud" formed by the proton vibrating in its interstitial "cage", the possibility of following the diffusional process even in the presence of other processes (such as trapping...).

Evidently, small or average diffusivities cannot be reached by this method. Only elements with a large enough incoherent scattering cross section can be studied; from this point of view, the best case is hydrogen, but deuterium, as well as Na, Cl, Ti, V, and Ni should also be mentioned.

II.2 Nuclear magnetic resonance. — NMR allows the observation of jumps in the lattice by the majority isotope.

1. *Review of NMR.* — Let $\hbar\mathbf{I}$ be the kinetic moment (spin) of the nuclei under consideration, and $\mu = \gamma\hbar\mathbf{I}$ their magnetic moment (dipolar); γ is the gyromagnetic ratio. In a magnetic field $\mathbf{H}_0$ imposed along the z axis, according to the classical theory, the moments μ precess around $\mathbf{H}_0$ with the Larmor frequency (Fig. IX.12):

$$\omega_0 = \gamma H_0. \tag{IX.29}$$

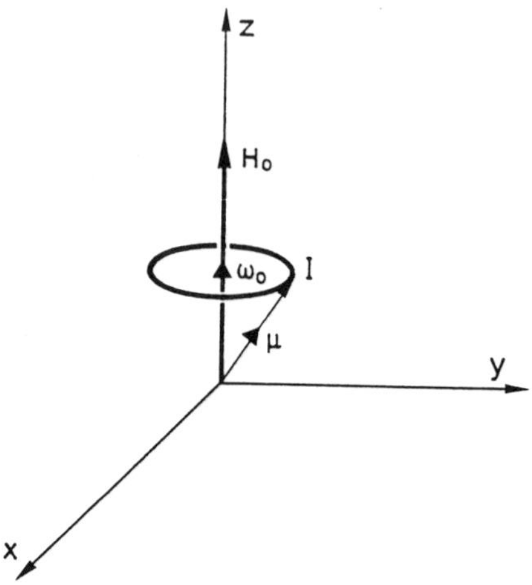

Fig. IX.12. — Larmor precession.

In the quantum theory, the components I_z and μ_z are quantized. The allowed values of I_z are $m_I = -I, -I+1, ..., I-1, I$. The energy of the magnetic dipole is then quantized (Fig. IX.13):

$$U = -\mu \cdot \mathbf{H}_0 = -\gamma\hbar I_z \cdot H_0 = -\gamma\hbar H_0 m_I,$$

or

$$U = -m_I \, \hbar \, \omega_0. \tag{IX.30}$$

If $I = 1/2$, the magnetic energy can take on only two values. The spins are distributed statistically on two levels. Since the difference $\Delta U = \hbar\omega_0$ is very small, the difference, Δn, in the populations of the two levels is very small, typically $\approx 10^{-5}$. It is responsible for the measured magnetization M_z.

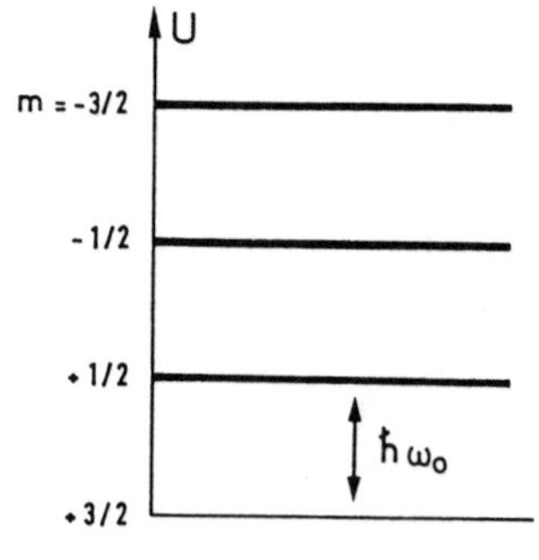

Fig. IX.13. — Principles of NMR.

In an NMR experiment, a radiofrequency field $\mathbf{H}_1$ of frequency ω ($\approx$ MHz) is applied in a direction perpendicular to $\mathbf{H}_0$ (Fig. IX.14). Since this field is equivalent to the superposition of two rotating fields with frequencies $+\omega$ and $-\omega$, there is a resonance when ω equals the Larmor frequency ω_0. This produces a maximum absorption of energy by quantum transitions $\Delta U = \hbar\omega_0$. The resonance is observed by varying ω, or, more conveniently, H_0. In reality, the resonance is not infinitely narrow, and a finite range of frequencies is observed. The form of the resonance curve is often represented by a Gaussian or a Lorentzian. Its width at half-maximum is denoted by $1/T_2$.

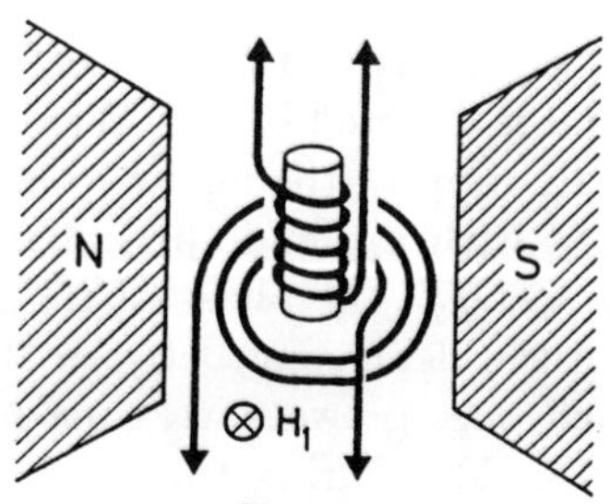

Fig. IX.14. — Schematic of an NMR experiment.

2. *The relaxation processes.* — The mechanism of the resonance tends to equalize the populations of each state characterized by the number m_I (Fig. IX.13). After interruption by the application of H_1, the relaxation process allows the re-establishment of the initial populations:

– *Spin-lattice relaxation* The "longitudinal" relaxation time, T_1, describes the evolution of the magnetization, M_z, toward equilibrium:

$$\mathrm{d}M_z/\mathrm{d}t = (M_0 - M_z)/T_1. \tag{IX.31}$$

For the two-level system, let P be the transition probability from one level to the other. Then, to equalize the populations of the two levels, the transitions take place in both directions:

$$\frac{\mathrm{d}\Delta n}{\mathrm{d}t} = 2\,P(\Delta n_0 - \Delta n),$$

where Δn is the difference in population of the two levels:

$$\Delta n = \Delta n_0\,[1 - \exp(-t/T_1)], \tag{IX.32}$$

with $T_1 = 1/2\ P$.

The return to equilibrium takes place either by absorption or emission of phonons with an exchange of energy $\hbar\omega$, or by a coupling of the spins with free electrons in metals through the process of inelastic scattering. The relaxation time, $(T_1)_e$, which is inversely proportional to the temperature (Korringa's relation), corresponds to the latter process.

– *Dipolar interactions. Spin-spin relaxation.* — Each spin precesses, in fact, in a field $\mathbf{H}_0+\mathbf{H}_l$. $\mathbf{H}_l$ is the local field created by neighboring dipoles, and varies as r^{-3}. Since these are randomly oriented, the local field varies from one point to another, thereby leading to a dispersion of the resonance frequency:

$$\Delta\omega_0 = 1/T_2 \sim \gamma\Delta H_l. \qquad \text{(IX.33)}$$

Typically, $\Delta H_l \approx 2\ G$, $T_2 \approx 100\ \mu$s and $\Delta\omega_0 \approx 10^4$ rad s^{-1}. This is characteristic of a "rigid" lattice and is denoted by T_2''. What happens then if the nuclei are mobile ? If they migrate rapidly, the field H_l sensed by a nucleus fluctuates randomly between $\pm\ \Delta H_l$, with a pseudo-period τ equal to the reciprocal of the jump frequency. If $\tau << T_2''$, the phase of the precessional motion varies in a random manner, and the dephasing between the spins, a function of the local field as seen by each one, grows with time much more slowly than in a fixed local field (Fig. IX.15). During the time τ, the spin is dephased by $\Delta\phi = \pm\gamma\Delta H_l\tau$ with respect to its precession in H_0. Since $\Delta\phi$ takes random values, in analogy with the equations for random walk in chapter II:

$$\langle\phi^2\rangle = n\ \Delta\phi^2 = n\gamma^2\ \Delta H_l^2\ \tau^2. \qquad \text{(IX.34)}$$

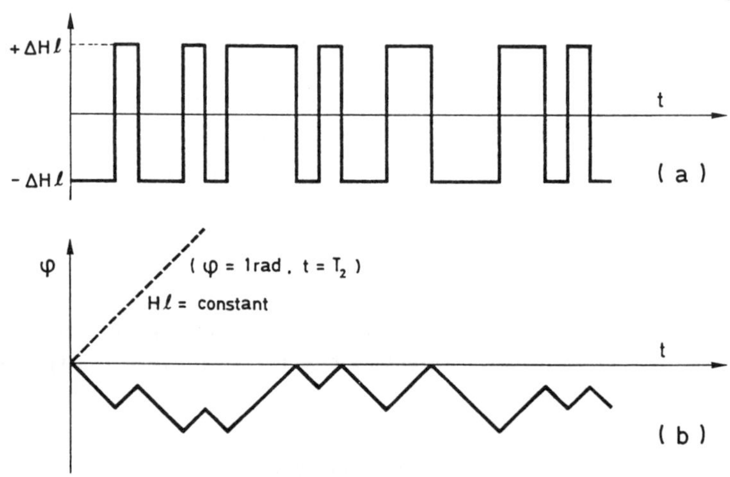

Fig. IX. 15. — The origin of diffusional narrowing: a) temporal fluctuations of the local field; b) variations of the phase change ϕ in a constant local field (dashed straight line) or a fluctuating local field (the jagged full line). The phase change is measured with respect to precession in the fixed field H_0.

To obtain a phase lag of 1 radian, n jumps are required, where:

$$n = 1/\left(\gamma^2 \Delta H_l^2\ \tau^2\right),$$

in a time:

$$T_2' = n\tau = 1/\left(\gamma^2\ \Delta H_l^2\ \tau\right), \qquad \text{(IX.35)}$$

different from the rigid lattice $T_2'' = 1/\Delta\omega_0 = 1/\left(\gamma\ \Delta H_l\right)$.

The width of the resonance becomes:

$$\Delta\omega' = \frac{1}{T_2'} = \left\langle \Delta\omega_0^2 \right\rangle \tau. \qquad \text{(IX.36)}$$

This is the ***diffusional narrowing***, which can be quite large. If $D \approx 10^{-8}$ cm^{2}s^{-1}, $\tau \approx 10^{-5}$ s, and $\Delta\omega' = 10^3$, the variation of $1/T_2'$ with temperature is shown in figure IX.16. Since τ follows an Arrhenius behavior, log $1/T_2'$ increases linearly with reciprocal temperature, until it reaches the saturation value $1/T_2' = \Delta\omega_0$, the value for the rigid lattice (jump frequency too slow to be observed).

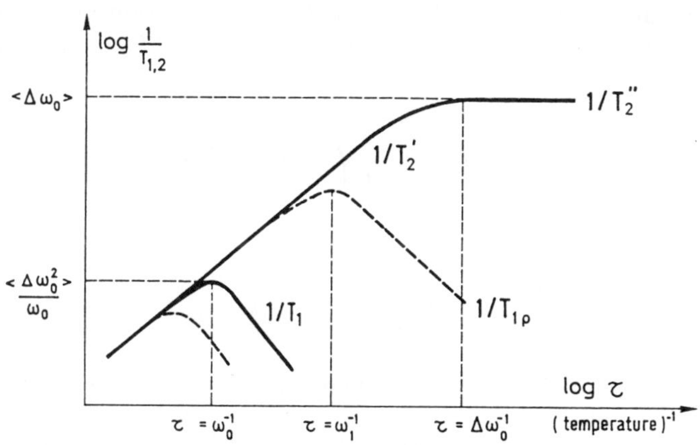

Fig. IX.16. — Schematic of the temperature dependence of the relaxation times measured by NMR. ω_0 is the Larmor frequency $(= \gamma H_0)$; $\left\langle \Delta\omega_0^2 \right\rangle = \gamma^2 \left\langle \Delta H_l^2 \right\rangle$; ΔH_l is the fluctuation of the local field.

– *Dipolar interactions. Spin-lattice relaxation.* — Dipolar interactions also allow coupling with the lattice. This contribution, denoted by $(1/T_1)_{\mathrm{d}}$, adds to that of the electrons $(1/T_1)_{\mathrm{e}}$, so that:

$$\frac{1}{T_1} = \left(\frac{1}{T_1}\right)_{\mathrm{d}} + \left(\frac{1}{T_1}\right)_{\mathrm{e}}. \qquad \text{(IX.37)}$$

The fluctuations of the local field modify the longitudinal relaxation time if they occur at a frequency of the order of ω_0, the frequency corresponding to the transitions between energy levels; $(1/T_1)_d$ then passes through a maximum for $\tau = \omega_0^{-1}$ if the fluctuations are due to atomic jumps of frequency τ^{-1}. This yields a peak analogous to an internal-friction peak (cf. Eq. (IX.10)); its form depends on the frequency spectrum $J(\omega)$ of the fluctuations in the local field:

$$\begin{aligned}\left(\frac{1}{T_1}\right)_d &= \langle \Delta\omega_0^2 \rangle \; J(\omega) \\ &\sim \frac{\langle \Delta\omega_0^2 \rangle}{\omega_0} \; \frac{\omega_0 \, \tau}{1 + \omega_0^2 \, \tau^2},\end{aligned} \tag{IX.38}$$

if $J(\omega)$ is a Lorentzian. From equation (IX.38), at high temperature $(1/T_1)_d \propto \langle \Delta\omega_0^2 \rangle \tau$; at low temperature, $(1/T_1)_d \propto \langle \Delta\omega_0^2 \rangle \tau^{-1}$. Figure IX.16 shows this peak as a function of log τ, i.e., as a function of reciprocal absolute temperature, since τ^{-1} is thermally activated. The slope of the two linear branches gives the activation energy for diffusion. The position in temperature and the height of the peak depend on the frequency ω_0 and therefore vary with the frequency of the oscillating radiofrequency field H_1.

– *Quadrupole interactions.* The nuclei are not only magnetic dipoles, but electric quadrupoles as well. The quadrupolar interaction can also contribute to the T_1 relaxation (quadrupole in a local electric field gradient) with a term $(1/T_1)_Q$.

In summary:

$$\frac{1}{T_1} = \left(\frac{1}{T_1}\right)_e + \left(\frac{1}{T_1}\right)_d + \left(\frac{1}{T_1}\right)_Q$$

and: (IX.39)

$$\frac{1}{T_2} = \frac{1}{T_2'} + \frac{1}{2} \frac{1}{T_1},$$

because the width of the $1/T_2$ line depends on the spin-spin relaxation process $1/T_2'$ and the mean lifetime of the quantum states $1/2\, T_1$ (transition probability from one state to another, Eq. (IX.32)). $T_2 \leq T_1$ because the phase coherence of the spins is limited by the lifetime T_1 of the spin states.

– *Slow atomic motions.* The residence time τ can be measured only if $\tau << T_2''$ (narrowing of the resonance) or if $\tau \approx 1/\omega_0$ (measurement of T_1). To study slower atomic motion, one can try to use lower Larmor frequencies so that $H_1 \approx H_l$. But the sensitivity is proportional to ω_0^2, so the following trick is needed: the spin system is demagnetized adiabatically while in a high field, the magnetic order is established by an excess of spins in the direction of the applied field $\mathbf{H}_0$; in zero field, only the local fields come into play.

Experimentally, the demagnetization is carried out in the frame rotating at the Larmor frequency, that of the oscillating field H_1 ($H_1 \approx H_l$). The resonance is thus carried out in a weak field; $T_{1\rho}$ is then measured, and goes through a maximum, not for $\tau = \omega_0^{-1}$ but for $\tau = \omega_1^{-1}$, with $\hbar\omega_1 = \gamma H_1$ (Fig. IX.16).

3. *Determination of the diffusion parameters.* — This is done starting with T_1, on the condition that the electronic contributions (in the case of metals) or the quadrupolar contributions can be subtracted, or starting with T_2', knowing T_1 and T_2. First τ is obtained, and then D is calculated from the relation:

$$D = a^2/6\tau \tag{IX.40}$$

for a cubic metal. Two corrections have to be made to compare this value with the tracer diffusion coefficient:

– correction for temporal correlation (see Ch. III, § III and Fig. III.8);

– correction for the isotopic mass, for one is comparing diffusivities of the majority isotope of mass m_0 and a tracer of mass m_1 (see the isotope effect, Ch. III. § III.3).

As shown in figure IX.17, the term $(1/T_1)_e$ can seriously limit the domain of D values or of temperature over which measurements can be carried out. In effect, $1/T_1$ or $1/T_2$ has to be larger than $(1/T_1)_e$.

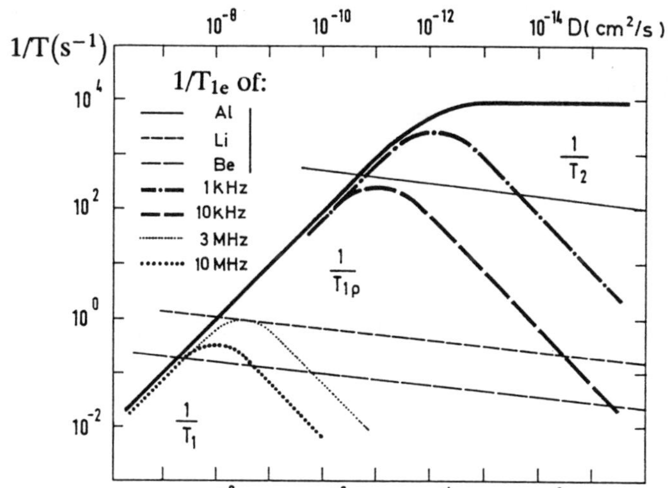

Fig. IX.17. — Limitation of the measurement of the diffusion coefficient from the relaxation times T_1 and $T_{1\rho}$ by the contribution of the electrons (straight lines) in three metals.

The difficult problem is the calculation of τ. The relations given above (Eq. (IX.38)) are approximate, and allow a first approximation to τ. As in the case of internal friction, it is necessary to use a precise model of the diffusion process, assuming one or more mechanisms with their specific temporal correlation effects. The random functions for a pair of identical nuclei i, j are calculated on the basis of this model:

$$\begin{aligned} Q_1(t) &= \mathbf{r}_{ij}^{-3} \sin \Theta_{ij} \cos \Theta_{ij} \exp(-i\phi_{ij}) \\ Q_2(t) &= \mathbf{r}_{ij}^{-3} \sin \Theta_{ij} \exp(-2i\phi_{ij}), \end{aligned} \tag{IX.41}$$

where $\mathbf{r}_{ij}(t)$ is the vector joining two carrier nuclei of the two dipoles with time-dependent polar coordinates r_i, Θ_i, ϕ_i. Then the pair-correlation functions (see Ch. II, § VII) are calculated:

$$\begin{aligned} J_1(t) &= \langle Q_1(0) \cdot Q_1(t) \rangle \\ J_2(t) &= \langle Q_2(0) \cdot Q_2(t) \rangle . \end{aligned} \tag{IX.42}$$

The Fourier transforms of these functions give the spectral densities of the fluctuations of the local field around the Larmor frequency ω_0 and its first harmonic $2\omega_0$:

$$\frac{1}{T_1} = \frac{3}{2} \gamma^4 \hbar^2 I(I+1) [J_1(\omega) + J_2(2\omega)] . \tag{IX.43}$$

Example: The atomic jumps lead to a series of discrete values of $\mathbf{r}_{ij}$ and thus of Q_1 and Q_2, where each orientation is given with a certain probability of its residence time. In the simplest case, t/τ jumps take place in the time t and the probability of N jumps is given by a Poisson distribution:

$$P_N(t) = \frac{(t/\tau)^N}{N!} e^{-t/\tau} .$$

Then the probability that the atom doesn't jump ($N = 0$) is:

$$P_0(t) = e^{-t/\tau},$$

from which:

$$J_1(t) = \langle Q_1^2(0) \rangle \exp(-t/\tau) \tag{IX.44}$$

where $\langle Q_1^2(0) \rangle$ depends on the geometry of the jumps. It follows that its transform is a Lorentzian:

$$J_1(\omega) \propto \frac{\tau}{1 + \omega_0^2 \tau^2}, \tag{IX.45}$$

which passes through a maximum at $\omega_0 \tau = 1$. In fact, this approximation is valid only for a liquid, in which the jump frequency is very high. The exact calculation of $J(t)$ and $J(\omega)$ requires a computer-simulation study of the atomic motions in each case. On the other hand, the asymptotic slopes of

T_1 and T_2 as functions of temperature (Fig. IX.16) are relatively insensitive to the model, a fact which allows an easy determination of the activation energy for diffusion in the case of a single diffusion mechanism.

Consider the case of diffusion via vacancies, with $\tau_v << \tau$:

– at very low temperature, $\tau_v > T_2''$. The spins can equilibrate after each jump of the vacancy;

– at intermediate temperatures, $\tau_v < T_2'' < \tau$. This is the most complex domain to analyze. During the time T_2'', the vacancy can jump one or several times; since the jump of a nucleus does not modify its spin, the vacancy leaves in its wake a streak of spins not in equilibrium. A computation by simulation is necessary:

– at high temperature, $\tau_v < \tau < T_2''$. This is the case of figure IX.15; the liquid model, i.e., equations (IX.44) and (IX.45) apply, with correction for temporal correlation.

The theoretical interpretation requires a description of the correlated movements of two atoms and a point defect instead of one atom and one defect, as in the problem of spatial correlation in the self-diffusion of a tracer.

Remarks:

1) Since NMR is sensitive to the majority nuclei, it is a method for studying volume *self-diffusion*. Nevertheless, *solute diffusion* can be studied in some cases by the quadrupolar interaction of the matrix nuclei with the solute for nuclear spins of 3/2. The broadening of the quadrupolar relaxation peak $1/T_1$ depends on the jump frequency of the solute.

2) The experiments are often carried out on powders, and sometimes on single crystals, taking advantage of the skin effect.

3) The comparison of diffusion coefficients measured by NMR and by tracers is the subject of a voluminous literature, especially for metals such as Li, Al, and Cu, ionic crystals (^{18}F in fluorides), and organic molecular crystals. The elements Ti, Cr, Ni, and Fe are not favorable for NMR.

4) Proton NMR allows the study of rotational and translational motions of molecules in organic molecular crystals.

II.3 Mössbauer effect.

Principle of the method. — The method uses two crystalline samples, one playing the role of *source*, the other of *absorber*. In the former, the nuclei emit γ rays, some of which are absorbed without atomic recoil in the latter (f is the fraction of γ rays absorbed recoil-free). The width of the absorption line Γ_0 is related to the lifetime τ_N of the excited nuclear level by the uncertainty relation:

$$\Gamma_0 \cdot \tau_N \gtrsim \hbar. \qquad \text{(IX.46)}$$

The source is moved relative to the absorber and the linewidth is recorded as a function of the relative velocity, which shifts the frequencies by the Doppler effect (Fig. IX.18).

If the atoms undergo diffusive jumps, the nucleus resides on one site only for a time $\tau_M < \tau_N$ and the resonance becomes wider:

$$\Gamma_M \cdot \tau_M \gtrsim \hbar \tag{IX.47}$$

with *broadening*:

$$\Gamma_M - \Gamma_0 = \Delta\Gamma \sim \hbar/\tau_M \tag{IX.48}$$

if $\tau_M << \tau_N$.

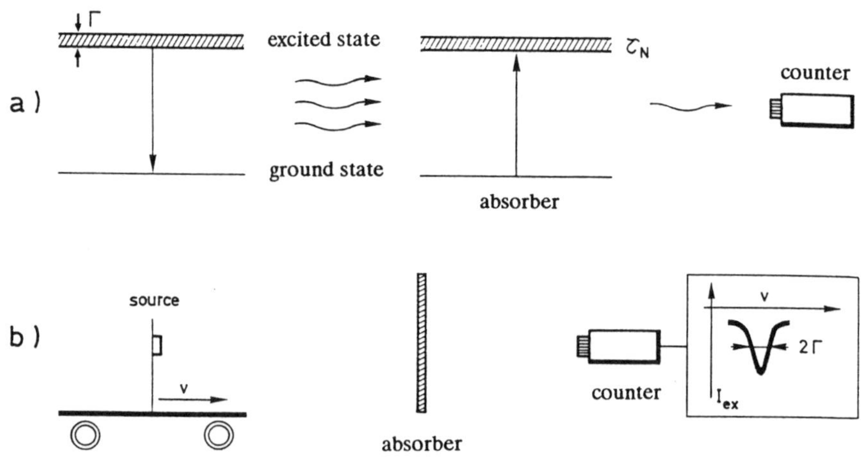

Fig. IX.18. — Mössbauer spectrometry: a) schematic of principles; b) arrangement of a source experiment (moving source).

Analysis of the process leads to an expression for the cross section for resonant absorption similar to equation (IX.25) with the self-correlation function $G_s(\mathbf{r}, t)$:

$$\sigma(\mathbf{k}_0, \omega) = (\sigma_0\Gamma/4\hbar) \int G_s(\mathbf{r}, t)\; e^{i\mathbf{k}_0\cdot\mathbf{r}}\; e^{-i\omega t} \cdot e^{-\Gamma_0|t|/2\hbar}\; d\mathbf{r}\; dt,$$

where σ_0 is the cross section if the incident energy $\hbar\omega'$ coincides with the energy of the absorber resonance $\hbar\omega_0$. Γ_0 is the natural width of the excited state of the nucleus. In an absorber experiment (moving absorber), the Mössbauer absorption peak described by $\sigma(\mathbf{k}_0, \omega)$ is broadened by this motion, as it is by the lifetime τ_N of the nucleus (see Eqs. (IX.46) and (IX. 47)).

To use this equation requires a model of the nuclear motion. For random jumps, the calculated broadening is:

$$\Delta\Gamma = (2\hbar/\tau_{\mathrm{M}}) \left[1 - \int P_1(\mathbf{r}) \exp(i\mathbf{k}\cdot\mathbf{r})\, \mathrm{d}\mathbf{r}\right], \qquad \text{(IX.49)}$$

where $P_1(\mathbf{r})$ is the distribution function of the $\mathbf{r}$ vectors after a jump; the factor in brackets then depends on the crystal structure and the correlation effects. For a polycrystal, it is assumed that the phases of the Fourier transform of $P_1(\mathbf{r})$ cancel on the average, so that approximately:

$$\Delta\Gamma \approx 2\hbar/\tau_{\mathrm{M}}. \qquad \text{(IX.49a)}$$

Neglecting correlation effects, τ_{M} is nothing but the reciprocal of the jump frequency, Γ. In a cubic crystal:

$$D = \frac{1}{6}\,\Gamma\lambda^2,$$

where λ is the jump length. From this:

$$\boxed{D = \frac{\lambda^2}{12}\,\frac{\Delta\Gamma}{\hbar}}, \qquad \text{(IX.50)}$$

neglecting the effects of temporal correlation.

Advantages of the method:

– a single sample can be used to explore the entire accessible domain of temperature. Very dilute solutions $(\approx 10^{-6})$ can be studied with the sample as source:

– the method complements radiotracer studies because it provides a local analysis of the atomic jumps.

Limitations of the method:

– the experiment in practice is constrained to $0.1\Gamma_0 < \Delta\Gamma < 100\,\Gamma_0$, which limits the measurement to 2 or 3 decades around $1/\tau_{\mathrm{M}}$;

– as $\tau_{\mathrm{M}} < \tau_{\mathrm{N}}$, only relatively large diffusion coefficients can be measured; for Fe, this condition gives $D \geq 10^{-9}$ cm^2 s^{-1},

– the nuclei under study must have a large value of the recoilless fraction, which limits the number of good Mössbauer isotopes to a few chemical species. The isotopes useful for diffusion measurements are: ^{57}Fe, ^{119}Sn, ^{151}Eu, and ^{161}Dy.

PART 3. — COMPUTER SIMULATION

It may appear surprising to introduce a paragraph on this subject into a chapter on experimental techniques, although the definition of "simulation" is: "experimentation on the functioning of a *model* in order to draw conclusions about the real system" (Mathieu, Kastler, and Fleury, Dictionary of Physics, Paris, 1983). In fact, the model can be constructed in the computer either in analog form, or, more commonly, numerically. Whatever the approach, it is neither theory nor experiment. It is a third original track, which allows "experiments" on a material that is perfectly characterized because it is constructed by the software or the hardware of the computer. Two general types of application will be distinguished: 1) purely statistical calculations; 2) studies of defects and diffusion mechanisms.

I. Statistical calculations.

To follow a random walk, or the deviations from a purely random walk, of a particle in a crystal lattice or in an amorphous structure, the individual paths can be sampled by drawing lots; this is the Monte Carlo method. The averages are taken over a sufficiently large number of paths to determine the characteristic macroscopic parameters (mean square displacement, correlation factor). To accomplish this, all the particles (atoms, eventually vacancies) must be tagged; the correlation factor is given by:

$$f = \sum_{i=1}^{N} R_i^2 \Big/ \sum_{i=1}^{N} n_i \, r^2, \qquad \text{(IX.51)}$$

where r is the jump distance (cubic lattice), n_i the number of jumps of particle i which effect the displacement R_i. The average is calculated over a very large number N of identical particles (N is larger, the smaller the number of sites on the chosen lattice).

A simple example has been given in chapter III, § IV. The number of lattice sites is generally limited to $p \times p \times p$, but this is repeated by the use of periodic boundary conditions.

In addition to these numerical methods, analog methods are also interesting. The lattice can be simulated by an electric network, the nodes of which represent atomic sites and the currents measure the number of atoms passing from one site to a neighboring one. This is an elegant method that was pioneered by Compaan and Haven (see Adda and Philibert, *op. cit.*, p. 413 ff).

II. Defect characteristics and diffusion mechanisms.

II.1 The goals of simulation. — The simulation methods can be used with three different objectives:

1) To determine the characteristics of defects in a material whose properties are as similar as possible to those of the real material. This assumes that "good" *inter-atomic potentials* are known. A large majority of computer programs uses central-force, i.e., two-body potentials, rather than the more realistic N - body potentials.

This approximation is reasonable in ionic crystals, since the ionic bond can be described by a Coulomb interaction. However, the Cauchy relation ($C_{12} = C_{44}$, for the elastic constants) is not in general obeyed, a sign that the interatomic forces are not purely central forces. Many-body effects are, however, included by the introduction of polarization terms; the polarization introduces for each ion a parameter that depends on the positions of all the surrounding ions. This is the "Breathing Shell" model, in which the ion is described as a point charge X, surrounded by a spherical shell of charge $-Y$; their relative eccentricity describes the polarization. The variations of the radius and form of the shell as functions of the positions of the neighboring ions allow the simulation of many-body effects (Fig. IX.19).

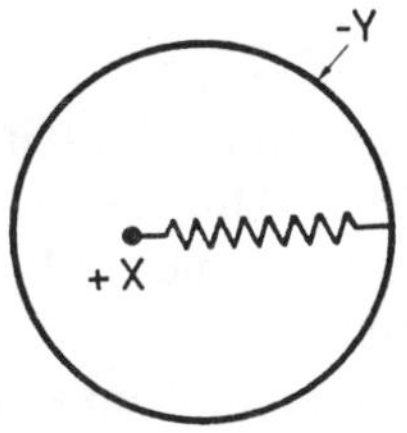

Fig. IX.19. — Schematic of "breathing shell model".

For metals, purely empirical potentials with an analytical form suitable for the calculation are used, such as Morse or Lennard-Jones potentials, or potentials of polynomial type (several polynomials for different distance intervals). However, without starting with true *ab initio* calculations, physically more realistic potentials can be used. For normal metals, these are based on the pseudo-potential model (a volume term due to the conduction electrons is superposed on an essentially screened Coulomb term). For transition metals, the interactions can be calculated in the tight-binding model; the method of moments then leads to the inclusion of the interactions of more than two bodies. All these potentials contain one or more terms with adjustable parameters.

In semiconductors with coordination number 4, pair potentials with two terms are used; one corresponds to the variation of the bond lengths between nearest neighbors, the other to their rotation, in a manner to describe the covalent bond (Keating potentials).

In all these cases, the short-range repulsion is described by a Born-Mayer potential of the form $\exp(-r/\rho)$ or another form that varies rapidly with distance. To limit the number of pair interactions that have to be calculated, a cut-off radius R is introduced, such that the potential is zero for $r \geq R$.

The parameters entering into the empirical or semi-empirical potentials are fitted to the properties of the material under study, the cohesion energy, elastic moduli and phonon spectra, dielectric constant,...

2) The second objective is a variant of the first: it consists of simulating an experimental process in order to allow a more correct interpretation of the measured macroscopic values: NMR, Mössbauer effect, neutron scattering.

3) The third objective consists of carrying out *exact calculations on model materials.* This is a qualitative approach in the sense that its results are felt to have a general importance for elucidating diffusion mechanisms. A central-force potential of Lennard-Jones type is generally used ([10]). It is a good representation of a rare-gas solid, which thus plays the role of the material being modeled. Studies of jump dynamics, dynamic correlation effects (a result of anharmonicity at high temperature), defects in defects (e.g., defects in low-dimensional structures such as surfaces or interfaces)... are accessible.

II.2 MODELS. — The calculation has to be limited to a relatively small number of particles, typically a few hundred. This number depends on the progress in the successive generations of computers, not only the size of the core memory, but the speed of calculation as well.

The limited dimensions of the crystal may render the study sensitive to surface effects. Different artifices are used to mitigate this:

– in molecular dynamics or Monte Carlo, *periodic boundary conditions.* The ensemble of particles is contained in a parallelepided-shaped box (commonly called the simulation box), which is repeated infinitely in three dimensions; a particle that leaves the box at one face is reintroduced at the opposite face. The interaction between particles can extend across the faces of the box.

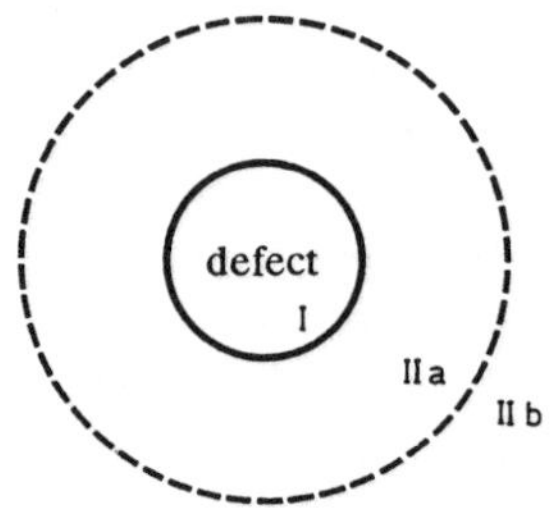

Fig. IX.20. — Separation of the crystal around a defect into several regions.

([10]) The Lennard-Jones potential has the form: $F(r) = A/r^n - B/r^m$, with $n > m$ (e.g., $n = 12$, $m = 6$).

The periodic boundary conditions fix the volume and the density of the system, as well as the concentration of defects (at least 1 for the volume of the elementary box);

– in a static calculation, the crystal is decomposed into two regions: an internal region I limited to a certain number of particles in which the interactions are calculated using the pair potentials, and an external region II, which is infinite in principle (see below, and Fig. IX.20).

II.3 METHODS. — Three methods of simulation are used: molecular dynamics, lattice statics, and the Monte Carlo method.

1) *Molecular dynamics.* — The paths of N interacting particles are calculated with the equations of motion of classical dynamics. If the initial positions and velocities of the N particles are specified, the path of the system in phase space is perfectly determined; it is located on a hypersurface of constant energy (microcanonical ensemble).

The force exerted on each particle by all its neighbors (including those contained in the reflections of the box) can be calculated at any instant:

$$\mathbf{F}_i = -\sum_{j \neq i} \nabla \Phi (\mathbf{r}_{ij}), \qquad \text{(IX.52)}$$

where $\Phi(\mathbf{r}_{ij})$ designates the interaction between particles i and j. The motion of the particle during the time interval δt is described by Newton's second law:

$$m (\mathrm{d}\mathbf{v}_i/\mathrm{d}t) = \mathbf{F}_i \quad i = 1,\ 2, ...N. \qquad \text{(IX.53)}$$

$3N$ coupled differential equations are obtained, with initial conditions on the positions and velocities, e.g., the N particles are atoms on the sites of a given lattice, with each atom having a velocity the components of which conform to a Maxwellian distribution (the corresponding atoms being chosen at random).

The paths are calculated for the interval δt. The corresponding forces are recalculated at time $t + \delta t$, and so forth by recursion.

The temperature is calculated at each time:

$$T(t) = \sum_{i=1}^{N} m v_i^2 / 3Nk, \qquad \text{(IX.54)}$$

as is the potential energy:

$$U = \frac{1}{2} \sum_{i=1}^{N} \sum_{j=1}^{N-1} \Phi(\mathbf{r}_{ij}) . \qquad \text{(IX.55)}$$

Thermal equilibrium is attained after some iterations (200 to 500 δt). Once equilibrium is attained, the system is allowed to evolve for a time long enough to yield statistically significant information.

The choice of δt is important. It must be less than 1/10 of the shortest vibration period of the system; typically, $\delta t \approx 10^{-15}$ or 10^{-14} s. A compromise must be found between the size of the crystal (i.e., the number of interactions to be calculated) and the number of iterations (sufficient for the calculation of averages). For example, if it takes the computer 1 s to calculate 10^4 pair interactions (crystal size of the order of 100 or 200 particles), the calculation becomes prohibitively long for more than 10^5 iterations, and no process lasting longer than $10^5 \times 10^{-14} = 1$ ns can be studied. It is thus quite improbable to see the spontaneous formation of a defect by thermal agitation; the defect has to be introduced at the beginning.

The calculation yields a table of positions and velocities as functions of time for N particles. Thus all observables that can be expressed in terms of these quantities can be calculated; and under the assumption that the system is ergodic, the spatial and temporal averages (e.g., jump frequencies) can also be calculated over a long enough path starting at the instant of thermal equilibration.

In addition to the temperature and the potential energy, one can also calculate the density of the particles of type α at every point:

$$\rho_\alpha(\mathbf{r}) = \left\langle \sum_i \delta\,(\mathbf{r} - \mathbf{r}_{i\alpha}) \right\rangle , \qquad \text{(IX.56)}$$

the mean square displacement, the position self-correlation function (Eq. (II.34)), the velocity self-correlation function (Eq. (II.45)), the diffusion coefficient (Eq. (II.46)), the structure factor, the dynamic scattering function $S_\alpha(\mathbf{k}, \omega)$ (Eq. (II.42)), the pair correlation function, etc... (see Ch. II, § VII).

Molecular dynamics yields an enormous quantity of raw data, which require intense thought to extract the most valuable information, tests for comparison with experiment, or models for new mechanisms. It is nevertheless limited in certain ways. It does not allow the direct calculation of entropies, nor of free energies, because these cannot be expressed as equilibrium averages of functions of position and velocity.

However, molecular dynamics allows the study of processes with energies less than kT; this is why it is especially applicable to the simulation of liquids, of diffusion in superionic conductors (Ch. V, § VIII), or of adatoms on close-packed surfaces (Ch. VII, part 2, § II). New methods of molecular dynamics calculations are under development for treating systems at constant pressure and constant enthalpy, a very promising development.

2) *Lattice statics.* — Contrary to molecular dynamics, this method rests on a static description of the lattice. One still considers a box containing N particles interacting via a certain potential, and searches for the configuration of minimum energy for a crystal containing one or more defects. The central problem is the relaxation of the atoms around a defect, a process which leads to the minimization of the energy of the system ([11]). Clearly, the relaxation (i.e., the displacement of the atoms relative to their normal positions on the lattice sites when no defect is present) is greatest in the immediate neighborhood of the defect, and decreases with increasing distance from it. Thus two regions are distinguished: an internal region (region I) where the displacements are large, and an external region (region II) which is but slightly perturbed (Fig. IX.20). The few hundreds of atoms typically contained in region I are described individually, and their interactions are described by the appropriate potential; symmetry considerations allow the reduction of the number of variables. In region II, the displacements are calculated by the application of the laws of elasticity (Navier's equation) or dielectric polarization (in ionic crystals). The energy takes the form:

$$E = E_{\mathrm{I}} + E_{\mathrm{II}} + E_{\mathrm{I/II}},$$

where $E_{\mathrm{I/II}}$ is the energy of interaction between the two regions. This expression is finally cast in a form which includes only the interactions among the ions of region I and between these and the ions of region II; all interactions among the ions of region II can be eliminated. A distinction between the regions IIa and IIb suggested by an imaginary boundary in region II permits the further simplification of the calculation according to the range of the different interactions.

Starting from a reasonable configuration, each component of the displacement of each atom i in region I is varied until the force on the atom i is cancelled. The forces exerted by region I on region II must similarly be brought to zero. From this, one obtains the energy and volume change for the formation of an elementary defect (vacancy, interstitial) or a group of defects ([12]). The choice and properties of the boundary between regions I and II is a difficult problem.

This method does not yield exact results in the sense of statistical mechanics except at the absolute zero of temperature. In effect, at finite temperatures the difference in potential energy ($\Delta\Phi = \Phi' - \Phi$) between the defect and ref-

---

([11]) For example, the energy to form a Schottky defect in MgO is calculated to be 40 eV without relaxation and 7 eV if relaxation is included.

([12]) Let E_0 and E_{A} be the potential energies of the perfect crystal and the crystal containing a defect on site A, respectively. The formation energy is then given by: $E_{\mathrm{f}} = E_{\mathrm{A}} - E_0 + E_{\mathrm{s}}$, where E_{s} is the potential energy of an atom located at a kink on the surface.

erence states is replaced by $\Delta\Phi_0 = \Phi_0' - \Phi_0$ calculated at 0 K, but with the value of the lattice parameter at temperature T. This makes it necessary that the harmonic approximation be fulfilled (this is called the quasi-harmonic approximation).

The method is also used for the calculation of the migration energies. The same minimization procedure is applied to the saddle-point configuration; the potential energy is minimized for a system in which the diffusing atom is held in a position defined in advance and the other atoms are free to relax. If E_A and E_C are the energies calculated for the defect in the ground state and for the diffusing atom at the saddle point, then the migration energy is given by $E_C - E_A$.

Note that in these methods it is a variation of energy at constant volume that is calculated, and not an enthalpy. In the quasi-harmonic approximation at a temperature T:

$$(\Delta H)_p = \Delta\Phi_0^{v} + T\alpha V \Delta P_0^{v},$$

where α is the coefficient of thermal expansion, and ΔP_0^{v} is the difference in pressure at constant volume. The second term on the RHS is important.

Lattice statics has already yielded numerous important results on ionic crystals, including oxides (e.g., the HADES ([13]) program offers a great flexibility as a function of the chosen potentials), and on alkali metals, two cases in which good, believable pair potantials exist. It is evidently limited to modeling processes involving energy variations well above kT.

3. *The Monte Carlo method.* — In the framework of statistical physics, the Monte Carlo method is not, properly speaking, a simulation, but rather an efficient technique for ***sampling configurations*** among all of those of a petit canonical ensemble (N, V, T). This allows the calculation of the equilibrium value of the quantity of interest X:

$$\langle X \rangle = Z^{-1} \int \exp(-\Phi/kT)\ X\ (\{q\})\ \mathrm{d}^{3\mathrm{N}}q. \qquad \text{(IX.57)}$$

Here Z is the partition function and q are the coordinates of the atoms. A randomly chosen configuration has every chance of having a negligible statistical weight because it corresponds to too large a potential energy Φ_i (the statistical weight is proportional to $\exp(-\Phi_i/kT)$). One starts at a "good" configuration, and progresses, by small successive displacements in the configuration space of the N-particle system, from configuration to configuration, rejecting those configurations that increase the energy by a value $\geq kT$. One obtains a random motion of the representative point in configuration space, a technique known by the name "Metropolis' Monte Carlo method" ([14]). The averages are evaluated

([13]) HADES = Harwell Automatic Defect Evaluation System.

([14]) Typically $M = 10^5 - 10^6$ moves are required for an adequate sampling of configuration space.

over the M configurations C_i by the relation:

$$\langle X \rangle = \frac{1}{M} \sum_{i=1}^{M} X\,[\{q\,(C_i)\}]\,. \qquad \text{(IX.58)}$$

The method allows the calculation of any quantity that can be expressed as an average over the configuration of the particles, for example, the free energies of formation and migration of point defects, based on the relation:

$$(F - F_0)\,/kT = -\mathrm{Log}\,\langle \exp\,[-\,(\Phi - \Phi_0)\,/kT]\rangle_0\,, \qquad \text{(IX.59)}$$

where the brackets $\langle\ \rangle$ represent a canonical average over the ensemble with index 0. Dynamic quantities, such as diffusion coefficients, cannot be calculated by this method.

This method has been nicknamed "calorimetry on a computer ". It can be coupled with molecular dynamics to study canonical ensembles at constant pressure.

APPENDIX I

Diffusion of gases.

Our discussion will be limited to continuous solid media. The reader is referred to specialized treatises on diffusion in *porous media* [15].

Two types of experiments can be carried out:

– experiments on ***desorption*** of gas from a solid charged with a uniform concentration of the gas. The equations given in chapter X § IV.4 for the elimination of vacancies at a surface are evidently applicable to this case. Interdiffusion experiments can also be carried out between a metal charged with gas and one not charged;

– experiments on ***permeation*** across a "membrane", which can be flat or cylindrical (a tube). If the gas pressures on both sides of the membrane are kept constant, a steady state is reached after a certain incubation time, and this is described by Fick's law:

$$J = D\frac{c_1 - c_2}{d} \quad \text{atoms (or molecules) cm}^{-2}\ \text{s}^{-1}$$

---

[15] See Cunningham, R. E., and Williams, R. J. J., *Diffusion in gases and porous media* (Plenum Publ. Corp.) 1980.

for a plane membrane of thickness d (Fig. IX.5), and

$$J = 2\pi D \frac{c_1 - c_2}{\mathrm{Log}\,(R_1/R_2)} \quad \text{atoms (or molecules) cm}^{-1}\ \text{s}^{-1}$$

for a hollow cylinder of unit length. (N.B. the changed meaning of flux and the changed units).

The permeation process can be governed by diffusion or by the reactions at the interfaces where the gas enters or leaves (e.g., hydrogen stays contained in steel bottles under pressures of 150-200 bars !). Two simple cases can be considered:

– the gas dissolves in the solid in the *molecular* form:

$$c = K p_{\mathrm{m}},$$

where K is Sieverts' constant (in molecules per cm^3 per unit pressure) and p_{m} is the partial pressure of the molecular species. It follows that:

$$J = \frac{KD}{d} \quad \Delta p_{\mathrm{m}} \text{ molecules cm}^{-2}\ \text{s}^{-1},$$

where Δp_{m} is the pressure difference across the membrane;

– the gas dissolves in *atomic* form; for a diatomic molecule:

$$c = K' p_{\mathrm{m}}^{1/2},$$

where K' is in atoms cm^{-3} and p_{m} is in atm. From this:

$$J = \frac{K'D}{d} \left[p_{\mathrm{m}(1)}^{1/2} - p_{m(2)}^{1/2} \right] \approx \frac{K'D}{d}\ p_{\mathrm{m}(1)}^{1/2},$$

if $p_{\mathrm{m}(2)}$ is negligible (vacuum, for example).

Thus, D cannot be obtained from a permeation experiment, but the "*permeation constant*" or "*permeability*", $P = KD$ is obtained instead. Since both D and either K or K' follow an Arrhenius law, the permeability does the same, with an enthalpy of activation equal to the sum of the activation enthalpy for diffusion and the enthalpy of solution of the gas.

In the experiments, great care must be taken to guard against surface layers of oxide or contaminant. If the gas dissolves in atomic form in all the layers of a lamellar material, it can be shown that:

$$J = \frac{p_{m(1)}^{1/2} - p_{m(2)}^{1/2}}{\displaystyle\sum_i \frac{d_i}{(DK')_i}},$$

i.e., a layer of very low permeability, even if its thickness d_i is very small, can dominate the expression for the flux.

For a plate of thickness l considered in III.2 (Pt.1), under the conditions $c = c_s$ at one face, $c = 0$ at the other, the quantity of gas that crosses the plate by diffusion in time t is given by:

$$Q(t) = \frac{Dc_s}{l}\left\{t + \frac{2\,l^2}{\pi^2\,D}\sum_{n=1}^{\infty}\frac{(-1)^n}{n^2}\left[1 - \exp\left(-\frac{n^2\,\pi^2\,Dt}{l^2}\right)\right]\right\}.$$

and $Q(t) \approx (Dc_s/l)\left(t - l^2/6D\right)$ when t is large.

The "induction time" is determined by the intersection of the linear part of the $Q(t)$ curve with the time axis $t_i = l^2/6D$. For t_i, the flux leaving the material reaches 0.6299 times the value for steady state, $J_\infty = Dc_s/l$. The "breakthrough time" $t_b = 0.494\,l^2/\pi^2 D$, determined by the intercept of the tangent at the inflection point of the $J(t)$ curve, is also used to determine D, although it is best to determine D from the total desorption curve $J(t)/J_\infty$ (see Tison, 1984). In case of trapping, an apparent D is measured.

Desorption of a gas by detrapping (see Fig. III.17).

In a certain temperature range, traps of type i, in concentration P_i, liberate gas. Let $\Theta_i(t)$ be the fraction to which the traps are filled at time t. The quantity of gas trapped per unit volume in the traps of type i is then $P_i\Theta_i$ and the quantity desorbed $Q_i(t) = P_i(1 - \Theta_i)$. If the detrapping kinetics are first order, the rate of desorption $\mathrm{d}Q_i/\mathrm{d}t$ is proportional to the quantity remaining:

$$\mathrm{d}Q_i/\mathrm{d}t = \text{Const.}\ (P_i - Q_i)\ \exp(-\Delta H_i/RT).$$

In general, detrapping is studied during heating at a constant rate $\mathrm{d}T/\mathrm{dt} = \Re$, and a desorption peak T_p is observed for each type of trap. If $\mathrm{d}Q_i/\mathrm{d}t$ passes through a maximum at $T = T_\mathrm{p}$, differentiation of the above equation yields:

$$\exp(-\Delta H_i/RT_\mathrm{p}) = \left(1/T_\mathrm{p}^2\right)\ \Re.$$

If experiments are carried out at various heating rates $\Re$, and log $\left(\Re/T_\mathrm{p}^2\right)$ is plotted *vs.* $1/RT_\mathrm{p}$, the slope of the resulting straight line is $-\Delta H_i$.

APPENDIX II

The Snoek effect.

The interstitials C, N and O in BCC metals give rise to anelastic effects (elastic after-effect and internal friction) which are interpreted on the basis of Snoek's model (1941).

The BCC structure contains two types of interstitial positions:

– *octahedral* sites of the type (1/2, 0, 0) and (1/2, 1/2, 0), located at the middle of the cube edges or at the face centers. These sites do not have cubic symmetry; there are 2 atoms along a <100> direction located $a/2$ from the center of the site, whereas 4 others in a plane perpendicular to <100> are at a distance $a\sqrt{2}/2$;

– *tetrahedral* sites, of the type (1/2, 1/4, 0); these sites have cubic symmetry with the 4 neighboring atoms at a distance of a $\sqrt{5}/4 = 0.56a$.

In general, the interstitials are not located on the tetrahedral sites, but on the octahedral ones. Since the latter do not have cubic symmetry, the interstitials distort the lattice, principally by pushing apart the two atoms located along the <100> direction. These interstitial sites can then be classed as one of three types, x, y, or z according to the direction of the principal distortion caused by the corresponding interstitial atom.

It is this asymmetric distortion that causes *the establishment of order under the action of a stress.* Normally, the numbers of atoms on each of the three types of sites are equal. If there is an excess of atoms in the z sites, an elongation of the crystal in the z direction results. Conversely, an elastic deformation which causes an elongation in a certain direction leads to a redistribution of the interstitials among the three types of sites (directional order induced by deformation).

Calculation of the relaxation time.

To calculate the relaxation time of this phenomenon, consider the interstitial jumps between the three possible types of sites. Let n be the number of interstitials per unit volume, and n_x, n_y, and n_z the number of interstitials on x, y, and z sites, respectively. The directional order is characterized by the parameter p:

$$p = n_z - n/3. \tag{A.1}$$

Consider the jump frequencies Γ_{xy} from an x site to a y site, Γ_{yz} from a y site to a z site, etc., the atom jumping a distance $a/2$. The rate of accumulation of atoms on the z sites is equal, on the average, to the rate of arrival of the atoms coming from the x and y sites, minus the rate of departures toward the x and y sites. Since a z site has two x and two y sites as nearest neighbors, the rate is:

$$\mathrm{d}n_z/\mathrm{d}t = \dot{n}_z = 2\Gamma_{yz}n_y + 2\Gamma_{xz}n_x - 2\left(\Gamma_{zx} + \Gamma_{zy}\right)\, n_z. \tag{A.2}$$

But when a stress is applied, the activation energies will not be the same for the different types of jumps. Let ΔH be the average height of the potential barrier between two interstitial sites, and δu the difference of potential energies between an x (or y) site and a z site. The height of the barrier is lowered by

$\delta u/2$ for the xz or yz jumps, and raised by $\delta u/2$ for zx and zy jumps:

$$\begin{aligned}\Gamma_{yz} = \Gamma_{xz} &= \nu \exp\left[-\left(\Delta H - \delta u/2\right)/kT\right] \\ &= \nu \exp\left(-\Delta H/kT\right) \exp\left(\delta u/2kT\right) \\ &= \Gamma_s \exp\left(\delta u/kT\right) \approx \Gamma_s \left(1 + \delta u/2kT\right),\end{aligned} \tag{A.3}$$

where $\Gamma_s = \nu \exp(-\Delta H/kT)$ is the probability of a jump in the absence of a stress. In the same way:

$$\Gamma_{zx} = \Gamma_{zy} \approx \Gamma_s \left(1 - \delta u/2kT\right). \tag{A.4}$$

Substitution of (A.3) and (A.4) in (A.2) yields:

$$\dot{n}_z = 2\Gamma_s \left(1 + \delta u/2kT\right)\left(n_x + n_y\right) - 4\Gamma_s \left(1 - \delta u/2kT\right) n_z,$$

or

$$\dot{n}_z = -6\Gamma_s \left[\left(n_z - \frac{n}{3}\right) - \frac{2}{9}\,\frac{n\,\delta u}{kT}\right], \tag{A.5}$$

where the approximation $n_z \approx n/3$ has been used to simplify the last term. When the order variable p defined in equation (A.1) is introduced, this equation takes the form:

$$\dot{p} = -\left(p - p_\infty\right)/\tau_R, \tag{A.6}$$

with the relaxation strength:

$$p_\infty = \frac{2}{9}\,\frac{n\,\delta u}{kT}, \tag{A.7}$$

and the relaxation time:

$$\tau_R = \left(6\Gamma_s\right)^{-1}. \tag{A.8}$$

Since each interstitial has 4 other interstitial sites as nearest neighbors, the total jump frequency of an atom equals:

$$\Gamma = 4\Gamma_s = \tau^{-1},$$

from which

$$\tau = \frac{3}{2}\,\tau_R. \tag{A.9}$$

Since in BCC interstitial solutions (Eq. (II.24)):

$$D = \frac{a^2}{24\tau}, \tag{A.10}$$

the diffusion coefficient is given by:

$$\boxed{D = a^2/(36\ \tau_R)}, \tag{A.11}$$

a relation which allows the exact determination of the diffusion coefficient of interstitial atoms in a dilute solid solution with BCC structure from a knowledge of the relaxation time.

Bibliography

I. General references

Adda, Y. and Philibert J. *op. cit.*, Ch. IV.
Bénière F., in *Mass Transport in Solids, op. cit.* (1983) Ch. 3, 21-41.
Rothman S. J., in *Diffusion in Crystalline Solids, op. cit.*, (1984), 1-56.
Non-Traditional Methods in Diffusion, G. E. Murch, H. K. Birnbaum, and J. R. Cost, eds., Met. Soc. AIME, Warrendale, Pa. (USA) 1984.

II. Specific references

Use of radiotracers
Mundy J. N., Rothman S. J. *et al.* in *Methods of Experimental Physics,* Vol. **21**, *Solid State; Nuclear Methods,* J. N. Mundy, S. J. Rothman, M. J. Fluss, and L. C. Smedskjær eds. (Academic Press, Inc.) 1983, p. 1-75.
Carslaw H. S. and Jaeger J. C., *Conduction of Heat in Solids* (Clarendon Press, Oxford) 1959.

High-Resolution Autoradiography
Aucouturier M., Lapasser G. and Asaoka T., *Metallography* **11** (1978) 5-21.

Microanalysis
Microanalysis and Scanning Electron Microscopy, École d'Eté de St.-Martin-d'Hères, English edition, (1978) (Les Editions de Physique, Orsay).

Ion probe analysis
Benninghoven A., Rüdenauer F. G., and Werner H. W., *Secondary Ion Mass Spectrometry,* (Wiley, New York) 1987.

Nuclear reactions, Rutherford backscattering

Nuclear Physics Methods in Materials Research, K. Bethge, H. Baumann, H. Jex, and F. Rauch, eds. European Physical Soc. (Vieweg & Sohn, Braunschweig) 1980.

Chu W.-K., Mayer J. W., and Nicolet M.-A., *Backscattering Spectrometry*, (Academic Press, New York) 1978.

Calvert J. M., Derry D. J. and Lees D. G., *J. Phys. D, Appl. Phys.* **7** (1974) 940-953.

Glow Discharges

Berneron R. and Charbonnier J. C., *Surface and Interface Analysis* **3** (1981) 134-141.

Sputtering

Atkinson A. and Taylor R. I., *Thin Solid Films* **40** (1977) 291-298.

Gupta D. and Tsui R. T. C., *Appl. Phys. Lett.* **17** (1970) 295.

Mehrer H., Maier K., Hettich G., and Rein G., *J. Nuclear Mater.* **69-70** (1978) 545-548.

Mundy J. N. and Rothman S. J., *J. Vac. Sci. Technol. A.* **1** (1983) 74-76.

Modulated multilayers

Cook H. E. and Hilliard J. E., *J. Appl. Phys.* **40** (1969) 2191-2198.

Greer A. L. and Spaepen F., in *Synthetic Modulated Structures* (Academic Press, New York) 1985, Ch. 11, p. 419-48.

Treatment of data

Aubin J. L., Ansel D. and Debuigne J., *J. Less-Common Met.* **113** (1985) 269-282.

Heijwegen C. P. and Visser G. J., *Scripta Metall.* **7** (1973) 223-228.

Kapoor R. K., Eagar T. W., *Met. Trans.* **21A** (1990) 3039-3047.

Complex impedance

Bauerle J. E., *J. Phys. Chem. Solids* **30** (1969) 2657-2670.

MacDonald J. R., *Impedance Spectroscopy* (Wiley, New-York) 1987.

Kleitz M., Bernard H., Fernandez E. and Schouler E. in *Advances in Ceramics*, Vol. **3** *Science and Technology of Zirconia*, A. H. Heuer and L. W. Hobbs, eds. (American Ceramic Society, Columbus) 1981, pp. 310-336.

Anelasticity

Nowick A. S. and Berry B. S., *Anelastic Relaxation in Crystalline Solids* (Academic Press, New York) 1972.

Philibert J. in *Defects in Solids, Modern Techniques,* A. V. Chadwick and M. Terenzi, eds. NATO ASI series, series B, 147 (Plenum, New York) 1986, pp. 349-375.

Induced anisotropy

Edler E. and Radeloff C., *J. Appl. Phys.* **40** (1969) 1526-1533.

Chambron W. and Caplain A., *Acta Metall.* **22** (1974) 357-366.

Caplain A. and Chambron W., *ibid.* **25** (1977) 1001-1011.

Neutron Scattering

Springer T. and Richter D., in *Methods of Experimental Physics* Vol. **23B**, *Neutron Scattering*, D. L. Price and K. Sköld, eds. (Academic Press, Orlando) 1987, p. 131-186.

Lechner R. E., Neutron Scattering Studies on Diffusion in Solids, in *Mass Transport in Solids, op. cit.* Ch. 8 (1983) p. 169-226.

Wakabayashi N., in *Methods in Experimental Physics*, Vol. 21 *op. cit.* (1983) p. 194-220.

Petry W. and Vogl G., *Vacancies and Interstitials in Metals and Alloys, op. cit.* (1987) pp. 323-348.

NMR and Mössbauer

Janot Ch., *J. Physique* **37** (1976) 253-265.

Janot Ch., *J. Physique Colloq.* **39** (1978) C2-238-C2-255.

Seeger A., Wolf D., and Mehrer H., *Phys. Status Solidi b* **48** (1971) 841-496.

Ailion D. C., in *Methods in Experimental Physics*, vol **21** *op. cit.* (1983) p. 439-482.

Petry W. and Vogl G., *loc. cit.*

Wolf D., *Spin Temperature and Nuclear Spin Relaxation in Matter* (Clarendon Press, Oxford) 1979.

Computer simulation

Doan N. V., in *Confolant* 77, *op. cit.*, Ch. IX, p. 285-303.

Bennett, Ch., Colloque Saclay (1976) *op. cit.*, p. 655-684.

Bennett, Ch., *Diffusion in Solids* (1975) *op. cit.*, p. 73-113.

Mackrodt W. C., *Mass Transport in Solids* (1983), *op. cit.*, Ch. V, p. 107-130.

Jacucci G., *ibid.* Ch. VI, p. 131-148.

Murch G. E., in *Diffusion in Crystalline Solids*, (1984) *op. cit. p.*379-430.

Computer Simulation of solids, C. R. A. Catlow and W. C. Mackrodt, eds. Lecture notes in Physics, vol. **166** (Springer Verlag, Berlin) 1982.

Diffusion of gases in solids

Cunningham R. E. and Williams R. J. J., *Diffusion in Gases and Porous Media* (Plenum Publ. Corp.) 1980.

Davanathan M. A. and Stachursky Z. O. J., *Proc. Roy. Soc. London* **A270** (1962) 90-102 (electrochemical permeation).

LE CLAIRE A. D., Permeation of Gases through Solids, *Diff. Def. Data* **33** (1983) 1-66.
TISON P. CEA report CEA-R-5240 (1) and (2), Service Documentation, CEA Saclay (1984).

CHAPTER X

The study of some diffusion-controlled processes

I. Diffusion in multi-phase systems and the formation of intermediate compounds.

In real systems, diffusion is often accompanied by the formation of new phases and/or chemical reaction; chemical reaction here means the formation of one or more compounds composed of the two base materials. Generally, the emphasis is placed either on the first aspect (multiphase diffusion, the growth of "intermetallics"), or on the second (oxidation, sulfidation, silicidation, etc.). However, it should not be forgotten that the kinetics observed for each aspect can be governed either by the rate of diffusion across the product phase ("diffusion control") or by the processes which take place at the interfaces ("interface control") and constitute the "chemical reaction" giving rise to the compound.

New phases can appear during welding, sintering, or coating, from surface deposits or other surface treatments,... from any operation in which two different materials are in contact at sufficiently high temperature. This is the reason for the technological importance of multi-phase diffusion: the formation of brittle intermetallics in coatings, of ferrites by the interdiffusion of Fe_2O_3 and MO, of silicides in VLSI, ... or the formation of compounds that serve as diffusion barriers, (protection of the fibers in a composite, protective oxides on a refractory alloy, etc.). The interest on a fundamental level is just as great, for these processes often raise questions for which no appropriate answers yet exist.

I.1 Nature of the phases formed by interdiffusion. — The nature, number, and composition of the phases can be predicted *a priori* in the case of diffusion in a binary system AB, where A and B can be two elements or two compounds. An isotherm T_0 of the phase diagram is a horizontal straight line which traverses the single-phase domains α, β... (Fig. X.1), extending over the composition interval in question. If a metallographic section of the diffusion couple is cut parallel to the diffusion direction, parallel bands corresponding to the α, β...phases will be observed.

Begin with pure A; owing to the diffusion of B, α solid solution forms and c_B grows to the value $c_{\alpha\beta}$ which corresponds to the point where the isotherm T_0 cuts the $\alpha \longleftrightarrow \alpha + \beta$ solvus line. The application of the phase rule to this

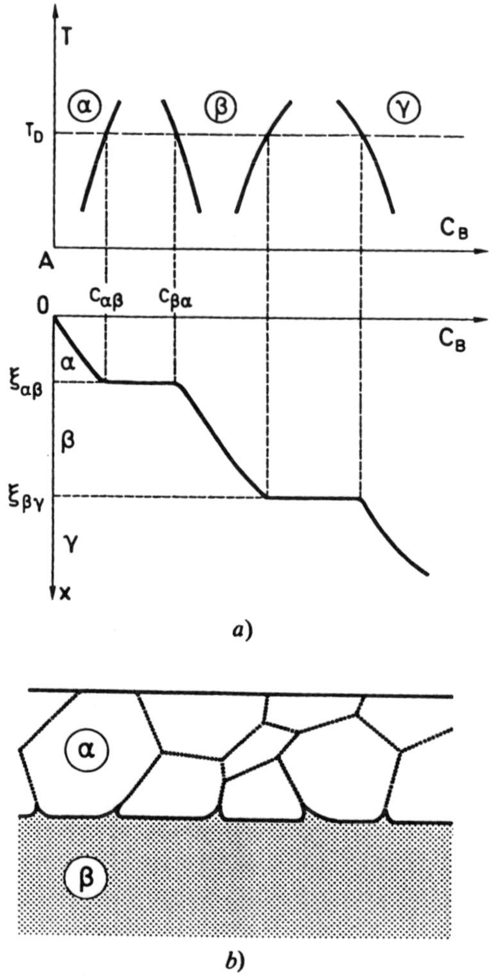

Fig. X.I. — a) Binary multiphase diffusion. b) Effect of grain boundary diffusion.

irreversible process predicts that diffusion takes place only in single-phase domains; the concentration changes abruptly from $c_{\alpha\beta}$ to $c_{\beta\alpha}$, the β solid solution B̲A having been concurrently enriched in A.

Assume now that an intermediate phase ζ can exist at the temperature under consideration. The supersaturation $c_{\alpha\beta} - c_{\alpha\zeta}$ of the α phase (Fig. X.2a) then involves the formation of the compound ζ. Finally, the $c(x)$ profile has the form of a series of curved segments, with discontinuities in the concentration at every interface; these discontinuities correspond to the intersection of the isotherm T_0 with the phase boundaries on the equilibrium diagram (Fig. X.1a). If such

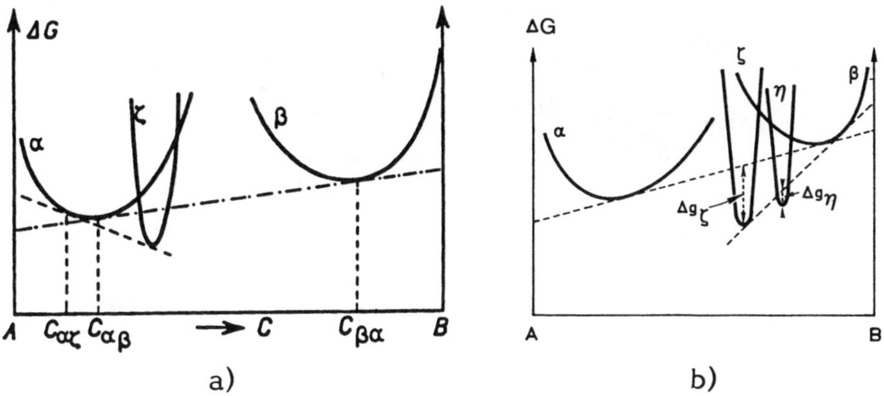

Fig. X.2. — **a) Formation of the α and β solid solutions and the intermediate phase ζ.** b) Successive formation of several compounds **ζ, η and the corresponding driving** free energies (after d'Heurle, 1988).

diffusion experiments are carried out at different temperatures T_1, T_2,..., the phase diagram can be constructed from these "horizontal" sections, a method very different from the classical technique for establishing a phase diagram.

Let us consider the simplest geometry, i.e., one-dimensional binary diffusion. If the temperature and pressure are fixed, the number of degrees of freedom of the system is $v = 2 - \phi$, where ϕ is the number of phases. For 2 phases, $v = 0$; thus the composition is fixed at a certain value of the abscissa, x_0. The interface between the two phases is fixed, it is thus a plane $(x = x_0)$. For a single phase, $v = 1$, and $c(x)$ varies along the abscissa in the range of concentrations corresponding to this phase.

Let us consider further the problem of ***planarity,*** i.e., of ***interface stability*** in the case of interdiffusion in a binary system. Imagine that a "bulge" of β phase develops at the α/β interface, and that the bulge has a certain radius of curvature R. The local equilibrium at the extremity of the bulge is then displaced according to the Thomson-Freundlich law, and the concentration of B in the α phase in thus higher than in the neighborhood of a plane interface (point M of Fig. X.3) (see, e.g., the section on coalescence, § IV.3 below). The perturbation under consideration thus displaces the interface in a direction opposite to the tendency to equilibrium, and expresses itself as the formation of a "precipitate" in the region otherwise under-saturated with B ($c_M > c_{M'}$, Fig. X.3a) which is thermodynamically impossible. The system will thus tend to resorb the bulge. However, planarity is not always attained, especially in the case of preferential diffusion along grain boundaries at low temperature; a phase can form ahead of the diffusion front in the form of "fingers" along the grain boundaries (Fig. X.1b).

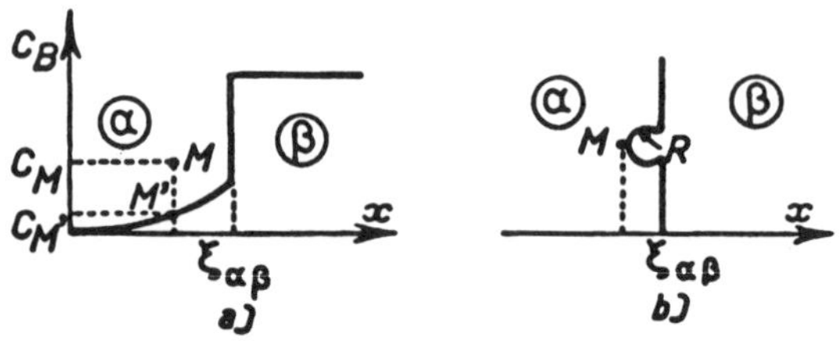

Fig. X.3 — Diffusion in a binary two-phase system. Stability of the interface.

I.2. EXPERIMENTAL STUDIES OF MULTIPHASE DIFFUSION. — Three types of sample are generally used:

- the "infinite" couple, in which the simultaneous presence of all the phases on the phase diagram is expected a *priori*;

- the "semi-infinite" couple, formed of a thin layer of A on the substrate B; the phases often appear successively in the course of time; the initial system A/B is replaced by a new system A_aB_b/B, until the source of A is exhausted, etc.;

- the thin film couple formed from thin-films of A and B.

The $c(x)$ profile is most often established on a transverse section by electron-probe microanalysis. In the case of a thin sample or a thin deposit, $c(x)$ is determined by SIMS. The latter technique allows the analysis of very thin phases or the measurement of compositions very close to the phase boundaries. Other techniques are also possible (nuclear-reaction analysis, Rutherford backscattering, see Ch. IX).

The Boltzmann-Matano analysis can be applied to the $c(x)$ profile of a multiphase binary diffusion couple because the function $c(x)$ is integrable and differentiable from the left and from the right at discontinuities. Hence the Matano formula allows the calculation of $\tilde{D}$ at every point. Kirkendall-effect experiments are possible. However, the calculation of the intrinsic diffusion coefficients requires the use of Darken's equations including the atomic volumes of the components (or, more exactly, the partial molar volumes); these partial molar volumes can vary strongly with composition (see Ch. VI, App. I).

In practice, the equilibrium conditions are not always realized:

1) certain phases expected from the phase diagram are sometimes absent from the diffusion zone, or they appear only after an "incubation time".

The growth generally follows a parablic law $\xi^2 = kt$ or $\xi^2 = k\,(t - t_0)$; the incubation times in the same system can differ from phase to phase, so that these appear in succession. For certain systems, and depending on the type of couple used ([1]), certain phases can just not appear or they regress.

([1]) E.g., infinite or semi-infinite couple, pure-A/pure-B couple or pure-A/AB-alloy couple,...

2) the *composition* of the phases can differ from those indicated on the phase diagram. The compositions of the terminal solid solutions are, in general, the expected ones. Departures from the expected compositions are more often encountered in intermediate compounds, especially intermetallics. The large volume change that accompagnies the formation of many of these compounds gives rise to internal stresses which may be the cause of these deviations. In the U/Cu and U/Al systems, for example, the concentration range of the compound is significantly reduced, coming closer to that predicted from the phase diagram if thin-layer samples (some hundreds of μm thick) are used instead of massive ones; the thin samples bend during the course of diffusion (Adda and Philibert, 1966).

3) extra phases can appear. In certain systems, interdiffusion at sufficiently low temperatures leads to the formation of an ***amorphous*** solid solution, e.g., Ni/Zr at 350°C. Two conditions are necessary; 1) a high enthalpy of formation for the compound AB, and 2) a very large difference between the diffusion coefficients of A and B in the amorphous phase $\left(\bar{D}_{\mathrm{A}} >> \bar{D}_{\mathrm{B}}\right)$. Note that these conditions are classical for certain cases of oxidation, e.g., the formation of amorphous SiO_2 by oxidation of Si below 1200°C.

It is interesting to know whether the growth of a compound takes place by diffusion of both species A and B (growth at two interfaces) or of only one species. To study this, inert markers are placed in the couple, either initially on the weld plane, or at some stage of the diffusion by implantation. The markers can be a rare gas or a heavy element, depending on the techniques for observation and analysis; these are sometimes used not only to measure the concentrations of the diffusing elements, but also to detect the markers. Another useful geometry is couples of the type $B/A/A'$, with two chemically similar elements (or isotopes of the same element) A and A' deposited on the substrate B; in this case, one observes if the order of the elements A and A' is preserved or reversed.

Remarks.

1) Even though parabolic growth kinetics $\left[\xi^2 \propto (t - t_0)\right]$ are observed for each phase in most cases, other kinetics can also be observed. A linear growth law, $x \propto t$, suggests that the growth is controlled by a process at the interface. A law $\xi \propto t^{1/n}$, with $n \approx$ 3-4, suggests growth governed by diffusion along boundaries between growing grains (see, e.g., Hirano and Iijima, 1985).

2) The rate constant in a parabolic or linear law does not always follow an Arrhenius dependence; in an extreme case, it can even decrease with increasing temperature.

I.3 THE KINETICS OF PHASE GROWTH. — The growth of the phases can be governed either by the diffusion of one or both species across each of the phases, by interface processes (chemical reactions among the species, point-defect reactions,...), or by a mixture of the two.

Interfacial reactions are especially dependent on the local supersaturation, from which, in the simplest case, first-order kinetics follow; the reactions rate being constant, the rate of growth of the phase remains the same. Such *linear kinetics* are often observed for very short times because the diffusion flux can be very high due to large gradients or a high value of $\Delta G/\xi$ (see below), so that diffusion is not the limiting process. Parabolic growth kinetics, on the other hand, are observed at longer times, when the growing phase is thicker, the gradient is less, and diffusion is the rate-limiting process.

The mixed kinetics observed in the intermediate case result from the addition of the "slowness" (harmonic sum of the rates) of each process. In the simplest case, this combination leads to an analytical expression for the growth law of the form:

$$\boxed{x^2 + Ax = B\,(t - t_0)},$$

especially if the growth takes place by a reaction at only one interface. Similar equations are obtained for oxidation (§ II.1 below) and the growth of a precipitate from a solid solution (§ IV.1).

The problem of diffusion-controlled growth is solved by writting the balance equation at the interfaces. For the $\alpha\beta$ interface in figure X.1, we have:

$$[c_{\alpha\beta} - c_{\beta\alpha}]\,\mathrm{d}\xi_{\alpha\beta} = (J_\alpha - J_\beta)_\xi\,\mathrm{d}t. \tag{X.1}$$

The left hand side represents the change of the number of B atoms arising from the growth of the β phase, $\xi_{\alpha\beta}(t)$; the right hand side represents the net flux, i.e., the difference in the fluxes in α and in β at the interface. This can be expressed in terms of Fick's law as:

$$\frac{\mathrm{d}\xi_{\alpha\beta}}{\mathrm{d}t} = \frac{1}{(c_{\alpha\beta} - c_{\beta\alpha})}\left\{-\left(D\frac{\partial c}{\partial x}\right)^{\alpha}_{\xi-} + \left(D\frac{\partial c}{\partial x}\right)^{\beta}_{\xi+}\right\},$$

the gradients being taken at the interface. Now the Boltzmann transformation $\lambda = x/\sqrt{t}$ can be applied. Since the concentrations at the interface remain constant, D and $\mathrm{d}c/\mathrm{d}\lambda$ at the interface are also constant:

$$\frac{\mathrm{d}\xi_{\alpha\beta}}{\mathrm{d}t} = \frac{(DK)_\beta - (DK)_\alpha}{c_{\alpha\beta} - c_{\beta\alpha}}\,\frac{1}{\sqrt{t}},$$

with $K = \mathrm{d}c/\mathrm{d}\lambda = (\partial c/\partial x)\sqrt{t}$. By integration:

$$\xi_{\alpha\beta} = P_{\alpha\beta}\sqrt{t}, \tag{X.2}$$

with:

$$P_{\alpha\beta} = 2\,\frac{(DK)_\beta - (DK)_\alpha}{c_{\alpha\beta} - c_{\beta\alpha}}. \tag{X.3}$$

The growth thus follows a parabolic time law; the rate constant $P_{\alpha\beta}$ can be positive, zero, or negative.

Now consider the three-phase system consisting of two terminal solutions α, β, and an intermediate phase ζ. The $\alpha\zeta$ and $\zeta\beta$ phase boundaries follow kinetic laws similar to equation (X.2), so that the thickness of the ζ phase also follows a parabolic law:

$$\Delta\xi_\zeta = (P_{\zeta\beta} - P_{\alpha\zeta})\sqrt{t} = \Delta P \cdot \sqrt{t}. \tag{X.4}$$

If the term ΔP is positive, the ζ phase grows with diffusion time, but if it is negative, any ζ phase that forms would be consumed by its neighbors as soon as it did.

Generally, ΔP increases with temperature. The appearance of certain phases can depend on the diffusion temperature. The $c(x)$ curve obtained from an A/B couple depends on the diffusion coefficients in all the phases; the same is true for ΔP through K, i.e., the gradients at the interfaces. If the original couple is made up of either pure A and a compound AB or two AB compounds, the $c(x)$ curve will be different; such modifications can change the sign of ΔP and cause a phase to appear that would be absent in a couple made up of pure A and pure B.

Note that it is sometimes more convenient to describe the growth of a compound, the composition range of which is rather narrow, by applying the Nernst equation. For the diffusing species i this is:

$$J_i = c_i F_i D_i / kT. \tag{X.5}$$

The driving force here is the gradient of Gibbs free energy. For a compound layer of thickness ξ, this is written $\Delta G_i/\xi$ where ΔG_i is the Gibbs free energy of formation of the compound per atom i. The growth law is then:

$$\mathrm{d}\xi/\mathrm{d}t = \Omega\bar{c}_i\left(\bar{D}_i/kT\right)\left(\Delta G_i/\xi\right), \tag{X.6}$$

where the bars indicate average values, and Ω is the volume of the compound per i atom. This indeed predicts a parabolic growth law $\left(\mathrm{d}\xi/\mathrm{d}t \propto \xi^{-1}\right)$.

If both species A and B contribute to the diffusion, the growth of the compound will take place at two interfaces. For the compound A_aB_b:

$$\mathrm{d}\xi/\mathrm{d}t = |J_\mathrm{A}|\,\Omega_\mathrm{m}/a + |J_\mathrm{B}|\,\Omega_\mathrm{m}/b, \tag{X.7}$$

where Ω_m is the molar volume of the compound. Since $c_\mathrm{A}=a/\Omega_\mathrm{m}$ and $c_\mathrm{B} =$

b/Ω_m, the Nernst equation yields:

$$d\xi/dt = |D_A F_A + D_B F_B|/kT. \tag{X.7a}$$

Let ΔG_f be the Gibbs free energy of the reaction:

$$aA + bB \longrightarrow A_a B_b.$$

Then

$$F_A = \Delta G_f / a\xi \quad \text{and} \quad F_B = \Delta G_f / b\xi.$$

From this we obtain:

$$\begin{aligned} d\xi/dt &= (\Delta G_f/kT)\,[D_A/a + D_B/b]\,/\xi \\ &= [(a+b)/ab]\,(\Delta G_f/kT)\left(\tilde{D}/\xi\right), \end{aligned}$$

with

$$\tilde{D} = N_B D_A + N_A D_B, \quad N_A = a/(a+b), \quad N_B = b/(a+b).$$

The parabolic growth law is then:

$$\boxed{\xi^2 = 2\,\frac{a+b}{ab}\,\frac{\Delta G_f}{kT}\,\tilde{D}t}. \tag{X.8}$$

I.4 PROBLEMS CONNECTED WITH NUCLEATION. — The conditions for the appearance of an intermediate phase or for the gradual appearence with time of several phases, as well as those for the different behavior of couples formed from pure components A and B or from compounds AB, depend not only on the growth kinetics, but also on the conditions of the nucleation of the intermediate phases.

In the course of interdiffusion, for example, one of the initial phases, say α, becomes supersaturated, followed by the "precipitation" of an intermediate phase ζ. However, the process of nucleating phase ζ can be different from the classical nucleation process in a homogeneous α solid solution; in the present case, we are dealing with the formation of a new phase along a plane interface of "infinite" extent.

Let γ_1, γ_2, and γ_3 be the interfacial energies per unit area of the initial interface and of the two interfaces created by the formation of a layer of thickness ξ of the new phase ζ, respectively. The change of the Gibbs free energy of the system per unit area is given by:

$$\Delta G = \xi \Delta g_v + \Delta\gamma, \tag{X.9}$$

with

$$\Delta\gamma = \gamma_2 + \gamma_3 - \gamma_1,$$

where Δg_v (< 0) is the free energy driving the transformation per unit volume (figure X.2a). It is evident from figure X.4 that:

– If $\Delta\gamma < 0$, ΔG is negative and decreases continuously as ξ increases; there is no critical condition for nucleation. This is especially the case for epitaxial growth ($\gamma_2 << \gamma_1$), the case of a diffusion couple in which one constituent is an oriented single crystal.

– If $\Delta\gamma > 0$, there exists a critical thickness $\xi^* = \Delta\gamma / |\Delta g_v|$, below which $\Delta G < 0$. If ξ^* is very small ($\leq$ an interatomic distance), the difference from the previous case is immaterial.

– If $\Delta\gamma > 0$ and $\xi^* \geq$ the interatomic distance, i.e., $|\Delta g_v|$ is quite small, a nucleation process for the new phase must still exist. This could be a heterogeneous process, with the nuclei localized in certain sites in the interface.

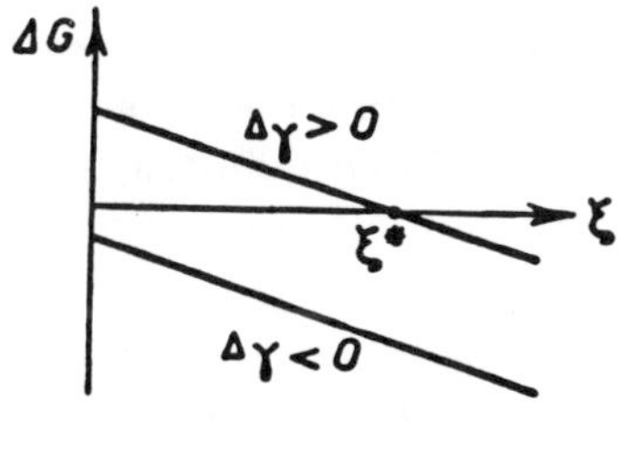

Fig. X.4. — Determination of the critical thickness from the sign of the change of the interfacial energy $\Delta\gamma$.

The differing specific volumes of the different phases introduce an extra mechanical-energy term into the balance equation for the Gibbs free energy. The sign of this term is the opposite of the sign of Δg_v; hence there is an increase in ξ^* and an enhanced trend toward a nucleation controlled process.

These nucleation problems can affect the order in which several intermediate phases appear, and explain why phases expected from the phase diagram are not always present. Once a phase ζ which does not require critical conditions for nucleation is formed, it is possible that the driving free energy for another intermediate phase ξ becomes very small, as can be seen on an isothermal diagram of the free-energy curves of different phases plotted as a function of the composition of the system (Fig. X.2b). The small value of the free energy results in a high value of ξ^* and eventually a difficulty in nucleating the ξ phase which can then not appear, or appear only after the temperature is increased sufficiently. One can thus observe a veritably explosive nucleation, which, however, can be interpreted perfectly well in the framework of classical nucleation theory. The activation energy for nucleation is proportional to $\gamma^3/\Delta g_v^2$; the nucleation probability thus varies rapidly with the temperature. (An example is the formation of $NiSi_2$ at 800°C from the NiSi/Si system which results from reactive diffusion of a Ni layer on Si at 500°C.)

These different cases have been found in the course of detailed studies of the formation of silicides, a subject reviewed by F. d'Heurle (1988).

Let us emphasize that in fact it is possible to predict neither the nature of the first phase that appears during the anneal of an A/B couple, nor the order in which the other phases expected from the equilibrium diagram will appear. The non- or late appearance of a phase can be due either to a difficulty in nucleation, or to a problem of materials balance in that the diffusion fluxes in the neighboring phases and the reaction rates at the interfaces are not such that the phase under consideration can grow. This is why in the case of two contiguous intermediate phases, it can be shown that one can appear only if the thickness of the other is above a certain critical value (see Philibert, 1988; Tarento and Blaise, 1988).

I.5 Ternary systems. — The prediction of the phases that appear in a ternary system is still more uncertain than in a binary, since the "diffusion path" (the curve c_A, c_B, c_C that represents the diffusion couple on the ternary phase diagram) cannot be predicted on the basis of general principles (see Ch. VIII, § IV).

Further, the interfaces between phases are not necessarily planar. The phase rule for fixed temperature and pressure $V = 3 - \phi$ indicates one degree of freedom for a two-phase region; $\bar{c}_A(x)$ and $\bar{c}_B(x)$ are variable in a two phase region, the proportion of the phases varying with x. However, a three-phase region is never found; in regions corresponding to the three-phase fields on the phase diagram, the diffusion path is virtual (dotted trace in Fig. X.5). The composition of the sample jumps from one boundary to the other. If the diffusion path crosses a two-phase region in the phase diagram, supersaturated zones will exist in the couple, depending on the part of the diffusion path under consideration. Hence one observes islets or isolated precipitates of a second phase (i.e., a true two-phase region), or non-isolated precipitates (i.e., a non-planar interface), as shown in figure X.6.

Figure X.5 shows some of the possible morphologies. Note the importance of the position of the diffusion path with respect to the tie lines in the two-phase regions of the equilibrium diagram; plane interfaces are observed when the diffusion path in the two-phase zone is parallel to the tie lines, in analogy to binary systems. The oxidation of binary alloys provides a typical example of this discussion (see, below § II.2).

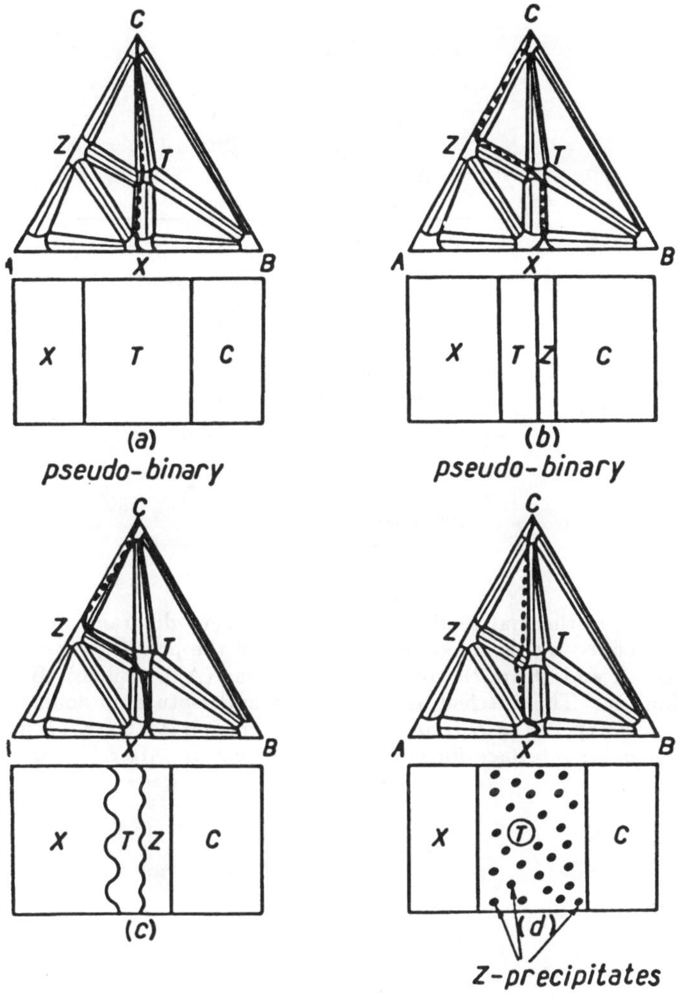

Fig. X.5. — Examples of possible morphologies in an X/C diffusion couple in a ternary system. The corresponding diffusion paths are traced on the isothermal section of the ternary phase diagram. After Van Loo *et al.* (1987).

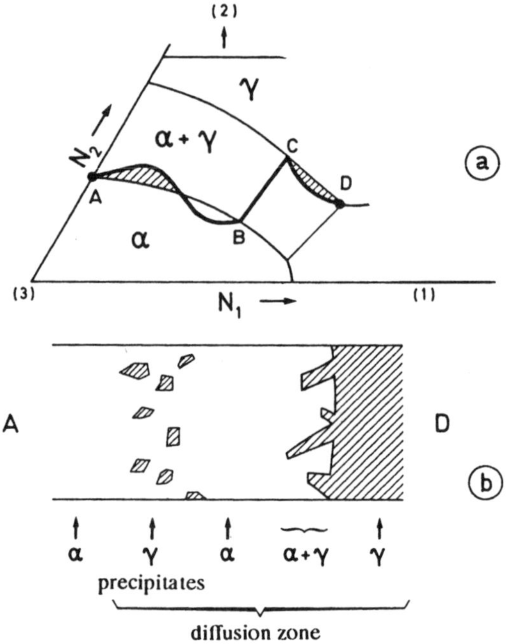

Fig. X.6. — a) Isothermal section of a ternary phase diagram. Virtual diffusion path ABCD, defined by the average composition of the phases perpendicular to the diffusion flux. A and D are the initial compositions (A/D couple), BC is a tie-line (α/γ) equilibrium. The hatched area denotes a supersaturated zone. b) Following this scheme, one expects an α zone containing γ precipitates for concentrations near A, and an irregular interface for concentrations near D. After Kirkaldy and Fedak (1962).

II. Oxidation

Oxidation is a particular case of the diffusion of an element from a gaseous atmosphere and an ensuing chemical reaction with the formation of a compound. The same considerations apply regardless of whether the diffusing element is in column VI (oxygen, sulfur) or in column VII (halogens) of the periodic table, or is one of the elemental constituents of the molecules in the gaseous atmosphere (methane, ammonia, borane, silane,...).

II.1 Oxidation of a pure metal. — Let X be the oxidizing element. It forms a layer of oxide M_aX_b, and under this, eventually a solid solution $\underline{M}X$ (Fig. X.7a). The oxide forms at the exterior interface MX/X or at the internal interface M/MX depending on whether the metal M or the oxidant X is the more rapidly

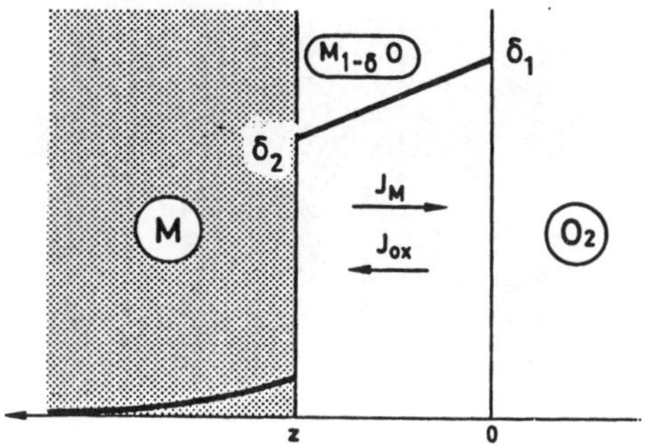

Fig X.7a. — Growth of a layer of oxide $M_{1-\delta}O$ on pure metal M.

diffusing species in the oxide ([2]).

Figures X.7b and c show in a schematic way some of the processes that come into play in two particular cases, chosen from among all the possibilities, i.e., according to the type of defect predominant in the metal (V_M or M_i) and in the oxide (V_M, M_i, V_X or X_i). Silicon is a slightly different case because the transport of the oxidizing agent probably takes place by the migration of O_2 molecules through the amorphous SiO_2 layer. Figures X.7b and c show that at the M/MX interface there exists a flux of defects which can involve either a local supersaturation in the oxide or a trapping at the interface, or "injection" into the M substrate ([3]).

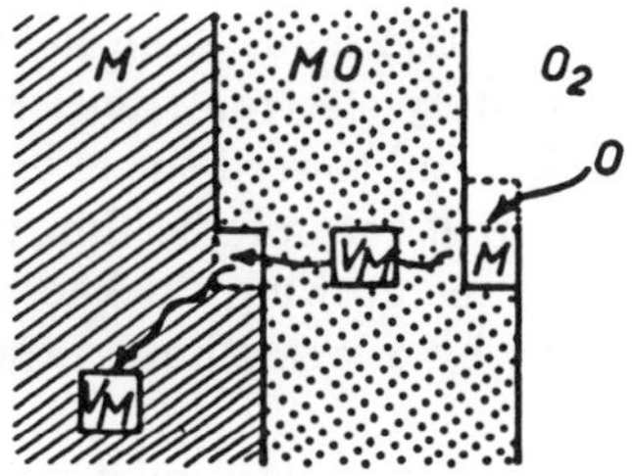

Fig. X.7b. — M is the more rapid diffuser and diffuses in the oxide via V_m. The oxide grows at the external interface. Cation vacancies are injected into the metal.

([2]) The diffusing species is identified experimentally with the aid of inert markers, or better, by isotopic marking: oxidation is carried out first in $^{16}O_2$ and then in $^{18}O_2$ or vice-versa, and one observes if the spatial order in the oxide layer is 16/18 or 18/16.

([3]) The formation of dislocation loops, indicating the injection of vacancies into metals or self-interstitials into silicon, is observed in transmission electron microscopic studies.

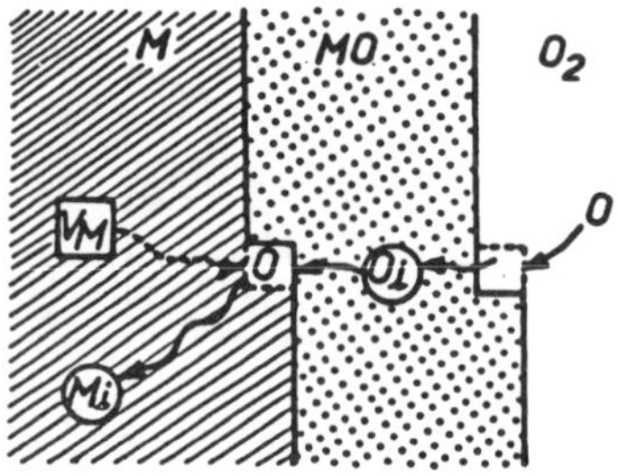

Fig. X.7c. — Oxygen as O_i (or O_2 in SiO_2) is the more rapid diffuser. The oxide grows at the internal interface. Self-interstitials are injected into the metal, or vacancies are removed from it at the interface.

It follows that the growth of the oxide layer depends on at least three processes, or groups of processes:

- one or more chemical reactions at the oxide/atmosphere interface,
- the diffusion of one or both species across the oxide film,
- a reaction at the metal/oxide interface.

Two cases can be considered, depending upon whether the interface reactions limit or do not limit the total kinetics.

1. *Oxidation controlled entirely by diffusion.* — For thin films ($\leq$ 10 nm), the model of Mott and Cabrera assumes electromigration of the diffusing species driven by a strong internal electric field. The field derives from an electric potential of several volts caused by electron transfer (by tunnel effect or thermoelectric emission) from the metal toward the external surface of the film. In the resulting kinetics, the growth rate of the film decreases exponentially as a function of the reciprocal of the growing thickness [exp $(1/z)$]. The growth law $z(t)$ is often approximated by a logarithmic expression, an approximation more or less justified.

For thick films ($\geq$ 10 nm), the growth follows a parabolic law, as already seen in chapter VI. The parabolic rate constant is in first approximation proportional to both the difference in composition across the oxide from the external surface to the metal/oxide interface and the chemical diffusion coefficient $\tilde{D}$ in the oxide (Eq. (VI.51)); it is generally given by Wagner's formula (see below).

It is instructive to derive the parabolic growth law, starting from the general flux equations. Consider the case of a substoichiometric oxide, $M_{1-\delta}O$, a p-type electronic semi-conductor. The fluxes to be considered are those of the cation vacancies ($V_M^{\alpha'}$) of charge $-\alpha\,|e|$ and electron holes h°.

Let n_i and u_i be the concentration (number of particles per unit volume) and the mobility of species i, respectively. Keeping in mind the Nernst field (diffusion of charged species):

$$\begin{cases} J_\mathrm{v} = -D_\mathrm{v}\,\dfrac{\mathrm{d}n_\mathrm{v}}{\mathrm{d}z} + n_\mathrm{v}u_\mathrm{v}E \\ J_\mathrm{h} = -D_\mathrm{h}\,\dfrac{\mathrm{d}n_\mathrm{h}}{\mathrm{d}z} + n_\mathrm{h}u_\mathrm{h}E. \end{cases} \tag{X.10}$$

Since there is no external electric field, there is no net electric current, and:

$$q_\mathrm{v}J_\mathrm{v} + q_\mathrm{h}J_\mathrm{h} = 0$$

or

$$-\alpha\,|e|\,J_\mathrm{v} + |e|\,J_\mathrm{h} = 0, \tag{X.11}$$

i.e.:

$$\alpha D_\mathrm{v}\,\frac{\mathrm{d}n_\mathrm{v}}{\mathrm{d}z} - D_\mathrm{h}\,\frac{\mathrm{d}n_\mathrm{h}}{\mathrm{d}z} = (\alpha n_\mathrm{v}u_\mathrm{v} - n_\mathrm{h}u_\mathrm{h})\,E.$$

Local electroneutrality requires that everywhere in the film:

$$n_\mathrm{h} = \alpha n_\mathrm{v}$$
$$\mathrm{d}n_\mathrm{h}/\mathrm{d}z = \alpha\,(\mathrm{d}n_\mathrm{v}/\mathrm{d}z)\,.$$

Using the Nernst-Einstein equation $u_i = q_iD_i/kT$ and noting that $u_\mathrm{h} >> u_\mathrm{v}$, $D_\mathrm{h} >> D_\mathrm{v}$, we obtain:

$$E \approx \frac{\alpha}{n_\mathrm{h}}\,\frac{D_\mathrm{h}}{u_\mathrm{h}}\,\frac{\mathrm{d}n_\mathrm{v}}{\mathrm{d}z} = \frac{kT}{n_\mathrm{v}\,|e|}\,\frac{\mathrm{d}n_\mathrm{v}}{\mathrm{d}z}. \tag{X.12}$$

The electric field has the same sign as the gradient of vacancy concentration; it thus reduces the flux of the positive species (the holes) and increases the flux of the negative species (the cation vacancies).

The vacancy flux is then:

$$J_\mathrm{v} = -\,(1+\alpha)\,D_\mathrm{v}\,(\mathrm{d}n_\mathrm{v}/\mathrm{d}z)\,, \tag{X.13a}$$

an expression in which the chemical diffusion coefficient $\tilde{D}$ appears:

$$(1+\alpha)\,D_\mathrm{v} = \tilde{D}. \tag{X.13b}$$

If $\tilde{D}$ does not depend on composition, the flux J_v is conserved, and $J_\mathrm{v}\mathrm{d}z$ can be integrated over the thickness z_0 of the film (4):

$$J_v z_0 = \tilde{D} \cdot \Delta n_v = \tilde{D} \cdot n\,(\delta_1 - \delta_2)\,,$$

where n is the number of cation sites per unit volume and δ_1 and δ_2 are the departures from stoichiometry at the surface and at the metal/oxide interface, respectively.

For $J_v dt$ vacancies transported, a volume of oxide equal to $\Omega J_v dt$ (Ω is the molecular volume of the oxide MO) is formed, of unit area and thickness dz_0 :

$$dz_0/dt = J_v\Omega = \tilde{D}\,(\delta_1 - \delta_2)\,/z_0,$$

since $n\Omega = 1$.

Hence the *parabolic growth law*:

$$\partial z_0/dt = K_p/z_0,$$

with :

$$z_0 = (2\ K_p t)^{1/2} \qquad \text{(X.14a)}$$

$$K_p = \tilde{D}\,(\delta_1 - \delta_2)\,. \qquad \text{(X.14b)}$$

The above integration was carried out assuming that $\tilde{D}$ is independent of composition. This restriction can be avoided by using the more general formula for the chemical diffusion coefficient (Eq. (VI.35)) which becomes here:

$$\tilde{D} = \frac{1}{2}\ D_{M^\bullet}\ \frac{\partial \text{ Log } P_{O_2}}{\partial\delta},$$

so that:

$$J_v = -\frac{1}{2}\ n D_{M^\bullet}\ \frac{\partial \text{ Log } P_{O_2}}{\partial z}.$$

Assuming that J_v is conserved, i.e., does not depend on z (vacancies uniquely created at the oxide/atmosphere interface and annihilated at the metal/oxide interface), the integral of the above equation, taken between 0 and z_0 gives $J_v z_0$:

$$J_v z_0 = -\ \frac{1}{2}\ n \int_{(P_{O_2})_2}^{(P_{O_2})_1} D_{M^\bullet}\ d \text{ Log } P_{O_2}$$

$$\frac{dz_0}{dt} = J_v\Omega = \frac{1}{z_0}\ \frac{1}{2} \int_{(1)}^{(2)} D_{M^\bullet}\ d \text{ Log } P_{O_2}.$$

---

(4) Local equilibrium and non-divergence of the flux is assumed; these two hypotheses are not necessarily compatible.

The parabolic rate constant is then:

$$\boxed{K_p = \frac{1}{2}\int_{(1)}^{(2)} D_{M^\bullet}\ \mathrm{d}\ \mathrm{Log}\ P_{O_2}}, \tag{X.15}$$

which is a particular case of Wagner's relation. Note that $D_{M^\bullet}$ is a function of P_{O_2}.

Remark: This relation is applicable to all processes with a reactive gas that lead to the formation of a compound: oxidation, nitriding, carburization, sulfidation, ... In general, for a compound MX:

$$K_p = \int_{(1)}^{(2)} D_{M^\bullet}\ \mathrm{d}\ \mathrm{Log}\ a_X,$$

if M is more rapidly diffusing species.

Wagner's oxidation constant. — We now calculate the flux across the oxide film in the general case, without the hypothesis $D_X << D_M$. Consider a binary oxide, or more generally a compound M_aX_b; the charges of the cations and the anions are:

$$q_+ = Z_+\,|e|$$
$$q_- = -\,|Z_-|\,|e| \quad Z_- < 0,$$

respectively, with the Z's denoting the valences. Then:

$$\frac{n_+}{n_-} = \frac{a}{b} = \frac{|Z_-|}{Z_+},$$

where $n_+ = an$, $n_- = bn$ indicate the numbers of the cations M and anions X per unit volume, and n is the number of "molecules" per unit volume. Note that $a \propto |Z_-|$ and $b \propto Z_+$. A molecular flux J_M/a corresponds to the cation flux J_M, and a molecular flux J_X/b corresponds to the anion flux J_X. Then:

$$J_{mol} = \frac{|J_M|}{a} + \frac{|J_X|}{b}. \tag{X.16}$$

However, the relation between the coefficients a, b and the valences $|Z_-|$ and Z_+ is not unique. To conserve the generality of the derivation, it is better to calculate the flux in equivalents ([5]):

$$J_{\mathrm{eq}} = Z_+ \, |J_+| + |Z_-|\,|J_-| \,. \tag{X.17}$$

The flux equations. Neglecting the cross-terms (independent sublattices), we have:

$$J_i = -\frac{L_i}{T}\,\frac{\mathrm{d}\tilde{\mu}_i}{\mathrm{d}z} = -\frac{L_i}{T}\left[\frac{\mathrm{d}\mu_i}{\mathrm{d}z} - q_i E\right], \tag{X.18}$$

with i denoting cations (+), anions (−), or electrons (e); E is the Nernst field. The zero current condition is:

$$\sum_i J_i \; q_i = 0.$$

From this we obtain, using the symbol ∇ for ease of notation:

$$E = \frac{L_+ \; q_+ \nabla\mu_+ + L_- \; q_- \nabla\mu_- + L_{\mathrm{e}} \; q_{\mathrm{e}} \nabla\mu_{\mathrm{e}}}{\sum_j L_j q_j^2},$$

$$J_i = -\frac{L_i}{T}\left\{\nabla\mu_i - q_i \; \frac{\sum_j L_j q_j \nabla\mu_j}{\sum_j L_j q_j^2}\right\}. \tag{X.19}$$

$\nabla\mu_{\mathrm{e}}$ can be eliminated by introducing the chemical potentials of the neutral species. From the equilibria:

$$\mathrm{M} \rightleftharpoons \mathrm{M}^{Z+} + Z_+ \; e_-$$
$$\mathrm{X} + |Z_-|\,e_- \rightleftharpoons \mathrm{X}^{Z-}$$

we obtain:

$$\begin{cases} \mu_{\mathrm{M}} = \mu_+ + Z_+ \; \mu_{\mathrm{e}} \\ \mu_{\mathrm{X}} = \mu_- - |Z_-|\,\mu_{\mathrm{e}} = \mu_- + Z_- \; \mu_{\mathrm{e}}. \end{cases}$$

([5]) An "equivalent" means the fraction of the compound that transports one positive and one negative unit charge. For example $\mathrm{Ni_{1/2}O_{1/2}}$, or $\mathrm{Al_{1/3}O_{1/2}}$ = 1/6 $\mathrm{Al_2O_3}$. For $\mathrm{M}_a\mathrm{X}_b$, the equivalent is given by $\left(b\,|Z^-|\right)^{-1}\mathrm{M}_a\mathrm{X}_b = \left(aZ^+\right)^{-1}\mathrm{M}_a\mathrm{X}_b$.

The calculation finally yields:

$$\left\{\begin{aligned} J_{\mathrm{M}} &= -\frac{L_+}{T}\left\{\left[1-\frac{L_+\,q_+^2}{\sum L_j q_j^2}\right]\nabla\mu_{\mathrm{M}} - \frac{L_-\,q_+\,q_-}{\sum L_j q_j^2}\nabla\mu_{\mathrm{X}}\right\} \\ J_{\mathrm{X}} &= -\frac{L_-}{T}\left\{\left[1-\frac{L_-\,q_-^2}{\sum L_j q_j^2}\right]\nabla\mu_{\mathrm{X}} - \frac{L_+\,q_+\,q_-}{\sum L_j q_j^2}\nabla\mu_{\mathrm{M}}\right\}. \end{aligned}\right. \tag{X.20}$$

The flux in molecules or equivalents can be calculated from equations (X.16) or (X.17). The fluxes being in the opposite directions, we take $|J_{\mathrm{X}}| = -J_{\mathrm{X}}$, and express J as a function of a single chemical potential, via the Gibbs-Duhem equation:

$$a\nabla\mu_{\mathrm{M}} + b\nabla\mu_{\mathrm{X}} = 0,$$

or:

$$|Z_-|\,\nabla\mu_{\mathrm{M}} + Z_+\;\nabla\mu_{\mathrm{X}} = 0.$$

After some manipulation, equation (X.17) yields:

$$J_{\mathrm{eq}} = \frac{1}{|Z_-|\,e^2}\left(\frac{L_+\,q_+^2 + L_-\,q_-^2}{\sum L_j q_j^2}\right) L_{\mathrm{e}}\,q_{\mathrm{e}}^2\,\frac{\nabla\mu_{\mathrm{X}}}{T}.$$

Note that because of the large mobility of the electrons, $L_{\mathrm{e}} >> L_+,\ L_-$, so that $L_{\mathrm{e}}\,q_{\mathrm{e}}^2/\sum L_j q_j^2 \approx 1$.

The phenomenological coefficients are related to the self-diffusion coefficients by:

$$L_+ = n_{\mathrm{M}}\,D_{\mathrm{M}}/k \quad\text{and}\quad L_- = n_{\mathrm{X}}\,D_{\mathrm{X}}/k,$$

which finally yields:

$$\begin{aligned} J_{\mathrm{eq}} &= n_{\mathrm{X}}\,[Z_+\,D_{\mathrm{M}} + |Z_-|\,D_{\mathrm{X}}]\,\frac{1}{kT}\nabla\mu_{\mathrm{X}} \\ &= -n_{\mathrm{M}}\,[Z_+\,D_{\mathrm{M}} + |Z_-|\,D_{\mathrm{X}}]\,\frac{1}{kT}\nabla\mu_{\mathrm{M}}. \end{aligned} \tag{X.21}$$

The quantities D_{M}, D_{X}, and $\nabla\mu_j$ depend on the composition, and thus on the depth z. However, if the flux does not produce local changes in the composition, the flux remains independent of z; then the integral of J_{eq} taken from 0 to z_0, the thickness of the oxide layer, equals $J_{\mathrm{eq}}\,z_0$ so that the preceding equation can be written:

$$J_{\mathrm{eq}} = K_{\mathrm{r}}/z,$$

with:

$$\boxed{K_r = n_X \int_{\mu'_X}^{\mu''_X} [Z_+ \, D_M + |Z_-| \, D_X] \frac{1}{kT} \mathrm{d}\mu_X} , \tag{X.22}$$

where μ'_X and μ''_X are the chemical potentials of X at the inner interface and at the external surface, respectively.

With $D_M >> D_X$, the previously derived equation is again obtained. For an oxide, for instance:

$$K_r = \frac{1}{2} \, Z_+ n_O \int_{p'}^{p''} D_M \, \mathrm{d} \; \mathrm{Log} \; P_{O_2}. \tag{X.23}$$

In these equations, the units of K_r are numbers of equivalents cm^{-1} s^{-1}, or moles of equivalents cm^{-1} s^{-1}. To convert to the experimentally measured rate constants for oxidation (measurement of gain in weight or thickness), the units must be changed; this procedure is detailed in appendix I. It can be seen that equation (X.22) leads to the parabolic rate constant of equations (VI.54) or (X.14b).

The rate constant for oxidation can be expressed in a number of different forms. It is sometimes given in terms of the electrical transport properties, the conductivity σ and the transport numbers of the mobile species, t_i (see Eq.(IV.48)). The reader can verify the equivalence of equation (X.22) and the general relation for an electronic and ionic conductor:

$$K_r = \frac{1}{|Z_-| \, e^2} \int_{\mu'_X}^{\mu''_X} (t_+ + t_-) \, t_e \, \sigma \, \mathrm{d}\mu_X, \tag{X.24}$$

where it has been assumed that $t_e \approx 1$.

Remarks:

1) If the diffusing species is not charged, the above model is not applicable, and the problem is that of simple diffusion in a gradient of chemical potential (see below, § II.1.2).

2) If the solubility of oxygen in the metal is not negligible (e.g., the oxidation of Ti or Zr), the problem is that of diffusion in a two-phase system treated in chapter I. Equations (I.41) of § VII give the analytic expressions for the diffusion profile in the underlying metal. Consider case c) of figure I.11. In the α solid solution, with c_I the concentration at the metal-oxide interface:

$$\frac{c - c_O}{c_I - c_O} = \frac{\mathrm{erfc}\,(x/2\sqrt{D_\alpha t})}{\mathrm{erfc}\;\gamma\varphi^{1/2}}.$$

The location of the interface on the abscissa, $\xi = 2\gamma\sqrt{D_\beta t}$, and $\phi = D_\alpha/D_\beta$. The denominator is then erfc $\left(\xi/2\sqrt{D_\alpha t}\right)$, and:

$$\frac{c - c_0}{c_\mathrm{I} - c_0} = \frac{\operatorname{erfc}\left(x/2\sqrt{D_\alpha t}\right)}{\operatorname{erfc}\left(\xi/2\sqrt{D_\alpha t}\right)}, \tag{X.25}$$

an expression in which the origin is taken at the exterior surface of the oxide: $x = x_\mathrm{I} + \xi$. This equation allows the measurement of the diffusion coefficient of oxygen in Ti or Zr. The equation is valid only if c_I remains constant during the process; as in all other interdiffusion problems, the variation of the molar volume with composition has to be taken into consideration (such a variation is possible for high solubilities of oxygen). Further, ξ has to be corrected for the expansion due to the formation of the oxide (Pilling and Bedworth ratio) so that the abscissa is homogeneous in time.

3) When the metal can form several oxides at the same temperature, successive layers corresponding to each of these oxides are observed, each being characterized by its own parabolic rate constant.

The same treatment applies to the oxidation of an oxide to a higher oxide (e.g., $Cu_2O \rightarrow CuO$).

4) At "intermediate" temperatures, transport along the grain boundaries of the oxide has to be taken account. If the film can be described as a two-phase medium (the grains and the boundaries), an effective diffusion coefficient can be introduced:

$$D_\mathrm{eff} = D + \alpha D' \delta/d,$$

where α is ≈ 2 for columnar grains, D is the volume diffusion coefficient, D' is the grain-boundary diffusion coefficient, δ is the grain-boundary width, and d is the grain diameter.

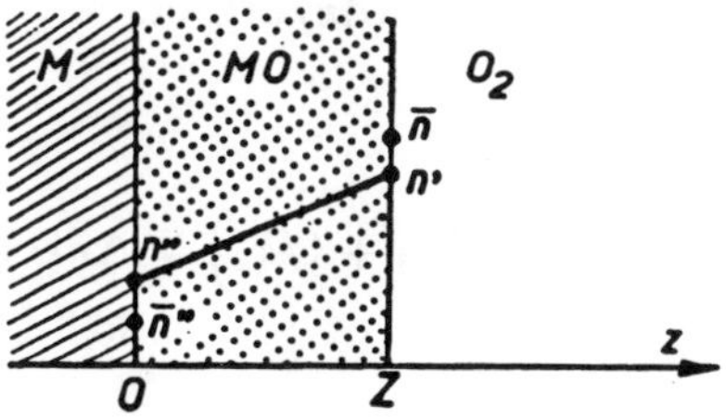

Fig. X.8. — Oxidation: deviations from equilibrium at the internal and the external interfaces.

2. *Oxidation controlled by diffusion and/or interfacial reaction.* — It is

convenient to carry out the derivation on the basis that oxygen is the diffusing element (Fig. X.8); clearly, similar equations would apply, *mutatis mutandis*, to the inverse case. The flux of oxygen across the oxide film is still written (see Eq. (X.13a) above):

$$\begin{aligned} J &= -\tilde{D}(\mathrm{d}n/\mathrm{d}z) \\ &= -\tilde{D}(n' - n'')/z, \end{aligned} \tag{X.26}$$

since steady state, i.e., div $J = 0$ implies $\tilde{D}(\mathrm{d}n/\mathrm{d}z) = \text{const}$. If $\tilde{D}$ is constant, the concentration n is a linear function of z.

The processes at the interfaces can be represented schematically by reactions obeying first order kinetics. At the external interface:

$$(\mathrm{d}n/\mathrm{d}t)_z = k_1(\bar{n} - n'), \tag{X.27}$$

where $\bar{n}$ is the concentration of oxygen absorbed at the surface corresponding to equilibrium with the oxygen partial pressure in the atmosphere, and n' is the actual oxygen concentration close to the surface.

At the internal interface, the kinetics are the same, if one again considers a departure from equilibrium:

$$(\mathrm{d}n/\mathrm{d}t)_0 = k_2\,(n'' - \bar{n}''). \tag{X.28}$$

These equations of continuity are written:

$$(\mathrm{d}n/\mathrm{d}t)_z = \mid J \mid = (\mathrm{d}n/\mathrm{d}t)_0\,, \tag{X.29}$$

permitting the elimination of the (possibly time-dependent) unknowns n' and n''. Since $\Delta\bar{n} = \bar{n} - \bar{n}'' = (\bar{n} - n') + (n' - n'') + (n'' - \bar{n}'')$,

$$\Delta\bar{n} = (\mathrm{d}n/\mathrm{d}t)_{0,\ z}\left(1/k_1 + 1/k_2 + z/\tilde{D}\right). \tag{X.30}$$

The materials balance at the internal interface, where the oxide forms, is written as:

$$\mathrm{d}z = \Omega\,(\mathrm{d}n/\mathrm{d}t)_0\,\mathrm{d}t.$$

This yields:

$$\mathrm{d}z\left(z/\tilde{D} + 1/k_1 + 1/k_2\right) = \Delta\bar{n}\Omega\ \mathrm{d}t, \tag{X.31}$$

which can be integrated to yield:

$$\boxed{z^2 + Az = B(t + \tau)}\,, \tag{X.32a}$$

with the parabolic rate constant:

$$B = 2\tilde{D}\Omega\ \Delta\bar{n} = 2K_{\mathrm{p}}, \tag{X.32b}$$

and the linear constant:

$$K_L = B/A = \Omega\, \Delta\bar{n}\,(1/k_1 + 1/k_2)^{-1}\,. \qquad \text{(X.32c)}$$

The boxed expression (X.32a) is generally known as the relation of Deal and Grove; it was established by them in 1965 for the understanding of successively linear and parabolic regimes observed during the oxidation of silicon. Actually, this relation has been known by metallurgists at least since Evans (1925) and has been derived in several books about the oxidation of metals. In the case of the oxidation of silicon, the linear constant is dominated by the reaction (or sum of reactions) at the internal interface ($k_1 >> k_2$), where the formation of silica takes place.
Remark: equation (X.31) shows that $(\mathrm{d}n/\mathrm{d}t)^{-1}$ is proportional to $z/\tilde{D} + k_1^{-1} + k_2^{-1}$, the harmonic sum of the rates due to each process individually. This is the law of "slowness"; the "slownesses", and not the rates, are additive when the processes are in series.

II.2 OXIDATION OF A BINARY ALLOY AB. — This is a well-known example of reactive ternary diffusion. Several configurations are possible; which one actually occurs depends on the relative stabilities of the oxides of A and B:

1) the solute element B is more easily oxidized than the base metal A under the given oxygen pressure. Two cases have to be considered, depending on the relative values of the products $c_B D_B$ and $c_{Ox} D_{Ox}$ in the alloy.

i) $c_B D_B > c_{Ox} D_{Ox}$: a single surface layer of the oxide of B or of mixed A and B oxides will form by diffusion of B toward the interface. (E. g., Fe-Cr alloys with Cr_2O_3, FeCrAl alloys with Al_2O_3.)

ii) $c_{Ox} D_{Ox} > c_B D_B$: *internal oxidation* (Fig. X.9). The oxygen that penetrates into the alloy is trapped in the form of particles of B oxide inside the alloy (e.g., AgSi, AgBe); the base metal eventually forms a surface-oxide layer (e.g., CuSi, CuBe). Such internal oxidation involves the diffusion of oxygen in the alloy and a more limited diffusion of the element B to form the particles of oxide. A practical measure of the penetration of the oxygen is the thickness of the internally oxidized region. This process takes place even if the solubility of oxygen is very small, e.g., iron alloys.

2) The solute element B is less readily oxidized than the base metal. In this case, either A oxide forms with globules of B dispersed in it, or a layer of A oxide forms, with the solid solution AB under it enriched in B (selective oxidation (Fig. X.10).

a) *Selective oxidation* of an AB alloy (Fig. X.10). — If only A is oxidized, the underlying metal is enriched in B. For example, in FeNi the concentration of the less easily oxidized Ni can become so great that the γ phase appears between the oxide and the center of the alloy.

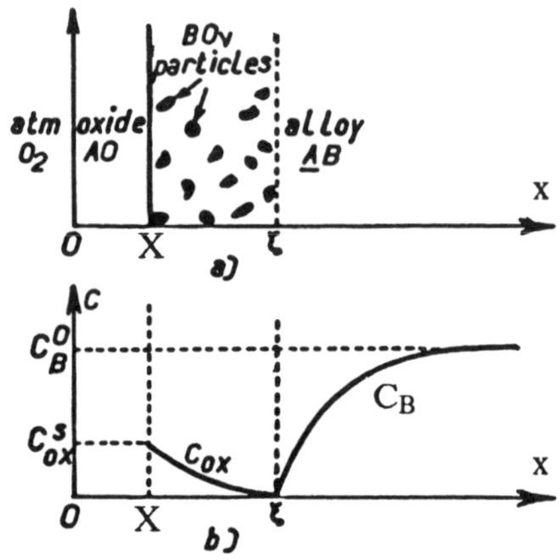

Fig. X.9. — Internal oxidation of an AB alloy and the accompanying concentration profile.

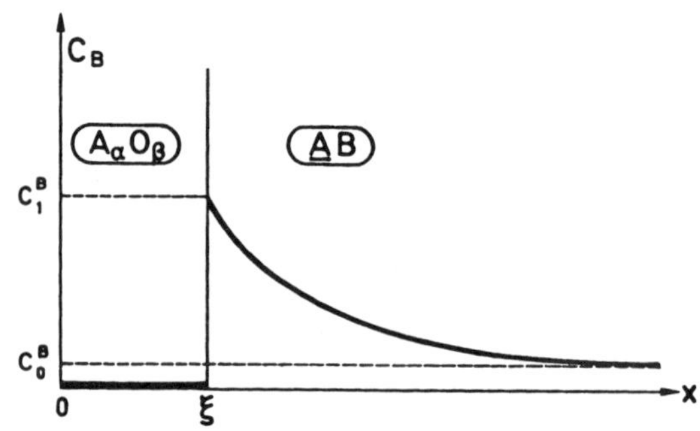

Fig. X.10. — Concentration profile for the selective oxidation of an AB alloy.

To determine the concentration profile of B, the method described in chapter I, § VII is used. The profile has the form:

$$c(x,\ t) = \mathrm{A} + \mathrm{B}\ \mathrm{erf}\left(x/2\sqrt{Dt}\right),$$

where D is the solute diffusion coefficient of B in A, assumed to be constant.

The motion of the oxide/metal interface follows the kinetics:

$$\xi = 2\gamma\sqrt{Dt}.$$

The initial conditions are:

$$t = 0,\ c = c_0 \text{ for all } x.$$

The boundary conditions are:

$$x = \xi^-, \quad c = 0 \text{ (no solubility in the oxide)}$$
$$x = \xi^+, \quad c = c_1, \quad \text{for } t > 0.$$

Assume a steady state, with $c_1 = \text{const}$. Then:

$$c(x,\ t) = c_0 + \frac{c_1 - c_0}{\text{erfc}\ \gamma}\left[1 - \text{erf}\left(x/2\sqrt{Dt}\right)\right].$$

The materials balance at the mobile interface is:

$$c_1 \frac{\mathrm{d}\xi}{\mathrm{d}t} = -D\left[\frac{\partial c}{\partial x}\right]_{\xi^+},$$

from which:

$$\frac{c_1 - c_0}{c_1} = \sqrt{\pi}\ \gamma \cdot \text{erfc}\ \gamma \cdot \mathrm{e}^{\gamma^2},$$

or:

$$\frac{c_1 - c_0}{c_0} = \frac{\sqrt{\pi}\ \gamma \cdot \mathrm{e}^{\gamma^2} \cdot \text{erfc}\ \gamma}{1 - \sqrt{\pi}\ \gamma\ \mathrm{e}^{\gamma^2}\ \text{erfc}\ \gamma}.$$

Finally:

$$\frac{c(x,\ t) - c_0}{c_0} = \frac{\sqrt{\pi}\ \gamma\ \mathrm{e}^{\gamma^2}}{1 - \sqrt{\pi}\ \gamma\ \mathrm{e}^{\gamma^2}\ \text{erfc}\ \gamma} \cdot \text{erfc}\left(x/2\sqrt{Dt}\right). \tag{X.33}$$

The enrichment factor at the interface is given by:

$$\frac{c_1}{c_0} = \frac{1}{1 - \sqrt{\pi}\ \gamma \cdot \text{erfc}\ \gamma \cdot \mathrm{e}^{\gamma^2}}. \tag{X.34}$$

This expression can be simplified if $\gamma >> 1$. Erfc (γ) can be replaced by its asymptotic expansion near zero:

$$\text{erfc}\ \gamma = \frac{\mathrm{e}^{-\gamma^2}}{\gamma\sqrt{\pi}}\left[1 - \frac{1}{2\ \gamma^2} + \cdots\right],$$

from which:

$$c_1/c_0 = 2\ \gamma^2. \tag{X.34a}$$

Example: with $\mathrm{d}\xi/\mathrm{d}t \approx 10^{-4}/\sqrt{t}$ cm s^{-1} and $D \approx 10^{-11}$ cm^2 s^{-1}, this becomes:

$$\gamma = 10^{-4}/\sqrt{10^{-11}} \approx 0.3 \times 10^2 >> 1,$$

and:

$$c_1/c_0 \approx 10^3.$$

b) *Internal oxidation of an* AB *alloy* (Fig. X.9). — The simultaneous equations for the diffusion of the species B and O have to be solved. A condition is imposed on the flux by the stoichiometry of the oxide BO_ν.

Take the origin at the surface. Let X be the thickness of the oxide layer, $\xi - X$ that of the internally oxidized layer. The diffusion equations for O and B are written ([6]):

$$\frac{\partial c_{\text{ox}}}{\partial t} = D_{\text{ox}} \frac{\partial^2 c_{\text{ox}}}{\partial x^2} \qquad X \leq x \leq \xi \tag{X.35a}$$

$$\frac{\partial c_{\text{B}}}{\partial t} = D_{\text{B}} \frac{\partial^2 c_{\text{B}}}{\partial x^2} \qquad x \geq \xi. \tag{X.35b}$$

The boundary conditions are: $c_{\text{ox}} = 0$ for $x \geq \xi$, all t,
$c_{\text{ox}} = c_{\text{ox}}^{\text{S}}$ for $x = X^+$, all t,
(c_{ox}^{S} is the equilibrium concentration of oxygen at the alloy-AB/oxide-AO interface).

$$c_{\text{B}} = 0 \quad \text{for} \quad x \leq \xi, \quad \text{all } t,$$
$$c_{\text{B}} = c_{\text{B}}^0 \quad \text{for} \quad x > 0, \quad t = 0$$

(c_{B}^0 is the initial concentration of B in the alloy).
Assuming that the kinetics of the process are governed by diffusion, and not, for instance, by the formation of oxide particles, we can write:

$X = (2Kt)^{1/2}$, K is the oxidation rate constant (cf. oxidation of a pure metal)

$\xi = 2\gamma(D_{\text{ox}}t)^{1/2}$, γ is a dimensionless constant.

To form the oxide BO_ν at depth ξ, the fluxes of O and B must obey the following condition:

$$-D_{\text{ox}} \left(\frac{\partial c_{\text{ox}}}{\partial x}\right)_{x=\xi^-} = \nu\, D_{\text{B}} \left(\frac{\partial c_{\text{B}}}{\partial x}\right)_{x=\xi^+}. \tag{X.36}$$

The solution of equations (X.35a and b) is then:

$$\begin{cases} c_{\text{ox}} = c_{\text{ox}}^{\text{s}} \cdot \dfrac{\text{erf }\gamma - \text{erf }\left(x/2\sqrt{D_{\text{ox}}\, t}\right)}{\text{erf }\gamma - \text{erf}\left(\sqrt{K/2\, D_{\text{ox}}}\right)} \\ c_{\text{B}} = c_{\text{B}}^{\text{O}} \cdot \left\{1 - \dfrac{\text{erfc }\left(x/2\sqrt{D_{\text{B}}\, t}\right)}{\text{erfc }\left(\gamma\, D_{\text{OX}}^{1/2}/D_{\text{B}}^{1/2}\right)}\right\}. \end{cases} \tag{X.37}$$

([6]) For $x < \xi$, the diffusion of B is not unidirectional.

The condition (X.36) becomes:

$$\frac{c_{\rm ox}^{\rm s}\ D_{\rm ox}^{1/2}\ {\rm e}^{-\gamma^2}}{{\rm erf}\ \gamma - {\rm erf}\ \left(\sqrt{K/2\ D_{\rm ox}}\right)} = \nu\ \frac{c_{\rm B}^0\ D_{\rm B}^{1/2}\ \exp\left(-\gamma^2\ D_{\rm ox}/D_{\rm B}\right)}{{\rm erfc}\ \left(\gamma\ D_{\rm ox}^{1/2}/D_{\rm B}^{1/2}\right)}. \qquad \text{(X.38)}$$

This system of equations can be considerably simplified for the case $\gamma < 1$, which corresponds to quite slow kinetics of internal oxidation. Then:

$${\rm e}^{-\gamma^2} \simeq 1$$

$${\rm erf}\ \gamma \simeq \frac{2}{\sqrt{\pi}}\ \gamma$$

$${\rm erf}\ \left(\sqrt{\frac{K}{2\ D_{\rm ox}}}\right) \simeq \frac{2}{\sqrt{\pi}}\ \sqrt{\frac{K}{2\ D_{\rm ox}}}.$$

Equation (X.38) becomes:

$$\boxed{c_{\rm ox}^{\rm s}\ D_{\rm ox} = \nu c_{\rm B}^0\ \frac{(\xi - X)\xi}{2\ t}\ \frac{1}{F\left(\xi/2\sqrt{D_{\rm B}\ t}\right)}}, \qquad \text{(X.39)}$$

where we have introduced the function:

$$F(u) = \sqrt{\pi}\ u\ {\rm e}^{u^2}\ {\rm erfc}\ u. \qquad \text{(X.40)}$$

The series expansion of this function is quite simple because:

$${\rm erfc}\ u = \frac{{\rm e}^{-u^2}}{u\sqrt{\pi}}\ \left[1 - \frac{2}{2\ u^2} + \frac{1}{2^2}\ \frac{3}{u^4} - \frac{1\cdot 3\cdot 5}{2^3\ u^6} + \cdots\right],$$

with $u = \gamma\sqrt{D_{\rm ox}/D_{\rm B}}$. In internal oxidation, $D_{\rm ox} >> D_{\rm B}$, and with $\gamma < 0.3$, $u > 2$, and the alternating series converges rapidly.

Knowing $D_{\rm B}$, we can use equation (X.39) to calculate the permeability $c_{\rm ox}^{\rm s}\ D_{\rm ox}$ from purely micrographic measurements (ξ and X). If there is no external oxide layer, X is set equal to zero. An approximate form of equation (X.39) can be derived very easily; see Exercice 60.

Note that the solution predicts the existence of a gradient of B in advance of the internal-oxidation front. These predictions have been confirmed experimentally by microprobe analysis.

III. Sintering.

Sintering here means the ensemble of processes that lead, by simple heat treatment at a high enough temperature, to the "welding" of two or more pieces of material in contact. In practice, sintering is used to fabricate parts in their final form (with only slight finishing necessary) and with the desired dimensions (allowing for shrinkage) without melting the constituents (powder metallurgy, ceramic fabrication). One generally starts with powders that are cold pressed to obtain a "green" compact which has 25-60% porosity by volume. The compact is subsequently heated to high temperature; the grains weld together, the empty spaces round off to form "pores" which are progressively eliminated. The elimination of the pores appears as a "densification" of the product. This ensemble of processes is by extension covered by the term "sintering".

The driving force which induces these processes is the decrease of the surface free energy of the aggregate. The two principal stages, welding of the particles by the formation of "necks" and the reduction of porosity, lead to the reduction of the solid/gas surface area. New surfaces form (the solid/solid interfaces), the total surface energy of which is much lower than of the original surfaces. Since the total surface area is larger when the particle size is smaller, it is advantageous to use very fine powders. The disappearance of free surface because of the sintering of powders $1\mu m$ diameter corresponds to a ≈ 1 cal./gram decrease in the free energy for a ceramic, or 0.35 cal./g for iron, all rather small values (≈ 100 J/mole).

The physical processes acting in the course of sintering must first be analyzed in order to predict the *kinetics* of the phenomenon. The possible processes are of three types:

- plastic deformation (viscous or non-viscous flow)([7]);
- evaporation/condensation;
- diffusion: surface, volume, or grain boundary.

The favored mechanism depends on the type of material, its form and grain size, and the conditions of temperature and pressure. Certain mechanisms can be coupled.

To establish analytical expressions for the kinetics, the sintering process has to be *modeled,* considering, for simplicity, that the material to be sintered is an aggregate of *spheres* or parallel *wires* of the same diameter. The case of spheres placed on a plane can also be used. "Necks" will form, circular or rectangular (very elongated) in cross section, depending on the geometry.

This discussion will be limited to the case of spheres (the Frenkel-Kuczynski model). Stage 0 corresponds to the quasi-instantaneous adhesion among the particles, stage 1 to neck growth, and stage 2 to an intermediate stage before stage 3; during stage 3, the pores become isolated (closed porosity) and spheroidal, and are resorbed (Fig. X.11). The classical analyses concern stages 1 and 3.

([7]) This process is very important in hot pressing (sintering under load).

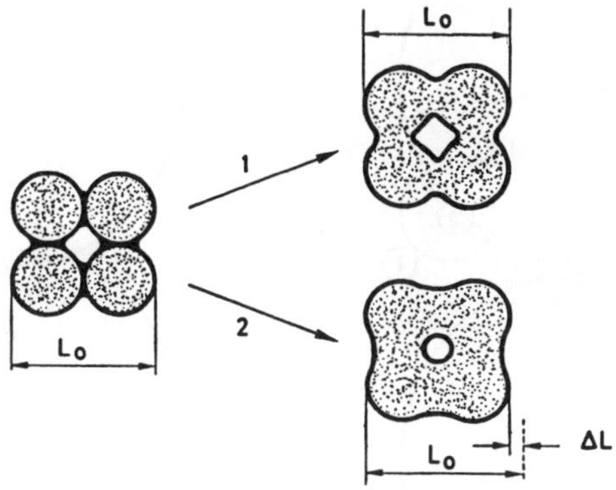

Fig. X.11. — Sintering of spherical particles: 1) change of the pore shape; 2) the same + shrinkage. Note that the change of the pore shape (spheroidization) does not necessarily imply shrinkage. The initially empty spaces become channels or isolated spheres (closed porosity).

III.1 STAGE 1 OF SINTERING IDENTICAL SPHERICAL PARTICLES. — Two geometrical models can be applied depending on the transport process and the nature of the sources and sinks of material: tangent spheres and intersecting spheres (Fig. X.12). Only certain processes lead to densification (intersecting spheres). The processes are listed in table I. Plastic deformation and viscous flow are not included in the table, even though the latter is in effect diffusional transport.

Table I. — *Sintering processes* (Fig. X.12)

Mechanism	Transport of material by	Source of Material	Sink of Material	Model
1	Evaporation/condensation	Surface	Surface of neck	Tangent spheres
2	Surface diffusion	"	"	
3	Volume diffusion	"	"	
4	" "	Dislocations	"	Intersecting spheres
5	" "	Grain boundaries	"	
6	Grain boundary diff.	"	"	

The area, S, of the external surface of the neck joining the two spheres, the radius of curvature of the neck, ρ, and the volume, V, included between this surface and the two spheres can be determined from geometrical calculations and elementary integrals, and can be expressed as a function of the radius, a, of the spheres and the radius, x, of the neck, with the assumption that

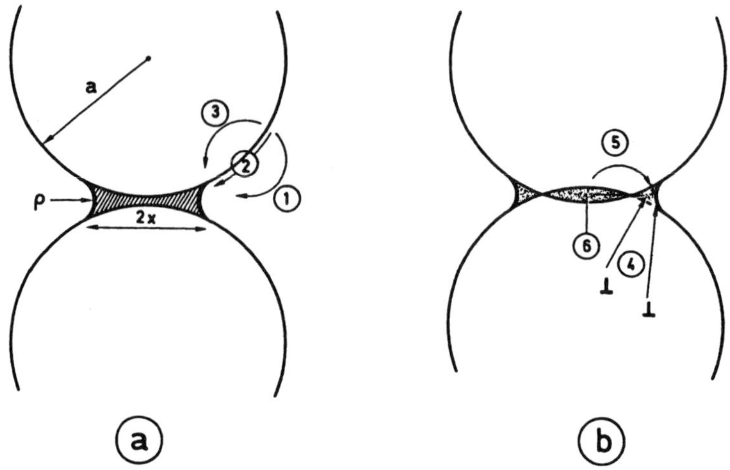

Fig. X.12. — Mechanisms of sintering for spherical particles: a) tangent spheres, b) intersecting spheres (see Tab. I). The arrows indicate the flux of material.

$\rho << x << a$:

Tangent spheres

$$\rho \approx x^2/2a \quad \text{(X.41a)}$$
$$S \approx \pi^2 x^3/a \quad \text{(X.41b)}$$
$$V \approx \pi x^4/2a. \quad \text{(X.41c)}$$

Intersecting spheres

$$\rho \approx x^2/4a \quad \text{(X.42a)}$$
$$S \approx \pi^2 x^3/2a \quad \text{(X.42b)}$$
$$V \approx \pi x^4/8a. \quad \text{(X.42c)}$$

Let us try to calculate the neck radius $x(t)$ which is a measure of the bonding between the particles, and thus of the mechanical strength of the aggregate.

a) *Evaporation/condensation* (Fig. X.12a). — The radius of curvature, ρ, of the neck region is small and negative; the vapor pressure over it is lower than over the spheres with radius a. The resulting pressure gradient leads to a transport of material in the vapor phase. The vapor pressure, p_1, above a surface, the principal radii of curvature of which are R_1 and R_2, is given by the Gibbs-Thomson relation:

$$\text{Log}\,\frac{p_1}{p_0} = +\frac{\Omega\gamma}{kT}\left[\frac{1}{R_1} + \frac{1}{R_2}\right], \quad \text{(X.43)}$$

where p_0 is the equilibrium vapor pressure above a plane surface, Ω is the atomic volume, and γ is the surface tension. For the surface of the neck, $R_1 = -\rho$, $R_2 = x$,

$$\mathrm{Log}\,\frac{p_1}{p_0} \approx -\frac{\Omega\gamma}{kT}\frac{1}{\rho} \qquad \text{since } x >> \rho.$$

The pressure above the spheres, $p \approx p_0$, since $a >> \rho$. The pressure difference is thus $p_0 - p_1 = \Delta p$, and

$$\frac{\Delta p}{p_0} \approx \frac{\Omega\gamma}{kT}\frac{1}{\rho}. \tag{X.44}$$

The rate, v, of condensation of the vapor, in g cm^{-2} s^{-1}, is given by Langmuir's equation:

$$v = \Delta p \left(\frac{M}{2\,\pi kT}\right)^{1/2}, \tag{X.45}$$

where M is the atomic mass. The condensation produces a volume change in the neck:

$$\mathrm{d}V = (vS \cdot \Omega/M) \cdot \mathrm{d}t.$$

Substituting the values of ρ, V, and S (Eqs. (X.41)) and integrating over time, we obtain:

$$\boxed{\frac{x^3}{a} = \beta_1\, t}, \tag{X.46}$$

with

$$\beta_1 = \frac{3\sqrt{\pi}\gamma\Omega^2\, p_0}{\sqrt{2}\, M^{1/2}\,(kT)^{3/2}}. \tag{X.46a}$$

Note that the rate of neck growth $\mathrm{d}x/\mathrm{d}t$ decreases with increasing time, following a $t^{-2/3}$ law.

b) *Diffusion.* — Just as differences in the radius of curvature induce a difference in the vapor pressure, they also induce a difference in the concentration of point defects, or more precisely, in their chemical potential. Herring's equation, written in its simplified form (Ch. VII, Eq. (VII.36a):

$$\mu - \mu_v = \mu_0 + \Omega\gamma\left[\frac{1}{R_1} + \frac{1}{R_2}\right] \tag{X.47}$$

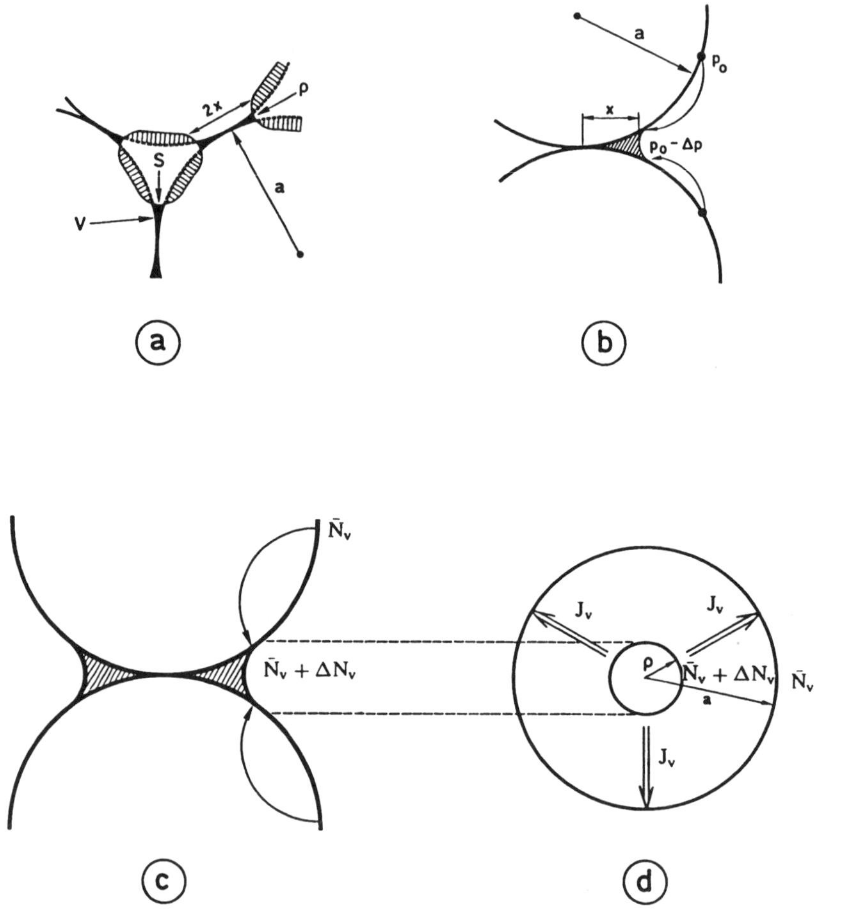

Fig. X.13. — Model of the tangent spheres: a) matter transport from the hatched regions towards the necks; b) evaporation + condensation; c) volume diffusion; d) vacancy flux (equivalent scheme).

allows the calculation of the vacancy concentration near a curved surface:

$$\text{Log}\,\frac{N_v}{\bar{N}_v} = -\frac{\Omega\gamma}{kT}\left[\frac{1}{R_1} + \frac{1}{R_2}\right], \tag{X.48}$$

where $\bar{N}_v$ is the equilibrium mole fraction of vacancies if the surface is plane. Near the neck surface, where the curvature is $-1/\rho$, the vacancy concentration is higher:

$$\frac{\Delta\, N_v}{\bar{N}_v} \approx \frac{\Omega\gamma}{kT}\,\frac{1}{\rho}. \tag{X.49}$$

Since $\rho << x << a$, this expression gives the approximate difference between the vacancy concentrations at the surface of the neck and at either the surface of the grains, or in their interior. The geometrical conditions are different for mechanisms (2) to (6).

(1) *Tangent spheres, volume diffusion* (mechanism 3). To express the vacancy flux, the geometry is first simplified, by reducing the problem to one of diffusion in a hollow cylinder with I.D.= ρ, O.D. $=a$ (Fig. X.13c and d). At steady state, the solution to Fick's equation is of the form $N_v = A + B$ Log r (Ch. I, App. I). Application of the boundary conditions $N_v = \bar{N}_v$ for $r = a$ and $N_v = \bar{N}_v + \Delta N_v$ for $r = \rho$ yields:

$$\frac{\mathrm{d}N_v}{\mathrm{d}r} = \frac{1}{r}\,\frac{\Delta N_v}{\mathrm{Log}\ (\rho/a)}.$$

The flux of vacancies leaving the interior cylinder is thus:

$$J_v = -D_v\left(\frac{\mathrm{d}n_v}{\mathrm{d}r}\right)_\rho = -D_v\frac{1}{\Omega}\left(\frac{\mathrm{d}N_v}{\mathrm{d}r}\right)_\rho = \frac{D\gamma}{kT}\,\frac{1}{\rho^2\ \mathrm{Log}\ (a/\rho)}, \tag{X.50}$$

with $D = \bar{N}_v D_v$, the self-diffusion coefficient. Thus a flux of material equal to J_v spreads itself out on the surface S and produces an increase in volume ΔV per unit time; using equation (X.41b) we obtain:

$$\frac{\mathrm{d}V}{\mathrm{d}t} = \Delta V = J\cdot S\Omega = \frac{D\gamma\Omega}{kT}\,\frac{4\,\pi^2\,a}{x\ \mathrm{Log}\ (a/\rho)}.$$

From equation (X.41a)

$$\frac{\mathrm{d}V}{\mathrm{d}t} = \frac{2\,\pi x^3}{a}\,\frac{\mathrm{d}x}{\mathrm{d}t}.$$

Setting these equations equal and integrating over t (neglecting the variation of the Log term with time), we obtain:

$$\boxed{\frac{x^5}{a^2} = \beta_3\ t}, \tag{X.51}$$

with:

$$\beta_3 = \frac{10\ \pi D\gamma\Omega}{kT\ \mathrm{Log}\ (a/\rho)}, \tag{X.51a}$$

an expression often used with Log $(a/\rho) = 4$ (i.e., $a/\rho \approx 55$), a value that corresponds well enough to the usual experimental conditions.

In fact, this mechanism is not very probable since the vacancies are more likely to be annihilated at internal sinks (grain boundaries) in the volume of the neck than at the surface of the spheres; see the intersecting-sphere model below.

(2) *Tangent spheres, surface diffusion* (mechanism 2). The results published by different authors are different because there is no analytical solution to the problem. Analysis by computer simulation leads to numerical results that are often complicated functions of time, indicating that a power law is only a first approximation:

$$\boxed{\frac{x^7}{a^3} = \beta_2\, t}, \tag{X.52}$$

with:

$$\beta_2 = \frac{80\ \gamma\Omega^{4/3}\ D_s}{kT}. \tag{X.52a}$$

(3) *Intersecting spheres.* In this model, a grain boundary forms between the two spheres and plays the role of sink for the point defects; the material can then diffuse from this boundary toward the neck, either in the volume (mechanism 5) or along the boundary itself (mechanism 6). Finally, a variant of mechanism 3 can be envisaged: volume diffusion, but with dislocations inside the grains as the sink for the point defects instead of the boundary (mechanism 4). In these three cases, the defect sinks are internal and the centers of the spheres move closer to each other during the process; this process leads to shrinkage of the compact, i.e., densification.

(3.1) *Volume diffusion* (mechanism 4). Repeating the calculations of § 1, but using equations (X.42a-c) this time, we obtain easily:

$$\boxed{\frac{x^5}{a^2} = \beta_4\, t}, \tag{X.53}$$

with:

$$\beta_4 = \frac{80\ \pi D\gamma\Omega}{kT\ \mathrm{Log}\ (a/\rho)}. \tag{X.53a}$$

an expression identical to that for β_3 except for a numerical factor.

(3.2) *Volume diffusion* (mechanism 5). The difference in vacancy concentration Δn_v between the surface and center of the neck leads to a gradient of order $\Delta n_v/x$, which gives rise to a flux of vacancies:

$$J_v \approx -\frac{D_v}{\Omega}\,\frac{\Delta N_v}{x}.$$

From this, a number of atoms per unit time JS arrives at the surface of the neck and changes the volume by $JS\Omega$. Following equation (X.42c) for the volume and equations (X.42a and b) for ρ and S, we arrive at:

$$\boxed{\frac{x^4}{a} = \beta_5\, t}, \tag{X.54}$$

with:

$$\beta_5 = \frac{16\ \pi D \gamma \Omega}{kT}.$$

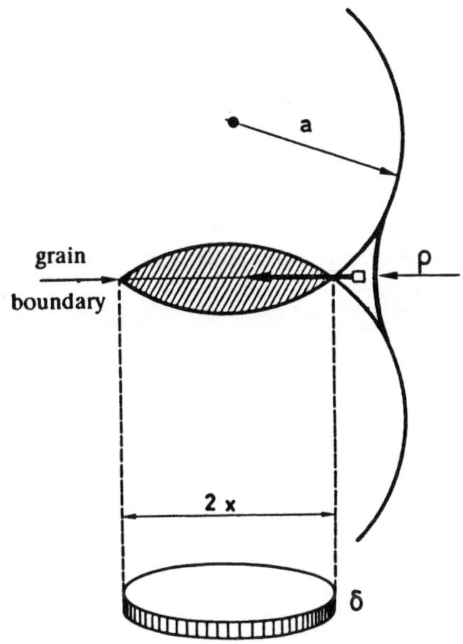

Fig. X.14. — Model with intersecting spheres. Grain boundary diffusion (□ denotes a vacancy).

(3.3) *Grain-boundary diffusion* (mechanism 6, Fig. X.14). The gradient $\Delta n_v / x$ gives rise to a vacancy flux along the boundary given by:

$$J_v \approx -\frac{D'_v \Delta N_v}{\Omega x},$$

where D'_v is the diffusion coefficient of the vacancies in the boundary. This diffusion is limited to a disk of radius x and thickness δ (Fig. X.14). The surface crossed by the vacancy flux J_v is thus $2\pi x \delta$. Finally, the volume change per unit time is given by:

$$\Delta V = \mid J_v \mid \Omega\ 2\ \pi x \delta, \tag{X.55}$$

from which $\Delta V = D'_v\ \delta\ 2\pi \Delta N_v = D'\delta\,(2\pi\gamma\Omega/kT)\,(1/\rho)$, where D' is the grain-boundary self-diffusion coefficient. With the help of equations (X.42a-c), the calculation can be carried out as before. After integration, we obtain:

$$\boxed{\frac{x^6}{a^2} = \beta_6\ t}, \tag{X.56}$$

with:

$$\beta_6 = \frac{96\ D'\ \delta\gamma\Omega}{kT}, \tag{X.56a}$$

and D' defined as above. The numerical factor is not precise because the expression for the gradient is only approximate. In reality, one is dealing with diffusion in a cylindrical geometry, and a term Log (a/ρ) should appear in the denominator of the expression for J_v, and in the denominator for β_6 as well.

In conclusion, it is found in every case that the diameter of the neck follows kinetics that are parabolic in time and a rate that decreases continually. The exponent of the time is always smaller than unity, and its value is characteristic of the mechanism:

$$\frac{x^p}{a^q} = F(\gamma,\ T)\ t,$$

or:

$$\boxed{\left(\frac{x}{a}\right)^p = F(\gamma,\ T)\ a^{q-p}\ t.} \tag{X.57}$$

Take two sets of particles with radii a_1 and a_2 respectively; the times, t_1 and t_2, required for the necks to grow in proportion to the radii (i.e., $(x_1/a_1) = (x_2/a_2)$) will be in the ratio:

$$\frac{t_2}{t_1} = \left(\frac{a_2}{a_1}\right)^{p-q} \tag{X.58}$$

the exponent $m = p - q$ assuming integer values typical of the mechanism. This expression is known as *Herring's scaling law.* Unfortunately, the same exponent m can correspond to different mechanisms:

Mechanism	p	q	m
1	3	1	2
2	7	3	4
3	5	2	3
4	5	2	3
5	4	1	3
6	6	2	4

Surface diffusion is operative only for very fine particles. It does not lead to densification, but it can affect the shape of the neck significantly (see below).
c) *Densification.* — Only mechanisms 4 to 6 lead to densification since in these mechanisms the centers of the spheres move closer to each other. The distance between the centers is approximatively $2(a/\rho)$; the rate of linear contraction thus varies as $2\ \mathrm{d}\rho/\mathrm{d}t = 2\mathrm{d}\left(x^2/4a\right)/\mathrm{d}t = (x/a)(\mathrm{d}x/\mathrm{d}t)$ and is therefore proportional to $t^{(2/p)-1}$. Upon integrating, a densification law in $t^{2/p}$ is obtained:

$$\frac{\Delta V}{V} = \frac{3\ \Delta L}{L} \propto t^{2/p}. \tag{X.59}$$

With volume diffusion mechanisms, the densification is proportional to $t^{2/5}$, as compared to $t^{1/3}$ for grain-boundary diffusion. (Recall that in this stage surface diffusion causes no shrinkage.)

Note that the relative rate of densification

$$(1/a)\ \mathrm{d}\rho/\mathrm{d}t = (1/2)\ (x/a)\mathrm{d}(x/a)/\mathrm{d}t$$

is proportional to $a^{2(q-p)/p}$, i.e., it varies inversely as a power of a. The densification rate is thus greater when the grain size is smaller.

Remarks:

1) Practical conditions rarely correspond to the Frenkel-Kuczynski model (there is a distribution of particle sizes, and still worse, the particles are nonspherical). The model is interesting because of the possibility of using it to test the role of geometrical and physical parameters under well controlled conditions.

2) The classical analysis followed in the preceding pages is very approximate; approximate functional forms must be assumed for the vacancy concentration gradient in order to obtain analytical solutions. However, a more fundamental objection is that the driving force for diffusion is not this gradient, but the gradient of chemical potential, the determination of which requires that the conditions of chemical and mechanical equilibrium be correctly written. The interplay of the surface tensions induces *compressive stresses*, $t_\mathrm{n}(r)$, normal to the plane of the boundary between the spheres.

Mechanical equilibrium at the cross-section of the neck is described by the equation:

$$\int_0^x 2\ \pi r t_\mathrm{n}(r)\ \mathrm{d}r = 2\ \pi x \gamma\ \cos\ \alpha,$$

where $2\alpha = \pi - \beta$; β is the angle between the two spherical surfaces coming together at the neck (typically, $\beta \approx 160°$). It follows that the diffusional flux is undoubtedly higher than that calculated from the simplified model.

Expressions for the radius of the neck were calculated on this basis by D. L. Johnson (1969); his results differed from those above in that the values of the exponent p were a little higher (4.12 instead of 4 for mechanism 5, 6.22

instead of 6 for mechanism 6) and the values of the numerical factor in β_i were different. From these, Johnson obtained an interesting expression for the rate of densification, but under much more restrictive assumptions: only mechanisms 5 and 6 are operative, and the angle $\beta = \pi$. For the first few percent of densification, with $Y = \Delta L/L_0$:

$$Y^{2.06}\dot{Y} = AY^{1.03} + B,$$

where:

$$A = \frac{2.63\ \gamma\Omega}{a^3\ kT}\ D,$$

$$B = \frac{0.70\ \gamma\Omega\ D'\ \delta}{a^4\ kT}.$$

Even though surface diffusion does not contribute to densification, it nevertheless almost always plays an essential role, if only to redistribute materials on the surface of the neck to maintain the equilibrium shape there. This process can go as far as increasing the radius of curvature ρ and thus changing the schematic form in figure X.12 ("overcutting"), a change calculable following Mullins' equations (see Ch. VII, App. II). This problem can only be solved numerically.

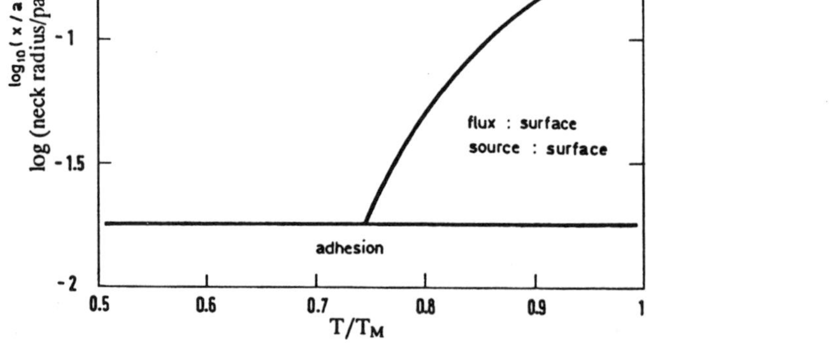

Fig. X.15. — Sintering map. Spherical particles of silver (dia = 20 μm) (after Ashby, 1974).

Further, it is not certain that the grain boundary always behaves as a perfect sink for vacancies. Finally, internal stresses can give rise to plastic deformation, which would not be negligible, at least in the first stages of neck formation.

Owing to the simultaneous operation of several mechanisms and the resulting coupling, the application of these models to practical cases raises some problems, even in simple geometries.

In spite of these restrictions, a comparative analysis of the principal processes that can play a role has lead Ashby and coworkers (1974, 1981) to establish *sintering maps* (Fig. X.15) where the domains in which the various mechanisms are dominant are represented in coordinates of Log (x/a) and T/T_{M}.

III.2 STAGE THREE OF SINTERING. — After a certain annealing time, the stage of closed porosity is reached. The pores are approximately cylindrical or spherical as a result of material transport processes tending to reduce the total surface area of the pores. The pores can be eliminated by various mechanisms which depend especially on recrystallization taking place. Recrystallization changes the position of the pores *vis à vis* the grain boundaries; hence the pores are eliminated by the diffusion of vacancies in the volume or along the grain boundaries. The processes can be easily modeled for either cylindrical or spherical pores; our discussion is limited to the later case (Fig. X.16).

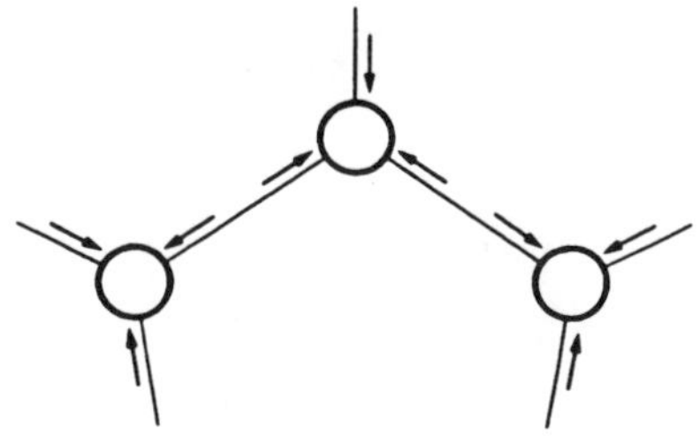

Fig. X.16. — Sintering: elimination of pores by a grain-boundary diffusion mechanism.

a) *Elimination of a spherical pore.* — A simplified model can be used: the vacancy source is the surface of a pore of radius r, the vacancy sink is the surface of a large sphere of radius $R > r$ (e.g., the surface of a grain). The solution of the diffusion equation for steady state in spherical coordinates is (see Ch. I, App. II):

$$c = B + A/\rho,$$

the boundary conditions are:

$$\begin{aligned} \rho = r, \qquad & N_v = \bar{N}_v + \Delta N_v \\ \rho = R, \qquad & N_v \approx \bar{N}_v, \end{aligned}$$

with (from Eq. (X.48)):

$$\Delta N_v = \bar{N}_v \, \frac{2\,\gamma\Omega}{kT} \, \frac{1}{r}.$$

It is easily found that:

$$\left(\frac{\mathrm{d}N_v}{\mathrm{d}\rho}\right)_r = \frac{1}{r^2} \, \Delta N_v \, \frac{rR}{R - r}.$$

The total flux of vacancies emitted by the surface of the pore is given by:

$$JS = -4\,\pi r^2 \cdot \Omega^{-1} \left(\frac{\mathrm{d}N_v}{\mathrm{d}\rho}\right)_r \cdot D_v .$$

Then the volume eliminated per unit time $JS\Omega = \Delta V$. The calculation yields, with $D = \bar{N}_v D_v$:

$$\Delta V = -D \, \frac{8\,\pi\gamma\Omega}{kT} \, \frac{R}{R - r}.$$

For a spherical pore:

$$\frac{\mathrm{d}V}{\mathrm{d}t} = 4\,\pi r^2 \, \frac{\mathrm{d}r}{\mathrm{d}t};$$

from which:

$$\frac{\mathrm{d}r}{\mathrm{d}t} = -\frac{1}{r^2} \, \frac{2\,D\gamma\Omega}{kT} \, \frac{R}{R - r}.$$

Neglecting r with respect to R and integrating, one obtains:

$$\boxed{r^3 - r_0^3 = -\left(\frac{6\,D\gamma\Omega}{kT}\right) t}, \qquad \text{(X.60)}$$

where r_0 is the initial radius of the pore.

This solution allows the variation of the porosity with time to be calculated on the condition that a geometrical model for the pore distribution is given (see below). Other processes can, however, be active; if the pore is situated on a boundary, the dominant process can be *grain boundary diffusion.*

In a simple model, the pore of radius r is on a facet of a grain, represented by a disk of radius R. The diffusion flux is then given in cylindrical coordinates, the boundary being simulated by a disk of thickness δ and large diameter R, at the center of which is the pore of radius of r. The appropriate solution of the diffusion equation $N_v = A + B$ Log ρ here yields:

$$(\mathrm{d}N_v/\mathrm{d}\rho)_r = \frac{1}{r}\,\frac{\Delta N_v}{\mathrm{Log}\,(r/R)}$$

$$J_v = -\frac{1}{\Omega}\,D'_v\,\frac{\Delta N_v}{\mathrm{Log}\,(r/R)}\,\frac{1}{r}.$$

The volume of material transported across the area $2\,\pi r\delta$ per unit time is:

$$\Delta V = 2\,\pi r\,\delta\Omega J_v.$$

A calculation similar to the preceding, neglecting the variation of Log r, i.e., with Log (R/r) = constant, yields:

$$\boxed{r^4 - r_0^4 = -\frac{8\,D'\,\gamma\Omega}{kT}\,\frac{1}{\mathrm{Log}\,(R/r)}\,t}, \qquad \text{(X.61)}$$

with the grain boundary diffusion coefficient $D' = N_v D'_v$. The kinetic law then has a $t^{1/4}$ dependence rather than $t^{1/3}$, a difference found in all transport kinetics.

b) *Calculation of the porosity.* — For this calculation, we follow Coble, who represented the grain by a cuboctahedron with edge a, and we assume that a pore is present at each vertex of this polyhedron. Since there are 24 vertices and each one is common to 4 polyhedra, the pore volume per polyhedron is:

$$V_{\mathrm{p}} = \frac{24}{4}\,\frac{4}{3}\,\pi r^3 = 8\,\pi r^3.$$

Since the volume of the cuboctahedron is $8\sqrt{2}a^3$, the porosity, P, is given by:

$$P = 8\,\pi r^3/8\sqrt{2}\,a^3 = \frac{\pi}{\sqrt{2}}\,\frac{r^3}{a^3}.$$

The initial porosity equals:

$$P_0 = \frac{\pi}{\sqrt{2}} \frac{r_0^3}{a^3},$$

from which, following equation (X.60):

$$\boxed{P - P_0 = -\frac{6\pi}{\sqrt{2}\,a^3}\left(\frac{D\gamma\Omega}{kT}\right)t}. \qquad \text{(X.62)}$$

In the case of grain boundary diffusion, there exists no simple analytical expression and it is better to have recourse to numerical simulation.

Conclusion: As remarked above, the process of sintering cannot be explained on the basis of a single mechanism. Further, in practice, sintering is carried out under conditions often far removed from the simple models described in this chapter. The role of sintering aids, indispensable in practical operations, is almost always poorly understood. In many cases, their solubility is small, their effect on the diffusion coefficients doubtful. It is reasonable that they often contribute to the formation of a second phase in the form of an intergranular film. Transport processes can occur in this film by dissolution and precipitation. Further, the diffusion of the sintering aids can cause recrystallization of the product by diffusion-induced grain boundary migration (see Ch. VII, part 1, § VI). The role of temperature gradients, especially during rapid or flash sintering, is also poorly understood.

IV. Precipitation and Aging.

IV.1 GROWTH OF A PRECIPITATE. — Consider the growth of a β phase in an AB solid solution supersaturated in alloying element B, starting from nuclei. The growth is governed by the slower of the two processes: arrival of B atoms by diffusion and attachment of B atoms to the β precipitate (or "interface reaction").

Assume that diffusion is the slower process. A concentration gradient is established around the precipitate particle; the particle is surrounded by a depleted (or enriched) zone if the β phase is rich (poor) in B (Fig. X.17a)). Let c_β and c_α be the concentrations of B at the α/β interface in the precipitate β and the matrix α, respectively. The materials balance at the interface of radius R is given, with the concentrations expressed in number of atoms per unit volume, by:

$$4\,\pi R^2 \cdot (c_\beta - c_\alpha)\,\mathrm{d}R = 4\,\pi R^2\,|J_\mathrm{B}|\ \mathrm{d}t$$

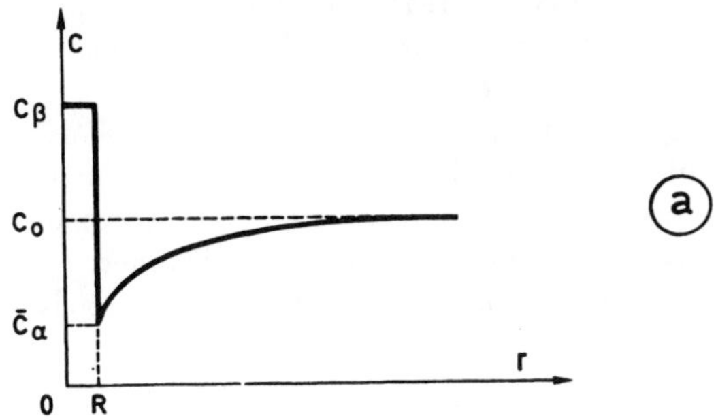

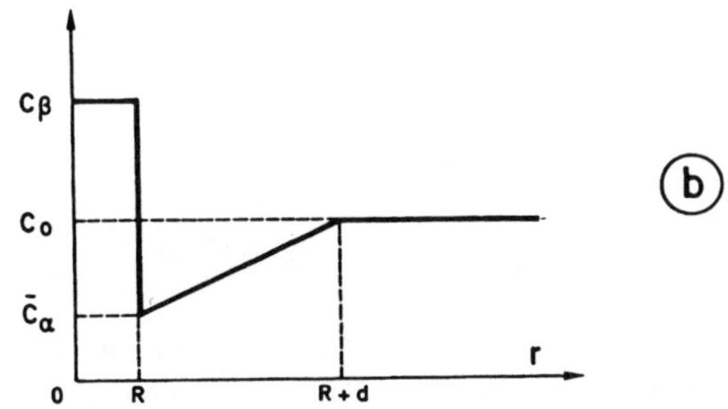

Fig. X.17. — Growth of a precipitate by diffusion: a) concentration profile $c(r)$; b) linear gradient.

with:

$$J_{\mathrm{B}} = -D\,(\nabla c_{\mathrm{B}})_R\,.$$

The coefficient D is the interdiffusion coefficient, not very different from the solute diffusion coefficient of B in α in a sufficiently dilute solution. Then:

$$\frac{\mathrm{d}R}{\mathrm{d}t} = \frac{D}{c_\beta - c_\alpha}\,(\nabla c_{\mathrm{B}})_R\,. \tag{X.63}$$

In order to know the growth rate of the precipitate, the concentration gradient at the interface has to be known. The problem can be solved easily under the following assumptions:

(1) The concentrations at the interface, c_α and c_β, are the equilibrium values. This is the assumption of ***local equilibrium***; capillarity effects, i.e., the variation of the equilibrium concentration with curvature, are neglected and $c_\alpha = \bar{c}_\alpha$.

(2) The precipitates are far enough away from each other so that the diffusion fields do not overlap; far from a precipitate, the concentration remains equal to the initial concentration c_0 (Fig. X.17a).

(3) The diffusion flux is *invariant in time*, or $\partial c/\partial t = 0$, so that the diffusion equation reduces to $\mathrm{div}(D\nabla c) = 0$ or to Laplace's equation, $\nabla^2 c = 0$, if D does not depend on concentration.

(4) If the supersaturation is small, $c_\beta - c_\alpha >> c_0 - c_\alpha$, Laplace's equation, independent of time, can be solved to determine the concentration profile in the matrix, assuming that R = constant. The gradient at the interface is obtained from this, and then $\mathrm{d}R/\mathrm{d}t$ can be obtained from equation (X.63). The inequality above implies in effect that $R(t)$ is a slowly varying function of time.

• *Case of a spherical precipitate.* Laplace's equation in spherical coordinates:

$$\frac{\mathrm{d}}{\mathrm{d}r}\left(r^2\,\frac{\mathrm{d}c}{\mathrm{d}r}\right) = 0 \qquad \text{(X.64)}$$

leads to

$$r^2\,(\mathrm{d}c/\mathrm{d}r) = \text{constant.} \qquad \text{(X.64a)}$$

With the boundary conditions:

$$\begin{array}{lll} r = R & c = \bar{c}_\alpha & \text{(local equilibrium)} \\ r = R_\infty & c = c_0 & \text{(initial concentration)}, \end{array} \qquad \text{(X.65)}$$

equation (X.64a) becomes:

$$\frac{\mathrm{d}c}{\mathrm{d}r} = \frac{1}{r^2}\,\frac{c_0 - \bar{c}_\alpha}{\dfrac{1}{R} - \dfrac{1}{R_\infty}}. \qquad \text{(X.66)}$$

With $R_\infty >> R$, this expression evaluated at the interface $r = R$ reduces to:

$$\left(\frac{\mathrm{d}c}{\mathrm{d}r}\right)_R = \frac{c_0 - \bar{c}_\alpha}{R}. \qquad \text{(X.67)}$$

The concentration profile $c(r)$ is a part of a hyperbola:

$$c(r) - c_0 = (\bar{c}_\alpha - c_0)\cdot(R/r)\,, \quad \forall t. \qquad \text{(X.67a)}$$

Introducing this value of the gradient in the balance equation (X.63), we obtain:

$$\frac{\mathrm{d}R}{\mathrm{d}t} = D\,\frac{c_0 - \bar{c}_\alpha}{c_\beta - \bar{c}_\alpha}\,\frac{1}{R}. \qquad \text{(X.68)}$$

Integrating equation (X.68) leads to

$$\boxed{R(t) = \left(2\ D\ \frac{c_0 - \bar{c}_\alpha}{c_\beta - \bar{c}_\alpha}\right)^{1/2}\ t^{1/2}}. \tag{X.69}$$

Growth in one dimension follows a $t^{1/2}$ law, i.e., *parabolic growth*, and *growth* of a volume a $t^{3/2}$ law.

Remarks: Other sets of assumptions also allow solution of the problem:

– Assumptions (3) and (4) above can be breached. The problem then becomes growth with a moving interface (the motion being governed by diffusion) of the type treated in chapter I, § 7, the only difference being that here the problem is in spherical coordinates, with the diffusion equation written as:

$$\frac{\partial c}{\partial t} = D\ \left(\frac{\partial^2 c}{\partial r^2} + \frac{2}{r}\ \frac{\partial c}{\partial r}\right). \tag{X.70}$$

– One can abandon assumption (3) but keep (4), i.e., $\mathrm{d}R/\mathrm{d}t \approx 0$ (stationary interface) and solve the time-dependent diffusion equation. In other words, it is assumed that the diffusion field around the particle is the same as the one that would have existed if the interface had been fixed at $r = R$ since $t = 0$. The effect of the displacement of the interface on the diffusion field is neglected in this approach.

– The diffusion field is approximated by a *linear profile* over a distance d from the interface (Fig. X.17b). The quantity d is determined by the condition of conservation of mass and the boundary conditions (X.65): $c(r,\ t) = c_0$ for $r \geq R + d$, with both R and d being functions of time:

$$\frac{4}{3}\ \pi R^3\ (c_\beta - c_0) = \frac{4}{3}\ \pi\ \left[(R+d)^3 - R^3\right] \cdot \frac{c_0 - \bar{c}_\alpha}{2},$$

from which

$$d = \gamma R,$$

with:

$$\gamma = \frac{1 - f(k)}{f(k)},$$

and:

$$f(k) = \left(\frac{4\ k}{1 - 4\ k}\right)^{1/3},$$

with k a measure of the supersaturation:

$$k = \frac{2\,(c_0 - \bar{c}_\alpha)}{c_\beta - \bar{c}_\alpha}. \tag{X.71}$$

The solutions corresponding to these three sets of assumptions are compared in table II to the first solution described here. The best approximations are the stationary interface for small supersaturations and the linear gradient for large supersaturations. The solution of the invariant field works only for very small supersaturations.

• *Case of a plate-like precipitate.* The method of chapter I, § VII leads easily to an exact solution of the time-dependent equation, but it is not possible to satisfy Laplace's equation and the initial and boundary conditions. The solutions corresponding to the other two approximations are compared with the exact solution in table II. The same conclusion is found as above: at low supersaturations the approximation of the stationary interface is best; at high supersaturations, the linear gradient.

• *Case of a cylindrical precipitate.* The solution of Laplace's equation in cylindrical coordinates $r\,(\mathrm{d}c/\mathrm{d}r) = 0$ leads, as for the case of spherical symmetry, to a gradient at the interface proportional to $1/R$, and hence to a parabolic growth law following equation (X.63).

• *Case of an ellipsoidal precipitate.* A general solution has been given by Ham (1959): in all cases, (i.e., whatever the eccentricity), the form is preserved in the course of growth (eccentricity independent of time), provided that the concentration is uniform over the entire interface and that surface diffusion is negligible. The growth law is *parabolic*, and the volume grows as $t^{3/2}$, whatever the eccentricity, including the extreme cases, disks and needles.

• *Reaction at the interface.* If the aggregation of the solute atoms onto the precipitates takes place slowly with respect to their rate of arrival by diffusion, a supersaturation at the interface results: $c_\alpha > \bar{c}_\alpha$. It can be assumed that the number of solute atoms attaching themselves to the precipitate per unit time is given by:

$$\frac{\mathrm{d}n_{\mathrm{B}}}{\mathrm{d}t} = 4\,\pi R^2 \cdot K \cdot (c_\alpha - \bar{c}_\alpha)\,,$$

analogous to the kinetics of a first-order reaction. This number must equal $4\,\pi R^2 \mid J_{\mathrm{B}} \mid$, a condition that fixes c_α. At steady state, following equation (X.67):

$$\mid J_{\mathrm{B}} \mid = D\left(\frac{\mathrm{d}c}{\mathrm{d}r}\right)_R = D\left(\frac{c_0 - c_\alpha}{R}\right),$$

and:

$$\frac{\mathrm{d}n_{\mathrm{B}}}{\mathrm{d}t} = 4\,\pi R D\,(c_0 - c_\alpha)\,.$$

Table II. — *Growth of precipitates* (after Aaron *et al.*, 1970)
$k = 2\,(c_0 - \bar{c}_\alpha)\,/\,(c_\beta - \bar{c}_\alpha) > 0.$

Assumptions	Spherical precipitates $R = \lambda_j \sqrt{Dt}$	Plate-like precipitates $\xi = \lambda_j \sqrt{Dt}$
Exact solution of eq. (X.70)	$\lambda_1 = 2\,\lambda$ $\lambda^2\,[1 - \lambda\sqrt{\pi}\,\exp \lambda^2 \cdot \operatorname{erfc}\lambda] = k/4$ $c(r,\,t) - c_0 = \dfrac{2\,\lambda\,(\bar{c}_\alpha - c_0)}{\exp(-\lambda^2) - \lambda\sqrt{\pi}\,\operatorname{erfc}\lambda} \cdot \left[\dfrac{\sqrt{Dt}}{r}\exp\left(-\dfrac{r^2}{4\,Dt}\right) - \dfrac{\sqrt{\pi}}{2}\operatorname{erfc}\left(\dfrac{r}{2\,\sqrt{Dt}}\right)\right]$ $\lim\limits_{(k\to 0)} \lambda_1 = \lambda_2$ and λ_3	$\lambda_1 = 2\,\lambda$ $\sqrt{\pi}\,\lambda\,\exp\lambda^2\,\operatorname{erfc}\lambda = k/2$ $\lim\limits_{k\to 0} \lambda_1 = \lambda_2$
Eq. (X.70) + dR/d$t \approx 0$ (stationary interface)	$\lambda_2 = \dfrac{k}{2\sqrt{\pi}} + \left(\dfrac{k^2}{4\,\pi} + k\right)^{1/2}$ $c(r,\,t) - c_0 = (\bar{c}_\alpha - c_0)\,\dfrac{R}{r}\,\operatorname{erfc}\left(\dfrac{r-R}{2\sqrt{Dt}}\right)$ $\lim\limits_{(k\to 0)} \lambda_2 = \lambda_3,$	$\lambda_2 = k/\sqrt{\pi}$
Laplace's equation (X.64) (invariant field)	$\lambda_3 = k^{1/2}$ $c(r) - c_0 = (\bar{c}_\alpha - c_0)\,\dfrac{R}{r}\,,\quad \forall t$	Not defined
Linear gradient	$\lambda_4 = (k/\gamma)^{1/2}$ $\gamma = \dfrac{1 - f(k)}{f(k)}$ and $f(k) = \left(\dfrac{4\,k}{1 - 4\,k}\right)^{1/3}$ $c(r,\,t) - c_0 = (\bar{c}_\alpha - c) \cdot \dfrac{R + d - r}{d}$	$\lambda_4 = \dfrac{k}{2\,(1 - k/2)^{1/2}}$ $\lim\limits_{k\to 0} \lambda_4 = k/2$

Elimination of c_α between these two equations yields $\mathrm{d}n_\mathrm{B}/\mathrm{d}t$:

$$\frac{\mathrm{d}n_\mathrm{B}}{\mathrm{d}t} = \frac{4\,\pi R^2\ KD}{KR+D}\ (c_0 - \bar{c}_\alpha)\,. \tag{X.72}$$

It is this expression which equals $4\,\pi R^2\,(c_\beta - c_\alpha)\ (\mathrm{d}R/\mathrm{d}t)$. Since $c_\beta > c_\alpha$, $c_\beta - c_\alpha \approx c_\beta - \bar{c}_\alpha$, and one again obtains equation (X.68) with the factor D/R replaced by $\dfrac{KD}{KR+D}$.

This equation is integrated to the form:

$$R^2 + AR = B\,(t - t_0)\,, \tag{X.73}$$

with the parabolic rate constant:

$$B = 2\ D\ \frac{c_0 - \bar{c}_\alpha}{c_\beta - \bar{c}_\alpha},$$

and the linear constant:

$$B/A = K\ \frac{c_0 - \bar{c}_\alpha}{c_\beta - \bar{c}_\alpha}.$$

If $KR << D$ (especially the case for small precipitates), the linear regime will dominate:

$$R = K\ \frac{c_0 - \bar{c}_\alpha}{c_\beta - \bar{c}_\alpha}\ t. \tag{X.73a}$$

For $KR >> D$ the growth process is governed by diffusion.

IV.2 DISSOLUTION OF A PRECIPITATE (Fig. X.18). — The problem is close to that for growth of a precipitate; the initial and boundary conditions are the same except that for $t = 0$, $R_0 > 0$.

- *Spherical precipitate.* The solution of Laplace's equation is:

$$c(r) - c_0 = (\bar{c}_\alpha - c_0)\ \frac{R}{r},$$

identical to equation (X.67a), except that here $\bar{c}_\alpha > c_0$ whereas the opposite is true for growth. After substitution of this result in the balance equation (X.63) and integration, we obtain:

$$R^2 = R_0^2 + kDt, \tag{X.74}$$

where k (< 0 here) has the same significance as above (Eq. (X.71)). Under the assumption of a stationary interface, the profile is analogous to the case of growth, but the kinetics are no longer simple.

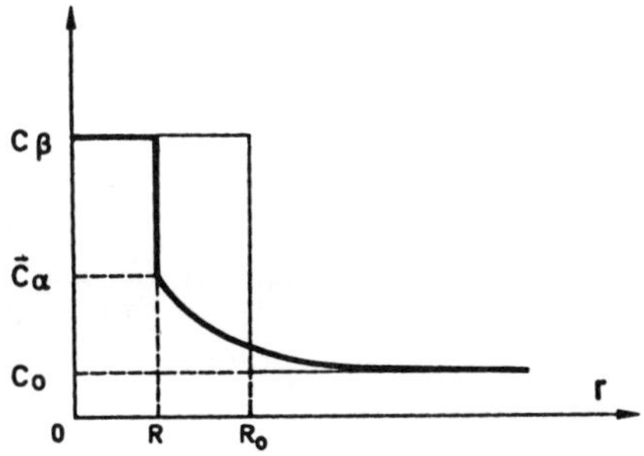

Fig. X.18. — Dissolution of a precipitate by diffusion.

• *Plate-like precipitate.* — The exact solution is easily written, and it leads to the expected parabolic kinetics. For the case of the stationary interface, the equations resemble those for growth, and again, Laplace's equation has no solution for the initial and boundary conditions.

These results are listed in table III. Note that the relationship between growth and dissolution is not independent of the form of the precipitate. Even though the kinetics for plate-like precipitates are parabolic for both cases, this is not true for the spherical case except in certain limited cases.

IV.3 Coalescence. — "Coalescence" is used here to mean the evolution of the shape and size of second phase particles or precipitates in the course of time. For a non-spherical precipitate, this process expresses itself by evolution of the shape towards that of a sphere, and for a population of precipitates, by an evolution of the size frequency distribution; the smallest particles dissolve, the largest grow, and, in the limit, all the precipitates are assembled in a

Table III. — ***Dissolution of a precipitate*** (after Aaron *et al.*, 1970):

$$k = 2\,\frac{c_0 - \bar{c}_\alpha}{c_\beta - \bar{c}_\alpha} < 0$$

Assumption	Spherical particle	Plate-like particle, $\xi = \xi_0 - \lambda_j\sqrt{Dt}$
Exact solution of the diffusion equation		$\lambda_1 = 2\,\lambda$ $\sqrt{\pi}\,\lambda\,\exp\lambda^2\,\mathrm{erfc}\,(-\lambda) = -k/2$ $\lim_{k\to 0}\lambda_1 = \lambda_2 > \lambda_4$
Stationary interface	No simple law	$\lambda_2 = k/\sqrt{\pi}$
Laplace's equation	$R^2 = R_0^2 + kDt,\ k < 0$	Not defined
Linear gradient	No simple law	$\lambda_4 = \frac{\bar{c}_\alpha c_0}{(c_\beta - c_0)^{1/2}(c_\beta - \bar{c}_\alpha)^{1/2}} = \frac{-k}{2(1-k/2)^{1/2}}$ $\lim_{k\to 0}\lambda_4 = -k/2$

single gross "precipitate". These processes are often lumped under the name "Ostwald ripening", the term "coalescence" being reserved for the evolution of the population.

In all cases, the driving force for these processes is the variation of the interfacial energy, correlated to the change of shape (this is why one sometimes refers to "capillary" effects). The change of shape or of size leads to a decrease in the interfacial energy per unit volume. In this discussion, precipitates and atoms of solute can be replaced by pores and vacancies, respectively.

To treat this problem, the chemical potential of the solute (or vacancy) has to be expressed as a function of the curvature of the interface. If local equilibrium in the neighborhood of the interface is assumed, the chemical potential of the solute (or vacancy) is the same on both sides of the interface (in the precipitate and in the matrix), and is given by the Gibbs-Thomson relation:

$$\mu(R) - \mu(\infty) = \frac{2\,\Omega\gamma}{R}, \qquad \text{(X.75)}$$

where $\mu(\infty)$ corresponds to a plane interface ($R = \infty$), γ is the interfacial energy (surface energy in the case of pores), and Ω is the atomic volume of the solute (vacancy).

The chemical potential decreases with increasing radius of curvature, R; hence the flux of solute atoms (vacancies) goes from the smallest precipitates (pores) to the largest.

If the solution is dilute enough, the mole fraction of solute (vacancies) in the matrix near the interface can be expressed simply via Henry's law:

$$\text{Log}\ \frac{N(R)}{N_\infty} = \frac{2\ \Omega\gamma}{kT}\ \frac{1}{R}, \tag{X.76}$$

or

$$N(R) = N_\infty \exp\ (\rho/R), \tag{X.76a}$$

with $\rho = 2\ \Omega\gamma/kT$. For a precipitate, $R > 0$, $N(R) > N_\infty$ and decreases with increasing R. For pores, the same equation applies to the vacancy concentration.

In a first approximation treatment, it is assumed that each precipitate evolves as if it were alone in the interior of a matrix in which the concentration of solute is determined by a certain average of concentrations (see below). This brings us back to the problem of the growth of a precipitate under conditions of small super- (or under-) saturation, already treated above:

$$\frac{\mathrm{d}R}{\mathrm{d}t} = D\frac{c_0 - c_\alpha}{c_\beta - c_\alpha}\ \frac{1}{R}, \tag{X.68}$$

where the concentration at the interface, c_α, is given by equation (X.76a):

$$c_\alpha = \bar{c}_\alpha \exp\ (\rho/R) \sim \bar{c}_\alpha \left(1 + \frac{\rho}{R}\right), \tag{X.77}$$

with $\bar{c}_\alpha$ the equilibrium concentration near a plane interface. With reasonable values, e.g., $\rho = 2\ \Omega\gamma/kT \approx 10\text{Å}$, the linearization is justified.

The concentration far from the precipitate, c_0, can be expressed by a formally analogous relation:

$$c_0 = \bar{c}_\alpha \exp\ (\rho/R_c)\,. \tag{X.78}$$

From this, and neglecting c_α with respect to c_β (which is always true for vacancies), we obtain:

$$\boxed{\frac{\mathrm{d}R}{\mathrm{d}t} = D\ \frac{\bar{c}_\alpha}{c_\beta}\ \frac{\rho}{R}\left[\frac{1}{R_c} - \frac{1}{R}\right]}. \tag{X.79}$$

This expression shows that the particle grows if $R > R_c$ and dissolves if $R < R_c$; R_c is thus a *critical radius for coalescence* determined by the condition of supersaturation of the matrix $c_0/\bar{c}_\alpha$ (Fig. X.19a). If the precipitate particles are far enough from each other, R_c equals the average radius. In effect, the conservation of the volume of the n precipitates implies:

$$\sum_{i=1}^{n} 4\,\pi R_i^2\,(\mathrm{d}R_i/\mathrm{d}t) = 0.$$

Then substituting equation (X.68) for the rate of growth or dissolution, always with $c_\alpha << c_\beta$, we obtain:

$$\sum_i 4\,\pi R_i\;D\;\frac{c_0 - c_\alpha}{c_\beta} = 0,$$

or, substituting equation (X.77):

$$\sum_i 4\,\pi R_i\;\frac{D}{c_\beta}\;[c_0 - \bar{c}_\alpha - \bar{c}_\alpha\,(\rho/R_i)] = 0,$$

or

$$(c_0 - \bar{c}_\alpha)\sum_i R_i = n\bar{c}_\alpha\rho,$$

so that

$$c_0 - \bar{c}_\alpha = \frac{\bar{c}_\alpha\rho}{\langle R\rangle},$$

where $\langle R\rangle$ is the mean radius $(1/n)\sum R_i$. Then:

$$c_0 - c_\alpha = \bar{c}_\alpha\left[\frac{1}{\langle R\rangle} - \frac{1}{R}\right]\rho,$$

and

$$\frac{\mathrm{d}R_i}{\mathrm{d}t} = D\,\frac{\bar{c}_\alpha}{c_\beta}\,\frac{\rho}{R_i}\,\left[\frac{1}{\langle R\rangle} - \frac{1}{R_i}\right]. \qquad \text{(X.80)}$$

The plot of $\mathrm{d}R/\mathrm{d}t$ *vs.* R goes through a maximum for $R = 2\langle R\rangle$ (Fig. X.19b). At any instant of time, particles with radius twice the average radius of the size distribution grow the fastest. In a long enough time, the number of particles, n, decreases, $\langle R\rangle$ increases, and the rate of growth decreases since, as seen from figure (X.19), the larger the mean radius, the lower the growth rate. The maximum growth rate, corresponding to a particle of radius $R = 2\langle R\rangle$ is then:

$$(\mathrm{d}R/\mathrm{d}t)_{\max} = D\,\frac{\bar{c}_\alpha}{c_\beta}\,\frac{2\,\Omega\gamma}{kT}\,\frac{1}{R^2}, \qquad \text{(X.81)}$$

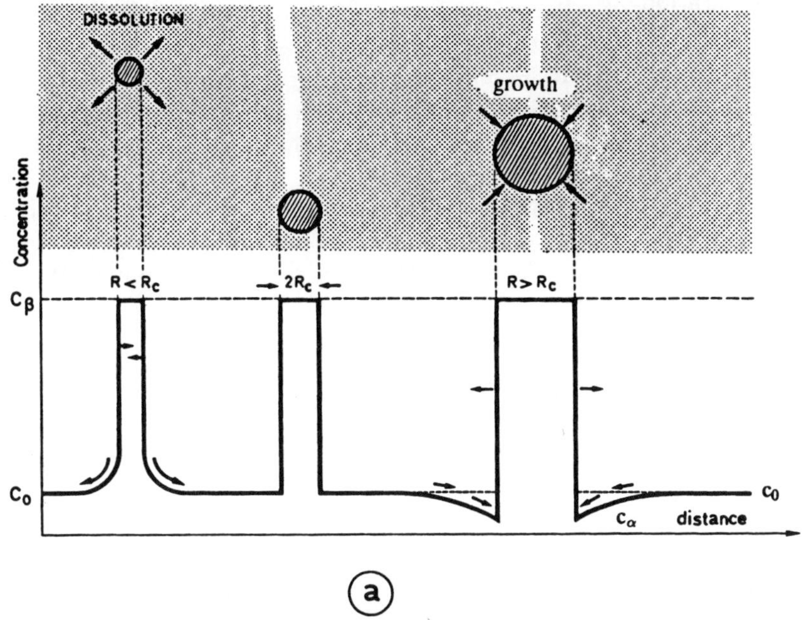

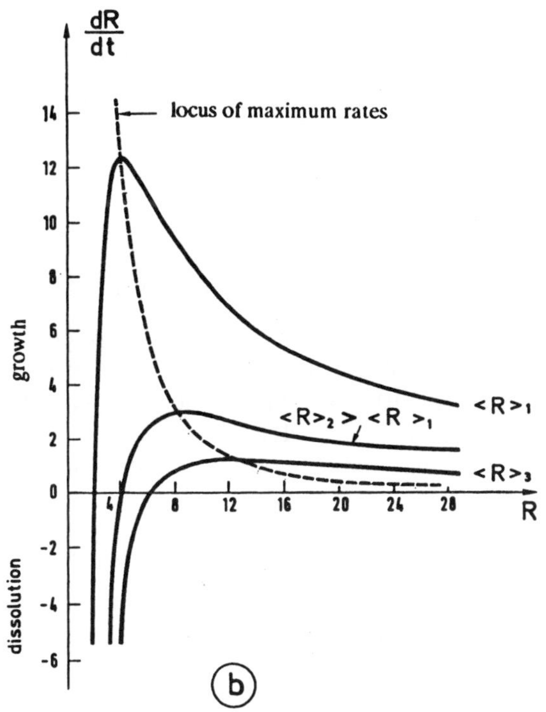

Fig. X.19. — a) Coalescence of spherical particles. R_c = the critical radius at the instant t. b) Kinetics of coalescence for spherical particles (after Greenwood, 1956).

and by integration

$$R^3 - R_0^3 = \frac{6\,D\,(\bar{c}_\alpha/c_\beta)\,\Omega\gamma}{kT}\,t. \tag{X.82}$$

Thus $t^{1/3}$ kinetics are observed for particles growing at the maximum growth rate.

A more detailed analysis of coalescence is beyond the scope of this book. Recall that Wagner as well as Lifshitz and Slyozov have succeeded in determining the asymptotic form of the steady-state size distribution, and deduced from it the asymptotic behavior of the mean radius. More recently, these solutions have been derived in a different way by Kahlweit (1975).

For spherical precipitates (isotropic surface energy) of pure solute formed with diffusion kinetics, these authors calculate:

– the number of precipitates per unit volume:

$$\frac{n(t)}{n(0)} = \frac{kT}{4\,D\gamma\Omega^2\,\bar{c}_\alpha}\,\frac{1}{t}; \tag{X.83}$$

– the mean radius:

$$\langle R(t)\rangle^3 - \langle R(0)\rangle^3 = \frac{8}{9}\,\frac{D\gamma\Omega^2\bar{c}_\alpha}{kT}\,t, \tag{X.84}$$

a formula of the same form as equation (X.82) with $c_\beta \approx 1/\Omega$,

– the critical radius: $R_c(t) = \langle R(t)\rangle$,

– the maximum radius: $R_{max} = 1.5\,\langle R\rangle$,

– the most probable radius: $R = 1.135\,\langle R\rangle$,

– the supersaturation:

$$c_\alpha - \bar{c}_\alpha = \left\{\frac{9\,\gamma^2\,\Omega\bar{c}_\alpha^2}{(kT)^2\,D}\,\frac{1}{t}\right\}^{1/3} \tag{X.85}$$

Remark: The nature of the diffusion coefficient introduced into these models presents no conceptual problem for vacancies. However, for the motion of a solute, we are dealing with coupled diffusion so that an effective diffusion coefficient has to be used (see App. II).

When diffusion occurs along the grain boundaries of the matrix, $t^{1/4}$ kinetics are predicted:

$$\langle R(t)\rangle^4 - \langle R(0)\rangle^4 = \frac{D'\delta\gamma\Omega^2 c'_\alpha}{kT}\,t,$$

where c'_α is now the solubility of the solute in the grain boundaries.

IV.4 Elimination of vacancies. — If the vacancies are eliminated by precipitation into pores, the kinetics are identical to those for precipitation. The fraction of vacancies remaining in solution can be calculated from equation

(X.67a) and the number of pores per unit volume.

If the vacancies are eliminated at sinks, the kinetics depend on the geometry. Fick's equation is solved for the following boundary conditions:

$$\begin{aligned} c_v &= c_0 \quad \text{at} \quad t = 0 \\ c_v &= 0 \qquad\quad t > 0;\ x,\ y,\ z \ \text{(sinks)}. \end{aligned} \tag{X.86}$$

The subscript v is omitted in the following.

Examples:

a) *Thin foil of thickness l.* (Fig. X.20). The sinks are the surfaces of the foil.

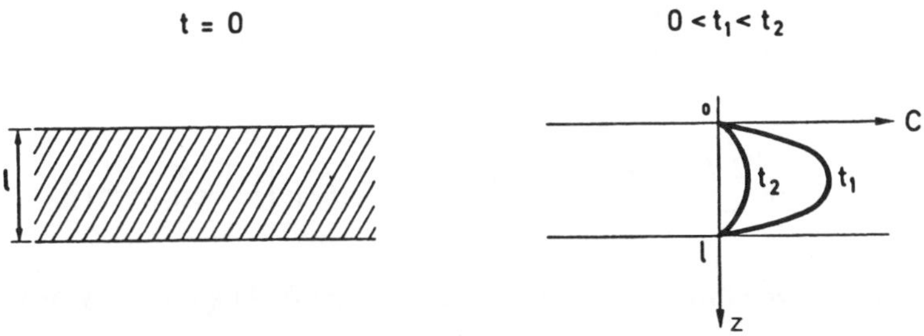

Fig. X.20. — Elimination of a supersaturation of vacancies at the surfaces of a foil.

The solution of the diffusion equation is obtained by the method of separation of variables:

$$\frac{c(z,\ t)}{c_0} = \frac{4}{\pi} \sum_{n=0}^{\infty} \frac{1}{(2\ n+1)} \exp\left(-\frac{(2\ n+1)^2\ \pi^2}{l^2}\ Dt\right) \sin\left(\frac{(2\ n+1)\ \pi z}{l}\right). \tag{X.87}$$

The terms with $n \geq 1$ decay rapidly, and the solution can be approximated for not too short times by:

$$\frac{c(z,\ t)}{c_0} = \frac{4}{\pi} \exp\left(-\frac{\pi^2}{l^2}\ Dt\right) \sin\left(\frac{\pi z}{l}\right).$$

The number of vacancies still present per unit area of foil at time t is given by:

$$\begin{aligned} \mathcal{N}_v(t) &= \int_0^l c(z,\ t)\ dz \\ \mathcal{N}_v(t) &= \frac{8\ c_0\ l}{\pi^2} \sum_{n=0}^{\infty} \frac{1}{(2\ n+1)^2} \exp\left(-\frac{(2\ n+1)^2\ \pi^2}{l^2}\ Dt\right), \end{aligned} \tag{X.88}$$

with $c_0 l = \mathcal{N}_v(0)$; for $t = 0$, the series for π, $\dfrac{\pi^2}{8} = \sum \dfrac{1}{(2\,n+1)^2}$ is found.

Neglecting short-time transients, we have approximately:

$$\boxed{\frac{\mathcal{N}_v(t)}{\mathcal{N}_v(0)} \sim \exp\left[-\frac{\pi^2}{l^2}\,Dt\right]}. \tag{X.89}$$

b) *Spherical geometry.* The same type of approximation leads to the same type of law for both spherical and cylindrical geometry. The spherical geometry is an approximation for a polycrystalline specimen in which the excess vacancies are eliminated at the grain boundaries (if spherical grains are assumed).

For a sphere of radius R, the solution of the diffusion equation is, with $c_1 = c(R)$:

$$\frac{c - c_0}{c_1 - c_0} = 1 + \frac{2\,R}{\pi r} \sum_{n=1}^{\infty} \frac{(-1)^n}{n} \sin\left(\frac{n \pi r}{R}\right) \cdot \exp\left[-\frac{n^2\,\pi^2\,Dt}{R^2}\right]. \tag{X.90}$$

If the vacancies are eliminated at the surface, $c_1 = 0$. The quantity of vacancies remaining at time t is given by:

$$\begin{aligned}\frac{\mathcal{N}(t)}{\mathcal{N}(0)} &= \frac{\int_0^R 4\,\pi r^2\,c(r,\,t)\,\mathrm{d}r}{\frac{4}{3}\,\pi R^3\,c_0} \\ &= \frac{6}{\pi R^2} \sum_{n=1}^{\infty} \frac{(-1)^n}{n} \int_0^R r \sin\left(\frac{n\pi r}{R}\right) \cdot \exp\left[-\frac{n^2\,\pi^2\,Dt}{R^2}\right] \mathrm{d}r \qquad \text{(X.91)} \\ &= \frac{6}{\pi^2} \sum_{n=1}^{\infty} \frac{1}{n^2} \exp\left[-\frac{n^2\,\pi^2\,Dt}{R^2}\right].\end{aligned}$$

If the time is not too short, the decay of the vacancy concentration is again dominated by the first term of the series:

$$\boxed{\mathcal{N}(t) \sim \mathcal{N}(0)\,\exp\left(-\frac{\pi^2\,Dt}{R^2}\right)}. \tag{X.92}$$

Hollow sphere. Steady state. A point sink is modeled by a sphere with capture radius R_c, so that every vacancy located at $r \leq R_c$ is considered trapped (or annihilated, depending on the nature of the trap). The problem becomes that of diffusion in a hollow sphere with outside radius R_∞ and inside radius R_c. The diffusion equation for steady state in spherical coordinates (Eq. (X.64)):

$$\frac{d}{dr}\left(r^2 \frac{dc}{dr}\right) = 0,$$

with boundary conditions $c(R_c^+) = c_c$, and $c(R_\infty) = c^0$, has for its solution:

$$c(r) = \frac{R_c\, c_c\, (R_\infty - r) + R_\infty\, c^0\, (r - R_c)}{r\,(R_\infty - R_c)}, \qquad \text{(X.93)}$$

from which:

$$\left(\frac{dc}{dr}\right)_{R_c^+} = \frac{R_\infty}{R_c}\,\frac{c^0 - c_c}{R_\infty - R_c}.$$

The total flux of vacancies arriving per unit time at the surface of the inside sphere is given by:

$$\Phi = 4\,\pi R_c^2 \cdot D \left(\frac{dc}{dr}\right)_{R_c^+}.$$

This is the number of vacancies initially contained in the volume $(4/3)\pi R_\infty^3$ and trapped per unit time. Thus:

$$\frac{1}{\mathcal{N}}\,\frac{d\mathcal{N}}{dt} = -\frac{\Phi}{\frac{4}{3}\,\pi R_\infty^3\, c^0},$$

or:

$$\boxed{\frac{d\mathcal{N}}{dt} = -K\mathcal{N}}, \qquad \text{(X.94)}$$

with:

$$K = D\,\frac{3\,R_c}{R_\infty^2\,(R_\infty - R_c)} \approx D\,\frac{3\,R_c}{R_\infty^3}. \qquad \text{(X.94a)}$$

The elimination of the vacancies at the sink follows *first-order kinetics*, corresponding to the "reaction" for non-saturable sinks:

$$\text{Vacancy} + \text{sink} \rightleftharpoons \text{sink}.$$

If the density of sinks is N_s per unit volume, R_∞ is defined by:

$$N_s\,\frac{4}{3}\,\pi R_\infty^3 = 1.$$

c) *Cylindrical geometry* (elimination at the surface). The calculation for the time-dependent regime is somewhat more complicated in cylindrical coordinates, because the solution of the diffusion equation involves Bessel functions. The solution is:

$$\frac{\mathcal{N}(t)}{\mathcal{N}(0)} = \frac{4}{R^2} \sum_{n=1}^{\infty} \frac{1}{\alpha_n^2} \exp\left[-D\alpha_n^2\, t\right], \tag{X.95}$$

with

$$J_0\left(R\alpha_n\right) = 0$$

for the boundary conditions: sinks on the surface of the cylinder R, $c(R) = 0$ for $t > 0$; $c = c_0$ for all r at $t = 0$. Here J_0 is the Bessel function of order 0. Since the first root of $J_0(x) = 0$ equals 2.405, the term dominating the decay is:

$$\boxed{\frac{\mathcal{N}(t)}{\mathcal{N}(0)} \sim \exp\left[-\frac{5.784\ Dt}{R^2}\right]}, \tag{X.96}$$

again leading to first-order kinetics.
Remark: The solutions discussed in paragraphs a), b) and c) also apply to the desorption of gas occluded in a sheet, sphere, or cylinder.

d) *Hollow cylinder.* The behavior of a dislocation as a linear sink for vacancies can be modeled by a hollow cylinder of outside radius R_∞, inside radius b, or, better, R_c. Here R_c is the capture radius that defines a cylinder equivalent to the dislocation line plus its stress field (see below, IV.5D, and Fig. X.21). In the interval $R_c \leq r \leq R_\infty$, the problem is purely one of diffusion, and every vacancy arriving at the boundary of the inside cylinder is assumed to be automatically trapped and then annihilated at a jog by very rapid diffusion along the dislocation core (the assumption of perfectly efficient sinks).

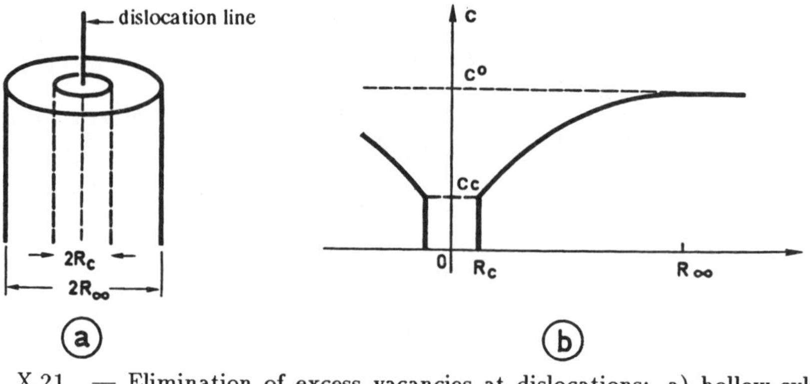

Fig. X.21. — Elimination of excess vacancies at dislocations: a) hollow-cylinder model; b) corresponding profile.

The boundary conditions are (with $c \equiv c_{\text{v}}$) :

$$\begin{aligned} c\,(R_\infty) &= c^0 \\ c\,(R_{\text{c}}^+) &= c_{\text{c}} \\ c\,(R_{\text{c}}^-) &= 0. \end{aligned} \tag{X.97}$$

Two cases should be considered:

1) *Transient regime.* Elimination of vacancies with initial concentration c^0. The solution of the diffusion equation is a series of terms with the form $\exp[-D\alpha_n^2 t]$, with preexponential factors containing Bessel functions. As above, for not too short times, the term with $n = 1$ will dominate the decay kinetics; for shorter times, more terms must be included. For the analysis of experimental results, an empirical relation:

$$\mathcal{N}(t) = \mathcal{N}(0)\ \exp\ [-At^m] \tag{X.98}$$

is more often used, with $m < 1$ for short times and $m \to 1$ for long times.

2) *Steady state.* Steady state is established if the concentration c^0 remains constant with time, e.g., irradiation under constant flux, or creep. In the latter case, $c^0 = \bar{c}$, the thermal equilibrium value.

The diffusion equation for cylindrical coordinates:

$$\frac{\partial c}{\partial t} = \frac{1}{r}\,\frac{\partial}{\partial r}\left(rD\,\frac{\partial c}{\partial r}\right)$$

is, for steady state with D constant:

$$\frac{\mathrm{d}}{\mathrm{d}r}\left(r\,\frac{\mathrm{d}c}{\mathrm{d}r}\right) = 0.$$

The general solution of this equation is of the type:

$$c = A + B\ \mathrm{Log}\ r,$$

and with the boundary conditions of equation (X.97):

$$c - c_{\text{c}} = \frac{(c^0 - c_{\text{c}})}{\mathrm{Log}\ (R_\infty/R_{\text{c}})}\ \mathrm{Log}\ (r/R_{\text{c}})\,. \tag{X.99}$$

The total flux of vacancies arriving at the dislocation per unit length of dislocation line is:

$$\Phi_v = 2\,\pi R_{\text{c}}\cdot D\,(\mathrm{d}c/\mathrm{d}r)_{R_{\text{c}}} = 2\,\pi D\,\frac{c^0 - c_{\text{c}}}{\mathrm{Log}\ (R_\infty/R_{\text{c}})}. \tag{X.100}$$

The loss of vacancies per unit time per unit length is given by:

$$\frac{1}{\mathcal{N}}\frac{\mathrm{d}\mathcal{N}}{\mathrm{d}t} = -\frac{\Phi_v}{\pi R_\infty^2\, c^0},$$

or

$$\boxed{\frac{\mathrm{d}\mathcal{N}}{\mathrm{d}t} = -K\mathcal{N}}, \tag{X.101}$$

with (for $c_c << c^0$) :

$$K = \frac{2\,D_v}{R_\infty^2\ \mathrm{Log}\ (R_\infty / R_c)}. \tag{X.101a}$$

Application to the climb of dislocations during creep. — The force on a dislocation due to the applied stress has a component that acts as a driving force for climb. If dislocation climb takes place by the absorption or emission of vacancies, it produces a local under- or supersaturation of vacancies. If these are not eliminated, dislocation climb is impeded. In other words, the under- or supersaturation exerts a certain effective force, called "chemical force", on the dislocation. The dislocation climb takes place if the vacancies diffuse away from (or towards) the regions where they are in thermodynamic equilibrium. To calculate the rate of climb, the expression for the vacancy flux Φ_v given by equation (X.100) is used, with:

$$c\,(R_\infty) = c^0 = \bar{c}_v,$$
$$c\,(R_c) = c_c = c_{ch}.$$

The first condition is somewhat arbitrary since the value for thermodynamic equilibrium far from the sources and sinks is used. In the second condition, c_{ch} is such that the chemical force due to c_{ch} just balances the force F_a resulting from the applied stresses:

$$F_a = -F_{ch} = -\frac{kT}{b^2}\ \mathrm{Log}\ \frac{c_{ch}}{c_{cap}} \approx -\frac{kT}{b^2}\ \mathrm{Log}\ \frac{c_{ch}}{\bar{c}_v}$$

where c_{cap} is the concentration at the capture radius in the absence of an applied force. To calculate the rate of climb, Φ_v has to be equated to the balance of absorption and emission of vacancies at jogs (see Edelin, 1978).

IV.5 SEGREGATION TO DISLOCATIONS. — Consider again the origin of the kinetics of the segregation of solute atoms or point defects (vacancies or interstitials) to dislocations.

The interaction energy between a solute atom or a point defect and a dislocation depends on the distance they are separated; the gradient of this energy is the origin of a drift term, which gives rise to a redistribution of solute atoms and point defects around the dislocations (Cottrell effect). It is important to know not only the equilibrium state of this segregation, but also the kinetics of its formation. This segregation can further lead to the precipitation of a solute-solvent compound phase in the dislocation core ("decoration" of the dislocations) or to annihilation of the point defects at the dislocation core. All these processes have important metallurgical consequences: the formation of Cottrell atmospheres, the Portevin-LeChatelier effect, aging after deformation, gettering of oxygen in silicon, annealing of defects created by irradiation or quenching.

A) *The dislocation-solute interaction.* — The interaction has its origin in both elastic and electrostatic effects (ionic crystals, semiconductors). For the first, we distinguish between size effects due to the hydrostatic component of the stress field around an edge dislocation and the inhomogeneity effect (or modulus effect) connected to the "hard" or "soft" character of the solute particle with respect to the elastic constants of the matrix. Both of these effects are active in the case of an edge dislocation, but only the modulus effect works in the case of a screw dislocation, the stress field of which does not have a hydrostatic component.

In metals, the size effect is dominant; small solute atoms and vacancies tend to segregate in the region of compressive stresses at the core of an edge dislocation, large solute atoms on the tensile side. Since the stress field is usually expressed as $(1/r)\, f(\theta)$, the energy of the elastic interaction (size effect, edge dislocation) is also written in polar coordinates:

$$W(r,\ \theta) = \frac{\mu b}{3\ \pi} \cdot \frac{1+\nu}{1-\nu}\ \Delta v\ \frac{\sin\ \theta}{r}, \tag{X.102}$$

where b is the Burgers vector, ν is Poisson's ratio, and Δv the difference between the atomic volumes of the solute and solvent. Briefly, for an edge dislocation:

$$W = \frac{A\ \sin\ \theta}{r}. \tag{X.102a}$$

Here A can be either less than or greater than zero, with $A < 0$ for attraction. Typically for a solid solution of C in α-Fe, $A \approx 10^{-29}$ Nm^2.

The equipotential lines (W = const.) are circles passing through the dislocation line; the lines of force, which are normal to the circles, are also circles, defined by cos θ/r = const. (Fig. X.22).

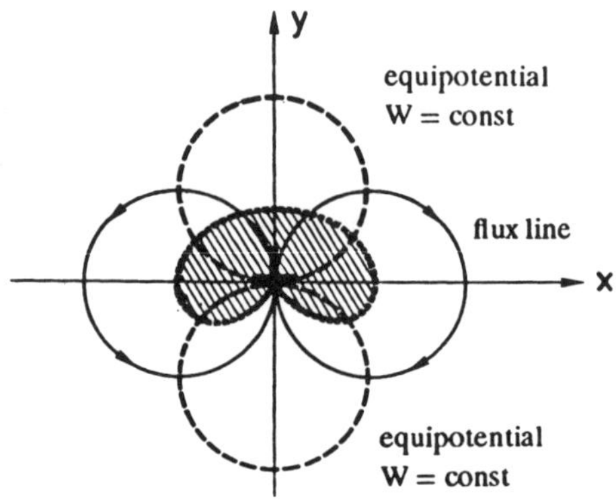

Fig. X.22. — Segregation to an edge dislocation for the case $\Delta v > 0$. Model of Cottrell and Bilby (1949). The hatched area is depleted in solute ($c = 0$), outside this zone $c = c_0$.

More generally,

$$W \propto \frac{1}{r^n} f(\theta), \qquad \text{(X.102b)}$$

with $n = 1$ for the size effect and $n = 2$ for the modulus effect. The modulus effect is important for vacancies, which behave like a particle with zero modulus; neglecting the angular dependence, and with B positive for attraction:

$$W \approx -B/r^n. \qquad \text{(X.102c)}$$

The flux of solute atoms or of point defects is given by the sum of the Fickian term and the drift due to the force field:

$$J = -D\,\nabla c - \frac{Dc}{kT}\,\nabla W, \qquad \text{(X.103)}$$

from which the radial and tangential components of the flux are:

$$\begin{cases} J_r = & -D\,\dfrac{\partial c}{\partial r} - \dfrac{Dc}{kT}\,\dfrac{\partial W}{\partial r} \\ J_\theta = & -\dfrac{D}{r}\,\dfrac{\partial c}{\partial \theta} - \dfrac{Dc}{kT}\,\dfrac{1}{r}\,\dfrac{\partial W}{\partial \theta}. \end{cases} \qquad \text{(X.103a)}$$

B) *Steady state.* — If the concentration c never exceeds the solubility limit, an *equilibrium distribution* can result such that the diffusion and drift terms cancel each other everywhere. The steady-state regime is found by setting:

$$J_r = J_\theta = 0 \qquad \forall r,\ \theta.$$

With the interaction described by equation (X.102a), equations (X.103a) yield:

$$\begin{cases} -D\,\dfrac{\partial c}{\partial r} + \dfrac{DcA\,\sin\theta}{kT\,r^2} = 0 \\ -\dfrac{D}{r}\,\dfrac{\partial c}{\partial \theta} + \dfrac{DcA\,\cos\theta}{kT\,r^2} = 0, \end{cases}$$

the solution of which is:

$$\boxed{c = c_0\,\exp\left(-A\,\sin\theta/kT\,r\right)}, \qquad \text{(X.104)}$$

where c_0 is the initial concentration. Equation (X.104) describes a *Maxwellian "atmosphere"* around the dislocation. This equation is valid only if c remains below the solubility limit and if the dislocation does not "saturate". (Since the segregation relaxes the elastic distortion, an accumulation of solute would reduce W).

C) *Kinetics.* — No analytical solution to the complete diffusion equation exists. Certain simplifications are necessary; the θ dependence is neglected, and a "capture radius" for the dislocation must be defined. The solutions can be calculated numerically; they predict kinetics in $t^{2/3}$ for short times and in $\exp(-t/\tau)$ for long times.

A completely different approach has often been followed since the work of Cottrell and Bilby (1949). It consists of *neglecting the Fickian* term in equation (X.103). The problem is further simplified by neglecting the dependence on θ. With $W = -B/r^n$ (see Eq. (X.102c)), the drift is given by:

$$J_r = -\frac{Dc}{kT}\,\frac{\mathrm{d}W}{\mathrm{d}r} = \frac{Dc}{kT}\,\frac{nB}{r^{n+1}},$$

from which the mean drift velocity is obtained as:

$$v_r = \frac{\mathrm{d}r}{\mathrm{d}t} = \frac{nBD}{r^{n+1}}\cdot\frac{1}{kT}.$$

Integrating from $t = 0$ to t, we obtain:

$$\frac{r^{n+2}}{n+2} = \frac{nBD}{kT}\,t,$$

or

$$r = \left[\frac{n(n+2)BD}{kT}\right]^{1/(n+2)} t^{1/(n+2)}. \qquad \text{(X.105)}$$

At the end of the time t, the solute atoms located at a distance $\rho \le r$ will have reached the dislocation core. The corresponding number of atoms per unit length of dislocation equals $\pi\ r^2 c_0$, or:

$$\boxed{\mathcal{N}(t) = \pi c_0 \left[\frac{n(n+2)BD}{kT}\right]^{2/(n+2)} t^{2/(n+2)}} \qquad \text{(X.106)}$$

The solution is the same within a numerical factor whether a purely radial flux is assumed, as above, or an interaction potential with angular dependence is used (circular flux line, Fig. X.22). This is why a purely radial interaction potential is generally used in problems involving sinks for point defects.

In the case of a dominant size effect, $n = 1$, $t^{2/3}$ kinetics are found; this is known as *Cottrell's and Bilby's law.* The time exponent 2/3 appears thus to be characteristic of the kinetics of segregation to dislocations. In fact, it depends on the nature of the interaction; for vacancies (elastic inhomogeneity, $n = 2$), an exponent of 1/2 is expected. Further, neglecting the Fickian term can be realistic only for short times, i.e., before a large concentration gradient is established.

The solution of Cottrell and Bilby, whether the angular variation of the potential is taken into account or not, leads to a region completely devoid of solute ($c = 0$), abutting on a region where $c = c_0$, with a border that moves as $t^{1/(n+2)}$ (Fig. X.22). Now the full diffusion equation is a second-order partial differential equation and requires two boundary conditions for its solution (e.g., in the case of a cylinder for $r = R$ (or ∞) and $r = R_c(\geq b)$). The equation without the Fickian term, with only the drift term included (see Eq. (X.103)):

$$\frac{\partial c}{\partial t} = \frac{D}{kT}\ \nabla(c\ \nabla W)$$

is first order; hence only one spatial boundary condition can be fulfilled. Since it is not possible to control the condition at the core-matrix interface ($r = R_c$), the existence of a perfect sink at $r = 0$ (or $r = R_c$) is accepted; this, in fact, corresponds to $c(r = 0) = 0$ (see the flux lines in Fig. X.22). This accumulation of solute at the dislocation core is known as a "*Cottrell atmosphere*", an ambigous expression since the solution hides precipitation in the core ! A true Maxwellian atmosphere can exist only if there are two terms; fickian and drift (Bullough and Newman, 1970).

Nevertheless, the Cottrell-Bilby solution (Eq. (X.106)) remains valid in practice for short times when the dislocation behaves as a true sink. Examples of this behavior are the beginning of the precipitation of the solute (the solute atoms, once at the dislocation core, migrate rapidly to the precipitate nuclei, Fig. (X.23)), or the climb of the dislocation by annihilation of vacancies at jogs. For longer times, the fickian flux due to the concentration gradient created by the dislocation can no longer be neglected.

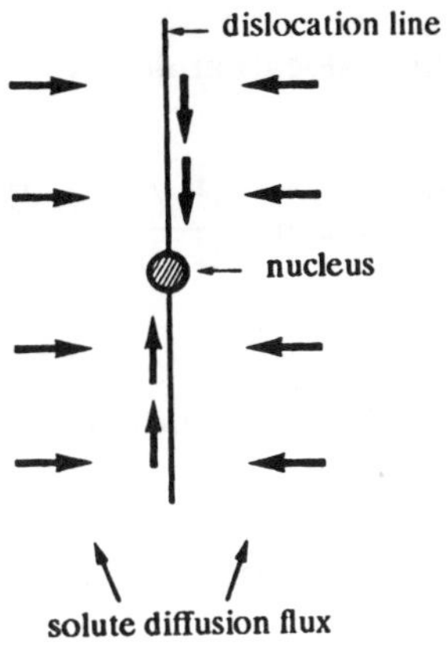

Fig. X.23. — Precipitation on dislocations.

The experimentally measured concentrations are often expressed in the empirical form:

$$c = c_0 \exp(-At^m) \qquad m \leq 1.$$

For long times, $m \to 1$, which yields the solution of first-order kinetics with $dc/dt \propto -c$, which in turn corresponds closely to an approximate solution of the diffusion equation for a cylinder (see above). For short times, the above equation can be approximated by:

$$c = c_0 [1 - At^m] .$$

The quantity of vacancies annihilated at dislocations per unit volume equals $c_0 At^m$; with $m = 2/3$, this is Cottrell's and Bilby's law.

The fact that the general solution of the diffusion equation in cylindrical coordinates does not predict the evolution of the concentration correctly can be attributed to a fault of the boundary condition chosen for $R = R_c$; the annihilation of the vacancies is probably governed by a local process that slows down the necessary climb of the dislocations (efficiency of the jogs). A more appropriate condition on the flux would be:

$$J(r) = K_c \{c - c_c(t)\} , \qquad r = R_c.$$

Here c_c, the concentration at the core, is different from $\bar{c}(\infty)$, the average concentration, owing to the stress field of the dislocation and its curvature; since the dislocation (assumed to be pinned by the Frank network) bends because of the climb, c_c varies with time. The other boundary condition should be written:

$$J(r) = 0, \qquad r = R_\infty$$

where $R_\infty \propto \rho_d^{-1/2}$ (ρ_d is the dislocation density). According to Bullough and Newman:

$$\frac{c_c(t)}{\bar{c}(\infty)} = \exp\left(\frac{b^2 W}{\ell kT}\,\theta - \frac{E}{kT}\right),$$

with

ℓ	= length of the dislocation segment;
θ	= angle of curvature;
E	= dislocation-line tension;
W	= vacancy-dislocation interaction energy.

D) *Precipitation on dislocations.* — For the case of precipitation from a supersaturated solution with preferential nucleation on the dislocations, the diffusion equation for the solute was numerically integrated by F. Ham (1959), with the following boundary conditions:

$r = R_\infty$ $\quad c = c_0$ (initial solute concentration)
$r \to 0$ $\quad c = 0$.

Ham first showed that at steady state, the problem reduced to pure fickian diffusion ([8]) (without a drift term) if the dislocation ($r \to 0$) were replaced by a cylinder of radius R_c :

$$R_c = 1.78\ A/4kT \quad \text{or} \quad 1.78\ B/kT \qquad \text{(X.107)}$$

where A and B are the constants from equations (X.102a and c) (edge and screw dislocations, respectively) and R_c is the effective capture radius; the cylinder with radius R_c has the same probability of capturing the solute atoms as the dislocation with its associated stress field.

The exact solution can then be compared with two approximate solutions:

– the diffusion equation solved with the drift term only (the Cottrell and Bilby solution),

– the diffusion equation without the drift term, but with the boundary condition $c = 0$ at $r = R_c$.

It can be shown (Fig. X.24) that the first is a good approximation for short times, the second for long times. Note that the Cottrell-Bilby solution predicts too small a rate of precipitation with increasing time, since the concentration gradient that becomes established induces a flux in the same direction as the stress field (this is the reverse of the Maxwellian atmosphere !). To a solute

([8]) By a change of variables $c = \psi(r, \theta) \exp\{A \sin\theta/(2r)\}$, a solution $\psi(r, \theta)$ of the diffusion equation is found.

atom which is located far from the dislocation and will not be captured for a long time, the dislocation appears as an equivalent cylinder of radius R_c without the associated stress field. In this case, a solution of the Cottrell-Bilby type is totally inadequate.

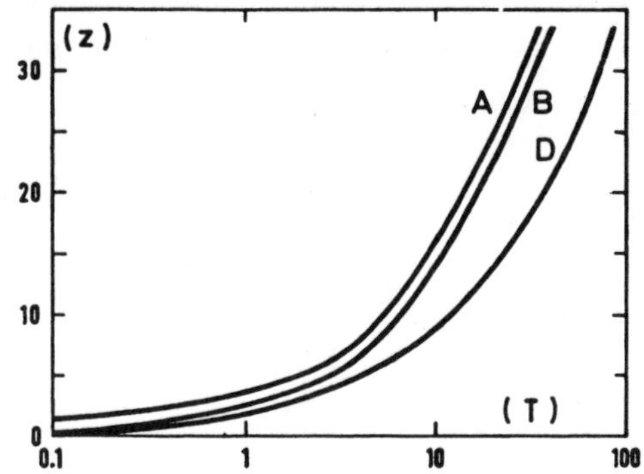

Fig. X.24. — Kinetics of precipitation on edge dislocations, in reduced units (see text): A: pure diffusion, dislocation simulated by cylinder of radius R_c; B: exact solution, D: Cottrell-Bilby solution (drift term only) (after Ham, 1959).

The curves of figure X.24 are plotted in reduced coordinates: Z = the number of atoms captured$/\pi R_c^2 c_0$ and $T = Dt/R_c^2$. For an edge dislocation, $Z = 2.3\ T^{0.762}$ for $0.5 < T < 100$. With the value of A for the αFe-C solution, $R_c = 14$ Å at ordinary temperatures; $T = 1$ corresponds to $t = 400$ s.

Numerical simulations on a two-dimensional lattice lead to values of 0.7-0.8 for the exponent n of the time, as long as the concentration c_0 far from the dislocation remains unaffected by the process.

When the precipitation takes place on a network of dislocations, there is competition owing to the overlapping of the diffusion fields. This of course does not occur at short times, and the Cottrell-Bilby solution remains valid. At long times, the precipitated fraction follows a time dependence:

$$1 - \exp(-t/\tau),$$

where τ can be calculated from R_c.

For a complete discussion, the reader should consult the review article by Bullough and Newman (1970).

V. The solidification of an alloy.

Owing to the differences in the solubilities of impurities or alloying elements in the solid and liquid phases of an alloy (Fig. X.25), the solid that forms from a liquid alloy is either richer or poorer in these elements, depending on whether the partition coefficient $k(T)$ is greater than or less than 1 ($\bar{k}_B$ is defined as the ratio of concentrations of B at the solidus and the liquidus of the phase diagram, $\bar{k}_B = \bar{c}_B^S / \bar{c}_B^L$. An effective partition coefficient is also defined: $k_B = c_B^S(0)/c_B^L(\infty)$; it is the ratio of the concentrations of B measured in the solid at the S/L interface and in the liquid far from the interface.

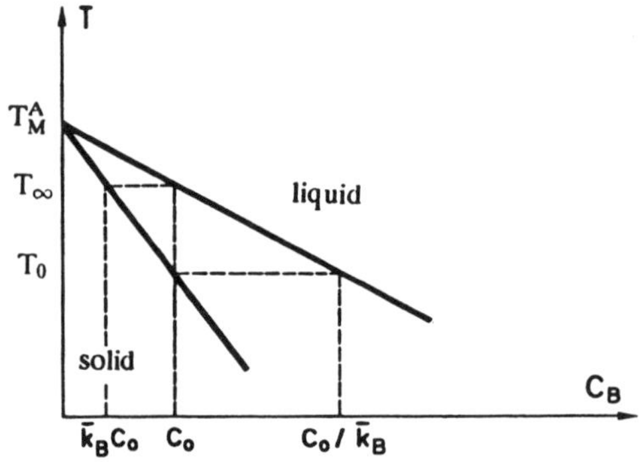

Fig. X.25. — Solid/liquid equilibrium in a binary AB alloy with partition coefficient < 1.

Consider the classic case of the unidirectional solidification of a binary alloy AB in a temperature gradient dT/dx. Three possible cases can be envisaged (Fig. X.26):

– the liquid is completely mixed; all the solute atoms rejected at the interface redistribute themselves in the liquid (owing especially to convection currents) (Fig. X.26a):

$$c_B^L(\infty) = c_B^L(0);$$

– no mixing in the liquid. Rejection of solute atoms at the advancing solidification front produces a local enrichment (or impoverishment) of the liquid near this interface. If the solidification rate is constant, a steady state is reached when the quantity of solute rejected by the solid equals that which diffuses down the concentration gradient in the liquid. The steady state signifies (Fig. X.26b):

$$c_B^S(0) = c_B^L(\infty) = c_0 \tag{X.108}$$

i.e.:

$k_B = 1$, even though $\bar{k}_B \neq 1$;

– partial mixing in the liquid; a steady state is obtained (Fig. X.26c) such that:

$$\bar{k}_B < k_B < 1 \quad \text{enrichment,}$$

or

$$\bar{k}_B > k_B > 1 \quad \text{depletion.}$$

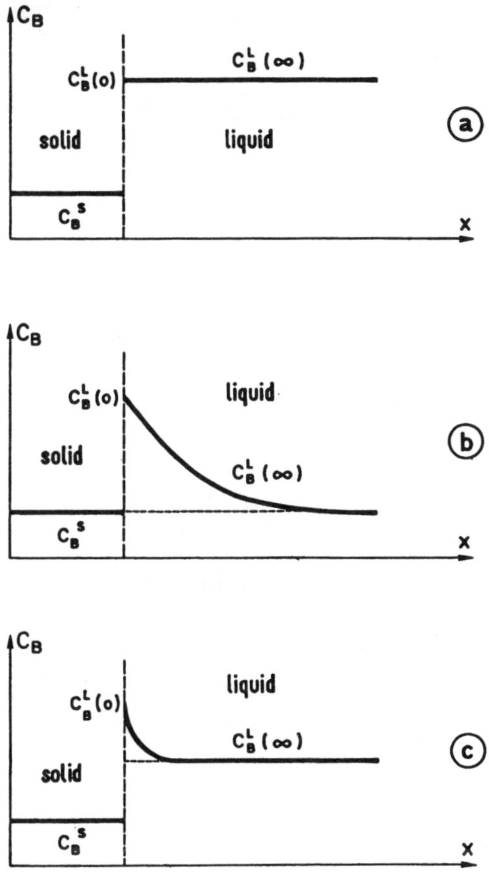

Fig. X.26. — Concentration profiles near the solid/liquid interface during solidification, partition coefficient < 1. a) complete mixing in the liquid; b) no mixing, steady-state diffusion; c) partial mixing.

In all cases, diffusion in the solid is assumed negligible $\left(D_{\mathrm{B}}^{\mathrm{L}}/D_{\mathrm{B}}^{\mathrm{S}} \approx 10^5\right)$.

The expression for the *concentration profile* in the liquid is easily obtained in the case of no mixing and at steady state, i.e., starting from the fraction solidified such that the rejected solute leads to a concentration $c_{\mathrm{B}}^{\mathrm{L}}(0) = c_0/\bar{k}_{\mathrm{B}}$ (local equilibrium at the interface, see Eq. (X.108)).

Let v be the rate of migration of the interface; for ease of writing, we set $c_{\mathrm{B}}^{\mathrm{L}} \equiv c$. The origin of x is taken at the interface. In this frame of reference, the diffusion flux occurs in a slab which moves at rate $-v$ from right to left. The variation of the concentration, δc, due to this motion during a time δt in the element δx is given by:

$$\delta c = \frac{1}{\partial x}\left[(vc)_{x+\delta x} - (vc)_x\right]\ \delta t \sim v\ \frac{\partial c}{\partial x}\ \delta t.$$

This term must be added to the Fickian term arising from the concentration gradient:

$$\frac{\partial c}{\partial t} = D\,\frac{\partial^2 c}{\partial x^2} + v\,\frac{\partial c}{\partial x}, \qquad \text{(X.109)}$$

where D is the diffusion coefficient of B in the liquid. For steady state:

$$D\,\frac{\mathrm{d}^2 c}{\mathrm{d}x^2} + v\,\frac{\mathrm{d}c}{\mathrm{d}x} = 0, \qquad \text{(X.110)}$$

with the boundary conditions:

$$\begin{aligned} c(\infty) &= c_0 \\ \bar{k}_{\mathrm{B}} \cdot c(0) &= c_{\mathrm{B}}^{\mathrm{S}} = c_0 . \end{aligned}$$

Equation (X.110) is easily integrated to obtain:

$$D\frac{\mathrm{d}c}{\mathrm{d}x} + vc = \text{Constant}.$$

When the constant on the RHS equals zero, the solution is $A\exp(-vx/D)$. Since c_0 is a particular solution of the full equation, the general solution can be written:

$$c = c_0 + A\exp\left(-\frac{v}{D}\,x\right),$$

and with the above boundary conditions:

$$\boxed{c = c_0\left[1 + \frac{1-\bar{k}}{\bar{k}}\exp\left(-\frac{v}{D}\,x\right)\right]}. \qquad \text{(X.111)}$$

Clearly, a variation of the liquidus temperature with distance x from the interface is associated with this variation in the composition of the liquid (given by Eq. (X.111)), i.e., a local increase $\Delta\,T_l$ if $\bar{k}_{\mathrm{B}} < 1$. If $\Delta\,T_l$ is high enough, then locally, ahead of the solidification front, $T < T_{\text{liquidus}}$. This is *constitutional supercooling,* so named because the temperature of the liquid is less than the temperature of its equilibrium with the solid. The important parameter that governs constitutional supercooling is the ratio v/D.

The liquidus temperature in the zone of constitutional supercooling is given by $T_{\mathrm{M}} - mc$, where m is the slope of the solidus (assumed linear) and T_{M} is the melting temperature of the pure solvent A (Fig. X.25). The temperature profile of the solidus can be compared to the imposed temperature profile (Fig. X.27). If the latter is represented by a constant gradient $\mathrm{d}T/\mathrm{d}x = G$, a critical condition for constitutional supercooling is obtained when this gradient is

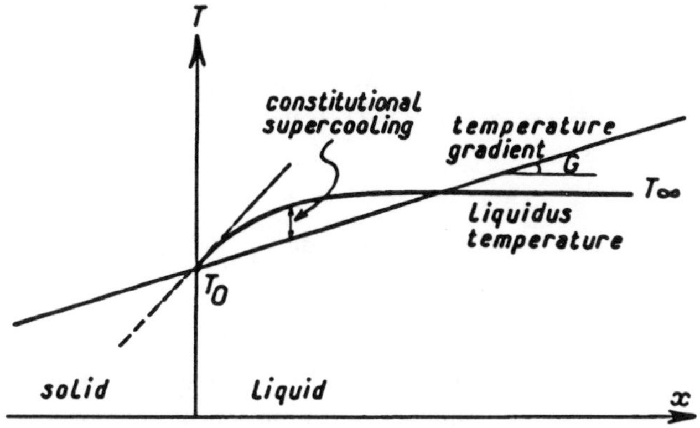

Fig. X.27. — Temperature gradient and liquidus profile ahead of a solid/liquid interface. The dashed line corresponds to the upper limit of the gradient for the generation of constitutional supercooling.

tangent to the temperature profile of the liquidus at $x = 0$ (the interface):

$$-m(\mathrm{d}c/\mathrm{d}x)_0 = G,$$

from which:

$$\boxed{\frac{G}{v} = \frac{mc_0}{D}\,\frac{1-\bar{k}}{\bar{k}}}, \qquad \text{(X.111a)}$$

which is the classic *equation of Tiller, Jackson, Rutter, and Chalmers* (see Chalmers, 1964). For temperature gradients G smaller than the value that satisfies this equation, there is constitutional supercooling in the zone where the liquidus profile is located above the temperature profile G.

Note that the solution of the diffusion equation which has just been determined is completely different from the solutions discussed above for the case of mobile interfaces (Ch. I., § VII); specifically, it differs from the solution for the case of enrichment in solute behind of the oxidation front of an alloy (see selective oxidation, this chapter, § II.2.a)). The reason for this is the following: in these problems, the entire process, including the phase transformation, is governed by diffusion, so that the interface moves as $(Dt)^{1/2}$, i.e., with a velocity $v \propto 1\sqrt{t}$. In the case of solidification, on the other hand, the process is governed by the diffusion of both matter and heat. It is the latter which, via the temperature gradient $\mathrm{d}T/\mathrm{d}x$, causes the interface to advance with constant v for a constant gradient.

Consider again the diffusion equation (X.110). If $v = h\sqrt{t}$, the change of variable $\lambda = x\sqrt{t}$ leads to:

$$\frac{\mathrm{d}^2 c}{\mathrm{d}\lambda^2} + \left(\frac{h}{D} + \frac{\lambda}{2\,D}\right)\frac{\mathrm{d}c}{\mathrm{d}\lambda} = 0,$$

or, with:

$$\frac{h}{D} + \frac{\lambda}{2\,D} = \frac{\lambda'}{2\,D},$$

to

$$\frac{\mathrm{d}^2 c}{\mathrm{d}\lambda'^2} + \frac{\lambda'}{2\,D}\,\frac{\mathrm{d}c}{\mathrm{d}\lambda'} = 0,$$

i.e., to the well-known Boltzmann-Matano equation, with the classical error-function solution if D is independent of c. In the case of solidification, the same change of variables does not allow the elimination of the variable t, since $v =$ constant instead of $ht^{1/2}$. Only the steady-state solution is simple, because the second order differential equation (X.110) has constant coefficients, unlike the Boltzmann-Matano equation.

VI. Diffusion under irradiation.

Irradiation by particles such as neutrons, protons, α particles, fission fragments, electrons, implanted ions... with energies above a certain threshold gives rise to the formation of Frenkel pairs (a vacancy and an interstitial) in the lattice. The number of these defects depends on the nature, energy, and flux of the incident particles.

VI.1 Defect concentrations. Balance equations. — The quasi-chemical approach proposed by Lomer (in his oft-quoted but unpublished Harwell report, 1954) allows the calculation of the average concentration of the defects by listing the events that can befall the point defects in the form of "chemical reactions":

Generation and recombination:

$$\text{Vacancy} + \text{Interstitial} \underset{G}{\overset{K_{iv}}{\rightleftharpoons}} 0.$$

Annihilation (creation) at fixed sinks (sources):

$$\text{Sink (or source)} + \text{Vacancy} \underset{S_v}{\overset{K_v}{\rightleftharpoons}} \text{Sink (or source)},$$

$$\text{Sink (or source)} + \text{Interstitial} \underset{S_i}{\overset{K_i}{\rightleftharpoons}} \text{Sink (or source)}.$$

The fixed sinks (dislocations, boundaries, surfaces), with concentration N_s, can also act as sources. Assuming that the fixed sinks are uniformly distributed, and neglecting the local spatial variations of concentration, we can write the balance equations with the above reaction-rate constants as:

$$\begin{aligned} \mathrm{d}N_v/\mathrm{d}t &= G - K_{iv}\ N_i\ N_v - K_v\ N_s\ N_v + S_v\ N_s \\ \mathrm{d}N_i/\mathrm{d}t &= G - K_{iv}\ N_i\ N_v - K_i\ N_s\ N_i + S_i\ N_s. \end{aligned} \tag{X.112}$$

The assumption of uniform distribution is reasonable for the dislocations, but not for the grain boundaries or surfaces; a supplementary diffusion term, $D_{v,i}\ \nabla^2 n_{v,i}$, has to be introduced in the equations to analyze the problem in the neighborhood of boundaries or surfaces. Finally, a number of events, such as the formation of multi-defects, divacancies, di-interstitials, defect-impurity complexes, etc... as well as the elimination of defects on dislocation loops formed by the condensation of the accumulated defects, which leads to a variable sink concentration, are left out of the detailed balance in the above equations.

Expressions for the reaction-rate constants.

• *The constants S_v and S_i.* — Without irradiation, $N_v = \bar{N}_v$ and $N_i = \bar{N}_i \approx 0$ (the barred quantities refer to values at thermal equilibrium); the concentrations do not vary with time. The balance equations reduce to:

$$\begin{aligned} K_v\ N_s\ \bar{N}_v &= S_v\ N_s \quad \text{or} \quad K_v\ \bar{N}_v = S_v \\ K_i\ N_s\ \bar{N}_i &= S_i\ N_s \qquad\quad S_i = 0, \end{aligned} \tag{X.113}$$

since $\bar{N}_i \approx 0$ in a metal.

• *Rate constant for mutual recombination K_{iv}.* — Assume that the vacancies are immobile (jump frequency $w_v << w_i$) and the mobile interstitials annihilate at the vacancies when they get close enough (within the recombination volume). The hollow-sphere model (see this chapter, § IV.4) applies to the diffusion of the interstitials. Equation (X.94a) yields:

$$\frac{\mathrm{d}n_i}{\mathrm{d}t} = -D_i\ \frac{3\ R_c}{R_\infty^3}\ n_i,$$

with the external radius of the sphere given by:

$$\frac{4}{3}\ \pi R_\infty^3\ n_v = 1 \text{ (density of sinks} = n_v).$$

Whence:

$$\frac{\mathrm{d}N_i}{\mathrm{d}t} = -4\ \pi D_i R_c N_i N_v/\Omega, \tag{X.114}$$

or, setting it equal to the second equation (X.112):

$$K_{iv} = 4\ \pi D_i R_c/\Omega.$$

Since $D_i = w_i a^2$ and $\Omega = a^3/4$ for an FCC structure, we can write:

$$K_{iv} = Z w_i,$$

with

$$Z = 16\pi R_c/a. \tag{X.115}$$

Z depends on the number of sites in the recombination volume ($\approx$ 50-150).

In the case where both species are mobile, the kinetics of the bimolecular reaction $v + i \rightleftharpoons 0$ yield the same result as long as $w_i >> w_v$.

• *Rate constant* $K_{i,v}$ *for annihilation on dislocations.* — To annihilate at a dislocation, the point defects must first diffuse to it and then migrate along the dislocation core until they reach a jog. If the latter mechanism is fast enough and the annihilation at jogs is efficient, the process is governed by volume diffusion, and the dislocation can be modeled by a hollow cylinder with inside radius R_c, the effective capture radius. Following the analysis of § IV, equation (X.101) and using the subscripts i, v to denote interstitials and vacancies respectively, we find:

$$\frac{\mathrm{d}\ N_{i,v}}{\mathrm{d}t} = -K_{i,v}\ N_{i,v}, \tag{X.116}$$

with:

$$K_{i,v} = \frac{2\ D_{i,v}}{R_\infty^2\ \mathrm{Log}\left(R_\infty/R_c^{i,v}\right)}$$ $= K_v\ N_s$ or $K_i\ N_s$ according to equations (X.112).

The capture radius for vacancies can be different from that for interstitials, because the elastic interaction is stronger for the latter (the so-called "bias").

With a dislocation density ρ_d such that $\pi R_\infty^2\ \rho_d = 1$:

$$K_{i,v} = \frac{2\pi\rho_d D_{i,v}}{\mathrm{Log}\left(R_\infty/R_c^{i,v}\right)}. \tag{X.116a}$$

If $\rho_d \approx 10^6$ to 10^{10} cm^{-2} and $D = a^2 w$, then approximately:

$$\begin{aligned} K_v N_s &\sim \rho_d a^2 w_v &= P_v w_v \\ K_i N_s &\sim \rho_d a^2 w_i &= P_i w_i, \end{aligned} \tag{X.116b}$$

and the balance equations become:

$$\begin{aligned} \frac{dN_v}{dt} &= G - Z w_i N_i N_v - P_v w_v \left(N_v - \bar{N}_v\right) \\ \frac{dN_i}{dt} &= G - Z w_i N_i N_v - P_i w_i N_i. \end{aligned} \tag{X.117}$$

VI.2 Steady state. — Since the interstitials are much more mobile than the vacancies, the concentration of the former evolves very rapidly during the transient stage; the evolution is controlled by the most efficient reaction for the annihilation of the interstitials. At the end of this stage, a pseudo steady state is attained such that $dN_i/dt = 0$:

$$N_i = \frac{G}{Z w_i N_v + P_i w_i}.$$

The differential equation dN_v/dt has an analytical solution which will not be discussed here. It can be shown that the concentrations of the vacancies and interstitials cross over during the transient; the vacancy concentration continues to grow during the pseudo-steady state, and the interstitial concentration remains constant or decreases. Finally, the true steady state is reached:

$$\begin{aligned} dN_v/dt &= 0 \\ dN_i/dt &= 0, \end{aligned}$$

which, according to equations (X.117) implies:

$$w_i N_i = \frac{P_v}{P_i}\, w_v \left(N_v - \bar{N}_v\right) \tag{X.118}$$

The ratio P_v/P_i represents the bias that characterizes the asymmetry of the dislocations vis-à-vis the vacancies and interstitials. When this expression for $w_i N_i$ is substituted into the balance equations for vacancies, we obtain, with $\alpha = P_i/P_v$:

$$N_v - \bar{N}_v = \frac{1}{2}\left\{\left[\bar{N}_v + \frac{\alpha P_v}{Z}\right]\left[-1 + \left[1 + \frac{4G\alpha}{Z w_v}\left(\bar{N}_v + \frac{\alpha P_v}{Z}\right)^{-2}\right]^{1/2}\right]\right\}. \tag{X.119}$$

This expression can be simplified in several cases.

The high-temperature regime. — The concentration of vacancies in thermal equilibrium is high enough so that:

$$\bar{N}_v >> \alpha P_v Z,$$

and

$$\bar{N}_v^2 >> 4\ G\alpha / Z w_v .$$

The term in curly brackets is approximately zero:

$$\begin{aligned} N_v &= \bar{N}_v + \frac{G\alpha}{Z w_v N_v} \approx \bar{N}_v \\ N_i &= \frac{G}{Z w_i \bar{N}_v} . \end{aligned} \qquad \text{(X.120)}$$

The medium- and low-temperature regimes. — If the temperature is low enough, the equilibrium concentration of vacancies is negligible:

$$\bar{N}_v << \alpha P_v / Z = P_i / Z.$$

Two cases are possible:
- high sink concentration (or not too low a temperature):

$$\frac{4\ G\alpha}{Z w_v} << \left(\frac{\alpha P_v}{Z}\right)^2 .$$

The kinetics are dominated by annihilation at dislocations; whence:

$$\begin{aligned} N_v &= \frac{G}{P_v w_v} \\ N_i &= \frac{G}{P_i w_i} . \end{aligned} \qquad \text{(X.121)}$$

- low sink concentration (or low temperature):

$$\frac{4\ G\alpha}{Z w_v} >> \left(\frac{\alpha P_v}{Z}\right)^2$$

from which:

$$\begin{aligned} N_v &= \left(\frac{G\ \alpha}{Z w_v}\right)^{1/2} \\ N_i &= \frac{1}{w_i}\left(\frac{G\ w_v}{Z\alpha}\right)^{1/2} . \end{aligned} \qquad \text{(X.122)}$$

The defect concentrations are then limited by mutual recombination. The regime changes with the temperature.

Figure X.28a shows the ensemble of these results, calculated with parameters that are reasonable for copper.

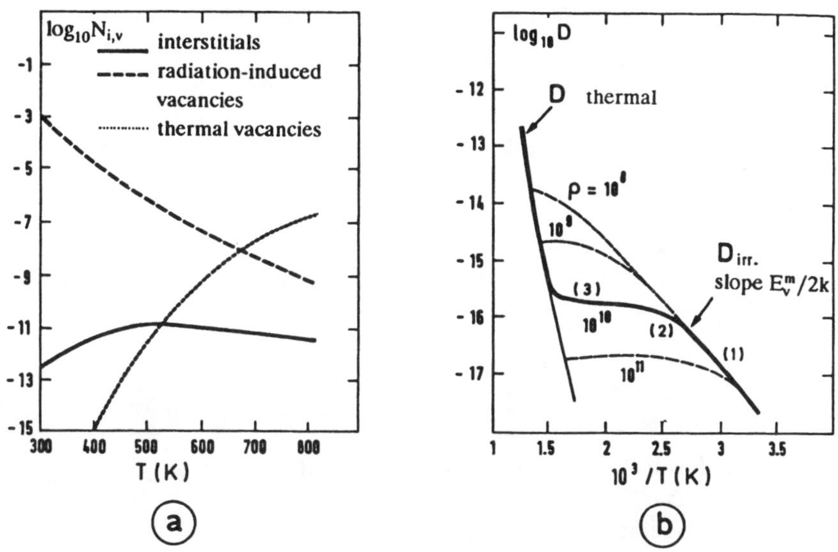

Fig. X.28. — Diffusion under irradiation (after Sharp). a) Concentration of vacancies and interstitials at steady state in irradiated copper, $G = 10^{-6}\ \mathrm{s}^{-1}$, $\rho_{\mathrm{d}} = 10^9\ \mathrm{cm}^{-2}$, $Z = 1$, E_v^{m}=20.3 kcal/mole, $E_v^{\mathrm{f}} = 26.7$ kcal/mole. b) Arrhenius plot for self-diffusion in copper at equilibrium and under irradiation. The same data as in a), except that ρ_{d} varies from 10^8 to $10^{11}\ \mathrm{cm}^{-2}$.

VI.3 TRACER SELF-DIFFUSION IN THE STEADY STATE. — The migration of tracer atoms can take place via both the vacancy and interstitial mechanisms, as opposed to the classic case of thermal equilibrium.

The diffusion coefficient under irradiation is given by:

$$D^* = N_i D_i + N_v D_v = \beta a^2 \left(N_i w_i + N_v w_v\right),$$

neglecting correlation effects and assuming that the structure is cubic (β is a numerical factor that depends on the structure).

The terms are often regrouped:

$$\begin{aligned} D^* &= N_i\ D_i + \left(N_v - \bar{N}_v\right) D_v \quad + \quad \bar{N}_v\ D_v \\ &= \qquad\qquad D^*_{\mathrm{irr}} \qquad\qquad + \quad D^*_{\mathrm{th}}, \end{aligned} \tag{X.123}$$

in a way that shows the enhancement of diffusion due to the irradiation and the thermal self-diffusion as separate terms.

At low temperature, under conditions such that mutual recombination is the dominant process of defect elimination, we have from equations (X.122) and (X.123):

$$D^* = D^*_{\text{irr}} = \beta a^2 \left(\frac{G w_v}{Z\alpha}\right)^{1/2} (1+\alpha)\,. \qquad \text{(X.124)}$$

The result $N_v D_v = N_i D_i$ and $D^*_{\text{irr}} = 2\ N_v D_v$ applies only if $\alpha = 1$, i.e., the dislocations are equally efficient for the elimination of vacancies and interstitials. The activation enthalpy for self-diffusion equals $\Delta H_v^{\text{m}}/2$ (ΔH_v^{m} is the migration enthalpy of a vacancy). The self-diffusion coefficient varies as the square root of the flux.

At high temperatures, where the thermal equilibrium vacancy concentration is dominant because the radiation induced defects are quickly eliminated at the dislocations:

$$D^* \approx \bar{N}_v w_v \beta a^2 = D^*_{\text{th}}, \qquad \text{(X.125)}$$

i.e., the normal self-diffusion coefficient is found, with the usual activation enthalpy $\Delta H_v^{\text{f}} + \Delta H_v^{\text{m}}$.

In the intermediate temperature range, however:

$$D^* = \beta a^2 G \left[P_v^{-1} + P_i^{-1}\right]. \qquad \text{(X.126)}$$

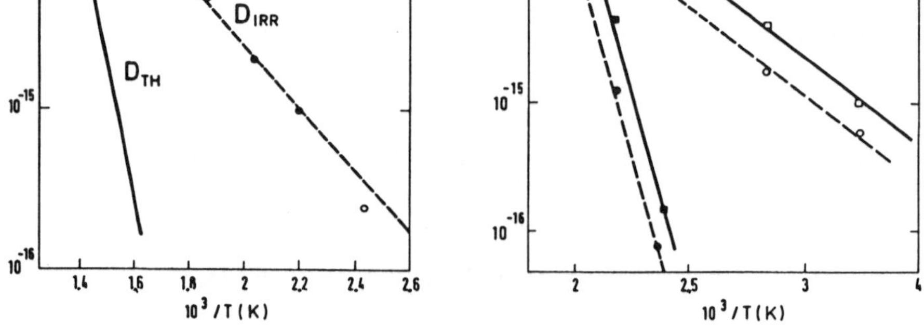

Fig. X.29. — Diffusion under irradiation. Self-diffusion in gold (left) and solute diffusion in aluminum (right). (After D. Acker, 1977).

Here the diffusion coefficient is quasi-independent of the temperature because both P_v and P_i are proportional to the dislocation density and do not contain the jump frequencies (see Eq. (X.116b)). The diffusion coefficient is proportional to the flux.

These predictions are shown schematically on the Arrhenius plot of figure X.28b, with reasonable defect parameters for copper. Figure X.29 shows actual experimental results for self-diffusion in gold, and solute diffusion in aluminum.

Remarks:

1) The transition temperature between the domains where mutual recombination and annealing to dislocations are respectively dominant increases with G. This is also why voids formed due to irradiation are observed at higher temperatures the higher the flux, for the same total dose.

2) If the sink density is almost zero, a steady state is predicted with:

$$N_i = N_v = (G/Zw_i)^{1/2}.$$

In fact, this condition is never found exactly, although such a quasi-steady state can be observed in the course of irradiation before the true state described by equation (X.119) is reached.

VI.4 CASCADE MIXING. — In the case of bombardment by heavy ions as in ion implantation, much of the radiation damage consists of large defect cascades. Since these cascades are at high temperatures for a brief time, and since recombination takes place inside the cascade, the atoms in each cascade are mixed. This results in a small temperature independent diffusion coefficient at low temperatures (down to liquid helium, Fig. X.30). The magnitude of this

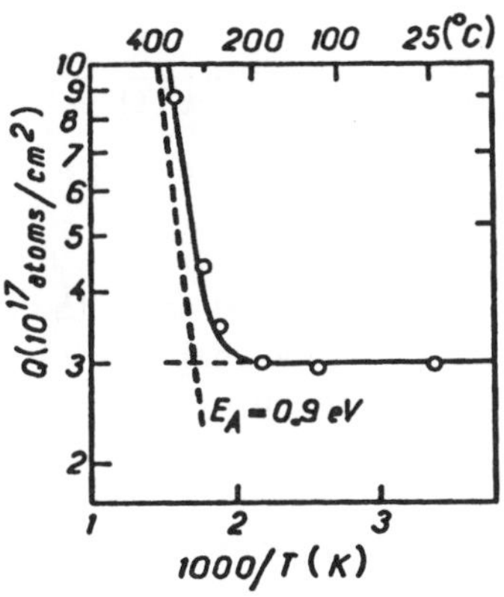

Fig. X.30. — Temperature dependence of ion beam mixing in Nb-Si. In this experiment, a 1440Å thick Nb layer, deposited on a single crystal of Si, was bombarded with a flux ϕ of 275 keV ^{28}Si ions (range < 1440Å); the quantity Q of Si mixed into the Nb film from the single crystal was measured by RBS. The dashed line (E_a= 0.9 eV) represents thermal mixing of Si and Nb. (Matteson, Roth, and Nicolet, 1979).

diffusion coefficient is approximately linear in the energy per ion deposited in unit length of the lattice. At higher temperatures, diffusion is again due to mobile point defects, but the cascades can act as sinks for the mobile defects. The subject has been reviewed by Averback (1986).

VII. Plastic deformation at high temperature.

Plastic deformation at temperatures $T \gtrsim T_M/2$ (T_M is the melting temperature) brings diffusion processes into play. Two phenomena are considered here, the analysis of which emphasizes mass transport: diffusional creep, and rupture by the growth of grain-boundary cavities during creep.

VII.1 DIFFUSIONAL CREEP. — According to an idea of Nabarro (1948), plastic deformation under certain conditions can take place not by the movement of dislocations, but by mass flow. In a non-isostatic stress field, the equilibrium concentrations of vacancies differ depending on the orientation of the surfaces or interfaces where they can be generated or annihilated. The difference in vacancy concentrations is the cause of a vacancy flux between different surfaces, and this flux gives rise to a mass flow in the opposite direction.

In fact, according to the phenomenological relations given in chapter VIII, the fluxes are proportional to the gradients of chemical potential, or, according to equations (VIII.18), to the gradients of the differences $(\mu_i - \mu_v)$ of the chemical potentials of the material species i and the vacancies. In a pure material A, the driving force for mass flow reduces to a single term $(\mu_A - \mu_v)$, written more simply in the following without the subscript A:

$$J_A = -J_v = -\frac{nD}{kT}\,\nabla(\mu - \mu_v) = -\frac{D}{kT\Omega}\,\nabla(\mu - \mu_v)\,, \qquad \text{(X.127)}$$

where D is the self-diffusion coefficient, n is the number of atoms per unit volume, and Ω is the atomic volume.

a) *Herring's model* (1950). — In the first stage of the calculation, we consider a single crystal of a pure material, cube-shaped with edge L, at temperature T, subjected to a stress system equivalent to a pure shear σ on the 45° planes. The traction $\mathbf{t}$ on each face is normal, its components are $t_n = \sigma$ on the z faces (tension) and $t_n = -\sigma$ on the x and y faces (compression) (Fig. X.31a).

At thermal equilibrium, near a surface to which a normal stress t_n is applied, the difference in the chemical potentials $(\mu - \mu_v)$ of the atoms and the vacancies is given by:

$$\mu - \mu_v = \mu_0 - t_n\Omega, \qquad \text{(X.128)}$$

where μ_0 is the chemical potential of the atoms in the absence of stress [9]. Recall that $\mu_v = kT$ Log $(n_v/\bar{n}_v)$, with $n_v = \bar{n}_v$ at thermal equilibrium in the absence of stress.

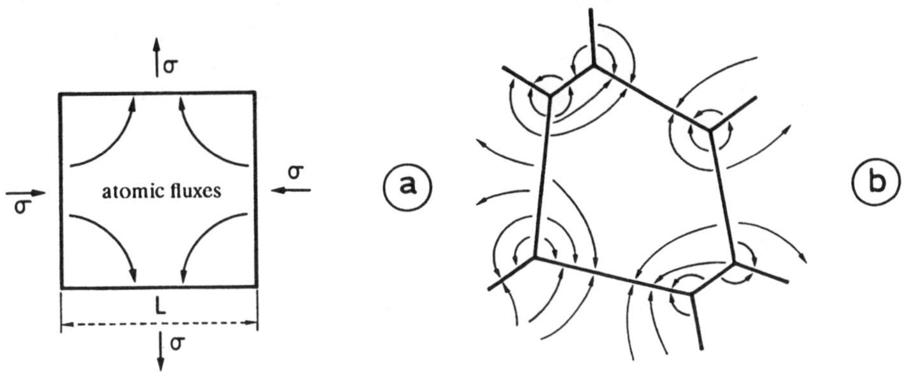

Fig. X.31. — Diffusional creep, Herring's model: a) single crystal under shear (equivalent normal stress $t_n = \pm\sigma$); b) Polyhedral grain in a polycrystalline aggregate, showing the mass flux. Same stress tensor as in a).

In the stressed crystal, there is a difference between the values of $(\mu - \mu_v)$ for the z faces on the one hand, and the x and y faces on the other; since the vacancies in this scheme are created and annihilated only at the surfaces, their chemical potential can be defined only there. It follows then that at every point in the volume div $J_v = 0$, or following (X.127):

$$\nabla^2 (\mu - \mu_v) = 0. \qquad \text{(X.129)}$$

Under these conditions, the chemical potential, or more exactly, the quantity $(\mu - \mu_v)$ can be calculated at all points of the crystal by solving equation (X.129) with the boundary conditions imposed on the surfaces by chemical equilibrium (Eq. (X.128)). Finally, the flux can be calculated throughout the crystal by applying equation (X.127).

In the general case, numerical methods have to be used, but approximate analytical solutions can be found quite easily. First of all, the condition of

(9) To form a vacancy, an atom situated on the surface is removed from the crystal under stress; in this operation, the Gibbs free energy changes by $-\mu + \mu_v$ (one atom lost, one vacancy gained). To conserve mass, an atom is removed from an identical but unstressed crystal and is placed on the surface of the stressed crystal; the corresponding change in the Gibbs free energy is $\mu_0 - t_n\Omega$. At equilibrium, the total change of the free energy is zero:

$$\mu - \mu_v = \mu_0 - t_n\Omega,$$

where t_n is the component of the traction normal to the surface (positive for tension).

mechanical equilibrium is simplified by assuming that the normal stresses t_n on the surfaces are uniformly equal to $\pm\sigma$. The flux of material then flows from the x and y surfaces at chemical potential $\mu + \sigma\Omega$ to the z surfaces at chemical potential $\mu - \sigma\Omega$. Then the gradient of chemical potential is of order $\sigma\Omega/L$, and the mass flux $J \propto D\sigma/kTL$. The number of atoms transported per unit time and per face is $\Phi \approx JL^2$, producing a volume change per face per unit time of $\Delta V = J\Omega L^2$ distributed over L^2, or a change in length $J\Omega$ per unit time, from which:

$$\dot{\varepsilon} = \frac{1}{L}\,\frac{\mathrm{d}L}{\mathrm{d}t} = \frac{1}{L}\,J\Omega$$

$$\boxed{\dot{\varepsilon} = B\,\frac{D}{kT}\,\frac{\sigma\Omega}{L^2}}\,. \tag{X.130}$$

A numerical calculation with fewer approximations is needed to determine the numerical factor B in the Herring-Nabarro equation (X.130). The essential result is that the ***strain rate is proportional to the stress ("viscous flow") and*** *to the reciprocal of the square of the crystal size.*

In the second phase of the calculation, the model is applied to diverse geometries and stress systems, e.g., a polycrystalline wire with "bamboo structure" under tensile load (the "zero creep" experiment for determining surface energy), or a polycrystalline sample. The latter is considered as an assembly of identical small crystals, with the grain boundaries playing the role of the surfaces in the above model (Fig. X.31b). Nevertheless, the model is only approximate; polyhedral grains must remain in contact with each other (compatible deformation). To achieve this, the flux at every point of the facets that make up the grain boundaries must adjust; a coupling which translates into *non-uniform normal stresses* t_n on these facets. Otherwise the flux would be higher near the edges (shorter distances, hence larger gradients), and as a result, the material arriving at the center of a facet under tension would be less than that arriving near the edges. The excess material at the edges gives rise to a compressive stress which locally reduces the normal tensile stress resulting from the applied load. The calculation of these stresses is simple only in the case of transport along the grain boundaries (see below). Further, the relative deformation of the contiguous grains gives rise to tangential stresses that have to be relaxed by grain boundary sliding; ***mass flow by diffusion and*** *grain-boundary sliding are necessarily coupled.*

For simplicity, Herring considered a cluster of spherical grains; this assumption leads to a numerical factor of $B = 16$ in equation (X.130), multiplied by 5/2 to take into account the relaxation of the tangential stresses.

Remarks:

1) The model was worked out for pure shear (equivalent normal stresses of $-\sigma$, $-\sigma$, σ). For a pure tensile or compressive stress σ, the equivalent shear

stress is $\sigma/3$; hence an extra factor of 1/3 should be included in equation (X.130). The factor B is of order 12 or 14 for different grain shapes under uniaxial tension.

2) In a solid solution or compound, the chemical composition remains constant during the deformation. Equation (X.130) applies on the condition that Ω is taken as the molecular volume and D is taken as the molecular diffusion coefficient as defined above in chapter VI, equation (VI.71).

b) *Coble's mechanism.* — The driving force for mass flow is the same as that in the Nabarro-Herring model, but the mass flow takes place by diffusion along the grain boundaries. This is the same as taking Herring's model for a cubic grain under a shear stress σ, equivalent to tensile and compressive normal stresses σ and $-\sigma$, respectively, on the grain faces, with the mass flow being restricted to an interfacial layer of thickness δ. Let D' be the grain-boundary diffusion coefficient. According to equation (X.127), the total flux in the interface is:

$$\begin{aligned} J'\delta &= -\frac{D'\delta}{\Omega kT}\,\nabla(\mu-\mu_{\mathrm{v}}) \\ &= \frac{D'\delta}{kT}\,\nabla t_{\mathrm{n}}. \end{aligned} \tag{X.131}$$

Since the deformation is uniform on a face, the quantity of material deposited per unit area and unit time should be constant:

$$\frac{\partial n}{\partial t} = -\delta \cdot \operatorname{div} J' = \beta > 0,$$

from which, in the x direction:

$$\frac{\mathrm{d}^2 t_{\mathrm{n}}}{\mathrm{d}x^2} = -\beta\,\frac{kT}{D'\delta}.$$

It follows that t_{n} *varies parabolically* from one edge of a face to the other. With the boundary condition $t_{\mathrm{n}} = 0$ at $x = 0$ and $x = L$, it is easily found that $t_{\mathrm{n}} = (\beta kT/2D'\delta)x(L-x)$. The condition of mechanical equilibrium:

$$\int_0^L t_{\mathrm{n}}\,\mathrm{d}x = \sigma L \tag{X.132}$$

can be used to determine the constant β:

$$\beta = \frac{12\,\sigma}{L^2}\,\frac{D'\delta}{kT},$$

from which:

$$t_{\mathrm{n}} = \frac{6}{L}\,x\left(1-\frac{x}{L}\right)\sigma. \tag{X.133}$$

To calculate the deformation, the rate of deposition dn/dt (number of atoms per unit area per unit time) is considered to produce a volume change $\Omega dn/dt$ (hence a change in the semi-diameter of the grain $d(L/2)/dt = \Omega dn/dt$), and a strain rate:

$$\boxed{\dot{\varepsilon} = \frac{C\sigma\Omega}{L^3}\frac{D'\delta}{kT}}, \qquad \text{(X.134)}$$

with $C = 24$. This equation has the same form as the one deduced by Coble in a different way and based on spherical grains. Differences in the factor C are here again due to the grain shape factor, to the applied stress system (tension or simple shear), and to taking into account the relaxation of the tangential stresses by grain boundary sliding. The essential result is that *the strain rate is proportional to the stress*, as in Herring-Nabarro creep, but to *the reciprocal of the cube of the grain size*, instead of the square as in Herring-Nabarro creep.

c) *The general case.* — In the general case of viscous flow, both volume and grain-boundary diffusion contribute to the mass transport. The strain rate retains the form of equation (X.130), but with the introduction of a diffusion coefficient that is a combination of the volume and grain-boundary self-diffusion coefficients, D and D' :

$$D_{\text{comp}} = D\left[1 + \frac{3\delta}{L}\frac{D'}{D}\right].$$

In effect, the ratio of the grain and grain-boundary volumes in which transport takes place is of order $3\delta/L$. The numerical factor in equation (X.130) remains equal to 14 for uniaxial tension and to 40 for shear.

For a compound $A_\alpha B_\beta$, the transport is given by the molecular diffusion coefficient, as defined in chapter VI, equation (VI.71):

$$D_{\text{mol}} = \frac{D_A\, D_B}{\alpha D_B + \beta D_A}.$$

Then, replacing D_A and D_B by the corresponding expressions for D_{comp}, we obtain:

$$D_{\text{comp}}^{\text{mol}} = \frac{1}{\alpha}\,\frac{D_A + \dfrac{3\delta}{L}D'_A}{1 + \dfrac{\beta}{\alpha}\,\dfrac{D_A + 3\dfrac{\delta}{L}D'_A}{D_B + 3\dfrac{\delta}{L}D'_B}}.$$

However, it seems physically more reasonable to couple the volume and grain-boundary fluxes separately, in order to conserve the chemical compositions of both the grain volume and the boundaries, and then to combine the molecular fluxes to obtain:

$$D_{\text{comp}} = \frac{D_{\text{A}}\ D_{\text{B}}}{\alpha D_{\text{B}} + \beta D_{\text{A}}} + (3\delta/L)\ \frac{D'_{\text{A}}\ D'_{\text{B}}}{\alpha D'_{\text{B}} + \beta D'_{\text{A}}}.$$

In all cases, the rate-controlling process for creep is the motion of the slower diffusing species along the easier path. This can be either the self-diffusion of A in either the volume or in the boundaries, or the self-diffusion of B in either the volume or the boundaries, depending on the relative values of the 4 diffusion coefficients (and thus the temperature) and the grain size.

For a more detailed discussion of the theoretical and experimental aspects of diffusional creep, the reader should consult the books by J. P. Poirier (1976, 1985).

d) *Grain-boundary sliding.* — In order to maintain cohesion between the grains, the viscous flow of a polycrystal by diffusional processes, analyzed in the preceding paragraphs, is necessarily accompanied by grain-boundary sliding. Conversely, sliding along non-planar grain boundaries requires a process of accomodation, in its nature either elastic, or diffusional, or plastic flow by dislocations (if the stresses are very high). We shall thus re-analyze the behavior of a polycrystal subjected to a shear strain, beginning this time from the point of view of grain-boundary sliding.

Consider the two-dimensional case for simplicity. In the principal plane of the shear, the polycrystal is simulated by a network of regular hexagons (Fig. X.32). Along the two orthogonal directions, the contours of the grains can be viewed as a periodically broken line, and these lines can be described by a Fourier series. Here they will be considered to be sinusoidal for simplicity.

The model is then the following: let two grains be separated by a wavy plane, the section of which is sinusoidal, and on which a shear τ_{a} is exerted (Fig. X.33). This involves the gliding of one grain with respect to the other, and the non-planarity of the boundary gives rise to normal stresses t_{n}. Transport of material by diffusion between the parts of the boundary in compression and tension allows the normal stresses to be relaxed.

The profile of the boundary is described by:

$$z = h\ \cos[(2\pi/\lambda)y]. \tag{X.135}$$

If the boundaries are the sole sources and sinks for vacancies, $\partial n_v/\partial t = 0$ everywhere in the grain volume, i.e., div $J_v = 0$. Since J_v is proportional to $\nabla(\mu - \mu_v)$, it follows that:

$$\nabla^2\,(\mu - \mu_v) = 0. \tag{X.136}$$

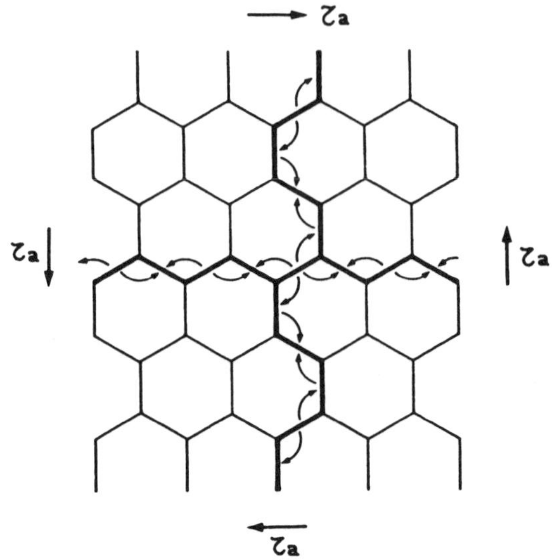

Fig. X.32. — Grain-boundary sliding in a crystal subjected to a shear τ_a, showing the mass flows required to maintain cohesion.

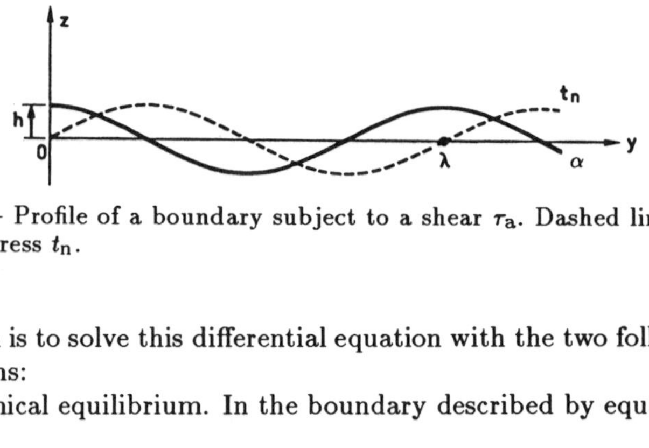

Fig. X.33. — Profile of a boundary subject to a shear τ_a. Dashed line: variation of the normal stress t_n.

The problem is to solve this differential equation with the two following boundary conditions:

1) Chemical equilibrium. In the boundary described by equation (X.135):

$$\mu - \mu_v = \mu_0 - t_n \Omega, \tag{X.137}$$

where t_n denotes the stress normal to the boundary arising owing to the glide (see Eq. (X.128)).

2) Mechanical equilibrium. The stress t_n has the same periodicity as the boundary. It is a maximum where the boundary resists the glide, i.e., where it is the most inclined to the $z = 0$ plane; t_n is thus 90° out of phase with $z(y)$:

$$t_n = t_0 \sin(2\pi/\lambda)y.$$

The stress t_n has a component $t_n \sin\alpha \approx t_n(-dz/dy)$ in the $z = 0$ plane (setting

$\sin\alpha \approx \tan\alpha$). At mechanical equilibrium, the tangential force thus produced equals that which results from the applied shear stress τ_a. This equality is, for an area of $\lambda \times 1$:

$$\tau_a\,\lambda = \int_{-\lambda/2}^{\lambda/2} t_n(-\mathrm{d}z/\mathrm{d}y)\mathrm{d}y.$$

Integration yields t_0, from which we obtain:

$$t_n = \frac{\tau_a\,\lambda}{\pi h}\,\sin\left(\frac{2\pi}{\lambda}\,y\right). \tag{X.138}$$

Equations (X.137) and (X.138) yield the value of the chemical potential in the plane of the boundary. The general solution must be periodic in y. Thus we are led to try as a solution of equation (X.136):

$$\mu - \mu_v = A\,\exp\left(-\frac{2\,\pi z}{\lambda}\right)\,\sin\left(\frac{2\,\pi y}{\lambda}\right) + \text{const.}$$

It is easy to determine that:

$$\mu - \mu_v = \mu_0 - \Omega t_0\,\exp\left(-\frac{2\,\pi z}{\lambda}\right)\,\sin\left(\frac{2\,\pi y}{\lambda}\right), \tag{X.139}$$

with $t_0 = \tau_a\,\lambda/\pi h$.

When the chemical potential is known, the diffusion equation is solved in the usual way, using the expressions for the mass flow as a function of $\nabla(\mu-\mu_v)$ and the equation of continuity.

Equation of continuity (material balance, Fig. X.34). Let an element Δy of the length of the boundary ($\cos\alpha \approx 1$) be displaced parallel to y by a distance $\dot{u}\,\Delta t$. This causes the "disappearance" of an element of material of section $\Delta y\dot{u}\,\Delta t\,\sin\alpha \approx \dot{u}\,\Delta t(\mathrm{d}z/\mathrm{d}y)\Delta y$. This quantity of material feeds a corresponding element in another part of the boundary by two diffusional fluxes:

– in the volume: $2J_z(y,0)\Delta y$, denoting by $J_z(y,0)$ the volume flux normal to the element under consideration and (approximately) parallel to the direction z, with its value dependent on y, and taken for the "plane" $z = 0$;

– in the boundary of the conventional thickness δ:

$$\delta\left[J'_y\,(y+\Delta y) - J'_y(y)\right] = \delta\cdot\Delta y\left[\frac{\mathrm{d}J'_y}{\mathrm{d}y}\right],$$

the flux "in" the boundary depends only on the space variable y. Then the balance equation is:

$$\dot{u}\,\Delta t(\mathrm{d}z/\mathrm{d}y)\,\Delta y = \Omega\,\Delta y\,\Delta t\,\left[2\,J_z(y,0) + \delta\left(\mathrm{d}J'_y/\mathrm{d}y\right)\right],$$

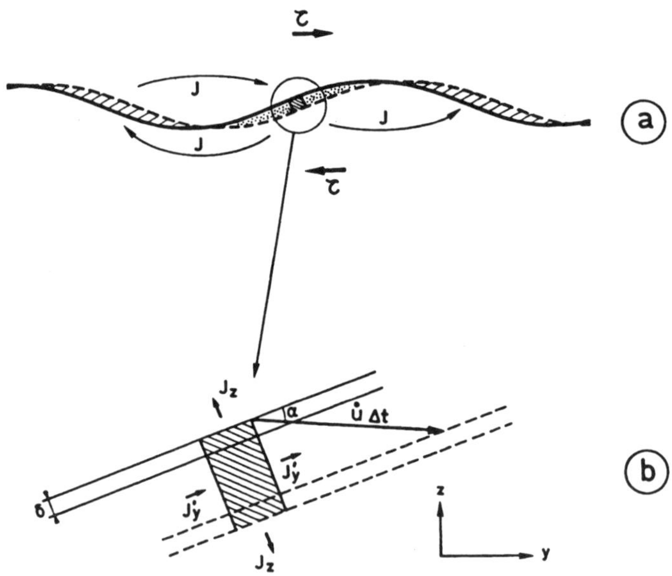

Fig. X.34. — a) Sliding of two grains along a wavy boundary and the corresponding mass flow; b) Detail of an element and material balance; $\dot{u}$ is the velocity of the glide.

and in the limit $\Delta t,\ \Delta y \to 0$:

$$J_z(y,0) + \frac{\delta}{2}\left(\frac{\mathrm{d}J_y'}{\mathrm{d}y}\right) = \frac{\dot{u}}{2\,\Omega}\,\frac{\mathrm{d}z}{\mathrm{d}y}. \qquad \text{(X.140)}$$

The mass flux is calculated as:

$$\begin{cases} J_z(y,0) = -\dfrac{D}{\Omega kT}\left[\dfrac{\partial\,(\mu-\mu_v)}{\partial z}\right]_{z=0} \\ J_y'(y) = -\dfrac{D'}{\Omega kT}\left[\dfrac{\partial\,(\mu-\mu_v)}{\partial y}\right]_{z=0}. \end{cases} \qquad \text{(X.141)}$$

Introducing the corresponding values in the equation of continuity (X.140), we find for the glide velocity:

$$\boxed{\dot{u} = \frac{2}{\pi}\,\frac{\Omega\tau_{\mathrm{a}}}{kT}\,\frac{\lambda}{h^2}\,D\left[1+\frac{\pi}{\lambda}\,\frac{D'\delta}{D}\right]}. \qquad \text{(X.142)}$$

This solution is valid for a slightly wavy boundary. Raj and Ashby (1971) have nevertheless applied it to a zig-zag boundary in hexagonal tiling with

$\lambda = 2L$ and $h = L/4$, where L is the grain diameter. The shear rate results from glide with velocity $\dot{u}$ along two families of orthogonal boundaries:

$$\dot{\gamma} = 2\dot{u}/L,$$

from which:

$$\boxed{\dot{\gamma} = \frac{40}{L^2}\frac{\Omega\tau_a}{kT} D\left[1 + \frac{\pi}{2}\frac{}{L}\frac{D'\delta}{D}\right]}, \qquad \text{(X.143)}$$

a formula identical to the one deduced from the viscous-creep models of Herring and Nabarro and of Coble with evidently the same numerical coefficient for the same applied stress system.

A more precise calculation, in which the profile $z(y)$ is described by a Fourier series does not yield a significantly different result (Raj and Ashby, 1971).

In conclusion, ***viscous diffusional creep and grain-boundary sliding are not*** independent processes; ***they are coupled.*** The rate of the resulting deformation can be equally well described as diffusional creep or grain boundary sliding with accomodation by diffusion.

VII.2 GROWTH OF VOIDS AT THE GRAIN BOUNDARIES DURING HIGH TEMPERATURE PLASTIC DEFORMATION. — Creep rupture can be caused by the growth and coalescence of voids along the grain boundaries. The voids grow in boundaries subjected to tensile stresses. The growth is due to diffusion of vacancies out of the boundary to the surface of the void, since the process of diffusional transport is faster than the process of plastic deformation for small pores near each other. In the usual conditions, transport by volume diffusion is negligible. Only the growth of the voids, not their nucleation, is considered here.

The model of Hull and Rimmer (1959). — Consider a boundary subject to a tensile stress normal to the plane of the boundary, t_n, and a spherical void. The spherical shape requires that: 1) the surface energy γ_s be independent of orientation; 2) the surface diffusion coefficient be high enough so that the same shape is maintained during growth.

The growth mechanism is the addition of vacancies to the void by diffusion along the boundary, the transfer of material between the void and the boundary giving rise to deformation in the direction of the tensile stress t_n. The driving force for the diffusion is the difference in chemical potential between the plane boundary and the surface of the void (Fig. X.35).

The difference in chemical potential of an atom and a vacancy on a curved surface is given (see Ch. VII, 2nd part, Eq. (VII.36)):

$$\mu - \mu_v = \mu_0 + \Omega\gamma_s\left(\frac{1}{R_1} + \frac{1}{R_2}\right), \qquad \text{(X.144)}$$

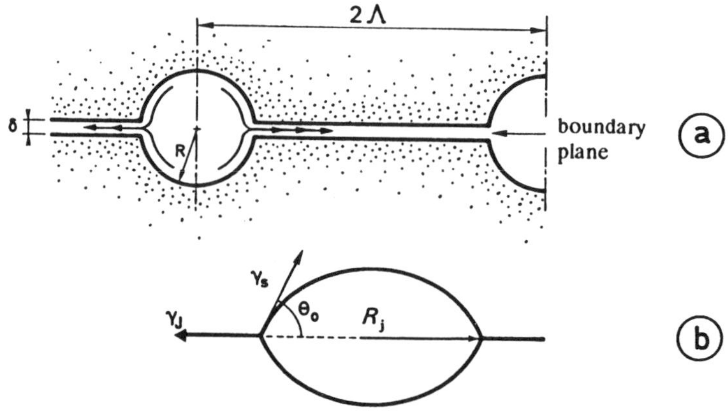

Fig. X.35. — Creep cavitation: a) growth by grain-boundary diffusion; b) equilibrium shape of a void on a boundary.

or, for a concave spherical surface with radius of curvature $-R$:

$$\mu - \mu_v = \mu_0 - 2\Omega\gamma_s/R, \qquad \text{(X.144a)}$$

where Ω is the atomic volume and γ_s is the surface energy.

On the boundary subjected to the normal stress t_n, according to equation (X.128):

$$\mu - \mu_v = \mu_0 - t_n\Omega. \qquad \text{(X.145)}$$

Let 2Λ be the distance between voids in the boundary plane. The average gradient of chemical potential is then:

$$\nabla(\mu - \mu_v) \sim \frac{\Omega}{\Lambda}(t_n - 2\,\gamma_s/R)\,,$$

which leads to a flux of material:

$$|\,J\,| \sim \frac{1}{kT}\frac{D'}{\Omega}\nabla(\mu - \mu_v) = \frac{D'}{kT}\frac{(t_n - 2\,\gamma_s/R)}{\Lambda}, \qquad \text{(X.146)}$$

where D' is the grain-boundary diffusion coefficient. In the simple model, it is assumed that $t_n = \sigma_\infty$, the normal stress applied to the sample.

According to equation (X.146), only voids with radius $R > R_c = 2\gamma_s/\sigma_\infty$ can grow. If a gas under pressure p is present in the void, the *critical radius* is reduced to $R_c = 2\,\gamma_s/(\sigma_\infty + p)$, since an extra term $p\Omega$ enters into the expression for the chemical potential (Eq. (X.144a)).

The total flux of atoms leaving the void diffuses in an intergranular disk of thickness δ, through a cross-sectional area $2\,\pi R\delta$. Then $2\pi R\delta\;|\,J\,|$ atoms move per unit time, leading to a volume change:

$$\frac{\mathrm{d}V}{\mathrm{d}t} = 2\,\pi R\delta \mid J \mid \Omega$$

$$= \frac{2\,\pi R\Omega D'\,\delta}{kT}\,\frac{(\sigma_\infty - 2\,\gamma_s/R)}{\Lambda}. \qquad \text{(X.147)}$$

Assume that the condition for growth $\sigma_\infty > 2\gamma_s/R$ is largely fulfilled; the second term in the parentheses can then be neglected. Then we deduce for a spherical void:

$$\boxed{\frac{\mathrm{d}R}{\mathrm{d}t} \approx \frac{D'\,\delta}{kT}\,\frac{\sigma_\infty\Omega}{2\,\Lambda R}}. \qquad \text{(X.148)}$$

The growth rate is inversely proportional to the radius of the void.

If there is gas in the void, substitute $(\sigma_\infty + p)$ for σ_∞; under the condition of exterior hydrostatic pressure p, substitute $(\sigma_\infty - p)$.

An approximate value of the time to rupture can be deduced from equation (X.148); the rupture takes place by coalescence of voids so that $R \approx \Lambda$. Integrating from $R = R_0$ to Λ, and using the fact that the critical radius $R_c << \Lambda$, we obtain:

$$t_R \sim \frac{kT\Lambda^3}{D'\,\delta\sigma_\infty\,\Omega}. \qquad \text{(X.149)}$$

The time to rupture by cavitation is inversely proportional to the normal stress.

Remarks: The model of Hull and Rimmer is approximate for several reasons:

1) The equilibrium of surface and interfacial tensions (Fig. X.35b) in fact imposes a lenticular, rather than a spherical, shape on the void, with an equatorial angle $2\,\theta_0$ such that $2\,\gamma_s \cos\theta_0 = \gamma_b$; hence the radius of curvature is given by $R = R_j/\sin\theta_0$ if R_j is the radius of the void.

2) If the surface diffusion coefficient D_s is not much greater than D', the shape of the void is not the equilibrium shape, but it is the flatter, the lower the value of D_s (tendency to the "microcrack" shape).

3) The condition of mechanical equilibrium is not compatible with a uniform normal stress over the plane of the boundary; for an applied stress σ_∞, the tensile stresses t_n are maximum midway between the voids.

The diffusion equations cannot be solved analytically in the general case. A finite-element solution has been given by Martinez and Nix (1982). The elements of a solution that takes into account remarks 1 and 3, but neglects 2 (assumption that $D_s >> D'$) are given in appendix III.

The models analyzed above are purely diffusional. They assume that the adjacent grains are perfectly rigid so that the material moves uniformly along the boundaries between the voids which "separate" the two contiguous grains. In reality, the plasticity of the material can play a role; the diffusing material can be accomodated by a separation of the grains localized near the voids. This has the effect of shortening the paths for mass flow, leading to an increase of the growth rate.

APPENDIX I

Rate constant for oxidation.

NOTE ON UNITS. — The rational parabolic rate constant K_r is given by equation (X.22) in equivalents cm^{-1} s^{-1}, or in moles of equivalents cm^{-1} s^{-1}, depending on the units chosen for n_X (atoms or moles of X per unit volume). Measurements of the oxidation kinetics are, on the other hand, carried out either by the measurement of the thickness z of the oxide layer, which obeys a law:

$$z^2 = 2\ K_p\ t, \qquad K_p\left(\mathrm{cm}^2{\cdot}\mathrm{s}^{-1}\right) \tag{A.1}$$

or by thermogravimetry:

$$\Delta m^2 = 2\ K_g\ t, \qquad K_g\left(\mathrm{g}^2{\cdot}\mathrm{cm}^{-4}{\cdot}\mathrm{s}^{-1}\right) \tag{A.2}$$

where Δm is the increase of mass per unit area.

To change from K_r to K_g, we note first that the equivalent of the compound $M_\alpha X_\beta$ is $(M_\alpha X_\beta)/(|\ Z_-\ |\ \beta)$. To the flux of moles of equivalents J_{eq} there corresponds a flux of oxide mass $J_{eq}\ M/(|\ Z_-\ |\ \beta)$, where M is the molecular weight of the compound.

Moreover, we can set:

$$z = m/\rho,$$

where m is the mass of compound per unit area, and ρ is its density in g $\cdot$ cm^{-3}. From the equation:

$$J_{eq} = K_r/z = K_r\ \rho/m, \tag{A.3}$$

the mass of compound formed per unit area is deduced as:

$$\frac{\delta m}{\delta t} = \frac{K_r\ \rho}{m} \cdot \frac{M}{|\ Z_-\ |\ \beta}. \tag{A.4}$$

In fact, only the change of mass due to the reaction of X with the metal is measured, i.e., a fraction $\beta M_X/M$ of the mass of oxide, with M_X the molecular weight of X. Transforming m and δm into a mass of X:

$$m_X\ \frac{\delta m_X}{\delta t} = \frac{K_r\ \rho M}{|\ Z_-\ |\ \beta} \cdot \left(\frac{\beta M_X}{M}\right)^2$$

$$\Delta m^2 = m_X^2 = 2\,\rho\,\frac{\beta}{|Z_-|}\,\frac{M_X^2}{M}\,K_r\cdot t. \tag{A.5}$$

For identification with (A.2):

$$\boxed{K_g\,(\mathrm{g^2{\cdot}cm^{-4}{\cdot}s^{-1}}) = \rho\,\frac{\beta}{|Z_-|}\,\frac{M_X^2}{M}\,K_r\,(\text{moles of equiv. cm}^{-1}\,\mathrm{s}^{-1})}\,. \tag{A.6}$$

To evaluate K_p, we change from flux of mass to flux of volume, i.e., a thickness per unit area, dividing by ρ :

$$\frac{\partial z}{\partial t} = J_{eq}\,M/\,|Z_-|\,\beta\rho, \tag{A.7}$$

and with $J_{eq} = K_r/z$:

$$z\,\frac{\mathrm{d}z}{\mathrm{d}t} = \frac{K_r\,M}{|Z_-|\,\beta\rho}$$

$$z^2 = \frac{2\,M}{|Z_-|\,\beta\rho}\,K_r\,t = 2\,K_p\,t. \tag{A.8}$$

$$\boxed{\underset{(\mathrm{cm^2{\cdot}s^{-1}})}{K_p} = \frac{M}{|Z_-|\,\beta\rho}\,\underset{(\text{moles eq. cm}^{-1}{\cdot}\mathrm{s}^{-1})}{K_r}}$$

Case of the oxide MO: $Z_+ = |Z_-| = 2,\ \beta = 1$

$$K_p = \frac{M}{2\,\rho}\,K_r,$$

with

$$\begin{aligned} K_r &= \frac{1}{2}\,c_X\,Z_+\int_{p'}^{p''}(D_M + D_O)\,\mathrm{d}\,\mathrm{Log}\,P_{O_2},\\ &= c_X\int_{p'}^{p''}(D_M + D_O)\,\mathrm{d}\,\mathrm{Log}\,P_{O_2}, \end{aligned}$$

from which

$$K_p = \frac{1}{2}\int_{p'}^{p''}(D_M + D_O)\,\mathrm{d}\,\mathrm{Log}\,P_{O_2}, \tag{A.9}$$

since $c_X\,M/\rho = 1$ in this case; the expression is identical to equation (VI.59).

APPENDIX II

Effective diffusion coefficient for coalescence.

During the coalescence of precipitates in an AB alloy, the fluxes of A and B must remain in the same ratio as their concentrations in the precipitate, or in the same ratio as their molar volumes, depending on whether the interface is coherent or not. The coupling of the fluxes is expressed by the relations of chapter VI, § IV. In that paragraph, following C. Y. Li *et al.* (1966), we introduced the coupling forces F_i in each expression for the flux $J_i (i = \mathrm{A},\ \mathrm{B})$. The principle of minimum entropy production leads to (Eq. (VI.66)):

$$J_\mathrm{A}\ \mathrm{d}F_\mathrm{A} + J_\mathrm{B}\ \mathrm{d}F_\mathrm{B} = 0 \tag{A.10}$$

A second condition on the flux allows the forces F_i to be determined and thus the fluxes J_i to be calculated:

1) For a moving incoherent interface (growth or shrinkage), the number of sites is not conserved. On the other hand, when the B atoms substitute for the A atoms during the growth of a precipitate rich in B for example, the fluxes across the interface must be such that:

$$J_\mathrm{A}\Omega_\mathrm{A} + J_\mathrm{B}\Omega_\mathrm{B} = 0 \tag{A.11}$$

(condition on the volumes). Here Ω_A is the molar volume of A in the matrix, Ω_B the molar volume of B in the precipitate β.

These two conditions on the fluxes yield:

$$\Omega_\mathrm{B}\ \mathrm{d}F_\mathrm{A} - \Omega_\mathrm{A}\ \mathrm{d}F_\mathrm{B} = 0,$$

from which:

$$\Omega_\mathrm{B}F_\mathrm{A} - \Omega_\mathrm{A}F_\mathrm{B} = 0. \tag{A.12}$$

It is easily deduced that:

$$J_\mathrm{B} = -\frac{c_\mathrm{A}\ c_\mathrm{B}\ D_\mathrm{A}\ D_\mathrm{B}\ \Omega_\mathrm{A}}{c_\mathrm{A}\ D_\mathrm{A}\ \Omega_\mathrm{A}^2 + c_\mathrm{B}\ D_\mathrm{B}\ \Omega_\mathrm{B}^2}\ \frac{1}{kT}\ [\Omega_\mathrm{A}\ \nabla\mu_\mathrm{B} - \Omega_\mathrm{B}\ \nabla\mu_\mathrm{A}], \tag{A.13}$$

where D_A and D_B are the self-diffusion coefficients of A and B, respectively, in the solid solution with composition c_A and c_B (expressed in numbers of atoms per unit volume).

$\nabla\mu_\mathrm{A}$ can be eliminated by use of the Gibbs-Duhem equation, yielding:

$$J_\mathrm{B} = -D_{\mathrm{B}_\mathrm{eff}}\ \nabla c_\mathrm{B} = -\frac{c_\mathrm{B}\ D_{\mathrm{B}_\mathrm{eff}}}{kT}\ \nabla\mu_\mathrm{B},$$

with the *effective diffusion coefficient*:

$$D_{B_{eff}} = \frac{D_A\, D_B\, \Omega_A\, (c_A\, \Omega_A + c_B\, \Omega_B)}{c_A\, D_A\, \Omega_A^2 + c_B\, D_B\, \Omega_B^2}.$$

This formula has been used, for example for the diffusion of carbon during the growth of cementite precipitates in ferrite. It applies equally well to precipitates of pure elements or to the formation of rare gas bubbles. For pores, the reader can check that with $D_B = D_v >> D_A$ and $c_A >> c_B = c_v$, D_{eff} is practically equal to the vacancy diffusion coefficient.

2) The coalescence of a stoichiometric compound $A_\alpha B_\beta$, only slightly soluble in the matrix M. It is assumed that the compound dissolves stoichiometrically in the matrix, i.e., $c_A \propto \alpha$ and $c_B \propto \beta$. The molecular flux is thus:

$$J_{mol} = \frac{J_A}{\alpha} = \frac{J_B}{\beta}.$$

The molecular diffusion coefficient has been calculated in chapter VI (Eq. (VI.66)):

$$D_{mol} = \frac{D_A\, D_B}{\alpha D_B + \beta D_A}.$$

The fluxes of $A_\alpha B_\beta$ and M must agree with the volume balance in order to allow migration of the interface:

$$J_M = -\frac{c_M}{kT}\, \frac{\Omega_{mol}\, D_M\, D_{mol}}{c_M\, D_M\, \Omega_M^2 + c_{mol}\, D_{mol}\, \Omega_{mol}^2} \cdot (c_{mol}\, \Omega_{mol} + c_M\, \Omega_M)\; \nabla \mu_M \tag{A.14a}$$

$$J_{mol} = \frac{\Omega_M}{\Omega_{mol}}\, J_M \quad \text{and} \quad \mu_{mol} = \alpha \mu_A + \beta \mu_B. \tag{A.14b}$$

Here Ω_{mol} denotes the variation of the volume when the "molecule" $A_\alpha B_\beta$ goes from the precipitate to the matrix, μ_A and μ_B are the chemical potentials of the elements A and B dissolved in the matrix M.

The above expression can be rearranged. Noting that:

$$c_{mol} = \frac{c_A}{\alpha} = \frac{c_B}{\beta} = \frac{(c_A\, c_B)^{1/2}}{(\alpha\beta)^{1/2}},$$

with

$$J_{mol} = -\frac{n_{mol}\, D_{mol}}{kT}\, \nabla \mu_{mol} = -\frac{(c_A\, c_B)^{1/2}}{kT}\, \overline{\overline{D}}\; \nabla \mu_{mol}, \tag{A.15}$$

we obtain after some calculation:

$$\overset{=}{D} = \frac{c_M (c_A\, c_B)^{1/2}\, \Omega_M^2\, D_A\, D_B\, D_M}{c_M\, \Omega_M^2\, D_M [c_B\, \alpha^2\, D_B + c_A\, \beta^2\, D_A] + c_A\, c_B\, \Omega_{mol}^2\, D_A\, D_B}. \tag{A.16}$$

APPENDIX III

Growth of voids at grain boundaries during high-temperature deformation.

The lenticular-void model (Raj and Ashby, 1975). — The radius of curvature of the voids is R. With $\theta_0 = \text{arc cos}\ (\gamma_b/2\gamma_s)$, their radius in the boundary plane is $R_j = R \sin \theta_0$ (Fig. X.33b). The voids are uniformly distributed over the boundary plane with 2Λ denoting the distance between neighboring voids.

The grains are subjected to a tensile stress σ_∞. This results in a stress component normal to the boundary t_n. The driving force for transport in steady state is the gradient of chemical potential in the boundary:

$$J'\, \delta = -\frac{D'\delta}{\Omega kT}\, \nabla (\mu - \mu_v), \tag{A.17}$$

with

$$\begin{aligned} \mu - \mu_v &= \mu_0 - 2\Omega\gamma_s/R, && \text{for } \rho = R_j, \\ \mu - \mu_v &= \mu_0 - t_n\Omega, && \text{for } R_j < \rho < 2\Lambda - R_j, \end{aligned}$$

where ρ is the radius vector, the origin being taken at the center of the void. At *steady state*, each element of area in the boundary should receive the same number of atoms which attach themselves there (thus causing a uniform deformation by "separation" of two contiguous grains, the so-called "jacking effect"):

$$\frac{\partial n}{\partial t} = \beta = \text{constant},$$

or

$$\text{div}\ (J'\, \delta) = -\beta, \tag{A.18}$$

and, following equation (X.150), with the Laplacian for cylindrical coordinates:

$$\nabla^2 (\mu - \mu_v) = \frac{1}{\rho}\frac{\partial}{\partial \rho}\left[\rho\, \frac{\partial (\mu - \mu_v)}{\partial \rho}\right] = \frac{\beta \Omega kT}{D'\, \delta}. \tag{A.19}$$

This differential equation is to be solved with the following boundary conditions:

$$\partial(\mu - \mu_v)/\partial\rho = 0 \quad \text{for } \rho = \Lambda,$$

for reasons of symmetry (this is halfway between two voids), and:

$$\mu - \mu_v = 2\,\Omega\gamma_s/R \quad \text{for } \rho = R_j,$$

to preserve continuity at the boundary of the void.

The solution of (A.19) is the following:

$$\mu - \mu_v = \beta\,\frac{kT\Omega}{4\,D'\,\delta}\,(\rho^2 - R_j^2) + \beta\,\frac{kT\Omega\Lambda^2}{2\,D'\,\delta}\,\mathrm{Log}\,\frac{R_j}{\rho} - \frac{2\,\gamma_s\Omega}{R}. \tag{A.20}$$

The stress t_n varies as $-(\mu - \mu_v)/\Omega$: it reaches its maximum value halfway between the voids. Its value is $2\gamma_s/R$ on the rim of the void, and is zero over the voids.

The condition of *mechanical equilibrium* requires that:

$$\pi\Lambda^2\,\sigma_\infty = \int_{R_j}^{\Lambda} t_n\,2\,\pi\rho\,\mathrm{d}\rho. \tag{A.21}$$

With t_n given by (A.20), the following expression is obtained for the parameter β:

$$\beta = \frac{\sigma_\infty - (2\,\gamma_s/R)\left(1 - \dfrac{R_j^2}{\Lambda^2}\right)}{\Lambda^2\left[\dfrac{3}{4} + \mathrm{Log}\,\dfrac{R_j}{\Lambda} - \dfrac{R_j^2}{\Lambda^2}\left(1 - \dfrac{1}{4}\,\dfrac{R_j^2}{\Lambda^2}\right)\right]}\,\frac{2\,D'\,\delta}{kT}. \tag{A.22}$$

The change $\mathrm{d}V$ of the void volume equals the volume of material deposited in the boundary, i.e., $(\partial n/\partial t)\pi(\Lambda^2 - R_j^2)\Omega \mathrm{d}t$. From which:

$$\begin{aligned}\frac{\mathrm{d}V}{\mathrm{d}t} &= \beta\pi\Omega\Lambda^2\,(1 - R_j^2/\Lambda^2)\\ &= 2\,\pi\Omega\,\frac{D'\delta}{kT}\left(1 - \frac{R_j^2}{\Lambda^2}\right)\sigma_\infty\cdot\\ &\quad\cdot\left[\frac{1 - (R_c/R)\,(1 - R_j^2/\Lambda^2)}{\dfrac{3}{4} - \mathrm{Log}\,\dfrac{\Lambda}{R_j} - \dfrac{R_j^2}{\Lambda^2}\left(1 - \dfrac{1}{4}\,\dfrac{R_j^2}{\Lambda^2}\right)}\right].\end{aligned} \tag{A.23}$$

The *critical radius* for growth is given by $R_c = 2\,\gamma_s/\sigma_\infty$.

To evaluate the time to rupture, the fraction of boundary area occupied by voids is calculated

$$A(t) = R_j^2/\Lambda^2.$$

From which, transforming equation (A.23), knowing that $R = R_j/\sin\theta_0$ and that:

$$V = \frac{2\,\pi}{3}\,\left(2 - 3\cos\theta_0 + \cos^3\theta_0\right)\,R^3,$$

$$\frac{dA}{dt} = \frac{32}{3\sqrt{\pi}}\,\frac{\Omega D'\,\delta}{kT}\,\sigma_\infty\,\frac{1}{(2\,\Lambda)^3}\cdot f(\theta_0)\cdot F(A). \tag{A.24}$$

and

$$t_R = \frac{3\sqrt{\pi}}{32}\,\frac{kT}{\Omega D'\,\delta}\,\frac{1}{\sigma_\infty}\,(2\,\Lambda)^3\,\frac{1}{f(\theta_0)}\int_{A_1}^{A_2}\frac{dA}{F(A)}. \tag{A.25}$$

The limits of integration are the values of A for R_c and for $R \le \Lambda$. Numerical integration shows that the value of the integral is not sensitive to the value of A_2.

Once again, the rupture time is found to vary inversely with the applied stress σ_∞. However, when the voids are no longer in equilibrium ($D_s \le D'$), the variation is as σ_∞^{-n} with $1 < n \le 3$ (Martinez and Nix, 1982).

The ratio of the variations in the void volume due to diffusion in the volume and along the boundaries is of order $2\Lambda D/3D'\,\delta$ (see above). With typical values $D = 5\times10^{-12}$ cm^2 s^{-1} and $D'\,\delta = 10^{-14}$ cm^3 s^{-1}, volume diffusion dominates only for a void density less than 10^5 cm^{-2}, i.e., intervoid distances of more than 50 μm. For a stress $\sigma_\infty = 10^{-4}\,\mu$ (μ = the shear modulus) and a void density of 10^6 cm^{-2}, the calculated time to rupture is about 1 day for a temperature of order 700°C.

Bibliography

§1 ADDA Y. and PHILIBERT J., *op. cit.* Ch. III.

CLARK J. B. and RHINES F. N., *Trans ASM* **51** (1958) 199-221.

KIRKALDY J. S. *Adv. Mater. Res.* **1** (1970) 56-100, H. Herman, ed. (Wiley, New York).

KIRKALDY J. S. and FEDAK D. G., *Trans. Metall. Soc. AIME* **224** (1962) 490-494.

HIRANO K. and IIJIMA Y., in *Diffusion in Solids, Recent Developments* (1985) *op cit.* p. 141-166.

VAN LOO F. J. J., VAN BEEK J. A., BASTIN G. F., and METSELAAR R.

ibid, p. 231-259.
VAN LOO F. J. J. and RIECK G. D., *Acta Metall.* **21** (1973) 73-84.
TARENTO R. J. and BLAISE G., *Acta. Metall.* **36** (1988) 1035-1041.
TENNEY D., UNNAM J., Numerical analysis for treating diffusion in single, two-and three-Phase binary alloy systems.
(1978) NASA Technical Memorandum 78636.
D'HEURLE F., *J. Mater. Res.* **3** (1988) 167-195.
PHILIBERT J., *DIMETA-88, op. cit.*, 995-1014.
DYBKOV V. I., *J. Phys. Chem. Solids* **47** (1986) 735-740.
§II ATKINSON A., *Rev. Mod. Phys.* **57** (1985) 437-470.
BENARD J., *L'oxydation des Métaux*, vol. 1 (Gauthier-Villars, Paris) 1962.
DEAL B. E. and GROVE A. S., *J. Appl. Phys.* **36** (1965), 3770-3778.
FROMHOLD A. T., *Theory of Metal Oxidation* 2 vols. (North Holland, Amsterdam) 1976, 1980.
KOFSTAD P., *High Temperature Corrosion* (Elsevier Applied Science, Amsterdam) 1988.
WAGNER C., in *Atom Movements*, ASM Seminar, Cleveland (1951) p. 153-173.
MAAK F., *Z. Metallk.* **52** (1961) 545-546.
RHINES F. N., JOHNSON W. A. and ANDERSON W. A., *Trans. AIME* **147** (1942) 205.
§III ASHBY M. F., *Acta. Metall* **22** (1974) 275-289.
CIZERON G., *L'industrie Céramique* (1968) 1-17.
SWINKELS F. B. and ASHBY M. F., *Acta Metall.* **29** (1981) 259-281.
Sintering, New Developments, M. M. Ristic, ed., Materials Science Monographs #4 (Elsevier Science Publishers) 1979.
JOHNSON D. L., *J. Appl. Phys.* **40** (1969) 192-200.
§IV AARON H. B., FAINSTEIN D., and KOTLER G. R., *J. Appl. Phys.* **41** (1970) 4404-4410.
BULLOUGH R. and NEWMAN R. C., *Rep. Prog. Phys.* **33** (1970) 101-148.
COTTRELL A. H. and BILBY B. A., *Proc. Phys. Soc.* (London) **A62** (1949) 49-62.
EDELIN G., in *Confolant 1977* (Les Editions de Physique) 1978, p. 107-147.
GREENWOOD G. W., *Acta Metall.* **4** (1956) 243-248.
HAM F. S., *J. Appl. Phys.* **30** (1959) 915-926.
HIRTH J. P. and LOTHE J., *Theory of Dislocations* (McGraw-Hill, New York) 1st ed. (1968) Ch. 18, p. 584 ff.
KAHLWEIT M., *Adv. Colloid Interface Sci.* **5** (1975) 1-35.
LI C. Y., BLAKELY J. M. and FEINGOLD A. H., *Acta Metall.* **14** (1966) 1397-1402.
Eléments de Métallurgie Physique, Vol. 4 (1978) Ch. 29, pp. 1096-1115.
§V *Eléments de Métallurgie Physique*, Vol. 4, Ch. 32.
CHALMERS B., *Principles of Solidification* (J. Wiley, New York) 1964.
§VI ACKER D., Thesis, University of Paris South, Orsay (1977).

Adda Y., Beyeler M., and Brebec G., *Thin Solid Films* **25** (1975) 107-156.

Averback R. S., *Nucl. Instr. Meth.* **b15** (1986) 675-687.

Matteson S., Roth J. and Nicolet M.-A., *Radiat. Eff.* **42** (1979) 217.

Rothman S. J., *Effect of Irradiation on Diffusion in Metals and Alloys*, in: *Phase Transformation and Solute Redistribution in Alloys during Irradiation*, F. V. Nolfi jr., ed. Applied Science Publishers (Barking, U. K.) 1981.

Sharp J. V., *Quoted in Eléments de Métallurgie Physique, op. cit.*, Ch. 26.

Sizmann R., *J. Nucl. Mater.* **69-70** (1978) 386-412.

§VII Hull D. and Rimmer D. E., *Philos. Mag.* **4** (1959) 673-687.

Martinez L. and Nix W. D., *Metall. Trans.* **13a** (1982) 427-437.

Raj R. and Ashby M. F., *Acta. Metall.* **23** (1975) 653-666.

Poirier J. P., *Plasticité à haute température des solides crystallins* (Eyrolles, Paris) 1976.

Poirier J. P., *Creep of Crystals* (Cambridge University Press) 1985.

Herring C., *J. Appl. Phys.* **21** (1950) 437-445.

Coble R. L., *J. Appl. Phys.* **34** (1963) 1679-1682.

Raj R. and Ashby M. F., *Metall. Trans.* **2** (1971) 1113-1127.

EXERCISES

Table of Contents

Introduction

These exercises are either direct applications of the text material, or complementary to it. The level of difficulty is highly varied. Exercises 1 to 32 refer essentially to chapter I. Exercises 33-39 bear on the material treated in chapters II and III, 40-44 to chapters IV and V, 45-48 to chapters VI and VIII, 49-55 to chapter VII, 56 to 60 to chapter IX, and the rest to chapter X.

I thank the colleagues who helped me to prepare these exercises, and the authors from whom data and calculations have been borrowed.

The solutions are available to teachers on request to the author.

J. PHILIBERT

1 – THE ERROR FUNCTION

A depth profile is represented by the complementary error function $c/c_0 = \text{erfc}\,[\,z/2(Dt)^{1/2}]$.

1) What is c_0?

2) What is the value of the reduced concentration c/c_0 at the depth $z = (Dt)^{1/2}$?

3) At what depth z is $c/c_0 = 0.5$?

Discuss.

Note that graph paper on which the error function plots as a straight line, called "probability paper", is available (e.g., Keuffel and Esser Co., New York, #356-23). The numbers along the ordinate are 50 [1 + erf (u)], but distance along the ordinate is linear in u.

2 – DOPING BY DIFFUSION

Silicon can be doped with boron in two steps :

1°) The boron is introduced by heat treatment in a gas of BCl_3 or B_2O_3. The vapor pressure is high enough so that the surface concentration reaches the maximum solubility of boron in silicon, c_s, at the heat treatment temperature.

Using the appropriate solution of Fick's equation, give an analytical expression for the quantity of boron, M, introduced during a treatment of duration t_1 at temperature T_1.

For $T_1 = 1100°\text{C}$, $t_1 = 6$ min, $c_s = 3 \times 10^{20}$ atoms cm^{-3}, $D = 4 \times 10^{-13}$ cm^2 s^{-1}, plot the penetration profile of boron over 5 decades of concentration.

Calculate the amount of boron introduced. Give the thickness of this layer in terms of $(Dt)^{1/2}$.

2°) A subsequent diffusion treatment is carried out to decrease the surface concentration and increase the depth of the doped region. Represent the doped zone of the previous paragraph as a thin layer at the surface, and assume that there is no loss of boron from the surface. What is the solution of the diffusion equation using the analytical results of the 1st paragraph? What conditions on the time t_2 and the temperature T_2 are necessary so that this solution can be used ?

With $T_2 = 1200°\text{C}$, $t_2 = 8$ or 80 min, $D = 3 \times 10^{-12}$ cm^2 s^{-1}, establish the corresponding diffusion profiles. Compare the penetration depths so obtained to that obtained in the first heat treatment at 1100°C. Discuss the result.

3°) What diffusion time at 1200°C is needed to obtain a concentration of 10^{18} atoms cm^{-3} at a depth of 8 μm? What is the concentration of boron at the surface after such a treatment?

3 – DOPING OF SILICON

On silicon uniformly doped with 10^{15} cm^{-3} atoms of arsenic, a very thin layer strongly doped with boron is deposited and capped by silicon nitride to prevent out-diffusion of boron.

After a first anneal, the concentration of boron on the surface of the silicon is 10^{18} cm^{-3}, and the junction depth is 0.41 μm.

1°) What is the form of the boron concentration profile?

2°) What is the quantity of boron introduced into this sample? [1]

3°) How long an anneal is necessary at 1100°C or at 980°C to obtain the same concentration profile ?

4°) The sample is subsequently annealed 36 h at 1000°C in nitrogen. What are the form of the concentration profile, the value of the surface concentration, and the junction depth after this second heat treatment ?

The values of the diffusion coefficient of boron in silicon are :

T, °C	980	1000	1100
D, $cm^2\ s^{-1}$	10^{-14}	1.6×10^{-14}	2×10^{-13}

4 – DOPING OF SILICON

Boron is diffused from a constant surface concentration into silicon doped uniformly with 10^{16} atoms cm^{-3} arsenic.

1°) What is the junction depth x_j with $c_s = 10^{18}$ cm^{-3} after diffusion for 4 h at 980°C, given that $D_B = 10^{-14}$ $cm^2\ s^{-1}$?

2°) What is the junction depth for $c_s = 10^{21}$ cm^{-3} and a 4 h anneal at 1100°C, given that $D_B = 2 \times 10^{-13}$ $cm^2\ s^{-1}$.

Examine this solution.

N.B.: The junction is assumed to be located at the depth where the concentrations of boron and arsenic are equal.

5 – DOPING OF SILICON

Boron is diffused into a thick substrate (1 mm). The initial distribution is given by table I; it is a Gaussian containing Q atoms. The initial condition $c(x,0)$ is given by :

$$c(x,0) = c_0 \exp[-(x-x_0)^2/A].$$

---

[1] The depth of the junction is determined by the condition that the concentrations of arsenic and boron are equal.

1°) What technology was used for the doping ?

2°) From the data, determine c_0, x_0, A, and Q.

3°) The sample is covered with a layer impermeable to the out-diffusion of boron, and an anneal is carried out at 1000°C.

3-1. Write the boundary and initial conditions.

Assume that $D_B = 2 \times 10^{-14}$ cm^2 s^{-1}, and is independent of doping.

3-2. Describe the evolution of the concentration profile of boron $c(x,t)$ as a function of annealing time t.

3-3. How should the initial concentration distribution be approximated in order to yield a simple analytical solution ? What is the order of magnitude of the approximation ?

3-4. Construct the $c(x)$ curve for $t = 4$ h.

Represent $c(x,0)$ and $c(x,t)$ graphically. Calculate the concentrations at $x = 0$, $x = 4000$ Å, $x = 8000$ Å etc... and every 4000Å until the concentration falls below 10^{15} cm^{-3}.

4°) The initial distribution is annealed at 1200°C. At the end of 5 min, the same profile is obtained as that after a 5 h anneal at 1000°C.

4-1. Calculate the diffusion coefficient of boron, $D(T)$, at 1200°C.

4-2. Expressing D in the Arrhenius form $D(T) = D_0 \exp(-H/kT)$, determine D_0 (cm^2 s^{-1}) and H (eV), given that $k = 0.8617\times 10^{-4}$ eV K^{-1}.

4-3. What would be the boron concentration profile after a 4 h anneal at 900°C? Calculate and plot the concentration profile on the figure with the same conditions as question 3-4.

Table I

$z(\mu m)$	0	1	2	3	4	5	6	7	8
c_B(cm^{-3})	4.4 10^{14}	5.3 10^{16}	1.5 10^{18}	1. 10^{19}	2. 10^{19}	1. 10^{19}	1.5 10^{18}	5.5 10^{16}	4.6 10^{14}

6 – CEMENTATION OF IRON

An experiment on the cementation of pure iron is carried out by diffusing carbon into iron platelets thick enough to be considered semi-infinite samples. The carbon concentration is maintained constant at 4 at.% at the surface, which is taken as the origin for the penetration depth.

The table below gives the carbon concentrations as a function of penetration depth for cementation experiments 2 h in duration and carried out at temperatures of 1000°C and 1250°C, respectively.

1°) Assuming that the diffusion coefficient of carbon is constant at a given temperature, calculate the preexponential factor D_0 and the activation energy Q for the diffusion of carbon in iron.

2°) Draw the penetration curve for carbon after a 1 h anneal at a temperature of 1100°C, the surface concentration being maintained at 4 at.% .

3°) If the cemented layer is limited to a carbon concentration ≥ 0.1 at.%, draw the curve of cemented layer thickness for anneals from 30 min to 16 h at 1000°C.

4°) After cementation for 2 h at 1250°C, the sample is slowly cooled. Describe the metallographic structures that would be observed at ambient temperature at different depths, assuming that there is no modification of the composition during cooling.

N.B.: In this type of diffusion experiment, a solution of Fick's second law of the form $c = A + B \text{ erf } (u)$, $u = x/2(Dt)^{1/2}$ can be used, with A and B determined from the initial and boundary conditions.

Table

Carbon concentrations at different depths

Diffusion at 1000°C (2 h)		Diffusion at 1250°C (2 h)	
Depth x(mm)	Concentration c(at.%)	Depth x(mm)	Concentration c(at.%)
0.2	3.45	0.4	3.55
0.4	2.90	0.8	3.20
0.6	2.40	1.2	2.70
0.8	1.95	1.6	2.30
1.0	1.50	2.0	1.95
1.2	1.15	2.4	1.60
1.4	0.85	2.8	1.30
1.6	0.65	3.2	1.05
1.8	0.45	3.6	0.80
2.0	0.30	4.0	0.65

7 – DECARBURIZATION OF IRON

Consider a sample of homogenous Fe-C alloy containing 1 wt. % C, 80 cm^2 cross-section, and long enough to be considered "semi-infinite". This sample, decarburized by heating for 20 h in vacuum at constant temperature, loses 0.376 g of carbon.

Calculate the diffusion coefficient and the carbon concentration 100 μm from the surface, assuming that :

- the diffusion takes place only in the direction perpendicular to the cross-section, and only at one face.

- the surface concentration of carbon is zero during the vacuum treatment.

The density of iron is 7.8 g cm^{-3}.

Given $D_0 = 2 \times 10^{-2}$cm^2 s^{-1}, and $Q = 20{,}000$ cal mole^{-1}, what is the annealing temperature ?

8 – CARBURIZATION OF STEEL IN A GASEOUS ATMOSPHERE

1°) The classic solution of Fick's equation for a constant surface concentration cannot always be used for calculations on carburization. This boundary condition can be a poor approximation because the supply of carbon depends on the rate of the chemical reaction for the deposition of carbon and its incorporation at the surface. Give the new boundary condition and the corresponding solution of Fick's equation, assuming that the kinetics of the reaction are first order.

2°) *Numerical example.* Carburization of γ iron at 1000°C. Assume that the diffusion coefficient of carbon is 3×10^{-7} cm^2 s^{-1} independent of concentration. The solubility of carbon in γ iron is 1.4 wt. % at 1000°C. Calculate the amounts of carbon that will have penetrated into the sample after 5 min and 1 h. Carry out the calculation for both constant and variable surface concentration. In the latter case, apply the theoretical results of 1°), taking 10^{-6}, 10^{-5}, and 10^{-4} cm s^{-1} as the values of the reaction-rate constant. Beginning at what value of the rate constant are the two solutions practically equivalent ?

Suggestion: to solve Fick's equation, carry out two successive changes of variable:

1) $c' = c - c_e$,

where c_e is the equilibrium surface concentration of carbon (i.e., the solubility), and

2) $\phi = c' - (1/h)(\partial c'/\partial x)$,

with $h = K/D$, where K is the rate constant of the first-order surface reaction. The solution is:

$$\frac{c - c_0}{c_e - c_0} = \text{erfc}\left[\frac{x}{2\sqrt{Dt}}\right] - \exp\,(hx + h^2 Dt)\ \text{erfc}\left[\frac{x\ +\ 2hDt}{2\sqrt{Dt}}\right].$$

9 – DEZINCIFICATION OF BRASS

A disk of brass is heated under vacuum for 16 h. The edges of the disk have been chrome plated to avoid all lateral evaporation.

Knowing that the composition of the sample is initially uniform, calculate the weight of zinc lost. Assume that $\sqrt{Dt}$ is small with respect to the thickness of the disk, and that the concentration of zinc on the surface is zero during the vacuum anneal.

Given:	Area of the disk	$50\ \text{cm}^2$
	Initial concentration	1 at. % Zn
	Diffusion coefficient	$D_{Zn} = 10^{-10}\text{cm}^2\ \text{s}^{-1}$
	Density of the brass	$8.9\ \text{g cm}^{-3}$
	Atomic masses Cu	63.54
	Zn	65.37

10 – DIFFUSION FROM A THICK LAYER

To study the solute diffusion of oxygen in TiC, a layer of TiO_2 of thickness $h = 500$ Å is deposited by radiofrequency sputtering on the polished surface of a TiC single crystal. After heating for 16 h, the oxygen depth profile is determined by a SIMS ion probe. The intensity of the signal is proportional to the oxygen concentration; the concentration $c\,(x,t)$ is thus expressed in arbitrary units and normalized to the concentration at $x = 0$. The following results are obtained:

x/h	0	0.1	0.2	0.4	0.6	0.8	1.0	1.2	1.4
$c(\times 10^2)$	100	99.5	99	96.5	92.5	87.5	80.5	73	66
x/h	1.6	1.8	2	2.4	2.8	3	4	5	
$c(\times 10^2)$	58	50	43	29	19	15	3.5	0.5	

1°) Using the solution for a plane source of finite thickness, calculate D.

Suggestion : use the values of c for $x/h = 1, 2, 3, 4$, and 5 in the corresponding expressions obtained as a function of the error functions erf(H), erf($2H$), erf($3H$), etc... Set:

$X = x/2(Dt)^{1/2}$, $H = h/2(Dt)^{1/2}$; equations of the type erf(nH) = F[erf(H)], n integer, are obtained which permit the estimation of D.

2°) For the case of diffusion from a layer of finite thickness h into a semi-infinite medium, show that the solution $c(x,t)$ of Fick's second equation approaches the solution for a thin layer as h becomes small.

Suggestion : Expand the error function in Taylor's series, change the variables to $X = x/2(Dt)^{1/2}$, $H = h/2(Dt)^{1/2}$; and neglect terms higher than first order.

3°) Using the thin layer solution, calculate D from the given data. Is this justified?

11 — SOLUTE DIFFUSION

A single crystal of TiC is heated for 24.4 h at 1610°C in a gas atmosphere with partial pressure of oxygen sufficiently low so that the solubility limit of oxygen in the carbide is never reached. The partial pressure of oxygen is assumed constant.

After the heat treatment, SIMS analysis yields the data in the table below, where c is the oxygen concentration in arbitrary units.

Determine the diffusion coefficient of oxygen in TiC for these conditions.

x (μm)	0	2.9	6.2	10	12.9	15.7	18.6	21.5	24.3
c	19.5	17	14.1	10.55	8.3	6.3	4.9	3.5	2.4
x (μm)	27.2	30							
c	1.7	1.15							

12 – DETERMINATION OF D BY THE METHOD OF MOMENTS

The analytic calculation of the diffusion coefficient by comparison of the two concentration profiles $c(x,0)$ and $c(x,t)$, measured before and after diffusion is possible only if $c(x,0)$ has a simple form, e.g., a Gaussian ; $c(x,0)$ can be the result of an implantation. However, the moments of the two profiles can be compared in all cases. The moment of order n is defined as:

$$M_n(t) = \int_0^\infty x^n c(x,t)\, dx \qquad (1)$$

For the calculation, assume that D is constant and that the initial and boundary conditions are:

$t = 0$, $c(x, 0)$ is the given distribution,

for all $t \geq 0$, there is no loss of diffusant through the surface.

1°) Differentiate equation (1) with respect to time, then integrate by parts, keeping in mind the initial and boundary conditions. Taking $n = 0, 1, 2, \ldots$ calculate $M_0(t)$, $M_1(t)$, $M_2(t), \ldots$ by integration over time, and find a recurrence relation between $M_n(t)$ and $M_{n-2}(t)$.

2°) Show that only the moments of even order permit a simple determination of D. Discuss the equations for M_2, M_4, and M_6.

3°) *Numerical calculation.* Doping of silicon by phosphorus. The phosphorus depth profiles, obtained by an ion probe after implantation (△) and after annealing 10s at 1100°C (o), are shown in the figure. Evaluate the moments of order 2, 4 and 6 numerically by the trapezoidal rule. Calculate D and compare the values obtained for $n = 2$, 4, or 6. Discuss the reasons for the observed disagreement.

(1) R. GHEZ, J. D. FEHRIBACH, and G. S. OEHRLEIN, *J. Electrochem. Soc.* **132** (1985) 2759-61.

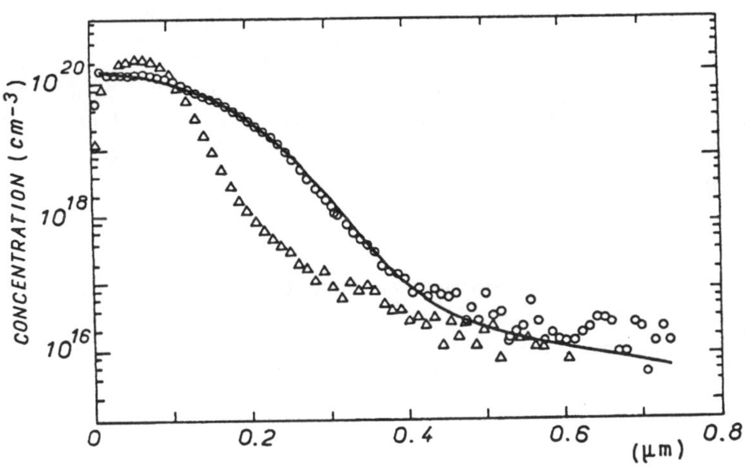

13 – DIFFUSION IN A THIN FILM

On a Fe-Ni substrate (which does not enter into the following), an aluminum film 5 µm thick is deposited, followed by a film of 50-50 NiCr alloy 300 Å thick. The interdiffusion of these two films is followed *in situ* during anneals at 250-300°C by measuring the intensity of the Kα Al, Kα Cr, and Kα

Ni X-rays by means of an electron-beam microanalyzer. The intensity of the chromium X-ray remains constant in time, while that of Al increases, and that of Ni decreases.

1°) Explain the process taking place in the sample.

2°) The ratio $I(t)/I_0$ of the intensities of the nickel X-ray emission at time t and at time 0 varies linearly with $t^{-1/2}$ when t is large enough. Explain this result on the basis of the appropriate solution of the diffusion equation. *Suggestion* : take the "thick layer" solution; justify.

3°) On a plot of $I(t)/I_0$ *vs.* $t^{-1/2}$ the slope at long times is 13.5 ± 1.5 $s^{1/2}$ for diffusion at 578 K. Calculate the diffusion coefficient knowing that the appropriate calibration yields $c/c_0 = 0.86\ I/I_0$. Justify the approximation made in this determination. What kind of diffusion coefficient is being measured here?

14 – INTERDIFFUSION

Consider a diffusion couple formed by welding two alloys 1 and 2. These two alloys were prepared from metals A and B, and their compositions, in atomic fractions, are:

- Alloy #1 (to the right of the weld) $N_{A1} = 40\%$ A
- Alloy #2 (to the left of the weld) $N_{A2} = 50\%$ A.

The diffusion couple is rapidly heated to a temperature T_1 and held there for 40 hours. After cooling, chemical analysis indicates that at a distance of 0.2 cm to the right of the weld, the composition is $N_A = 42.5\%$ A.

1°) Give the shape of the diffusion profile, i.e., the variation of N_A with x or with $\lambda = x/2\sqrt{Dt}$.

2°) Calculate the diffusion coefficient D.

3°) Assuming that the diffusion coefficient D stays constant, find the time required to reach the same concentration ($N_A = 42.5\%$ A) at a distance twice as great from the interface.

15 – INTERDIFFUSION

Consider some A/AB diffusion couples made up of thick samples of pure metal A and a homogeneous alloy dilute in B, concentration C_0. Each couple is treated at a constant temperature, 800°C, 850°C, etc., for 16 h. After these treatments, the distance x_i between the weld plane and the isoconcentration plane $c(x,t) = 0.3\ C_0$ is determined (say, by a micrographic method).

Temperature, °C	800	850	900	1000
x_i, cm	1.25×10^{-2}	1.56×10^{-2}	2.1×10^{-2}	3.21×10^{-2}

Assuming that the diffusion coefficient is independent of concentration, calculate D at each temperature, as well as the activation energy and the preexponential term D_0.

16 – HOMOGENIZATION OF AN ALLOY

Segregation in the form of elongated bands in the rolling direction is shown by micrographic methods in a steel plate after hot rolling.

1°) Explain the origin of this segregation and the possibility of revealing it metallographically.

2°) For simplification, this segregation can be described by a sinusoidal profile:

$$\Delta C^{\circ} = C^{\circ} - \bar{C}^{\circ} = \frac{\Delta C^{\circ}_{\mathrm{m}}}{2} \sin\,(\pi x/\ell)$$

where the indices ° refer to the initial state (before annealing).

An anneal at the temperature T can reduce this segregation by diffusion of the alloying elements. A solution of the diffusion equation is sought by the method of separation of variables:

$$\Delta C = \Delta C^{\circ}\ \tau(t).$$

Determine the function $\tau(t)$. Assume that the diffusion coefficient is independent of concentration.

3°) Numerical example for the diffusion of manganese:

Width of the bands: 50 μm,

Concentration C° minimum 11 wt.% , maximum 15 wt.% .

Calculate the annealing time necessary at 1300°C to reduce the concentration difference to $\Delta C = 0.5\%$, given that $D_0 = 2.7$ cm^2 s^{-1}, $Q = 66$ kcal mole^{-1} = 276 kJ mole^{-1}.

4°) The concentration profile is given more exactly by a Fourier series:

$$\Delta C^{\circ} = \frac{\Delta C^{\circ}_{\mathrm{m}}}{2}\,\frac{4}{\pi} \sum_{n=0}^{\infty} \frac{\sin\,[\{2n + 1\}\,\pi x/\ell]}{2n + 1} \quad (t = 0)$$

Evaluate the error in the previous evaluation of the annealing time for the same degree of homogenization.

17 – CALCULATION OF THE DIFFUSION COEFFICIENT BY THE BOLTZMANN-MATANO METHOD

1°) Show that the interdiffusion coefficient $\tilde{D}$ can be calculated without a preliminary determination of the Matano interface; the area that appears in the numerator of the Matano formula is replaced by a linear combination of the two hatched areas of the figure.

Suggestion: integrate $x\,dc$ by parts, or do an algebraic calculation based on the areas defined on the figure and the Matano interface.

2°) Assuming that the molar volume varies with the composition, derive from the former result, the new expressions for $\tilde{D}$ (known as Den Broeder's formula (1)). Suggestion : use equation (A.16), chapter VI, for the change of variable.

(1) F. J. A. DEN BROEDER, *Scripta Metall.* **3** (1969) 321-325.

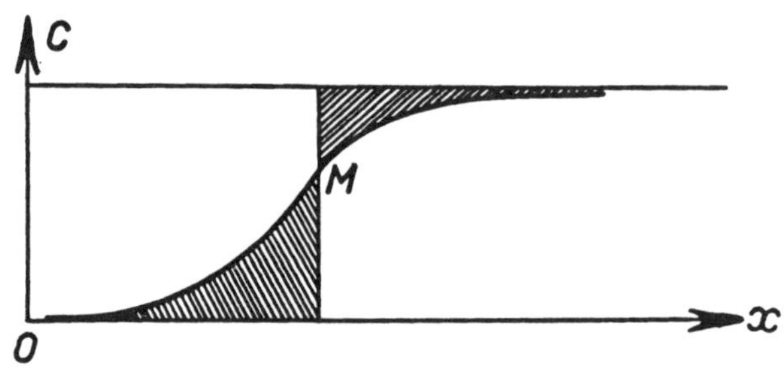

18 – THE INTERDIFFUSION OF IRON AND NICKEL

The diffusion couples were made by joining together a nickel and an iron pellet 4 mm thick, and annealing in vacuum for 48 h at 1200°C. The concentration profile was measured on a transverse section of the couple by X-ray spectrometry with a Castaing electron microprobe. The results are given in the table below.

1°) Determine the position of the Matano interface.

2°) Calculate the interdiffusion coefficient $\tilde{D}$ over the entire concentration range 10 % - 90 % and establish the plot of $\tilde{D}(c)$.

TABLE

distance (μm)	50	80	120	160	180	200	220	240	260	280
wt. % Ni	99.5	98	96	92	89	87	82.5	79	75	71
distance	300	320	340	360	380	400	420	440	460	480
wt. % Ni	66.5	59	51	40.5	30	19	12.5	4	1	≈ 0

19 – CALCULATION OF THE DIFFUSION COEFFICIENT BY THE MATANO METHOD (semi-infinite geometry).

Boron was diffused into a single crystal of silicon. The boron concentration profile as a function of depth as determined by SIMS is given in the figure.

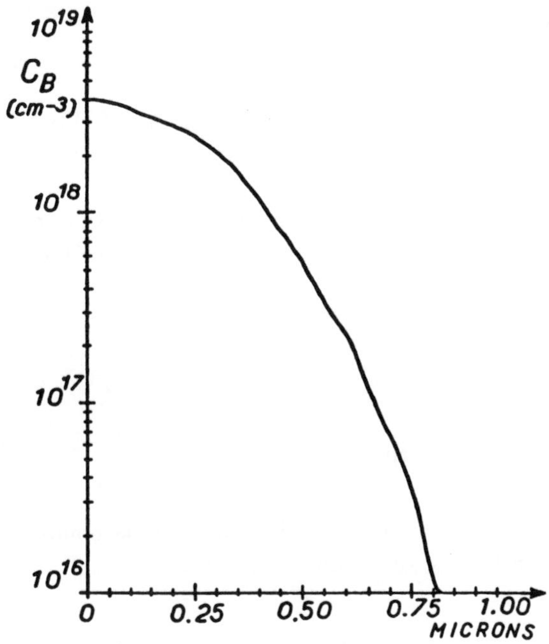

1°) Determine the variation of $\tilde{D}t$ with boron concentration C. (The experimental profile has been smoothed so this function decreases monotonically).

2°) Using log-log coordinates, show $\tilde{D}t$ graphically as a function of C.

3°) Can the temperature at which the heat-treatment was carried out be deduced from this profile?

20 – CALCULATION OF THE INTERDIFFUSION COEFFICIENT BY A NUMERICAL METHOD

The interdiffusion profile $c(x)$ is represented by an expansion in Legendre polynomials $P_n(u)$, $-1 \leq u \leq 1$.

Calculate $\tilde{D}$ by the Matano formula as a function of the P_n, their integrals Q_n, and their derivatives P'_n.

Can this method be applied to the entire $c(x)$ curve?

Suggestion: Change variables to $u(x)$, with u varying over the interval $[-1, +1]$ when x varies over $[x_1, x_2]$, and set:

$$c(u) = \sum_{n=0}^{M} A_n P_n[u].$$

21 – INTERDIFFUSION COEFFICIENT AT THE ENDS OF THE DIFFUSION COUPLE (Hall's method)

The ends of the concentration profile for interdiffusion, $c(x)$ at either high or low concentrations, can be approximated by an error function:

$$c = (c_0/2)\,[1 + \text{erf } u],$$

$$\text{with } u = h(x/\sqrt{t}) + k.$$

1°) Show that the Matano formula under these conditions leads to the following expression for the diffusion coefficient:

$$\tilde{D}(c) = \frac{1}{4h^2t}\left\{1 \pm k\pi \exp\left(u^2\right)\left(1 \pm \text{erf } u\right)\right\},$$

with the sign + if $c \to 0$, − if $c \to c_0$.

N.B.: The origin on the abscissa is taken as the Matano interface, and the concentration profile $c(x)$ is represented in such a manner that c is an increasing function of x.

2°) Suggest a practical method for calculating $\tilde{D}(c)$ from the experimental $c(x)$ curve.

22 – INTERDIFFUSION, APPLICATION OF HALL'S METHOD

An Al-Fe diffusion couple, annealed 72 h at 900°C, was analyzed with an electron microprobe. The experimental values of the aluminum concentration as a function of distance are given in the table below. Calculate $\tilde{D}(c)$ by Hall's method.

x, μm	0	30	40	60	80	110	130	150	180
wt.% Al	0.10	0.39	0.42	0.57	0.93	1.32	1.98	2.46	3.18
x, μm	200	220	240	260	280	300	> 500		
wt.% Al	3.39	3.96	4.26	4.32	4.44	4.56	4.65		

23 – INTERDIFFUSION BETWEEN A TWO-PHASE ALLOY AND A PURE METAL

The attached concentration-distance curve is from a diffusion couple made up of a thick sample of uranium and a two-phase uranium-nickel alloy with an average nickel concentration c_0= 1.3 wt.% , that had been annealed at constant temperature (≈ 950°C) for 16 h. The origin on the abscissa corresponds to the position of the weld plane after diffusion.

1°) What does the concentration c_α indicated on the graph represent?

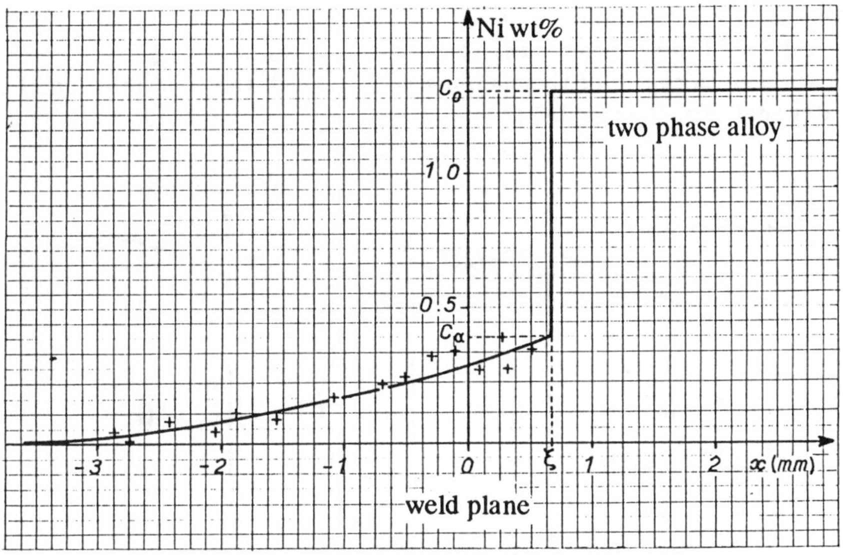

2°) Determine the coordinate of the Matano interface; then calculate the diffusion coefficient by the Matano method for the concentration $c = 0.2$ wt.%.

3°) Assuming that the diffusion coefficient is constant, find the solution $c(x)$ of Fick's second equation for the boundary conditions implied by the concentration-penetration curve.

4°) Assume that the boundary of the two-phase region is displaced by a distance ξ during the time t, and that this process is diffusion-controlled (in other words, assume that the rate of dissolution of the second phase is very fast and that the slower phenomenon governs the kinetics). Thus we can set $\xi = 2\gamma\sqrt{Dt}$, where γ is a constant.

a) Write the equation connecting the conservation of matter at the plane $x = \xi$ during the time dt to the displacement ($d\xi$) of the boundary of the two-phase region.

b) From the equations obtained above, deduce the diffusion coefficient D.

Calculate the value of $\tilde{D}$ via this equation from the data on the curve. Compare this value with the one calculated by the Matano method.

24 – TWO-PHASE DIFFUSION

Consider a system of metal A clad by metal B, heated for 9 days at a temperature θ at which 2 phases, α and β, form during diffusion (see the figure below, where c indicates the atomic fraction of B).

Assume that at the temperature θ, $D_\alpha = D_\beta = 4 \times 10^{-12}$ cm^2 s^{-1}, that the diffusion coefficients do not vary with concentration, and that the system is infinite in the x direction. Calculate:

1°) The displacement of the plane of the discontinuity α/β.

2°) The thickness of the cladding (B) in which the concentration of A is greater than 1 at.% ($c < 0.99$).

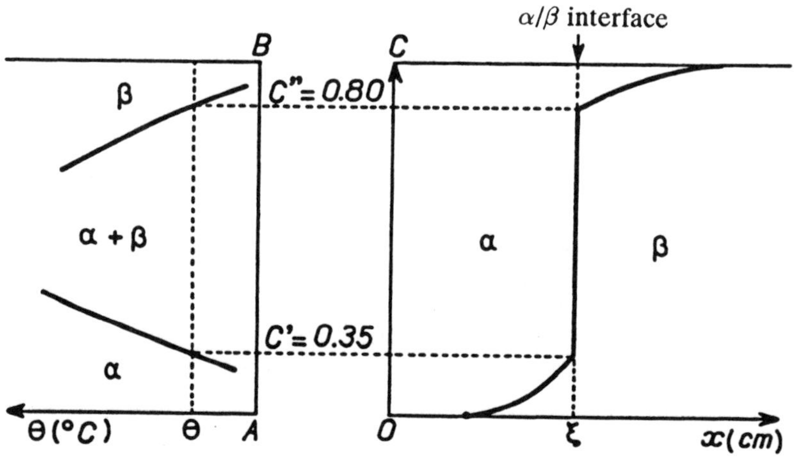

25 – INTERDIFFUSION OF THIN FILMS

A layer of gold 0.1 μm thick is deposited on a sheet of silver 0.1 mm thick, and the diffusion couple thus obtained is heated at a constant temperature θ.

The diffusion coefficient at this temperature is 2×10^{-9}cm^2 s^{-1}, independent of concentration. Assuming that the evaporation of the two metals is negligible, calculate:

1°) The surface concentration of gold in wt. % on each face of the sample after 10 h of treatment.

2°) The time necessary to obtain an alloy homogenous to ± 1 % by wt.

Neglect the difference between the densities of silver and gold.

26 – INTERDIFFUSION, THIN LAYER

A thin layer of metal B is deposited on the surface of a sheet of metal A, 0.1 mm thick, and the couple thus formed is heated at 700°C for 20 h. Assuming that the diffusion coefficient is independent of concentration and that the activation energy and preexponential are 40 kcal/mole and 4×10^{-2} cm^2 s^{-1}, respectively, what should be the thickness of the layer of B so that the concentration of B is 2.4 wt. % , within 10 % , at 10 μm from the surface initially in contact with the deposit ? (Neglect the difference in density between the two metals).

27 – THIN LAYER (CYLINDRICAL COORDINATES)

A thin layer of nickel is deposited on the surface of a 2 mm dia. copper wire. The thickness of the deposit is chosen to obtain an alloy with 1 wt.% nickel after complete homogenization. To obtain such an alloy, the sample is heated at a constant temperature where the diffusion coefficient is $D = 10^{-9}$ cm^2 s^{-1}.

Calculate the annealing time necessary to obtain an alloy that is homogeneous within ± 0.5 % .

28 – OUT-DIFFUSION OF HYDROGEN

A metallic sphere 1.3 cm in diameter contains a weight fraction of hydrogen equal to 4×10^{-6}, distributed homogenously. It is heated for 1000 s at constant temperature in vacuum. If the diffusion coefficient is 5×10^{-5} cm^2 s^{-1} at this temperature, calculate the volume of hydrogen at S.T.P. that escapes from the sphere. (The density of the metal is 8 g cm^{-3}).

29 – DIFFUSION OF GAS THROUGH A MEMBRANE (PERMEATION)

Consider a membrane of thickness l through which a gas diffuses in a manner such that the concentrations on the entrance and exit surfaces remain constant and equal to c_1 and 0, respectively.

The diffusion equation with these boundary conditions can be solved by two methods. They yield:

Separation of variables:

$$c(x,t) = c_1 \left[1 - (x/l) - (2/\pi) \sum_{n=1}^{\infty} \frac{1}{n} \sin \frac{n\pi x}{l} \exp\left(-\frac{n^2\pi^2 Dt}{l^2} \right) \right] \quad (1)$$

Laplace Transformation:

$$c(x,t) = c_1 \left\{ \sum_{n=0}^{\infty} [-1]^n \operatorname{erfc}\left[\frac{x + 2nl}{2\sqrt{Dt}} \right] - \sum_{n=0}^{\infty} [-1]^n \operatorname{erfc}\left[\frac{2\,[n+1]\,l - x}{2\sqrt{Dt}} \right] \right\} \quad (2)$$

However, the quantity measured in most experiments is the exit flux, $J(l,t)$, or the integrated flux, i.e., the quantity of gas which has come out in time t, $Q(t) = \int J\,[l,t]\,dt$.

1°) Calculate $J(l,t)$ for both solutions. Calculate $Q(t)$ for the first solution.

2°) Using the first solution, give an approximate expression for these two quantitites for long times.

3°) Starting with the second solution, propose an approximate expression for $J(l,t)$ for short times.

4°) What is the value of $J(l) = J_\infty$ for steady state?

5°) Compare the two approximate solutions over the interval $0 < \tau < 1$, where the reduced time $\tau = Dt/l^2$.

6°) Describe the variation with time of $J(l,t)$ and $Q(t)$ in the framework of the second approximate solution (short times).

7°) The tangent at the inflection point of the $J(l,t)$ *vs.* t curve cuts the time axis at t_1, called the "breakthrough time". Give an expression for t_1 (hint: its form is l^2/D times a numerical factor).

8°) Extrapolation of the linear part of $Q(t)$ cuts the time axis at t_2. Give an expression for t_2.

9°) What is the value of $J(t_2)/J_\infty$?

30 – PERMEATION OF HYDROGEN

The permeation of hydrogen through slabs of Cr-Mo steel has been measured electrochemically. The hydrogen is introduced by cathodic charging. The flux leaving the slab is measured in a medium of 0.1N NaOH by the anodic current under an imposed potential corresponding to the oxidation of the hydrogen, $H \rightarrow H^+ + e^-$ (see figure below).

The results of an experiment are given in the table below. The thickness of the slab was $l = 0.75$ mm.

t (hours)	0.5	1	2	3	4	5	6	7	8
i (μa cm^{-2})	0.01	0.12	0.30	0.47	0.62	0.73	0.81	0.87	0.90
t (hours)	9	10	11	12	13	14	15		
i (μa cm^{-2})	0.905	0.91	0.92	0.925	0.93	0.93	0.93		

1°) Establish the relationship between the measured anodic current density, i, and the flux leaving the slab, expressed in cm^{3}of H_2 at S.T.P. per unit area and unit time.

2°) Plot the curve $J(t)$ or $J(t)/J_\infty$, where J_∞ is the steady-state flux, as well as the curve $Q(t)$ of the integrated flux ($Q(t)$ = the quantity of hydrogen having diffused out per unit area since time zero).

3°) Evaluate the diffusion coefficent, D, from these curves by several methods. Compare the values obtained.

4°) Compare the experimental $J(t)$ curve with the theoretical curve given by the simplified expression valid for short times:

$$\frac{J}{J_\infty} = \frac{2}{\sqrt{\pi\tau}} \exp\left[-\frac{1}{4\tau}\right],$$

with $\tau = Dt/l^2 \leq 0.30$, and $J_\infty = Dc_1/l$.
Discuss.

5°) Calculate the hydrogen concentration c_1 at the entry face. Assuming that the hydrogen is located at the tetrahedral sites in α iron, express this concentration in atoms of hydrogen per tetrahedral site. The lattice parameter of α iron is 2.86 Å.

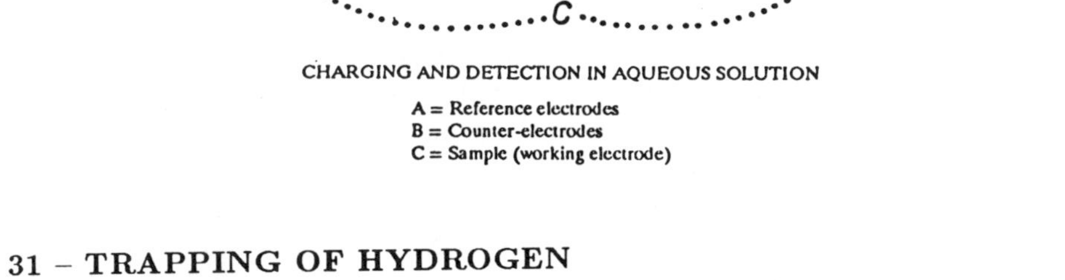

CHARGING AND DETECTION IN AQUEOUS SOLUTION

A = Reference electrodes
B = Counter-electrodes
C = Sample (working electrode)

31 – TRAPPING OF HYDROGEN

It is known that the diffusion of hydrogen does not always follow Fick's laws because of the trapping of hydrogen atoms by certain defects. The latter are modeled by spheres of diameter d, present in number N per unit volume, in α iron which has a lattice parameter a_0. All the lattice sites on the surface of these defects are considered to behave as traps for hydrogen.

1°) Calculate the number of such traps per unit volume, n_p. To be trapped, an atom of hydrogen must:

a) be located in a site neighboring a trap (in view of the geometry of the model, this number of sites is taken to be n_p).

b) jump into the trap. What is the probability of this latter event?

2°) Knowing the concentration (number per unit volume) of hydrogen, C_0, calculate the number of hydrogen atoms on sites neighboring a trap, and then the number that are trapped per unit time. Let kC be this number. Express k as a function of N, d, D, a_0 and C_0.

3°) *The balance equation.* Write the balance equation for trapping, equating the number of atoms trapped per unit time and unit volume to the variation of concentration. Integrate the resulting differential equation. What precisely is the nature of k according to this equation? Comment.

4°) *Numerical exercise.* The solution $C(t)$ of this equation allows the experimental determination of k. At ordinary temperatures, $k = 0.25\ \text{s}^{-1}$ in iron. From this, deduce the dimension d of the defects, if their density is $N = 10^9\ \text{cm}^{-3}$ and if $a_0 = 2.86$ Å and $D = 5 \times 10^{-5}\ \text{cm}^2\ \text{s}^{-1}$.

N. B.: The result of question #2 is: $k = ANd^2D/a_0$, where A is a numerical factor.

32 – TRAPPING

Consider the diffusion of an interstitial solute in a dilute solution with extended defects in the matrix where the solute atoms can segregate and be trapped with a binding enthalpy $-\Delta H_b$. In the matrix, the atoms of the diffusing species always find nearest neighbor sites which are vacant, as the solution is very dilute whereas, in the extended defects, the probability of finding a vacant nearest neighbor site is smaller than one because of the segregation.

Derive the analytical expression for the solute diffusion coefficient and discuss its temperature dependence.

Suggestion. Trapping sites are characterized by the energy 0 (vacant) or $-\Delta H_b$ (occupied by a solute atom). Describe the occupancy probability with Fermi-Dirac statistics and derive D as the product of the site availability times the jump frequency and the jump distance squared.

Numerical Application. Calculate the temperature of the transition between the two regimes defined above, given $\Delta H_b = 0.2$ or 0.5 eV and solute concentration 1 or 100 at.ppm.

33 – APPLICATION OF THE NERNST-EINSTEIN EQUATION TO SELF-DIFFUSION

Rederive Fick's equation for the self-diffusion of a tracer, starting with the Nernst-Einstein equation.

Suggestion: the driving force arises from the entropy of mixing.

34 – INTERDIFFUSION

Formulation based on the Nernst equation. Show that in order to use the Nernst formulation to calculate the interdiffusion flux, the gradient of Gibbs free energy $(\partial g/\partial x)_T$ should not be used as the driving force. Show that the objections to this choice disappear if the driving force is taken as the gradient of $(\partial g/\partial c)_T$.

35 – THE NERNST-EINSTEIN EQUATION (ELECTROMIGRATION)

The linearization of the exponentials for the jump frequencies to the right, Γ_{12}, and to the left, Γ_{21}, in a potential gradient leads to the Nernst-Einstein equation:

$$v = FD/kT,$$

if $v = (\Gamma_{12} - \Gamma_{21})\lambda$, where λ is the interplanar spacing. Taking the expansion one order higher, evaluate the driving force necessary to observe a significant deviation from the linear Nernst-Einstein law.

Take the case of an electric field applied at ambient temperature or at 1000°C, and particles carrying either the electronic charge, or 50 times that ("effective charge"). The interplanar spacing is $\lambda = 3$ Å.

36 – NON-FICKIAN BEHAVIOR

1°) Write the balance equation for the concentration C_n among three neighboring lattice planes $n-1$, n, $n+1$, placed in a concentration gradient. Assume that the jump frequency is constant.

2°) Using a Taylor's expansion of C_n, deduce the diffusion equation as a function of successive derivatives of $c(x)$.

37 – QUANTUM EFFECTS

What are the conditions on the mass of a diffusing particle and on the

temperature so that the jump frequency is affected by quantum effects? Carry out an appropriate numerical calculation.
Suggestion: use a dimensionally exact relation between Planck's constant and the energy kT. Recall that $\hbar = 1.05 \times 10^{-34}$ J s.

38 – DIFFUSION OF AN INTERSTITIAL ELEMENT IN THE BCC STRUCTURE

An interstitial element in a BCC structure can occupy either octahedral (O) or tetrahedral (T) sites.

1°) Identify these sites and give the number of each per unit cell.

2°) The diffusion can take place via three types of jumps between first-nearest-neighbor interstitial sites: O → T, T → O, and T → T. The diffusion coefficient is thus expressed as a function of the corresponding three jump frequencies W_{TT}, W_{OT}, and W_{TO} :

$$D = a^2 \frac{W_{OT}\,[W_{TO} + 2W_{TT}]}{12\,[W_{TO} + 2W_{OT}]},$$

where a is the lattice parameter of the BCC lattice (1).

Starting with this expression, find the classical formula in which only the octahedral sites are taken into consideration. Express D as a function of the residence time on the O sites.

(1) S. Ishioka and M. Koiwa, *Phil. Mag.* **A52** (1985), 267-277.
R.H. Condit and D.N. Beshers, *Trans. AIME* **239** (1967) 680.

39 – DIFFUSION IN THE HEXAGONAL CLOSE-PACKED STRUCTURE

1°) Diffusion in the basal plane. Choose orthogonal x and y axes. Based on the formula $D_x = 1/2 \sum \Gamma_s x_s^2$, where x_s is the projection of the length of a type s jump on the x axis, and an analogous formula for D_y, calculate D_x and D_y and show that they are equal.

2°) Calculate D_z, the z direction being taken along the six-fold axis. Under what condition will $D_z = D_x$ $(= D_y)$?

40 – SELF-DIFFUSION IN BERYLLIUM

Beryllium is a metal with HCP structure having parameters $c = 3.58$ Å and $a = 2.28$ Å. The self-diffusion coefficients are measured both parallel ($D_{\|}$) to and perpendicular ($D_{\perp}$) to the six-fold axis in a single crystal.

1°) The results of a first experiment are:

$D_{\perp} = 2 \times 10^{-9}$ cm^2 s^{-1}, $D_{\parallel} = 10^{-9}$ cm^2 s^{-1}.

Calculate from this the jump frequency Γ_1 in the basal plane and the jump frequency Γ_2 out of the basal plane, neglecting correlation effects.

2°) Further experiments carried out at various temperatures yield:

$D_{\perp} = 0.5 \exp(-37600/RT)$

$D_{\parallel} = 0.6 \exp(-39400/RT)$,

in cm^2 s^{-1}, with activation energies in cal mole^{-1}.

Calculate the jump frequencies Γ_1 and Γ_2, taking correlation effects into account, at two temperatures: 727° and 1060°C. From this, deduce the activation energy and preexponential factor corresponding to each kind of jump.

41 – SELF-DIFFUSION IN COPPER

To study self-diffusion in copper, a radioactive copper isotope is electrodeposited on single crystals of that metal. Heat treatments are carried out at diverse temperatures for a constant time equal to 49 h. The thickness of the deposited layer is assumed to be negligible with respect to $\sqrt{Dt}$. After diffusion, the distance at which the activity is e times less than that on the surface is determined by sectioning (e is the base of natural logarithms). The results are:

T(°C)	600	715	850	1063
x(μm)	4.5	22	94	511

1°) Obtain graphically both the activation energy, rounding it off to an integer, and the preexponential factor, rounded off to a simple number, for self-diffusion in copper.

2°) Calculate the jump frequency for copper atoms at 850°C, knowing that self-diffusion in copper takes place via single vacancies.

3°) The enthalpy and entropy for vacancy formation are respectively $\Delta H_f = 1.17$ eV and $\Delta S_f = 1.5\,k$, where k is Boltzmann's constant. What is the equilibrium concentration of vacancies at 850°C? What is the jump frequency for vacancies at the same temperature?

4°) Under irradiation of a single crystal, the atomic fraction of vacancies in copper is 10^{-4}. Calculate the self-diffusion coefficient under irradiation at 850°C. What is the distance, under these conditions, in which the activity drops to $1/e$ times the value on the surface after 49 h of diffusion?

Given: the correlation factor, $f_0 = 0.8$,
the lattice parameter, $a_0 = 3.6$ Å.

42 – ISOTOPE EFFECT

Cobalt diffuses very rapidly in $\beta - \mathrm{Zr}$, so it is thought to diffuse interstitially. The isotope effect has been measured with the isotope pair ^{57}Co and ^{60}Co to be $E = 0.23 \pm 0.03$, apparently independent of temperature (1).

Is this result consistent with a simple interstitial mechanism, or a mechanism in which several atoms move in each jump (such as a dissociated interstitial or a mixed dumbell)?

Suggestion: compare $E(1)$, the value given above, with $E(2)$ and $E(3)$. The atomic mass of Zr is 91.22.

(1) C. HERZIG. J. NEUHAUS, K. VIEREGGE, and L. MANKE, *Materials Science Forum* **15-18** (1987) 481.

43 – ISOTOPE EFFECT

From the definition of the isotope effect E, show that the diffusion coefficient is proportional to $m^{-E/2}$ rather than to $m^{-1/2}$.

Suggestion: rewrite E in terms of the differences ΔD and Δm.

44 – DIFFUSION OF A DOPANT IN SILICON

Let $n_i(T)$ be the concentration of intrinsic carriers, and let n_x be the concentration of dopant X.

1°) Show that if $n_x << n_i(T)$, the diffusion profile of X follows a simple form, such as $D_x =$ constant, but if $n_x > n_i(T)$, the diffusion kinetics become complex.

2°) Application to silicon (band gap = 1.12 eV). What is the critical concentration of the element X for passing from one regime to the other at 900°C?

N.B.: Recall that in a semi-conductor:

$$n_i^2 = 1/2 \left[\frac{kT}{\pi \hbar^2}\right]^3 (m_e m_h)^{3/2} \exp\left(-E_g/kT\right),$$

where m_e and m_h are the effective masses of electrons and holes, respectively, and E_g is the energy gap.

45 – INTERDIFFUSION IN A CONCENTRATION GRADIENT: A/B

The relation between the self-diffusion coefficient, D_i, and the intrinsic diffusion coefficient, $\bar{D}_i$, of the same chemical species i (= A or B) was derived in the text to be:

$\bar{D}_i = D_i\phi$, where ϕ is the thermodynamic factor

$$\phi = 1 + \frac{\partial \log \gamma_i}{\partial \log N_i},$$

γ_i is the activity coefficient, and N_i is the mole fraction, $N_{\mathrm{A}} + N_{\mathrm{B}} = 1$.

1°) Show that ϕ can be written in the form

$$\phi = \frac{N_{\mathrm{A}}\,[1 - N_{\mathrm{A}}]}{RT}\, G'',$$

where G'' is the second derivative of the Gibbs free energy:

$$G'' = \mathrm{d}^2G/\mathrm{d}N^2.$$

2°) Assume that the AB solid solution is regular. Show that:

$$\phi = 1 - 2\Delta H_{\mathrm{m}}/RT,$$

where ΔH_{m} is the enthalpy of mixing.

Recall that the entropy of mixing of a regular solution is the same as that of an ideal solution:

$$\Delta S_{\mathrm{m}} = -R\,[N_{\mathrm{A}} \log N_{\mathrm{A}} + (1 - N_{\mathrm{A}}) \log (1 - N_{\mathrm{A}})],$$

and the enthalpy of mixing varies parabolically with the composition:

$$\Delta H_{\mathrm{m}} = \Lambda N_{\mathrm{A}}\ (1 - N_{\mathrm{A}}).$$

3°) With the same model, derive the expression $\phi = 1 - T_{\mathrm{s}}/T$, where T_{s} is the spinodal temperature.

46 – INTERDIFFUSION AND THE KIRKENDALL EFFECT

In order to verify experimentally the theories of the Kirkendall effect for the iron-nickel system, a diffusion couple was prepared, made up of a pellet of nickel 4 mm thick, a nickel foil calibrated at 200 μm (the foil-disk weld was intended to serve as a reference plane), inert markers (very fine sillimanite wires) and an iron pellet 4 mm thick. After annealing 48 hours at 1200°C in vacuum (sealed in a quartz capsule), micrographic examination established that the markers were displaced by 70 μm towards the iron.

The nickel-concentration profile as a function of penetration (obtained from electron-probe microanalysis) is given in the table below. The ordinates are given in atomic % , the abscissas in μm, with the nickel pellet-foil weld taken as the origin.

1°) Determine the position of the Matano interface.

2°) What is the location x_K of the Kirkendall plane, and what is the corresponding atom fraction of nickel?

3°) What is the value of the interdiffusion coefficient, $\tilde{D}$, for this concentration?

4°) What are the intrinsic diffusion coefficients $\bar{D}_{Ni}$ and $\bar{D}_{Fe}$ for the concentration corresponding to the Kirkendall interface? (Recall that the displacement of the Kirkendall interface is proportional to the square root of the time.)

5°) Calculate the thermodynamic factor for Fe and for Ni for these conditions, knowing that the self-diffusion coefficients of Ni and Fe in an alloy of the corresponding composition are:

$$D^*_{Ni} = 1.20 \times 10^{-10} \text{ cm}^2 \text{ s}^{-1}, \quad D^*_{Fe} = 1.5 \times 10^{-10} \text{ cm}^2 \text{ s}^{-1}.$$

What can be concluded from this?

6°) Carry out the same calculation as in 4 using Heumann's method (see text, e.g., Eq. (VI.22)).

Table

Distance (μm)		50	80	120	160	180	200	220	240	260
Ni, wt. %		99.5	98	96	92	89	87	82.5	79	75
280	300	320	340	360	380	400	420	440	460	480
71	66.5	59	51	40.5	30	19	12.5	4	1	≈ 0

47 – INTERDIFFUSION AND THE KIRKENDALL EFFECT

A diffusion couple is prepared from two thick samples of gold and of silver. A foil and wires of tungsten are placed at the weld plane. The couple is treated for 9 h at 915°C. After electron-probe microanalysis, the table of results (see below) is obtained, permitting construction of the $C_{Au}(x)$ curve.
Micrographic observation shows that the tungsten wires were displaced 65 μm with respect to the tungsten foil in the direction of positive x (toward the silver). The coordinate of the Kirkendall (i.e., marker) plane is $x_K = 510$ μm after diffusion. Determine:

1°) the location of the Matano interface;

2°) the value of the gold concentration, C_{Au}, at the Kirkendall interface;

3°) the coordinates of the Kirkendall and Matano interfaces; compare;

4°) the chemical diffusion coefficient, $\tilde{D}$, for this concentration;

5°) the intrinsic diffusion coefficients $\bar{D}_{Au}$ and $\bar{D}_{Ag}$ for this concentration;

6°) calculate the thermodynamic factor, given the self-diffusion coefficients:

$$D^*_{Au} = 0.9 \times 10^{-9} \text{ cm}^2 \text{ s}^{-1}, \quad D^*_{Ag} = 2.25 \times 10^{-9} \text{ cm}^2 \text{ s}^{-1}.$$

Table

x (μm)		0	25	50	75	100	125	150	175	200
C_{Au} (at. %)		100	99.5	99	98.5	97.5	96.5	95	94	92
225	250	275	300	325	350	375	400	425	450	475
89	87	84	80	76	71	66	61.5	56	51.5	45
500	525	550	575	600	625	650	675	700	725	750
39	33	26	20	14	9	5	2	1	0.5	≈ 0

48 – DIFFUSION IN A SOLID SOLUTION WITH MODULATED COMPOSITION

Using the expression of Cahn and Hilliard for the free energy of a heterogeneous solid solution, calculate the difference in Gibbs free energy between a homogeneous binary solution of composition c_0 and a solid solution with a harmonic variation of the composition in one direction $c - c_0 = A \cos \beta x$. Determine the conditions for the amplification of this modulation according to its wavelength Λ.

Numerical exercise: Consider a regular solid solution. In a nearest-neighbor model one defines the parameter

$$\omega = V_{\text{AB}} - (V_{\text{AA}}+V_{\text{BB}})/2,$$

where the V_{ij} are the i-j bond energies.

Recall that in this case the enthalpy of mixing per unit volume is written:

$$\Delta H_{\text{m}} = \mathcal{N} \zeta \, N_{\text{A}} \, N_{\text{B}} \, \omega,$$

with ζ the coordination number, $\mathcal{N}$ the number of atoms per unit volume, N_{A} and N_{B} the mole fractions of A and B, respectively, $(N_{\text{A}} + N_{\text{B}} = 1)$; the entropy of mixing is the same as in an ideal solution. In the same model, Cahn's gradient factor is given by:

$$K = \zeta a_0^2 \, \omega/6,$$

with a_0 the atomic diameter.

Given that $z = 12$, $a_0 = 2.5$ Å, and k (Boltzmann) $= 1.4 \times 10^{-23}$ JK^{-1}.

1°) Given that the critical temperature for miscibility of a 50-50 solution is 700°C, find ω.

2°) Calculate Λ_c for demixing this solution at 500°C.

3°) If the size effect is $\eta = 5\%$, discuss the values of the coherence-energy term.

Take E (Young's modulus) $= 210$GPa, and ν (Poisson's ratio) $= 0.3$.

49 – DIFFUSION IN A BICRYSTAL

A recent publication presented the results of a study of the diffusion of chromium in niobium bicrystals with a plane tilt boundary and without detectable precipitates (1). By sectioning a bicrystal, the authors prepared samples in the form of disks, with surfaces normal to the boundary plane and perpendicular to the tilt axis. A layer of pure chromium was electroplated on the surface, and the samples were annealed in quartz tubes sealed under vacuum for 100 h at 1273°C.

After annealing, the samples were sectioned along a plane normal to the surface of the disk and to the boundary plane. On this section, after careful polishing, the authors measured the local concentration of chromium by means of an electron-probe microanalyzer.

1°) describe the diffusional processes, after having sketched schematically the geometry of the system under study. Knowing that the solute diffusion coefficient of chromium in niobium is given by an Arrhenius expression with $D_0 = 0.30\ \text{cm}^2\ \text{s}^{-1}$, $Q = 83500$ cal mole^{-1}, in what diffusional regime did this experiment take place? What solution of the diffusion equation would you choose? Is the approximate formula valid?

2°) On the polished section, the concentration of chromium was measured point-to-point along a straight line perpendicular to the boundary at a distance $z = 15.5\ \mu$m from the surface on which the chromium had been deposited. The results are given in the following table (the origin of x is the boundary plane).

x (μm)	−27	−17	−11.5	−2	0	5	11.5	19
Cr (wt.%) (±0.02 %)	≈ 0	0.02	0.075	0.135	0.19	0.095	0.065	≈ 0

Draw the corresponding graph. How is the grain-boundary diffusion coefficient D' calculated from these results?

3°) Similar measurements were carried out at varied depths z. How would $\bar{C}_z$ be calculated from them? The results are given in the following table (to be completed for $z = 15.5\ \mu$m from the answer to 2°).

$\bar{C}_z$ (wt. %–μm)	6.03	4.82		3.29
z(μm)	6	11.5	15.5	18.5

Which grain-boundary diffusion parameter can be determined? What is its value?

4°) Compare the correlation coefficients obtained from linear regressions of $\bar{C}(z)$ *vs.* $z^{6/5}$ or z^2.

(1) X. R. Qian and Y. T. Chou, *Philos. Mag.* **A52** (1985) L13-L18.

50 – SELF-DIFFUSION ALONG GRAIN BOUNDARIES

To study the diffusion of nickel in the grain boundaries of nickel oxide, NiO, in the temperature range between 500°C and 800°C, polycrystalline nickel oxide samples were prepared by oxidation of pure nickel foils. The material thus obtained had equiaxed grains with an average diameter of about 10 μm. A thin layer of tracer (^{63}Ni) was deposited on the surface. After diffusion annealing, the penetration profile was determined by a sectioning technique based on RF sputtering. The anneals were carried out at 4 different temperatures (500, 600, 700 and 800°C) under a pressure of one atmosphere of pure oxygen. The data are given in the table I below (1):

Table I

T(°C)	t(s)	D (cm^3 s^{-1})	$D'\delta$ (cm^3 s^{-1})
800	6.6×10^2	2.1×10^{-14}	1.4×10^{-16}
700	7.2×10^3	1.2×10^{-15}	
600	6.05×10^5	3.7×10^{-17}	1.8×10^{-18}
500	8.6×10^5	4.5×10^{-19}	5.6×10^{-20}

D and D' are the diffusion coefficients of nickel in the volume and the grain boundaries, respectively. δ is the width of the grain boundary, taken as $\delta = 7 \times 10^{-8}$ cm in the numerical exercises.

1°) Show that these experiments are indeed in the Harrison B kinetics regime.

2°) The experimental values for average activity as a function of depth, found at 700°C, are given in table II.

Table II

z (cm) $\times 10^4$	0.1	0.12	0.14	0.16	0.18	0.22	0.24	0.30	0.50
$\bar{c} \times 10^3$	550	340	160	100	29	30	25	15	9
0.68	0.90	1	1.15	1.4	1.6	2.1	2.6	3.1	3.6
5.2	4.6	4	3.8	3.1	3	2.2	1.6	1.05	0.85

Obtain $D'\delta$ and the grain boundary diffusion coefficient D'.
Is the condition $\beta > 10$ fulfilled? Justify *a posteriori* the relation used to calculate D'.

3°) For a given temperature, how does the slope of the concentration profile vary with the time of diffusion?

Plot the straight lines that should be obtained for the diffusion "tails" for times of $t_1 = 10^5$ s and $t_2 = 10^6$ s at 700°C. (The straight lines are forced to pass through the point obtained by extrapolating to $z = 0$ the "tail" of the curve defined in the second question.)

4°) Can an activation energy for grain-boundary diffusion be defined? Compare it to the activation energies for volume diffusion.

5°) Study the graph of log $\bar{c}$ *vs.* x^2. Does the "diffusion tail" corresponding to grain-boundary diffusion appear linear? To what precision? What can be concluded from this?

(1) A. ATKINSON and R. I. TAYLOR, *Philos. Mag.* **A43** (1981) 979-998.

51 – DENUDED ZONE NEAR GRAIN BOUNDARIES

A denuded zone, i.e., a zone in which precipitates do not appear, is classically observed around grain boundaries in light metals after quenching and annealing. This is attributed to the "pumping" of the solute element by the grain boundaries, where it forms fine intergranular precipitates. The supersaturation is thus insufficient to produce precipitation near the grain boundaries (see figure).

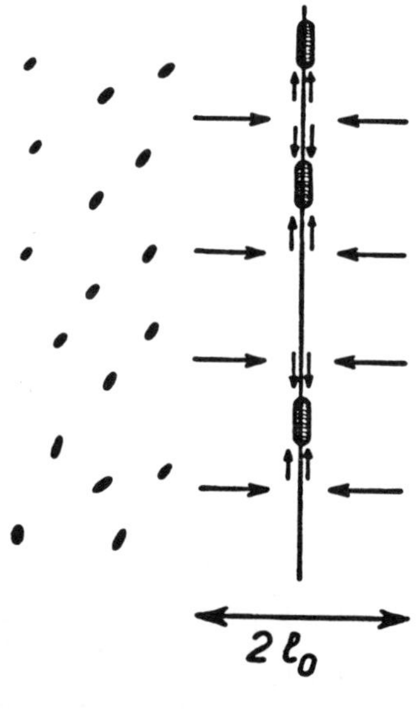

Such a study was carried out on an Al-Li alloy (2.5 wt.% Li). The alloy was annealed at 500°C and quenched, then aged for different times at 200°C to cause precipitation of the metastable phase $\delta' - Al_3Li$.

1°) The width of the denuded zone, l_0, was measured as a function of the annealing time:

t (hours)	3	12	24	48
l_0 (cm) × 10^5	1.0	2.5	3.3	4.6

Can a value of the diffusion coefficient of lithium at 200°C be deduced from these data?

2°) An analytical transmission electron microscope with electron beam about 20 nm in diameter was used to carry out a series of "point" analyses in a direction perpendicular to the boundary on a thin foil. The results for a treatment of 48 h at 200°C are given in the table below.

Table of Analyses (1)

x, distance to the boundary, (nm)	0	38	100	145	205	250	300
Li (at. %)	0 – 0.45	0.62	1.4	2.85	2.2, 3.15	2.85	3.5, 4.9

350	405	460	500	560	750	760	810	850	910
3.62	3.15, 4.4	5.05	4.9	6.3	6.15	7.4	7.25	7.9	7.4

What is the theoretical profile that these points should fit? Deduce the value of the diffusion coefficient, and compare it to the value obtained in the first question. Plot the experimental points on a graph with the theoretical profile obtained from the best value of D.

(1) SAINFORT P., Thesis, University of Grenoble, 1985.

52 – ZONE IMPOVERISHED BY INTERGRANULAR PRECIPITATION

According to the classical model of intergranular precipitation, the precipitates grow by the diffusion of solute toward the grain boundary, the solute being drained off along the boundary to the precipitates (Figure).

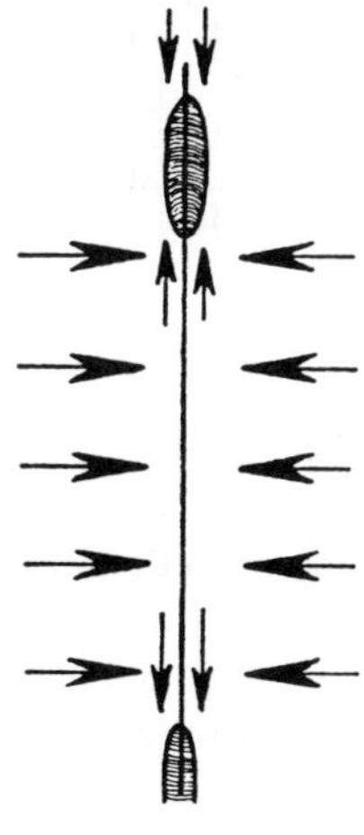

To verify this model, experiments on precipitation were carried out on an Al-Zn alloy containing 4.5 at.% Zn. After a homogenizing anneal of 2 h at 450°C, the samples were air cooled to the aging temperature ($110 < \theta < 160$°C) where they were held for a time t and then quenched.

Thin layers were prepared from these samples for observation in transmission in an electron microscope equipped with an X-ray microanalyzer, with 100 Å resolution under the best conditions. The analyses for zinc in a "traverse" of a grain boundary are given in figure below for aging 28 h 37 min at 125°C.

Which classical solution of the diffusion equation should the concentration profile normal to the boundary follow? Propose a method for extracting the value of the diffusion coefficient, D, from the experimental results. Give the value of D so determined. Show the theoretical profile on the figure and compare it to the experimental one. Why is there a difference between these two profiles near the minimum?

Propose a measure of the width of the impoverished zone. How does it vary with the holding time and the aging temperature?

(1) A. W. Nichols and J. P. Jones, *J. Phys. Chem. Sol.* **44** (1983) 671-676.

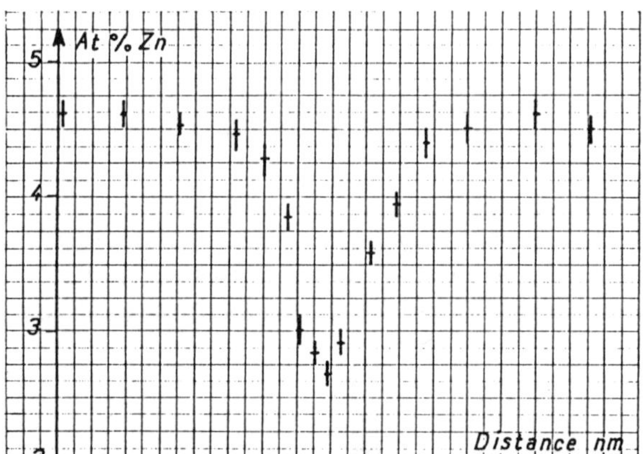

53 – GRAIN-BOUNDARY DIFFUSION

1°) Given that the ratio of the diffusion coefficients in the volume and the grain boundaries, D/D', is small enough, show that there exists a simple relationship between the expressions for the average concentrations $\bar{c}_S(z,t)$ and $\bar{c}_W(z,t)$ given by Suzuoka's and Wipple's solutions, respectively.

Suggestion: differentiate $\bar{c}_W$ with respect to the penetration z.

2°) Figure VII.7 shows that the "tails" on the penetration profiles corresponding to grain-boundary diffusion, calculated according to Suzuoka's solution, extrapolate to values of $\bar{c}$ at $z = 0$ that are almost independent of the diffusion time. Show that if this result is exact, it cannot be true for Whipple's solution.

Suggestion: to simplify the calculation, assume that $\log \bar{c}$ is linear in z rather than in $z^{6/5}$.

54 – THE ACCUMULATION METHOD

Consider diffusion along the grain boundaries that traverse a film with columnar structure. Let h be the film thickness and d the grain diameter, measured in the plane of the film.

Assume that the conditions correspond to the regime C and that the film is thin enough so that a steady state is established between the two surfaces of the film. The entry surface is maintained at a constant concentration c_0, while on the exit surface, the concentration at the grain-boundary intersections remains practically zero if surface diffusion along the exit surface is sufficiently rapid. The quantity of diffusing substance accumulating on the exit surface, $Q(t)$, is measured as a function of time. $Q(t)$ is expressed in number of atoms per unit area (cm^{-2}) or in volume per unit area (cm).

1°) Derive an equation for $Q(t)$ that has crossed the film in steady state. Can the grain-boundary diffusion coefficient be deduced from this?

2°) The same question for dislocations threading throughout the film.

3°) How can the grain-boundary diffusion coefficient D' be obtained, instead of the product $D'\delta$?

Numerical example: Aluminum diffuses through polycrystalline films of silicon covered with a 20 nm thick deposit of aluminum (1). The silicon film thickness $h = 350$ nm, the grain size $d = 40$ nm. The quantity of aluminum accumulating on the exit surface is followed by Auger spectrometry. This method is further combined with sputtering to determine the concentration profile, $\bar{c}(z)$, which is shown to be linear. The results are given in the table below:

Temperature (°C)	Time (hours)	$Q(t)$ (cm)	$d\bar{c}/dz$ (cm^{-1})
400	1	1.1×10^{-7}	100

Calculate D' by the two methods (see questions 1° and 3°). Compare and discuss. Lattice parameters: Si – 5.43×10^{-10} m, Al – 4.04×10^{-10} m.

(1) HWANG J. C. M., HO P. S., LEWIS J. E. and CAMPBELL D. R., *J. Appl. Phys.* **51** (1980) 1576.

55 – SURFACE OR GRAIN BOUNDARY SEGREGATION

Consider a solid solution AB with concentration C_0 of the solute B, and assume B is segregating to the surface (or to the grain boundaries) according to the Gibbs equilibrium law. Calculate the segregation kinetics i.e. the variation with time of the surface concentration $\Gamma_s(t)$, (in number of B atoms per unit area) according to one of the following assumptions:

1°) the solute concentration in the alloy, just below the segregated layer $C(x = 0,\ t)$ remains constant, or eventually nil (see the figure).

2°) local equilibrium is maintained at the surface, i.e. $\Gamma_s(t) = k \times C(x = 0,\ t)$ at any time.

Discuss the conditions for the validity of these assumptions.

Suggestion for question 2°) : use the Laplace transform for the diffusion equation and boundary conditions.

Answer:

$$\frac{\Gamma_s(t) - \Gamma_s(\infty)}{\Gamma_s(0) - \Gamma_s(\infty)} = \exp\left(\frac{Dt}{k^2}\right)\ \operatorname{erfc}\left[\sqrt{\frac{Dt}{k^2}}\right].$$

Give an approximation for short times and show it is identical with the case 1°) solution.

Numerical application: Alloy Ag-0.5 at.% Pb.

How long a time is necessary to obtain a surface coverage equal to 0.5 on a (111) surface at 300°C ?

Given: lattice parameter: 4.17 Å.

Pb diffusion coefficient at 300°C: 2.3×10^{-15} cm^2 s^{-1}.

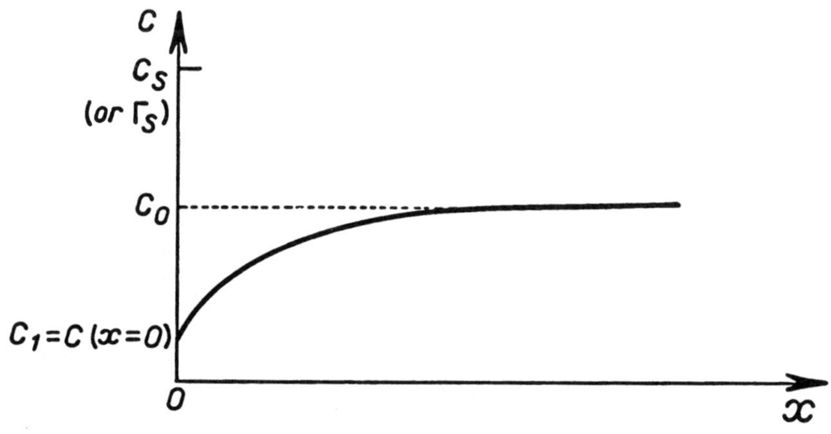

56 – DIFFUSION OF A TRACER

An available grinding method is good down to a section thickness of 0.1 μm. Knowing that to measure the diffusion coefficient precisely, a drop in the tracer activity of at least two orders of magnitude is needed, that at least 12 experimental points are necessary, and that they should cover a penetration depth at least $4\sqrt{Dt}$, what is the minimum value of D that can be determined with this method?

The experiment is carried out on nickel. What is the minimum temperature at which D can be determined if the heat treatments are limited to 10 days? For self-diffusion in nickel, $D_0 = 1.9\ \text{cm}^2\ \text{s}^{-1}$, $Q = 68$ kcal/mole.

57 — THE SECTIONING-COUNTING METHOD: CORRECTION OF THE PENETRATION DEPTH

In a sectioning and counting experiment, each section i has a definite thickness t_i. But because of the gaussian decrease of the activity, the center of gravity of the activity of each section i is shifted by δz_i with respect to z_i, the coordinate of the center of the section. To calculate D, log A_i must thus be plotted as a function of $(z_i + \delta z_i)^2$ *instead of* z_i^2. Calculate the correction δz_i.

58 — THE RESIDUAL-ACTIVITY METHOD

The Gruzin-Seibel equation relates the radioactivity A_n measured after the removal of n successive layers to the concentration $c(z)$ at the depth z attained after these n sections:

$$\mu A_n - \frac{\partial A_n}{\partial z} = \text{constant} \times c(z),$$

where μ is the coefficient of linear absorption of the radiation. For hard radiation, the first term of this equation is negligible. For this case, express A_n/A_0 as a function of z for two classic conditions: 1) the instantaneous or thin-layer source, 2) constant surface concentration.

Compare the two corresponding graphs using simple criteria such as the depth z for which $A/A_0 = 1/2$, and the slope at the origin.

59 — CALCULATION OF D_0 AND Q BY LINEAR REGRESSION

The following values of the diffusion coefficient, D, were measured at four temperatures:

T (°C)	850	950	1000	1050
D (cm^2 s^{-1})	8.9×10^{-11}	2.6×10^{-10}	5.5×10^{-10}	8.6×10^{-10}

Calculate D_0 and Q by the method of least squares, and give the corresponding confidence limits.

60 — INTERNAL FRICTION

Consider the Snoek peak due to atoms in interstitial solution in α iron. The position of the Snoek peak for an Fe-C alloy is determined for different frequencies: 10°C for a period of 17.8 s and 48°C for a period of 0.476 s.

1°) Calculate:

– the corresponding activation energy.

– the diffusion coefficient of carbon at 48°C, assuming that the carbon atoms occupy only octahedral sites (lattice parameter of α iron $a_0 = 2.861$ Å).

– the temperature of the Snoek peak for a period of 2 s.

2°) For this period (2 s), the Snoek peak for nitrogen is observed at 18°C. Are the carbon and nitrogen peaks separated enough to allow the calculation of the carbon concentration from the peak height in the presence of nitrogen?

61 — OXIDATION OF NICKEL

A nickel foil is oxidized in air at 1000°C. NiO is a metal-deficient oxide: $Ni_{1-\delta}O$. At 1000°C, the mole fraction of cation vacancies, δ, equals 10^{-4} in equilibrium with air, and 4×10^{-6} in equilibrium with nickel. The vacancies are assumed to have a charge $-2\,|e|$.

1°) Graph the variation of the concentration of vacancies and electron holes across the oxide film. Justify the condition on the vacancy flux $\mathrm{d}J/\mathrm{d}x = 0$; recall further that the diffusion coefficient of the vacancies is independent of their concentration at low concentrations. Assume a film thickness of 20 μm.

2°) Calculate the electric field across the film; describe the origin of the field. Graph its variation across the film.

3°) Show that owing to this field, there exists an internal space charge, ρ. Calculate it, using Poisson's equation. Express the charge ρ as a difference in concentration of carriers of charge $\rho \equiv |e|\ \Delta n$. Compare Δn to the concentrations of vacancies and holes. Discuss the validity of the assumption of electroneutrality for films of thickness 20 μm and 200 Å respectively. Given: lattice parameter of NiO: 4.18 Å, dielectric constant = 9.

62 — GROWTH OF AN OXIDE LAYER

The growth of a NiO layer on nickel between 700 and 1100°C follows a parabolic law: $\xi^2 = \gamma t$. From the Wagner diffusion model one predicts $\gamma = 6.4\ D$, where D is the coefficient of self-diffusion of Ni in NiO.

From thermogravimetric measurements at 1000°C one determines $\gamma = 5.8\ 10^{-13}\ \mathrm{cm^2{\cdot}s^{-1}}$. The oxide layer is polycristalline with columnar grains spreading over the whole thickness with an average diameter equal to 3 μm.

Are these two kinds of results consistent?

Given at 1000°C D(Ni) in NiO $= 2.8 \times 10^{-14}\ \mathrm{cm^2 \cdot s^{-1}}$
$D'\ \delta$ $= 1.4 \times 10^{-16}\ \mathrm{cm^3 \cdot s^{-1}}$

63 — INTERNAL OXIDATION

Consider a binary alloy <u>A</u>B, in which the element B, present at low concentration (n_B B atoms per unit volume), is more easily oxidized than the solvent A and forms particles of the oxide BO_ν by internal oxidation.

1°) Give a simplified derivation of the kinetic equation for the depth $\xi(t)$ attained by internal oxidation. For this, assume that the oxygen gradient between the surface (C_{ox} = constant) and the oxidation front ($C_{ox}(\xi) = 0$) is linear. The materials balance is effected by comparing the quantity of oxygen brought in by diffusion during a time (dt) and the quantity needed to oxidize element B completely. (That element is not supposed to diffuse over long distances). The expression thus obtained is close to the exact expression obtained in chapter X, equation (X.39).

2°) How does the size of the oxide particles vary with depth?

64 — DECARBURIZATION OF A STEEL DURING OXIDATION

During an austenitizing anneal of a steel in an oxidizing atmosphere, a layer of oxide scale forms on the surface of the metal. At the metal/oxide interface, the carbon reacts according to the reaction C + FeO → Fe + CO. If the scale is sufficiently porous for the carbon monoxide to escape, the steel will be decarburized in the region just under the scale (1).

1°) Solve the diffusion equation to determine the depth profile of the carbon $c(x)$. Take the initial concentration as c_0 and the concentration at the metal/scale interface as zero.
Suggestion: Assume that the metal/scale interface moves parabolically: $x^2 = 2K_pt$.

2°) Determine the depth ξ of the zone decarburized to a carbon concentration $c_1 = \gamma c_0$ with $\gamma < 1$. Can this depth be written in the form $\xi = 2\alpha\sqrt{Dt}$, where D is the diffusion coefficient of carbon in austenite? Obtain the thickness $d = \xi - x$ of the decarburized zone as a function of the annealing temperature and time.
Suggestion: For $\xi \approx 2\alpha\sqrt{Dt}$, see the numerical example.
Numerical example:
$c_0 = 0.85$ wt.%
$\gamma = 0.92$
Temperature range: 900°C< T <1300°C
Diffusion of carbon in austenite: $D_0 = 0.25$ cm² s⁻¹, $Q = 35{,}000$ cal/mole.
Parabolic rate constant for oxidation $K_p\,(\text{mm}^2\,\text{s}^{-1}) = 57.1\exp(-21720/T)$.

(1) N. BIRKS and C. H. MEIER, *Introduction to high temperature oxidation of metals*, Ch. 9, Edward Arnold (London) 1983.

65 — OXIDATION OF SILICON

A wafer of (111) silicon slightly doped with boron $\left(10^{15}\ \text{cm}^{-3}\right)$ is oxidized in a dry oxygen atmosphere at 1 atm. pressure.

A first oxidation takes place at 1000°C for a time t_1; the thickness of SiO_2 formed is x_1. Windows are opened in this layer by removing the entire thickness of oxide. In some of these windows, phosphorus is implanted at low energy, so that the concentration of phosphorus near the surface remains almost constant for a shallow depth.

These wafers are then annealed for 2 h at 920°C under the same conditions as before. The oxide thickness outside the windows is x_2; the oxide thickness over the windows without phosphorus is x_3; and over the windows implanted with phosphorus, x_4 (Figure).

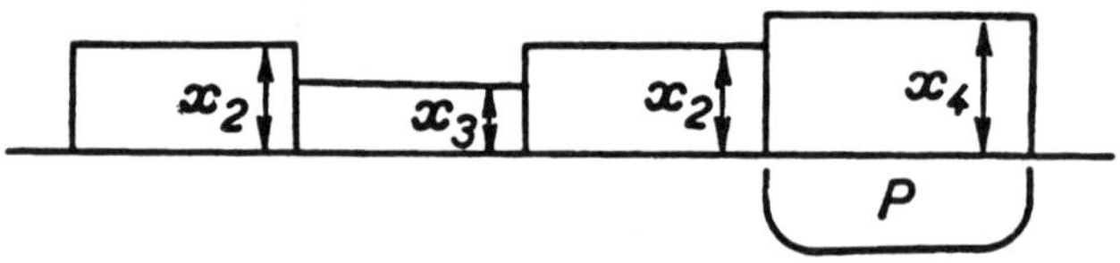

Given $x_2 = 800$ Å. Determine x_1, t_1, and x_3.
What can be said about x_4? Recall that the oxidation of silicon follows the kinetics given by Deal and Grove: $x^2 + Ax = B(t + \tau)$, where x is the thickness of the oxide film and τ takes into account the initial accelerated stage of oxidation.

T (°C)	920	1000
A (μm)	0.15	0.10
B ($\mu\text{m}^2\ \text{h}^{-1}$)	4×10^{-3}	6.3×10^{-3}
τ (h)	1.15	0.42

66 — OXIDATION OF DOPED SILICON

Take a wafer of silicon doped uniformly to $c_0 = 10^{18}\ \text{cm}^{-3}$ atoms of boron. The wafer is oxidized in dry oxygen at 1000°C. The dopant dissolves in the oxide film during the oxidation.

1°) Write the flux of dopant at the Si/SiO_2 interface.

2°) Carry out a balance of the quantity of dopant that accumulates in a layer (dx) of oxide in time (dt.). Use the segregation ratio S between the dopant concentrations in the SiO_2 and Si, and the ratio M of the corresponding thickness of SiO_2 and Si.

3°) What is the profile of the dopant on the Si side for $S = 0.4$ if $M = 2.5$?

4°) Give the form of the $c(x, t)$ profiles as a function of S for long oxidation times. If ξ is the coordinate of the interface, when is $c(\xi, t) > c_0$, and when is $c(\xi, t) < c_0$?

67 — IONIC CURRENT (IMPURITIES) IN SiO_2

Instabilities in the operation of MOS (metal-oxide (SiO_2)-semiconductor) diodes are connected with the migration of impurity (Na^+ ion) contaminants in the SiO_2 film.

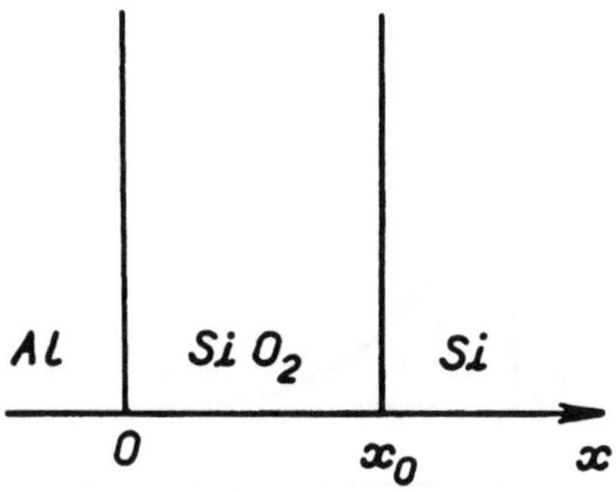

An oxide layer is grown at the surface of a plane, semi-infinite sample of silicon, and a thin layer of aluminum is deposited on it. This oxide is contamined by mobile Na^+ ions. There is no other charge carrier in the oxide.

1°) Considering that at equilibrium the flux of Na^+ ions in the x direction is zero, show that:

$\left(\frac{2kT}{q} \cdot \frac{dE}{dx}\right) - E^2 = c_1$, where c_1 is a constant of integration, and E is the electric field.

2°) Integrate this equation for $c_1 > 0$, $c_1 = 0$, $c_1 < 0$. Let c_2, c_3 and c_4 be the constants of integration in the three cases.

3°) Deduce $N(x)$, the concentration of Na^+ ions; $V(x)$, the potential such that:

$$E = -dV/dx, \text{ and the total charge } Q = \int_0^{x_0} N(x)\, dx$$

for $c_1 \geq 0$ (verify that $N(x) \geq 0\ \forall x$). Deduce the physical limitations on the integration constants.

4°) Give the general form of the curves for $N(x)$ and $E(x)$. Why do these present extrema at certain interfaces? Discuss this in terms of the values of c_2 and c_3. Can these constants be determined physically, and from what data?

5°) If the potential between the metal and the semiconductor is zero, give the exact form of $N(x)$. Deduce the value of c_1 as a function of Q.

68 — SELECTIVE OXIDATION

Consider the oxidation of an alloy AX, in which the element X is less oxidizable than the base metal A. In this case, a layer enriched in the element X forms just behind the metal/oxide interface. Assume that the interface acts as a perfect filter; the initial concentration of X in the alloy is c_0, the concentration in the oxide is zero, the concentration in the metal is $c(x,t)$. Let V be the velocity of the interface motion due to oxidation.

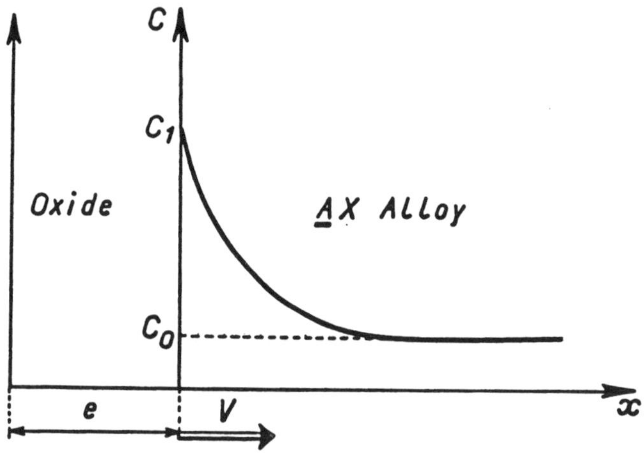

The diffusion equation for parabolic oxidation kinetics was solved in chapter X. Apply this result to the two following cases (c_0 and c_1 designate the initial concentration and the concentration at the metal/oxide interface, respectively, of the element X).

1°) Oxidation of an Fe-Ni alloy at 850°C. Take $V = 7.1 \times 10^{-5}\ t^{-1/2}$, $D = 1.5 \times 10^{-11}$ cm^2 s^{-1}. Calculate c_1/c_0. For what initial concentration c_0 does a layer of γ-FeNi form under the oxide layer (solubility of Ni in α-Fe at 850°C: 1 wt %)?

2°) Oxidation of an Fe-As alloy at 800°C. For an alloy with $c_0 = 0.075$ %, microprobe measurement yields $c_1 = 3.5$ %. Knowing that a layer 580 μm thick foms in 24 h, what is the value of the diffusion coefficient of As? (Note that the ratio of the density of the metal to the oxide $\approx$ 1.7).

69 — DIFFUSION BRAZING

In this process, also called "transient liquid phase bonding", a foil of brazing material, made of an alloy AB (near the eutectic) with liquidus temperature less than those of the alloys to the brazed, is placed between the two alloys to be brazed, A and A'. The entire assembly is raised to a temperature T, ΔT above the liquidus of the braze.

For simplicity, both of the pieces to be brazed will be considered to consist of pure metal A. On the basis of the AB phase diagram, describe the progress of the brazing process during an isothermal heat treatment at the temperature T. Establish the equation describing the evolution of the thickness $2h(t)$ of the liquid zone. Determine the time t_f necessary to obtain, at the temeprature T,

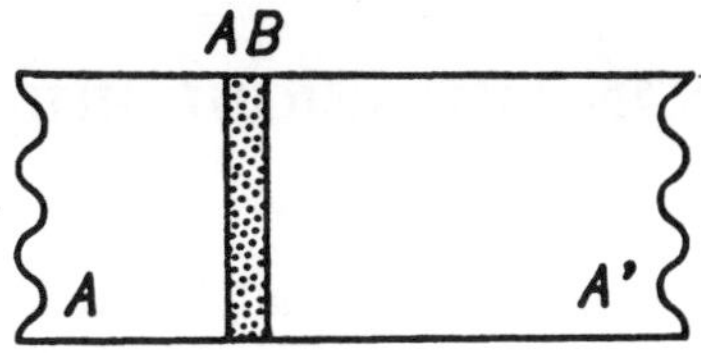

complete solidification of the brazed joint.

Numerical example: Brazing nickel with a Ni-15 at% boron braze. At 1100°C, the diffusion coefficient of B in Ni is $D = 10^{-6}$ cm^2 s^{-1}.
c_s (concentration of boron for the Ni-B solidus) $= 5 \times 10^{-4}$ at.% . $2\,h_0$ (thickness of the film of brazing material) $= 50$ μm. Evaluate the role of grain boundary diffusion taking D_{eff} according to Hart's formula for the diffusion of boron in nickel; grain size $= 50$ μm, $D'/D = 10^6$, $\delta = 5$ Å.

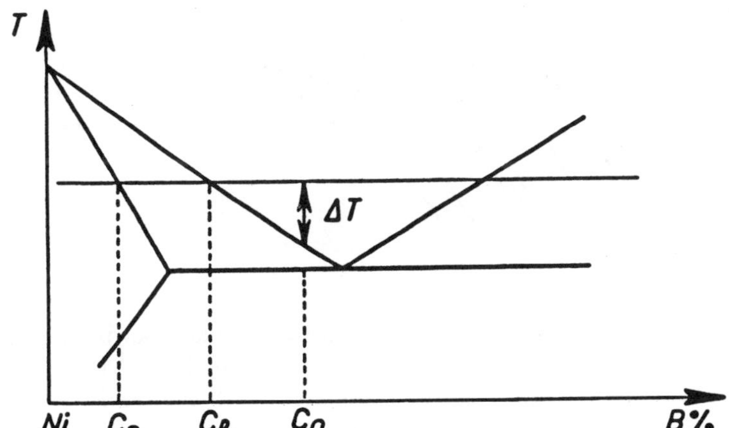

Reference. J TUAH-POKU, M. DOLLAR, T. B. MASSALSKI, *Met. Trans.* **19A** (1988) 675-686.

70 — KINETICS OF COALESCENCE

1°) Show that the evolution of the average concentration in the matrix (Eq. (X.83)) can be deduced from the growth law (Eq. (X.82)) by application of the Gibbs-Thomson relationship.

2°) Two types of measurement were carried out on samples treated to produce coalescence of precipitates:

1 – determination of the average radius of the particles by microscopy;

2 – determination of the concentration of the solute in the matrix by direct analysis or by an indirect technique (electrical resistivity, magnetic moment). Both of these quantities are found to follow a $t^{1/3}$ law. Are these data sufficient to verify the model for coalescence?

71 — SEGREGATION IN THE NEIGHBORHOOD OF A MICRO-CRACK

Near the tip of a crack in a mechanically stressed material there exists a stress field. Atoms of impurity can migrate in the field and segregate near the tip or at the sides of the crack.

1°) For a tensile stress normal to the plane of the crack, calculate the field of isostatic pressure, starting from the classical expressions for the stress field (see a textbook on fracture mechanics). The head of the crack is considered as a wedge located along the z axis. Plane strain is assumed.

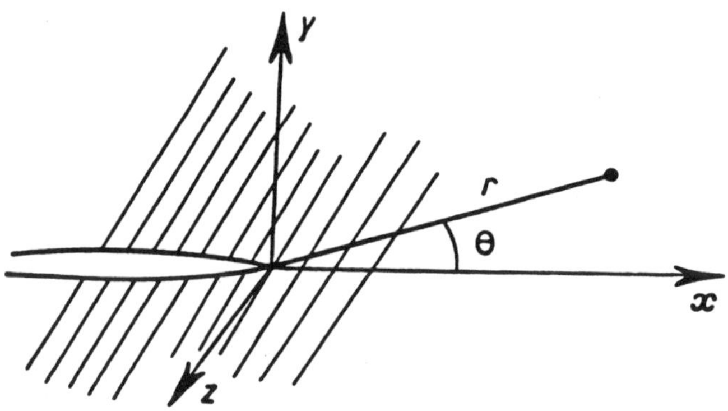

2°) Consider an impurity atom, labeled by its polar coordinates r, θ. It is represented as a sphere and characterized by its size effect ΔV (< 0 or > 0). Write the elastic interaction energy of this atom with the crack.

3°) Write the drift flux induced by the elastic field, and then the diffusion equation in $\partial c/\partial t$, neglecting the Fickian flux. State precisely the initial and boundary conditions for the differential equation for $c(t,\ r,\ \theta)$ for the problem under consideration.

4°) Solve the equation by the method of characteristics for the two cases $\Delta V > 0$ and $\Delta V < 0$.

5°) For both cases, graph the characteristic curves in the xy plane, as well as some equipotential curves and some atomic trajectories. Deduce for each case where the atoms will segregate (assuming that this place behaves as a perfect sink).

6°) Calculate the total number, $N(t)$, of atoms segregated in time t, per unit length of crack (i.e., along the z axis). Plot the number of atoms segregated along the crack, $n(x, t)$.

Note that $N(t)$ is the area under this curve.

7°) It is assumed that the crack propagates by fits and starts. During the time it is immobile, $N(\Delta t)$ atoms segregate along the edge. Then the crack advances brusquely. If an average crack velocity, v, is defined, it moves ahead by $v/\Delta t$ jerks per unit time. Calculate the rate of covering of the crack as a function of $N(t)$ and v, in number of atoms per unit area of crack.

8°) *Numerical exercises*

1) Segregation of hydrogen in α-Fe at ambient temperature. Given:
Size difference $\Delta V = 4.28 \times 10^{-31}$ m^3
Diffusion: $D_0 = 2.3 \times 10^{-7}$ m^2 s^{-1}, $Q = 6,680$ J mole^{-1}.
Solubility in atom fractions:
$c = 1.85 \times 10^{-3} (p_{H_2})^{1/2} \exp(-28{,}600 \text{ (J)}/RT)$
$p_{H_2} = 690$ torr
$K_I = 37$ MPa m$^{1/2}$
Average propagation velocity: $v = 4 \times 10^{-5}$ m s^{-1}, $\Delta t = 1$ s.

2) Segregation of sulfur in a mild steel at 500°C.
Size effect $\Delta V = -0.6\Omega$ with $\Omega = 11.22 \times 10^{-24}$ cm^3.
Diffusion of sulfur: $D_0 = 1.68$ cm^2 s^{-1}, $Q = 2.13$ eV.
Solubility $c_0 = 2 \times 10^{18}$ atoms cm^{-3}
$K_I = 30$ MPa m$^{1/2}$
$v = 10^{-5}$ cm s^{-1}
$\Delta t = 1$ s.

Take Poisson's ratio $\nu = 0.33$

References

C. A. Hippsley, H. Rauh, and R. Bullough, *Acta Met.* **32** (1984) 1384-1394.

C. A. Hippsley and C. L. Briant, *Scripta Met.* **19** (1985) 1203-1208.

TABLE I

THE ERROR FUNCTION

(ADDA and PHILIBERT, *op. cit.* p. 1111-1112).

u	Θ	*u*	Θ	*u*	Θ	*u*	Θ
0.01	0.0112833	0.51	0.5292437	1.01	0.8468105	1.51	0.9672768
0.02	0.0225644	0.52	0.5378987	1.02	0.8508380	1.52	0.9684135
0.03	0.0338410	0.53	0.5464641	1.03	0.8547842	1.53	0.9695162
0.04	0.0451109	0.54	0.5549392	1.04	0.8586499	1.54	0.9705857
0.05	0.0563718	0.55	0.5633233	1.05	0.8624360	1.55	0.9716227
0.06	0.0676215	0.56	0.5716157	1.06	0.8661435	1.56	0.9726281
0.07	0.0788577	0.57	0.5798158	1.07	0.8697732	1.57	0.9736026
0.08	0.0900871	0.58	0.5879229	1.08	0.8733261	1.58	0.9745470
0.09	0.1012806	0.59	0.5959365	1.09	0.8768030	1.59	0.9754620
0.10	0.1124630	0.60	0.6038561	1.10	0.8802050	1.60	0.9763484
0.11	0.1236230	0.61	0.6116812	1.11	0.8835330	1.61	0.9772069
0.12	0.1347584	0.62	0.6194114	1.12	0.8867879	1.62	0.9780381
0.13	0.1458671	0.63	0.6270463	1.13	0.8899707	1.63	0.9788429
0.14	0.1569470	0.64	0.6345857	1.14	0.8930823	1.64	0.9796218
0.15	0.1679959	0.65	0.6420292	1.15	0.8961238	1.65	0.9803756
0.16	0.1790117	0.66	0.6493765	1.16	0.8990962	1.66	0.9811049
0.17	0.1899923	0.67	0.6566275	1.17	0.9020004	1.67	0.9818104
0.18	0.2009357	0.68	0.6637820	1.18	0.9048374	1.68	0.9824928
0.19	0.2118398	0.69	0.6708399	1.19	0.9076083	1.69	0.9831526
0.20	0.2227025	0.70	0.6778010	1.20	0.9103140	1.70	0.9837904
0.21	0.2335218	0.71	0.6846654	1.21	0.9129555	1.71	0.9844070
0.22	0.2442958	0.72	0.6914330	1.22	0.9155339	1.72	0.9850028
0.23	0.2550225	0.73	0.6981038	1:23	0.9180501	1.73	0.9855785
0.24	0.2657000	0.74	0.7046780	1.24	0.9205052	1.74	0.9861346
0.25	0.2763263	0.75	0.7111556	1.25	0.9229001	1.75	0.9866717
0.26	0.2868997	0.76	0.7175367	1.26	0.9252359	1.76	0.9871903
0.27	0.2974182	0.77	0.7238216	1.27	0.9275136	1.77	0.9876910
0.28	0.3078800	0.78	0.7300104	1.28	0.9297342	1.78	0.9881742
0.29	0.3182834	0.79	0.7361035	1.29	0.9318987	1.79	0.9886406
0.30	0.3286267	0.80	0.7421010	1.30	0.9340080	1.80	0.9790905
0.31	0.3389081	0.81	0.7480033	1.31	0.9360632	1.81	0.9895245
0.32	0.3491259	0.82	0.7538108	1.32	0.9380652	1.82	0.9899431
0.33	0.3592785	0.83	0.7595238	1.33	0.9400150	1.83	0.9903467
0.34	0.3693644	0.84	0.7651427	1.34	0.9419137	1.84	0.9907359
0.35	0.3793819	0.85	0.7706680	1.35	0.9437622	1.85	0.9911110
0.36	0.3893296	0.86	0.7761002	1.36	0.9455614	1.86	0.9914725
0.37	0.3992059	0.87	0.7814398	1.37	0.9473124	1.87	0.9918207
0.38	0.4090093	0.88	0.7866873	1.38	0.9490160	1.88	0.9921562
0.39	0.4187385	0.89	0.7918432	1.39	0.9506733	1.89	0.9924793
0.40	0.4283922	0.90	0.7969082	1.40	0.9522851	1.90	0.9927904
0.41	0.4379690	0.91	0.8018828	1.41	0.9538524	1.91	0.9930899
0.42	0.4474676	0.92	0.8067677	1.42	0.9553762	1.92	0.9933782
0.43	0.4568867	0.93	0.8115335	1.43	0.9568573	1.93	0.9936557
0.44	0.4662251	0.94	0.8152710	1.44	0.9582966	1.94	0.9939229
0.45	0.4754818	0.95	0.8208908	1.45	0.9596950	1.95	0.9941794
0.46	0.4846555	0.96	0.8254236	1.46	0.9610535	1.96	0.9944263
0.47	0.4937452	0.97	0.8298703	1.47	0.9623729	1.97	0.9946637
0.48	0.5027498	0.98	0.8344235	1.48	0.9636541	1.98	0.9948290
0.49	0.5116683	0.99	0.8385031	1.49	0.9648979	1.99	0.9951114
0.50	0.5204999	1.00	0.8427008	1.50	0.9661052	2.00	0.9953223

TABLE I *(continued)*

u	Θ	*u*	Θ	*u*	Θ	*u*	Θ
2.01	0.9955248	2.51	0.9996143	3.01	0.9999793	3.51	0.99999930905
2.02	0.9957195	2.52	0.9996345	3.02	0.9999805	3.52	0.99999935766
2.03	0.9959063	2.53	0.9996537	3.03	0.9999817	3.53	0.99999940296
2.04	0.9960858	2.54	0.9996720	3.04	0.9999829	3.54	0.99999944519
2.05	0.9962581	2.55	0.9996893	3.05	0.9999839	3.55	0.99999948452
2.06	0.9964235	2.56	0.9997058	3.06	0.9999849	3.56	0.99999952115
2.07	0.9965822	2.57	0.9997215	3.07	0.9999859	3.57	0.99999955527
2.08	0.9967344	2.58	0.9997364	3.08	0.9999867	3.58	0.99999958703
2.09	0.9968805	2.59	0.9997505	3.09	0.9999876	3.59	0.99999961661
2.10	0.9970205	2.60	0.9997640	3.10	0.9999884	3.60	0.99999964414
2.11	0.9971548	2.61	0.9997797	3.11	0.9999891	3.61	0.99999966975
2.12	0.9972836	2.62	0.9997888	3.12	0.9999898	3.62	0.99999969358
2.13	0.9974070	2.63	0.9998003	3.13	0.9999904	3.63	0.99999971574
2.14	0.9975253	2.64	0.9998112	3.14	0.9999910	3.64	0.99999973636
2.15	0.9976386	2.65	0.9998215	3.15	0.9999916	3.65	0.99999975551
2.16	0.9977472	2.66	0.9998313	3.16	0.9999921	3.66	0.99999977333
2.17	0.9978511	2.67	0.9998406	3.17	0.9999926	3.67	0.99999978990
2.18	0.9979505	2.68	0.998494	3.18	0.9999931	3.68	0.99999980528
2.19	0.9980459	2.69	0.9998578	3.19	0.9999936	3.69	0.99999981957
2.20	0.9981372	2.70	0.9998657	3.20	0.9999940	3.70	0.99999983285
2.21	0.9982244	2.71	0.9998732	3.21	0.9999944	3.71	0.99999984517
2.22	0.9983079	2.72	0.9998803	3.22	0.9999947	3.72	0.99999985663
2.23	0.9983878	2.73	0.9998870	3.23	0.9999951	3.73	0.99999986726
2.24	0.9984642	2.74	0.9998933	3.24	0.9999954	3.74	0.99999987712
2.25	0.9985373	2.75	0.9998994	3.25	0.9999957	3.75	0.99999988629
2.26	0.9986071	2.76	0.9999051	3.26	0.9999960	3.76	0.99999989477
2.27	0.9986739	2.77	0.9999105	3.27	0.9999962	3.77	0.99999990265
2.28	0.9987377	2.78	0.9999156	3.28	0.9999965	3.78	0.99999990995
2.29	0.9987986	2.79	0.9999204	3.29	0.9999967	3.79	0.99999991672
2.30	0.9988568	2.80	0.9999250	3.30	0.9999969	3.80	0.99999992300
2.31	0.9989124	2.81	0.9999293	3.31	0.9999971	3.81	0.99999992881
2.32	0.9989655	2.82	0.9999334	3.32	0.9999973	3.82	0.99999993421
2.33	0.9990162	2.83	0.9999372	3.33	0.9999975	3.83	0.99999993921
2.34	0.9990646	2.84	0.9999409	3.34	0.9999977	3.84	0.99999994383
2.35	0.9991107	2.85	0.9999443	3.35	0.9999978	3.86	0.99999995208
2.36	0.9991548	2.86	0.9999476	3.36	0.9999980	3.88	0.99999995915
2.37	0.9991968	2.87	0.9999507	3.37	0.9999981	3.90	0.99999996521
2.38	0.9992369	2.88	0.9999536	3.38	0.9999982	3.92	0.99999997039
2.39	0.9992751	2.89	0.9999563	3.39	0.9999984	3.94	0.99999997482
2.40	0.9993115	2.90	0.9999589	3.40	0.9999985	3.96	0.99999997860
2.41	0.9993462	2.91	0.9999613	3.41	0.9999986	3.98	0.99999998183
2.42	0.9993793	2.92	0.9999636	3.42	0.9999987	4.00	0.9999999845828
2.43	0.9994108	2.93	0.9999658	3.43	0.9999988	4.10	0.99999999330
2.44	0.9994408	2.94	0.9999679	3.44	0.9999989	4.20	0.99999999714
2.45	0.9994694	2.95	0.9999698	3.45	0.9999989	4.30	0.99999999881
2.46	0.9994966	2.96	0.9999716	3.46	0.999999008	4.40	0.99999999951
2.47	0.9995226	2.97	0.9999733	3.47	0.999999076	4.50	0.99999999980
2.48	0.9995472	2.98	0.9999750	3.48	0.999999141	4.60	0.99999999992
2.49	0.9995707	2.99	0.9999765	3.49	0.999999201	4.80	0.99999999999
2.50	0.9995930	3.00	0.9999779	3.50	0.999999257	5.00	0.99999999999843253

INDEX (*)

(*) Page numbers in italics refer to exercises.

Composition : Mathor-Tex ; Composciences,
« Le Tropic », Avenue des Indes, 91940 LES ULIS - France.
Imprimé en France. – EMD, 53 Lassay-les-Châteaux.
N° 2061. Dépôt légal : septembre 1991.